Parasitology and Vector Biology

Parasitology and Vector Biology

Edited by **Cherilyn Jose**

SYRAWOOD
PUBLISHING HOUSE
New York

Published by Syrawood Publishing House,
750 Third Avenue, 9th Floor,
New York, NY 10017, USA
www.syrawoodpublishinghouse.com

Parasitology and Vector Biology
Edited by Cherilyn Jose

International Standard Book Number: 978-1-68286-351-0 (Hardback)

Printed in China.

Contents

Preface

This book was inspired by the evolution of our times; to answer the curiosity of inquisitive minds. Many developments have occurred across the globe in the recent past which has transformed the progress in the field.

Parasitic and vector borne diseases are a major threat to health of humans and other organisms in present day scenario. It is necessary to study these diseases to understand the cause and to ensure better control measures. This book on parasitology and vector biology provides readers with in-depth knowledge of various parasites, their hosts, and the relationship between them. Different approaches, evaluations, methodologies and advanced studies on helminthology, parasitism, DNA sequence markers, and host-pathogen interactions have been included in this book. It will help students and researchers in gaining a comprehensive insight into the field.

This book was developed from a mere concept to drafts to chapters and finally compiled together as a complete text to benefit the readers across all nations. To ensure the quality of the content we instilled two significant steps in our procedure. The first was to appoint an editorial team that would verify the data and statistics provided in the book and also select the most appropriate and valuable contributions from the plentiful contributions we received from authors worldwide. The next step was to appoint an expert of the topic as the Editor-in-Chief, who would head the project and finally make the necessary amendments and modifications to make the text reader-friendly. I was then commissioned to examine all the material to present the topics in the most comprehensible and productive format.

I would like to take this opportunity to thank all the contributing authors who were supportive enough to contribute their time and knowledge to this project. I also wish to convey my regards to my family who have been extremely supportive during the entire project.

Editor

Antiparasitic effects of the water extract from *Chenopodium ambrosioides* L. (Chenopodiaceae) against some gastrointestinal nematodes in West African Long Legged goats

Sahidou SALIFOU[1], Dadjo Florian DAGA[2], Sabbas ATTINDEHOU[1]*, Rodolph DEGUENON[1] and Cyprien Félix BIAOU[1]

[1]Laboratoire de Recherche en Biologie Appliquée, Université d'Abomey-Calavi, 01 BP 2009 Cotonou, Bénin.
[2]Département de Production Animale, Faculté des Sciences Agronomiques, Université d'Abomey-Calavi, 01 BP 526 Cotonou, Bénin.

An *in vivo* test was undertaken in order to evaluate the efficacy of an aqueous extract of leaves of *Chenopodium ambrosioïdes* against nematode parasites in goats. The study used fifteen naturally infected West African Long Legged goats (*Capra hircus*) from Northern Benin. Three levels (1, 2 or 4 ml/kg body weight/3 days treatment) of concentration of the herbal drug were administrated to three experimental groups and in comparison with two control groups (one without treatment and the other treated with levamisole as a reference drug). Results showed significant (P<0.001) decrease of faecal strongyle eggs in experimental groups. After three days of treatment, the reduction in the rate of egg shedding was over 70% in animals treated with herbal medicine, whereas egg shedding rates remained the same in the non treated group. The rates of reduction in faecal eggs reached almost 100% in few days (5 to 6 days) among treated animals without a significant dose-dependent effect. Moreover, helminthological autopsy performed on the study animals after treatment showed an almost total absence of worms. *C. ambrosioïdes* aqueous extract exerted a kind of parasite clearing effect in the abomasums and in the small intestine. This indicates the effectiveness of the herbal extract against gastrointestinal parasitic. The lowest dosage of 1 ml/kg body weight for three days treatment was found to be appropriate and prescriptive. No acute toxicity was recorded and the autopsy of internal organs revealed only rare cases of congestion and petechiae on the liver and lungs in animals treated with 4 ml/kg body weight.

Key words: Goats, gastrointestinal nematodes, controlling, herbal medicine, *Chenopodium ambrosioïdes*.

INTRODUCTION

In tropical regions, gastrointestinal parasites are an important impediment for livestock production. Their control is also a real economic constraint since the drugs used to control parasites infection can be costly. Benzimidazoles, avermectins, imidothiazoles and pyrimidines are the main and common anthelmintic moleculars used to protect small ruminants, but most of them are subject to drug resistance (Chandrawanthani, 2004; Gaskin, 2006). Due to these issues, numerous farmers in poor areas are using endogenous recipes. In recent years, there has been a resurgence of interest in herbal medicine. In Africa, many farmers use plants to control animal diseases (Djoueche et al., 2011); thus, African flora may provide pharmaceuticals that are currently unknown to the scientific community and that may be cheaper and more environmentally sound. A dynamic of rationalization of endogenous practices is underway and the medicinal properties of many tropical

*Corresponding author. E-mail: sabbastino@yahoo.fr.

Table 1. Study animal groups and corresponding treatments.

Group	Administrated treatments (oral route)	
	Medicinal drug and dosage/kg body weight	Duration of treatments
CG_{WT}: Control group (no treatment)	None	-
CG_{Lev}: Control group treated	Levamisole (50 mg)	Single dose
EG 1 ml/kg = Experiment group no. 1	Plant extract (1 ml)	
EG 2 ml/kg = Experiment group no. 2	Plant extract (2 ml)	One dose daily for three days
EG 4 ml/kg = Experiment group no. 3	Plant extract (4 ml)	

Figure 1. Morning faeces recovery device ↑ (diaper) from animals.

plants are being checked. This study aims at evaluating the efficacy of an aqueous extract of the leaves of *Chenopodium ambrosioïdes* against strongyle infection in West African Long Legged (WALL) goats. The plant is common in many parts of the world like Europe, Asia and Africa. It is used as medicinal herb (Quinlan et al., 2002; Ruffa et al., 2002; Efferth et al., 2002; MacDonald et al., 2004; Patrício et al., 2008). It is well known in rural areas of Benin as "trouzouman" (in "fongbé", a national language) where it is deemed effective against some human parasitic diseases such as pinworms, tapeworms, roundworms and hookworms.

MATERIALS AND METHODS

Experimental design

The subjects of this study were 15 over one year old Sahelian goats naturally infested with strongyles (*Haemonchus contortus* and *Trichostrongylus* species). These study animals were selected among thirty animals purchased at Zongo livestock market in Cotonou after testing their stools. Only the animals with a faecal eggs count (FEC) over 1000 eggs/g of faeces were used in the study. The subjects were divided into five homogeneous groups (based on body weight and future treatment) of three animals (Table 1).

The plant material consisted of powdered leaves of *C. ambrosioïdes*. The leaves were harvested in December 2011 at Hêvié, district of Abomey-Calavi, Department of Atlantic, Benin, where the plant grows naturally. These leaves were sun-dried, then ground, bottled and kept dry within the Applied Biology Research Laboratory (LARBA) of the Abomey-Calavi Polytechnic School. Two substances were used as drugs in the study: the herbal drug and the control drug. The herbal drug was the aqueous extract of the powder of *C. ambrosioïdes*. It was obtained by infusing, during 30 min, 50 mg of powder in 1000 ml of distilled and deionised hot water. The filtrate was used as the herbal drug at the latest 24 h after. Levamisole bolus® (a conventional anthelmintic) was used as a control drug.

In vivo test of the anthelmintic activities of the drugs

The *in vivo* anthelmintic activity of the leaves of *C. ambrosioïdes* was assessed in farm conditions. For 15 days of adaptation, the animals were fed *ad libitum* with forage and received daily 1000 g of concentrate. The experiment was undertaken from December 2011 to January 2012 at the Doguici farm, located at Attogon, district of Allada. The parasitological status of the animals was reconfirmed according to the selecting criteria (FEC ≥ 1000) first. From the beginning of the test (day 0) to the third day (day two), animals received the appropriate treatment (Table 1) and their general conditions and behaviour were carefully monitored by the booking of animal's posture, temperature, respiratory frequency, rumination, aspect and frequency of faeces; any toxicological abnormality. Each night of the experiment, each study animal was outfitted with a diaper-like collection sheet (Figure 1) in order to recover, individually and efficiently, the morning faeces of the animals. Faecal material was collected daily for each animal and was stored at 4°C in sealed labelled vials. A total of 150 samples were then taken and examined for faecal eggs counts. The parasitological examination method of Mc Master (Gaskin, 2006) was used to assess the FEC over 10 days (three adaptation days and seven experiment days). The herbal medicine's efficacy was evaluated by calculating the reduction rates of faecal eggs shedding (R_{FES}) with the following formula.

$$R_{FES}(n) = 100 \times (FEC_0 - FEC_n)/FEC_0$$

where $R_{FES(n)}$ = faecal eggs shedding rate of day "n", FEC_n = faecal eggs count of day "n" and FEC_0 = faecal eggs count before treatment.

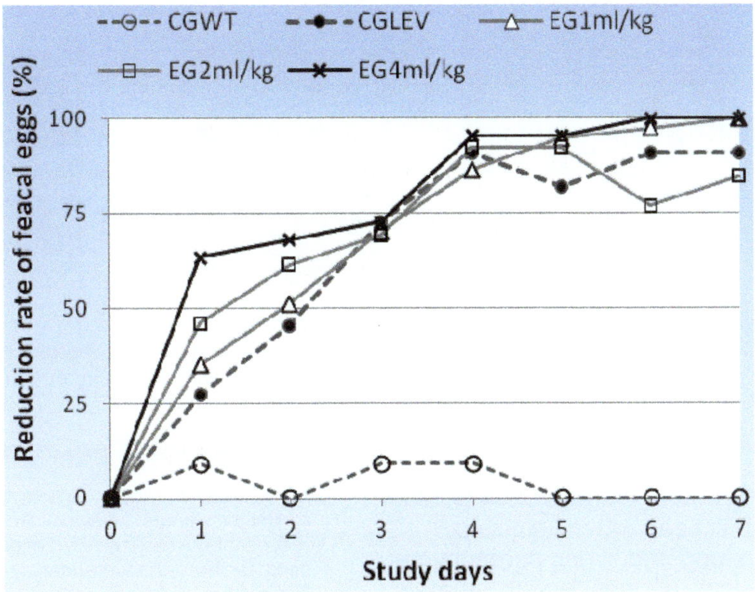

Figure 2. Effect of *C. ambrosioïdes* extract on the faecal eggs excretion.

Helminthological and toxicological autopsy

At the end of the experiment (day 10), the animals were slaughtered at the abattoir of Cotonou. Their digestive systems were collected and examined tracking lesions and present adult nematodes or cestodes worms. The abomasums and intestines of each gut are separately and carefully inspected after being incised along their long axis. The luminal content was recovered in a container for examination. The mucosa was also inspected for fixed worms. Worms were then collected in glass bottles containing 70° ethanol for nematodes and 5% formalin for other helminthes and were transferred to the LARBA. Identification of parasites was done later based on criteria provided by Zajac et al. (2006). The worm burden for each animal was determined for each species of parasite.

Statistical analysis

The FEC recorded data and the calculated eggs shedding reduction rates (R_{FES}) were performed under Stata 11 following a Poisson distribution model.

RESULT AND DISCUSSION

In vivo anthelmintic effects of *C. ambrosioïdes* aqueous extract

The daily reduction (percent) of faecal eggs excreted in each treatment condition is as shown in Figure 2. There was a significant decrease in eggs excreted with all treatment when compared with the control group (P<0.001). Reduction of faecal eggs excretion (increase of R_{FES}) was seen spontaneously (72 h) at levels over 70% (required level for biological significance according to Githiori et al., 2003) in the experimental groups ($EG_{1\ ml/kg}$, $EG_{2\ ml/kg}$ and $EG_{4\ ml/kg}$). Reduction in rates approached 100% by the sixth day in the $EG_{4\ ml/kg}$ group. No significant difference was found between the results for the $EG_{4\ ml/kg}$ and CG_{Lev} groups. With reference to the results of similar tests with other herbal plants, *C. ambrosioïdes* can be considered very efficient in decreasing the excretion of faecal parasites eggs. For example, roots' extracts of *Halothamus somalensis* have induced only 50% reduction within three weeks (Dawo and Tibbo, 2005). A high R_{FES} of 96.2% was recorded in sheep as maximum with a four plants formulation (Zaman et al., 2012) within two weeks. The efficacy of *C. ambrosioïdes* was not dose-dependent like several other cases of herbal medicines since the levels of success were not significantly different between the dosages. This means that the lower concentration could be a suitable and prescriptive one.

Acute clinical toxicity

Only the presence of pasty stools indicated a perturbation of the digestive transit in animals treated with the two highest doses of extract.

Autopsy results

At helminthological autopsy, the parasites burden in the animals was near zero with treated animals while the un-treated group harboured more than 350 adult of the main gastrointestinal nematodes (*H. contortus, Trichostrongylus*

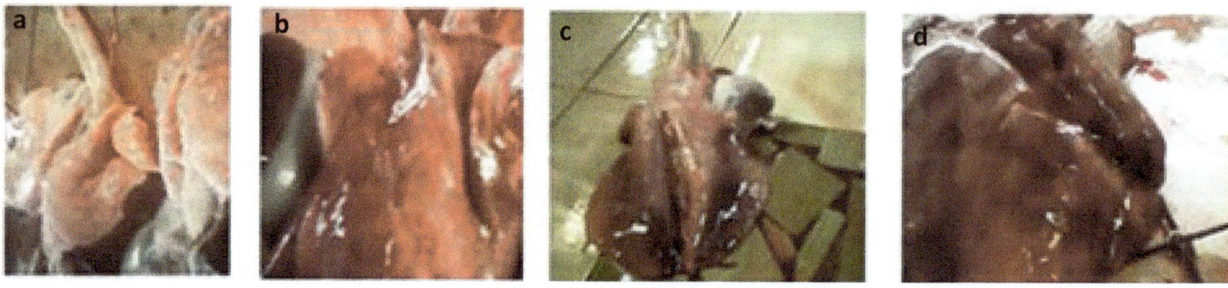

Figure 1. Pathological impact of *C. ambrosioïdes* extract on lungs with Wall goats.
a. Normal lungs taken from animal in group CGWT; b. Normal lungs taken from animal in group EG 1 ml/kg; c. Mild pulmonary congestion with animal group EG 2 ml/kg; c. Marked pulmonary congestion with animal in group EG 4 ml/kg.

spp, and *Trichurus* spp). The herbal extract had a potent deworming effect similar to that seen with levamisole.

The intestinal mucosa, the lung and liver's parenchyma showed a few petechial lesions (Figure 3) with animals treated with 4 ml/kg body weight dosage. This implies a dose-related toxicity, but no immediate distress or life-threatening effects were observed. These results confirm however the toxicological risks of the plants. Therefore, using a minimum dosage must be recommended.

Conclusion

This study shows that an aqueous extract of the leaves of *C. ambrosioïdes* is a potent anthelmintic in WALL goats. Ease of formulation and the efficacy of low dose promote its use in farmers as an alternative to synthetic molecules. Further studies are needed to clarify the active ingredients and the mode of action as well as long term dosage utilization.

ACKNOWLEDGEMENTS

The study was made possible through the support of the NGO Doguici who put provided access to their unit of livestock production facilities. They authors thank the villagers in Kpoguetomey, Attogon, District of Allada for their cooperation. This study is dedicated to the memory of the Late Mother Anne-Marie KPOGUE, a resident of the village and resource person who actively contributed to the monitoring of the study animals.

REFERENCES

Chandrawanthani P, Yusoff N, Wan LC, Ham A, Waller PJ (2004). Total anthelmintic failure to control nematode parasites of small ruminants on government breeding farms in Sabah, East Malaysia. Vet. Res. Commun. 28(6):479-489.

Dawo F, Tibbo M (2005). Anthelmintic effect of Halothamus somalensis in Arsi-Bale goats. Livestock Res. Rural Dev. 17(6).

Djoueche CM, Azebaze AB, Dongmol AB (2011). Investigation of plants used for the ethnoveterinary control of gastrointestinal parasites in Bénoué Region, Cameroon. Tropicultura 29(4):205-211.

Efferth T, Olbrich A, Sauerbrey A, Ross DD, Gebhart E, Neugebauer M (2002). Activity of ascaridol from the anthelmintic herb *Chenopodium anthelminticum L.* against sensitive and multidrug-resistant tumor cells. Anticancer Res. 22(6C):4221-4224.

Gaskin JM (2006). Mastering the McMasters method of fecal helminth egg count quantification. Presented at the Annual Conference, Florida Goat Association. Gainesville, FL.

Githiori JB, Höglund J, Waller PJ, Baker RL (2003). Evaluation of anthelmintic properties of extracts from some plants used as livestock dewormers by pastoralist and smallholder farmers in Kenya against *Heligmosomoides polygyrus* infections in mice. Vet. Parasitol. 118(3-4):215-226.

MacDonald D, VanCrey K, Harrison P, Rangachari PK, Rosenfeld J, Warren C, Sorger G (2004). Ascaridole-less infusions of *Chenopodium ambrosioides* contain a nematocide(s) that is (are) not toxic to mammalian smooth muscle. J. Ethnopharmacol. 92(2-3):215-221.

Patrício FJ, Costa GC, Pereira PVS, Aragão-Filho WC, Sousa SM, Frazão JB, Pereira WS, Maciel MCG, Silva LA, Amaral FMM, Rebêlo JMM, Guerra RNM, Ribeiro MNS, Nascimento FRF (2008). Efficacy of the intralesional treatment with *Chenopodium ambrosioides* in the murine infection by *Leishmania amazonensis*. J. Ethnopharmacol. 115(2):313-319.

Quinlan MB, Quinlan RJ, Nolan JM (2002). Ethnophysiology and herbal treatments of intestinal worms in Dominica, West Indies. J Ethnopharmacol. 80(1):75-83.

Ruffa MJ, Ferraro G, Wagner ML, Calcagno ML, Campos RH, Cavallaro L (2002). Cytotoxic effect of Argentine medicinal plant extracts on human hepatocellular carcinoma cell line. J. Ethnopharmacol. 79(3):335-339.

Zajac M, Conboy G (2006). Veterinary Clinical Parasitology, 7th ed. Black Well Publishing Company, U.K.

Zaman MA, Iqbal Z, Khan MN, Muhammad G (2012). Anthelmintic activity of an herbal formulation against gastrointestinal nematodes of sheep. Pak. Vet. J. 32(1):117-121.

2

A comparative study on the prevalence of intestinal helminthes among rural and sub-urban pupils in Gwagwada, Nigeria

Auta, T.[1] , Wartu, J. R.[2], Bawa, J. A.[1] and Jabbi, A. M.[3]

[1]Department of Biological Sciences, Federal University, Dutsin-Ma, Katsina State, Nigeria.
[2]Department of Microbiology, Kaduna State University, Kaduna, Nigeria.
[3]Department of Biological Sciences, Police Academy, Wudil, Kano State, Nigeria.

A comparative study on infection with intestinal helminths among pupils in rural and sub-urban primary schools was carried out in Gwagwada district, Chikun Local Government Area of Kaduna State, Nigeria. Faecal samples were collected from 244 pupils in eight rural and sub-urban public schools and examined for eggs of intestinal helminths using the filtration technique and microscopically examined for intestinal parasites. The overall prevalence in both rural and sub-urban pupils was 67.2%, with the sub-urban pupils having a prevalence of 70.8% and the rural pupils a 58.9% prevalence. The five intestinal helminths observed in the study were *Ascaris lumbricoides* (37.0 and 29.8%), *Trichuris trichiura* (4.1 and 4.7%), hookworms (9.6 and 5.3%), *Taenia* species (32.9 and 11.1%) and *Schistosoma mansoni* (21.9 and 1.8%) in rural and sub-urban pupils, respectively. There was no significant association of the prevalence with location. Mass deworming campaign should be embarked upon immediately and sanitation facilities should be provided to curtail these alarming infections.

Key words: Helminths, rural, sub-urban, pupils, Gwagwada, Nigeria.

INTRODUCTION

Soil-transmitted helminth infections are among the most common infections worldwide and affect the poorest and most deprived communities. Latest estimates indicate that more than 880 million children are in need of treatment for these parasites (WHO, 2014). In sub-Saharan Africa, intestinal helminths are the most common and diseases with a very higher negative public health and socio-economic impacts (Enimien et al., 2014; Ojurongbe et al., 2014).

Parasitic helminths are endemic in Nigeria, due to poor environmental sanitation, pollution, and contamination of water and soil. Children in Nigeria are highly exposed and very vulnerable to these infections (Damen et al., 2010). In rural and sub-urban settlements in Nigeria, intestinal helminthes have been a major problem. This is as a result of their poor socio-economic status and lack of basic amenities such as pipe borne water and other sanitary facilities (Okon and Oku, 2001) and it may also be due to their illiteracy and careless behaviour(s). Generally, parasitic infections abound in Nigeria not only because a large number of the population still live(s) in insanitary surroundings with constant faecal pollution of soil, food and drinking water, but also as a consequence of the tropical environment, suitable for easy parasitic growth and spread (Adeyeba and Essiet, 2001).

A typical intestinal helminth has an intestinal lifecycle stage. Intestinal helminths of health relevance are nematodes (roundworms), cestodes (tapeworms) and trematodes (flukes). The most common of these helminths in Nigeria are *Ascaris lumbricoides, Trichuris trichiura* hookworms (*Necator americanus* and *Ancylostoma duodonale*), *Strongyloides stercoralis* and Schistosomes (Odening, 1976; Brooker et al., 2006). Other helminths of human health relevance but not very common in the tropics include: *Taenia* species, *Hymenolepis nana, Dicrocoelium hospes* and *Enterobius vermicularis* (WHO, 1994). The lifecycle of most helminths may involve more than one host, interspersed with a free living phase. Intestinal worms lay eggs inside the host body before they are passed out with faeces onto the ground or water as the case may be. The deposited eggs remain in the water/soil until they are ingested by the host. As they are ingested, they hatch into larvae and moult into adults which then start laying new eggs (Parker et al., 2003).

Due to the public health implications of these intestinal parasites, this study was carried out with the aim of comparing the intestinal helminthes infestation among pupils in rural and sub-urban primary schools in Gwagwada district, Chikun Local Government Area of Kaduna State, Nigeria.

MATERIALS AND METHODS

Study area

The study was carried out in Gwagwada educational district from January and September, 2006. Four public primary schools, each were randomly selected, from both the rural and sub-urban communities of the educational district, making a total of eight schools. The schools selected were Gwagwada, Dutse, Gazamari and Bakin Kasuwa (sub-urban schools) and Bakin Liza, Bashishi, Gwazunu and Lukuru (rural schools). Fifteen percent of the pupils (both male and female) in each school and class were randomly selected for the study. Approval was obtained from the local government education secretary, and consent from the district education officer, the head teachers of each school and parents/guardian of the pupils was also obtained before sample collection.

Faecal sample collection

Clean, labelled wide-mouth sample plastic bottles with covers were given out to the selected pupils for the study and were properly instructed on how to collect their early morning faecal sample without contamination using the applicator stick attached to the sample bottle cover. The bottles were then collected from the pupils as they resume for morning classes and the faecal samples were immediately preserved with formalin and were finally taken to the laboratory for analysis. Sample collection was carried out in accordance to internationally best practices (Odoba et al., 2012; Pham-Duc et al., 2013).

Paper and pen questionnaires were administered to selected pupils through personal interview and information on sex, age, class and school name were collected.

Laboratory examination of faecal samples

The faecal samples were examined for parasites using the formaldehyde-ether concentration technique as described by Cheesbrough (1992). An aliquot of 1 g of faeces was suspended in 10 ml of 10% formaldehyde solution and mixed with a glass rod. The suspension was passed through a funnel covered with a gauze pad, to remove debris into a centrifuge tube. Three millilitres of ether was added and the suspension thoroughly mixed. The tubes were centrifuged for 3 min at 4000 rpm. Four layers were formed at the end of the centrifugation. The first layer was the ether with fats dissolved in it, the second was the debris, the third was the formaldehyde solution and the fourth was the sediment of eggs and/or larvae.

The centrifuge tubes were decanted, leaving only the sediment. The sediment was examined by sampling a drop with a pipette and depositing it on a glass slide. The slide was covered with a slide cover slip and examined microscopically using X10 and X40 objectives of the microscope as describe by Cheesbrough (1992). The eggs/larvae were identified using Atlas of Medical Helminthology and Protozoology as well as Medical Parasitology (Jeffrey and Leach, 1975; Arora and Arora, 2010).

Statistical analysis

The data obtained in the study are shown in Tables 1, 2, 3, 4 and 5, interpreted in percentages and analysed with respect to rural and sub-urban schools, age, sex, class and type of helminth infection. Chi square was used to test for association of prevalence with rural and sub-urban pupils across class of study, sex and age using Excel Microsoft (2010).

RESULTS

The prevalence of intestinal helminths among the rural and sub-urban pupils and their distribution with schools are presented in Table 1. Out of a total of 244 pupils, in both rural and sub-urban schools examined, 164 pupils were found to be positive with at least one of the helminth parasites, which gave a prevalence of 67.2%. Among the rural schools, 43 (58.9%) pupils were found to be positive out of the 73 pupils examined, while 121 (70.8%) pupils were recorded positive out of 171 pupils examined among the sub-urban schools. The highest prevalence (76.6%), among all the schools was recorded in Dutse, a sub-urban area, and was followed by Lukuru (73.3%), a rural area, while the lowest prevalence (41.6%) was obtained in Gwagwada, the district headquarters.

The results, as shown in Table 2, indicate the sub-

Table 1. Prevalence in rural and sub-urban schools.

Rural			Sub-urban		
School	No. examined	No. positive (%)	School	No. examined	No. positive (%)
Bakin Lizza	15	8 (53.3)	Gwagwada	77	53 (68.8)
Bashishi	18	9 (50.0)	Dutse	64	49 (76.6)
Gwazunu	25	15 (60.0)	Gazamari	11	7 (63.6)
Lukuru	15	11 (73.3)	Bakin Kasuwa	19	12 (63.2)
Total	73	43 (58.9)	Total	171	121 (70.8)

Table 2. Prevalence by sex in rural and suburban schools.

School	Male		Female	
	No. examined	No. positive (%)	No. examined	No. positive (%)
Rural	45	26 (57.7)	28	17 (60.71)
Suburban	99	73 (73.74)	72	48 (66.67)

Males χ^2= 3.7; df = 1; p = 0.05 and Females χ^2= 0.3; df = 1; p = 0.05

Table 3. Prevalence by age group in rural and sub-urban schools.

Age (Years)	Rural		Sub-urban	
	No. examined	No. positive (%)	No. examined	No. positive (%)
7-9	19	13 (68.4)	49	38 (77.6)
10-12	17	11 (64.7)	43	32 (74.4)
13-15	14	8 (57.1)	41	27 (65.9)
>16	23	11 (47.8)	38	24 (63.2)

χ^2= 2.1; df = 3; p = 0.05.

urban pupils had the higher prevalence in both males (73.7%) and females (66.7%) than the male and female pupils in rural areas. However, there was no significant association with sex between rural and sub-urban prevalence (P>0.05). When compared across age groups (Table 3), the group of 7 to 9 years in sub-urban schools had the highest prevalence (77.6%), while the group >16 years in rural schools had the lowest prevalence (47.8%). Though, there was no significant association of prevalence with age between the two areas, there was a decline in prevalence as the age increased, in both areas.

The highest prevalence (80.0%) was obtained in sub-urban schools among Primary 4 pupils while the lowest prevalence of 47.8% was among Primary 6 pupils of sub-urban schools. Though no significant association was found, there was slight decline in prevalence as one went higher the classes (among rural schools) from Primary 1 to 3, a slight increase in Primary 4 and a continuous decline in Primary 5 and 6. The prevalence among sub-urban pupils, across classes showed a staggered pattern (Table 4).

Ascaris lumbricoides was the most prevalent helminth, with 37.0 and 29.8% among both rural and sub-urban pupils, respectively (Table 5). *Schistosoma mansoni* had the lowest prevalence (1.8%) among sub-urban pupils, but slightly higher among the rural pupils (21.9%).

DISCUSSION

The overall high prevalence (67.2%) of intestinal helminth infections among both rural and sub-urban pupils is similar to the high prevalence (80.9%) by Damen et al. (2011) reported among almajiris, 59.1% among primary school pupils by Usip et al. (2013). This could be attributed to carelessness and unhygienic behaviour(s) among these pupils both at home and in school. Lack of sanitation facilities in these schools might have also contributed to the high prevalence. In all the selected schools and environs, it was observed that there were no toilet facilities or, where present, they were not functional or no adequate water supply. The pupils mostly defecate in open field both at home and in school.

Table 4. Prevalence with class between rural and sub-urban schools.

Class	Rural		Sub-urban	
	No. examined	No. positive (%)	No. examined	No. positive (%)
Primary 1	16	11 (68.8)	35	23 (65.7)
Primary 2	8	5 (62.5)	29	22 (75.9)
Primary 3	12	7 (58.3)	27	20 (74.1)
Primary 4	10	6 (60.0)	30	24 (80.0)
Primary 5	13	7 (53.8)	27	21 (77.8)
Primary 6	14	7 (50.0)	23	11 (47.8)

χ^2= 1.3; df = 5; p = 0.05.

Table 5. Prevalence with type of helminthes between rural and sub-urban schools.

Type of helminth	Rural		Sub-urban	
	No. examined	No. positive (%)	No. examined	No. positive (%)
Ascaris lumbricoides	73	27 (37.0)	171	51 (29.8)
Trichuris trichiura	73	3 (4.1)	171	8 (4.7)
Hookworm	73	7 (9.6)	171	9 (5.3)
Taenia spp.	73	24 (32.9)	171	19 (11.1)
Schistosoma mansoni	73	16 (21.9)	171	3 (1.8)

Consequently, the faeces could contaminate nearby household and school environments with eggs from geohelminths, and this might have resulted to the high counts from this study. The pupils could have picked up the eggs from the contaminated soils during extracurricular activities and through eating with dirty, contaminated hands, thus leading to infection which affects the health, well-being and the academic performance of the pupils.

The very high prevalence ratio of infection obtained in this study is similar to that reported by Leykun (2001) in a study in Ethiopia among school children and that of Sebastián and Santi (2001) among Naporuna school children in the Low-Napo region, North-Eastern Ecuador. A high prevalence (77%) of intestinal geohelminthiasis among school children in riverine communities of Nigeria was reported by Ariyo et al. (2007).

Majority of the parents/guardians of these school pupils are either farmers or nomadic cattle rearers, having little or no education on good hygienic behaviour and so, children are not under adequate adult supervision of their sanitation habits even at home.

The insignificant association in prevalence between sex and locality could be due to the common lack or poor sanitation facilities and portable drinking water. The higher prevalence recorded among males in both rural and sub-urban pupils matches with the reports of Ajibola (2013), Ajibola and Hassan (2013), Alemu et al. (2011), Leykun (2001) and Sehgal et al. (2010), but contrary to the findings of Amaechi and Okore (2013). This could be attributed to the activities the males get involved in, which predispose them to infection, such as swimming, fishing, farming, football and other sports which females are mostly not involved in. The slight decrease of infection with increase in age may be due to increase in hygiene and behavioural changes that mostly come with increases in age and this also applies to the prevalence across the classes. In both rural and sub-urban pupils, the highest prevalence of *A. lumbricoides* corresponded with the reports of Ajibola (2013), Ajibola and Hassan (2013), Ekpenyong et al. (2008), Emeka (2013), Usip et al. (2013), Damen et al. (2011) and Ojurongbe et al. (2014), but are contrary to that of Odoba et al. (2012) that reported hookworm to have the highest prevalence among childrenin Zaria.

Conclusion

In conclusion, there was a very high prevalence of intestinal helminths among pupils in these rural and sub-urban areas, which reveals that intestinal helminthes are endemic in these areas. *Ascaris lumbricoides* has a very high prevalence among pupils. The health status of rural and sub-urban pupils was not significantly different, because the available social, health and sanitary facilities in these areas are not different in any way. There is a need for the state/local government and Non-Governmental Organizations (NGOs) to embark on urgent mass deworming campaigns among all pupils, and potentially all people living in these areas. The need for health/hygiene education and awareness among pupils, parents/guardians, teachers and food vendors will contribute immensely for improving their behaviour. Government agencies, at state and local levels, should provide more facilities, such as boreholes, toilet facilities, good drainage system and other basic facilities that will promote more healthy living conditions.

ACKNOWLEDGEMENT

The contributions and cooperation of all pupils, teachers, school heads, district education officer and Chikun Local Government education secretary are highly recognised and appreciated.

Conflict Interests

The authors declare that there is no conflict of interests regarding the publication of this article.

REFERENCES

Adeyeba O, Essiet U (2001). Prevalence of Helminth and protozoal infection among a Religious sect that walk Barefooted in Iseyin, Nigeria. Nig. J. Parasit. 22:85-94.

Ajibola B (2013). Epidemiology and Risk Factors of Soil Transmitted Helminths in Two Communities of Yewa North Local Government Area, Ogun State. 8th Annual Conference of the Zoology Society of Nigeria (ZSN), Lagos. P 76.

Ajibola B, Hassan AA (2013). Prevalence and Epidemiological Risk Factors of Soil Transmitted Helminths Infections among School Children in two communities of Ogun State. 8th Annual Conference of the Zoology Society of Nigeria (ZSN) Lagos 2013. P 77.

Alemu A, Atnafu A, Addis Z, Shiferaw Y, Teklu T, Mathewos B, Birhan W, Gebretsadik S, Gelaw B (2011). Soil transmitted helminths and *Schistosoma mansoni* infections among school children in Zarima town, Northwest Ethiopia. BMC Infect. Dis. 11:189.

Amaechi EC, Okore OO (2013). The Prevalence and intensity of Urinary Schistosomiases among Pupils in Umuorie Community, Abia State, Eastern Nigeria. Faculty of Science University of Ibadan, Nigeria International Conference on Science and Sustainable Development in Nigeria. P 14.

Ariyo F, Fagbenro-Beyioku AF, Faweya T, Monye P, Oyewole F, Oyibo WA, Okoro C, Olufunlayo TF, Sanyaolu A, Soremekun B, Ukpong M (2007). Helminthic reduction with albendazole among school children in riverine communities of Nigeria. J. Rural Trop. Public Health 6:6-10.

Arora DR, Arora B (2010). Medical Parasitology.3rd ed. Delhi: CBS publishers.

Brooker S, Clements AC, Bundy DA (2006). Global epidemiology, ecology and control of soil-transmitted helminth infections. Adv. Parasitol. 62:221-61.

Cheesbrough M (1992). Medical Laboratory Manual for Tropical Countries, 2nd ed. Cambridge University press. P 203.

Damen JG, Lar P, Mershak, P, Mbaawuga EM, Nyary BW (2010). A Comparative Study on the Prevalence of Intestinal Helminthes in Dewormed and Non-Dewormed Students in a Rural Area of North-Central Nigeria. Lab. Med. 41(10):585-589.

Damen JG, Luka J, Biwan EI, Lugos M (2011). Prevalence of Intestinal Parasites among pupils in rural North Eastern, Nigeria. Niger. Med. J. 52(1):4-6.

Ekpenyong EA, Eyo JE (2008). Prevalence of Intestinal Helminths infections among schooling children in tropical semi urban communities. Anim. Res. Int. 5(1):804-810.

Emeka LI (2013). Prevalence of Intestinal Helminthic Infection among School Children in Rural and Semi Urban Communities in Nigeria. J. Dent. Med. Sci. 6(5):61-66.

Enimien OJ, Fana SA, Emmanuel WB (2014). Intestinal Helminthic Infection in Numan (Northeast Nigeria). Int. J. Innov. Appl. Stud. 5(2):102-105.

Jeffrey HC, Leach RM (1975). Atlas of Medical Helminthology and Protozoology, 2nd ed. London; Churchill Livingstone.

Leykun J (2001). Soil-Transmitted Helminth Infections and *Schistosomiasis mansoni* in School Children from Chilga District, Northwest Ethiopia. Ethiop. J. Health Sci. 11(2):79-87.

Odenning G (1976). Conception of terminology of host in Parasitology. Adv. Parasitol. 14:1-93.

Odoba MB, Otalu O, Balogun JB (2012). Prevalence of Helminth Parasite Eggs in Pupils and Playing Grounds of some Selected Primary Schools in Zaria, Nigeria. World J. Life Sci. Med. Res. 2(5):192.

Ojurongbe O, Oyesiji KF, Ojo JA, Odewale G, Adefioye OA, Olowe AO, Opaleye OO, Bolaji OS, Ojurongbe TA (2014). Soil transmitted helminth infections among primary school children in Ile-Ife Southwest, Nigeria: A cross-sectional study. Int. Res. J. Med. Med. Sci. 2(1):6-10.

Okon O, Oku E (2001). Prevalence of intestinal parasites among school children in two contrasting communities in Cross River state, Nigeria. Niger. J. Parasitol. 22:117.

Parker GA, Chubb JC, Ball MA, Roberts GN (2003). Evolution of complex life cycles in helminth parasites. Nature 425:480-484.

Pham-Duc P, Nguyen-Viet H, Hattendorf J, Zinsstag J, Phung-Dac C, Zurbrügg C, Odermatt P (2013). Ascaris lumbricoides and Trichuris trichiura infections associated with wastewater and human excreta use in agriculture in Vietnam. Parasitol. Int. 62:172-180.

Sebastián MS, Santi S (2001). Control of intestinal helminths in schoolchildren in Low-Napo, Ecuador: impact of a two-year chemotherapy program. Rev. Soc. Bras. Med. Trop. 33(1):69-73.

Sehgal R, Reddya, G V, Jaco J, Verweijb, Atluri, V, Rao S (2010). Prevalence of Intestinal Parasitic Infections among School Children and Pregnant Women in a Low Socio-Economic Area, Chandigarh, North India. Rev. Inf. 1(2):100-103.

Usip L, Esiest P, David NC (2013). The prevelance of human intestinal helminthes and the efficacy of anthelmentic levamisole drug in Abak Local Government Area of Akwalbom State Nigeria. Basic Res. J. Med. Clin. Sci. 2(5):52-58.

World Health Organization (WHO) (1994). Children in the tropics intestinal parasitoses. Technical Report Series No. 210. WHO, Paris. pp. 14-20.

World Health Organization (WHO) (2014). Intestinal Worms. Available at: http://www.who.int/intestinal_worms/epidemiology/en/

3

Parasitological examination of Ava stream used in irrigation in Enugu State, South-Eastern Nigeria: An implication for helminth transmission

D. E. Nwele[1], A. C. Uhuo[1], E. C. Okonkwo[2], G. A. Ibiam[3], C. S. Onwe[1], J. I Ugwu[3], J. C. Onwe[3] and A.C. Nwuzo[2]

[1]Department of Applied Biology, Faculty of Biological Sciences, Ebonyi State University, Abakaliki, Nigeria.
[2]Department of Applied Microbiology, Faculty of Biological Sciences, Ebonyi State University Abakaliki, Nigeria.
[3]Department of Medical Laboratory Science, Faculty of Health Sciences, Ebonyi State University Abakaliki, Nigeria

The parasitological examination of Ava stream in Abakpa Nike, Enugu state, Nigeria, was investigated to identify the species and load of parasites in the stream. Five different sites were sampled; these sites were chosen based on the different anthropogenic activities going on around these sites. One litre sample each was collected from these sites and were preserved by the addition of 10% formaldehyde (2 ml/L), and were transported to the laboratory. The samples were examined using the flotation technique. The species of parasites identified include *Schistosoma haematobium*, *Ascaris lumbricoides*, *Hymenolepsis nana*, *Enterobius vermicularis*, *Trichuris trichuria*, *Diphybotrium latum*, and *Taenia* species. Samples of Ogburugbu, recorded the highest egg count/L, as well as the highest percentage occurrence, and were followed closely by Ugwuagor, while Agbalike site recorded the least occurrence. This research has helped to identify the possible health risk of consuming raw vegetables and fruits irrigated with wastewater and organic manure sources and the unhygienic activities of the people.

Key words: Parasites, examination, anthropogenic, stream, irrigation.

INTRODUCTION

Wastewater is any water that has been adversely affected in quality by anthropogenic influence (Cornish et al., 1999). It comprises of liquid waste discharged by domestic residences, commercial properties, industry, and/or agriculture and can encompass a wide range of potential contaminants and concentrations. Around 90% of wastewater produced globally remains untreated, causing widespread water pollution, especially in developing countries (Tchobanoglous et al., 2003). Increasingly, agriculture is using untreated raw and wastewater for irrigation since there is no alternative to farmers due to scarcity of water resources (IWMI, 2010).

Raw and wastewater reuse for irrigation, as practiced with Ava stream, is a common practice to overcome water scarcity especially in countries with limited water resources (Mara and Cairncross, 1989). Defecation around water bodies as well as leakage of sewage materials containing human excreta is a common practice around Ava stream due to poor standard of living of those living around this area. However, there is a potential risk of transmission of diseases through excreted organisms in the case of irrigation with untreated wastewater, as well as those natural water receiving human excreta (Feachem, 1983).

The health hazards associated with direct and indirect wastewater use in irrigation agriculture are of two kinds: the rural health and safety problem for those working on the land or living on or near the land where the water is being used, and the risk that contaminated products from the wastewater use area may subsequently infect humans and animal through consumption or handling of the foodstuff or secondary human contamination by consuming foodstuff from animals that used the area (WHO, 1989). Therefore, in order to prevent the transmission of diseases, it has been recommended that only treated wastewater be used for crop irrigation (IWMI, 2010).

The objectives of this study were to (1) determine the level of parasite contamination of Ava stream and (2) assess the health risk of consuming vegetables and other fruits (uncooked) irrigated with the stream.

MATERIALS AND METHODS

The study area is Abakpa Nike, Eastern part of Enugu State, South-Eastern Nigeria. It covers about 8,422km^2, with an estimated population of about 120,835 people, which fall in the group of low income earners and majority of whom are farmers, traders and students, with few civil servants.

Ava stream forms the major source of water for people living in Abakpa metropolis. It flows East-South of Enugu State capital city. The stream is utilized for both domestic and agricultural purpose. Due to low economic status of the people living in this area, majority of their residential buildings lacks basic sanitary facilities and at such, faecal matters are deposited in and around the stream, as most people choose bush method as alternative method of defecation. Cultivation of vegetables of different kinds such as those used uncooked, in making African salad (spinach, tomato, pepper, lettuce, *Solanum* species, and *Telferia* species), and those eaten after cooking (pumpkin, bitter leaf, curry leaf, okra, etc.) form the major farming activities around the stream since the stream serves as the only source of irrigation water.

Experimental

Water samples from Ava stream, were collected from five (5) different sites: Ogwuagor, Ogburugbu, Agbalike, Silas, and Ifoo, respectively. These sites were chosen based on personal observation of the anthropogenic activities such as washing and bathing, scooping of the water into irrigation farms, promiscuous discharge of human and animal waste direct into the water, use of animal and poultry dropping for organic fertilization of the vegetable within and around the stream. Samples of 1 L were collected from each site and preserved by the addition of 10% formaldehyde (2 ml/L) in sterile bottles and are transported to Applied Biology Laboratory, Faculty of Biological Sciences, Ebonyi State University Abakaliki. In the laboratory, the samples were allowed to settle for 8 h to sediment overnight.

The supernatant was carefully removed and discarded without disturbing the sediment, using a siphon. The sediment was then transferred to centrifuge tubes. The walls of the sedimentation container was washed thoroughly using a spray bottle with detergent solution and the rinsing was added to the sediments in the centrifuge tubes. All the recovered materials were centrifuged at 1000 g for 15 min. The supernatant was removed and discarded and the sediments transferred to one tube. Flotation technique which is ideal for the recovery of helminth eggs was used by suspending the pellet in an equal volume of aceto-acetic buffer (pH 4.5) (Cheesbrough, 1999).

The mixture was re-centrifuged at 1000 g for 15 min and the pellets was transferred to microscope counting cell for final examination using objectives 10 and 40× magnification. The total number of eggs/L(N) present in the original sample of the water was determined from the equation:

$$N= AX/PV$$

Where N is the number of eggs per litre of sample, A is the number of eggs counted in the counting slide or the mean of counts from two or three slides, X is the volume of the final products (ml), P is the volume of the microscopic counting cell (0.3 ml), and V is original wastewater sample volume.

RESULTS

The analysis of this survey shows that Ava stream is contaminated with helminth eggs of different classes of Nematode, Trematode and Cestodes, respectively. Among the Trematodes encountered were the eggs of *Schistosoma haematobium*; Nematodes encountered were *Ascaris lumbricoides*, *Enterobius vermicularis*, and *Trichuris trichuria*. Among the Cestodes implicated were the eggs of *Taenia* spp., *Hymenolepsis nana*, and *Diphyllobotrium latum.* The quantification of these helminth eggs in Ava water samples from the five different sites is shown in Table 1. As shown in Table 1, the Ava stream contained different groups of helminthes, including those mentioned earlier. There was significant difference between the total egg count/L in samples of the five different sites sampled ($P< 0.05$). This difference was the most significant between Ogwuagor and Ogburugbu.

Table 2 shows the percentage of occurrence of different helminth eggs from the five different sites sampled. The effluents of Ogburugbu had the greatest percentage followed by Ogwuagor. Silas and Ifoo shared equal percentage occurrence while Agbalike has the least occurrence.

Of all the parasitic eggs isolated in the Ava stream from the five sites sampled, *A. lumbricoides* was the most frequently recovered (Table 1).The mean total egg count/L in the five different sites was much higher in Ogburugbu (mean= 2.4 egg/L) than the other four sites. This was closely followed by Ogwuagor with the mean total egg count/L of 2.2 eggs/L. However, the difference between the mean total egg count/L in Ogburugbu and that of Ugwuagor was 2.4 eggs/L.

DISSCUSION

Water is a vital need of both plants and animals and this has necessitated to a higher demand to maintain our day to day needs. The Enugu State Government has developed water production and sanitation scheme whose sole responsibility is to service water to the entire people of Enugu and environs. This notwithstanding has led to scampering of water by villagers who do not have access

Table 1. Quantification of parasite eggs in Ava water samples from five different sites of sample collection.

Parasite	Sites of collection and number of parasites identified				
	Ogwuagor (n=11)	Ogburugbu (n=12)	Agbalike (n=5)	Silas (n=6)	Ifoo (n=6)
Schistosoma haematobium	1	4	1	0	0
Ascaris lumbricoides	3	3	1	4	1
Hymenolepsis nana	3	2	0	1	1
Enterobius vermicularis	2	0	2	0	2
Trichuris trichuria	0	0	1	0	0
Diphylobotrium latum	0	2	0	0	1
Taenia spp.	2	1	0	1	1

Table 2. Percentage occurrence of helminth eggs from the five different sites sampled.

Site	Volume of sample examined (V, L)	No. present (%)
Ogwuagor	1	11 (1.1)
Ogburugbu	1	12 (1.2)
Agbalike	1	5 (0.5)
Silas	1	6 (0.6)
Ifoo	1	6 (0.6)

to either good portable water sources to service their domestic needs. Ava people lives at the Eastern part of Enugu State. This has affected their easy access to portable water, may be because of their bed rock. Ava water is one of the most accessible water sources among their inhabitants as this is seen to assist their fast need of water. As pertaining to food and water, this stream has served a dual purpose of agricultural irrigation and drinking sources not only, but washing, bathing and even serving their domestic animals. This has transposed to the emergence of other uncalled act of defecating within and around the stream, dumping refuse and other ill transmitting activities which has yielded propagation of parasitosis among their users and domestic animals. This was evident in earlier work of El kettani et al. (2008) which showed prevalence of intestinal helminthiasis in a group exposed to raw wastewater to be 4.7%. The use of domestic animal dung as organic fertilizers are not left out in their agricultural practices, but this is seen to do more harm than good, because these waste animal dungs serve as a vehicle of parasite transmission especially their infective stages as well as other human pathogens. Of these, helminth infections pose the greatest risks and are a serious public health concern (WHO, 1989). However, the result of this study showed that Ava stream contained different helminth eggs. These parasite species identified were classified into three groups: Cestodes, Trematodes, and Nematodes. The commonest among the identified species however, was *A. lumbricoides* with mean concentration of 2.4 eggs/L. This result agrees with other studies which showed the frequent presence and high concentration of *Ascaris* eggs in wastewater and raw sewage worldwide as reported by Stott (1997), Valbuena (2002) and Coelho et al. (2001).

In addition, the resistance of *Ascaris* to external conditions (Crompton, 1989), allows the eggs to remain viable longer than other helminth eggs (Arfaa, 1987; Erdogrul and Sener, 2005).

The stream used in irrigation in this study does not respond to the WHO standard of <1 helminth egg/L (WHO, 1989), since this study implicated 1 egg/L of water. Consequently, it is not recommended for irrigation of vegetables especially those which are eaten raw. This finding is in accordance with the earlier work of Hajjami et al. (2013) which recorded mean concentration of 8.98 eggs/L. There was a great variation in parasite eggs identified in the five different sites sampled, with Ogburugbu, recording the highest number of parasite eggs/L. This was probably due to the rate of human activities going on around these sites as earlier discussed in this work. The consumption of raw vegetables and fruits irrigated with this stream also plays an important role in the transmission of parasitic contaminations (Arther et al., 1981; Daryani et al., 2008). More reason could be attributed to the fact that population around this area were economically poor, they have a very low standard of living and as such, they lack basic sanitary facilities and therefore, all faecal and other human waste are discharged directly into the stream. The occurrence of the egg of *S. haematobium* indicates the presence of infected individual who possibly could have passed their infected urine into the stream few hours before the time of sample collection.

An earlier survey conducted by Emenyonu et al. (2010) on the effect of waste water use on vegetable crop production in Imo State, Nigeria, shows that wastewater use has negative effect on vegetable crop production some of which they mentioned to include bad odour as well as parasitic infection of the irrigated crops and vegetables. Similarly, studies in different countries by World Health Organization (WHO, 1989), have served to highlight the impact of these practices on the environment and health, especially when these wastewaters, are used in irrigation agriculture.

CONCLUSION/RECOMMENDATION

Ava stream is highly loaded with eggs of helminth parasites and this could mean a serious danger to people, animals and the environment in general. Poor sanitary hygiene as observed within Ava stream illustrates the sanitary and environmental risks associated with the use of this stream as shown by the result of this study. These risks would even be higher if crops and vegetables irrigated with this stream are consumed raw. Both people working around Ava stream and the consumers of the vegetables and other agricultural products produced and irrigated with the stream have the greatest risk of helminthic infections.

However, it is recommended that government should issue warning to owners of residential houses around this stream, forcing them to build good toilets or have their houses sealed by her sanitation agency, this will discourage defecation around the stream. Also, inhabitants of these areas should be educated on the dangers of urinating inside the stream as well as direct discharge of sanitary waste into the stream. A survey should be carried out to identify those individuals passing blood in urine. Lastly, agricultural produce around this stream should be washed with clean water and properly cooked before consumption.

REFERENCES

Arther RG, Fitzgerald PR, Fox JC (1981). Parasite ova in anaerobically digested sludge. J. Water Pollut. Contr. Fed. 53:1333-1338.

Arfaa F (1987). Medical helminthology, Vol. II. Round worms. Tehran, Danesh Pajooh press, (In Persian).

Cheesbrough M (1999). District Lab. Practices in Tropical Countries. University press. Cambridge (UK). pp. 216-239.

Coelho LM, Oliveira SM, Milman MH, Karasawa KA, Santos RD (2001). Detection of transmissible forms of enteroparasites in water and vegetables consumed at school in Sorocaba, Sao Paulo state, Brazil. Rev. Soc. Bras Med. Trop. 34:479-482.

Cornish GA, Mensah E, Ghesquire P (1999). An Assessment of surface water quality for irrigation and its implication for human health in perr-urban zone of Kumasi Ghana.Report OD/TN/95, Sept. HR Wallington, UK. p. 460.

Crompton DWT (1989). Biology of the *Ascaris.* In: Crompton DWT, Neisheim MC, Pawlowski Z. S, eds. *Ascaris* and its prevention and control. London, Taylor and Francis Publication, pp. 9-69.

Daryani A, Ettehad GH, Sharif M, Ghorbani L, Ziaei H (2008). Prevalence of intestinal parasites in vegetables consumed in Ardabil, Iran. Food Contr. 19:790-794.

El Kettani S, Azzouzi E, Boukachabine K, El Yamani M, Maata A, et al. (2008) Intestinal parasitosis and use of untreated wastewater for agriculture in Settat, Morocco. East Mediterr Health J. 14:6.

Emenyonu CA, Odii MA, Ohajianya DO, Henri-Ukoha A, Onyemauwa SC, Ben-Chendo GN , Munonye OU (2010). Effects of Waste Water Use on Vegetable Crop Production in Imo State, Nigeria. Researcher 2(10):47-56.

Erdogrul OR, Sener H (2005). The contamination of various fruit and vegetable with Enterobius vermicularis, *Ascaris* eggs, Entamoeba histolytica cysts and *Giardia lamblia* cysts. Food Contr. 16:557-560.

Feachem R (1983). Sanitation and Disease: Health aspect of excreta and wastewater management. New York, John Willy and Sons (for the World Bank).

Hajjami K, Ennaji MM, Fouad S, Oubrim N, Cohen N (2013). Wastewater Reuse for Irrigation in Morocco: Helminth Eggs Contamination's Level of Irrigated Crops and Sanitary Risk (A Case Study of Settat and Soualem Regions). J. Bacteriol. Parasitol. 4:163.

International Water Management Institute (IWMI), (2010). Wastewater use in agriculture; Not only an issue where water is scarce. Water Issue Brief p.4.

Mara D, Cairncross S (1989). Guideline for the safe use of wastewater and excreta in Agriculture. Geneva World Health Organization.

Stott R (1997). A survey of the microbial quality of wastewaters in Ismailia, Egypt and the Implications of wastewater reuse. Water Sci. Technol. 35(11-12):211-217.

Tchobanoglous G, Burton FL, Stensel HD (2003). Wastewater engineering (Treatment and Disposal Reuse)/Metcalf and Eddy, Inc. (4th ed). Mcgraw-Hill Book Company.

Valbuena D (2002). Detection of intestinal helminths and contamination-indicator bacteria In treated and untreated wastewater. Interciencia 27(12):710-714.

WHO, (1989). Health guidelines for the use of wastewater in agriculture and aquaculture. Wld. Hlth. Org. Tech. Rep. Ser. No. 778.

4

Challenges of schistosomiasis prevention and control in Ethiopia: Literature review and current status

Tadesse Hailu Jember

Department of Microbiology, Immunology and Parasitology [DMIP], College of Medicine and Health Science, Bahir Dar University, P. O. Box 79, Bahir Dar, Ethiopia.

The effect of schistosomiasis on human health is high especially in resource poor countries like Ethiopia. Intestinal schistosomiasis is the major causes of morbidity in most parts of the country. School age children are the vulnerable segments of the population. Lack of sanitation, potable water, environmental modification, health education, mass treatment, awareness and manageable water harvesting programs like irrigation are the contributing factors for schistosomiasis. Proper health impact assessment for new irrigation schemes and other water resources projects will provide a solid basis for the incorporation of health safeguards at design and construction plans in Ethiopia.

Key words: Schistosomiasis, irrigation schemes, sanitation, migration.

INTRODUCTION

Intestinal parasites adversely affect the health of humans in many parts of the world. They continue to be global problem, particularly among children in developing nations (WHO, 1981). The most prevalent and important helminths in developing countries are the soil-transmitted group and schistosomes (WHO, 2004). Intestinal helminthiasis is intimately related to poverty and most infections are entrenched particularly in tropical regions where the biophysical environment and cultural practices favor transmission (Cheesbrough, 2000). For instance, of the global burden of schistosomiasis, an estimated 85% is found in sub-Saharan Africa (WHO, 2006). Schistosomiasis also known as bilharziasis is a parasitic disease caused by trematode fluke of the genus *Schistosoma*. Schistosomiasis is mainly found in Asia, Africa and South America in areas where fresh water snails serve as the intermediate hosts (Gryseels et al., 2006). Schistosomiasis is endemic in 74 tropical countries worldwide, affecting over 200 million people while 500 to 600 million people are at risk of becoming infected (Rozendaal, 1997; Steinmann et al., 2006; Siddiqui, 2011).

Schistosomiasis is a disease caused by one of six *Schistosoma* species, namely *Schistosoma haematobium*, *Schistosoma guineensis*, *Schistosoma intercalatum*, *Schistosoma mansoni*, *Schistosoma japonicum*, and *Schistosoma mekongi* (Davis, 2009). A number of animal species as *Schistosoma margrebowiei* or *Schistosoma bovis* may also occasionally infect humans (McManus and Loukas, 2008).

In Ethiopia, infection of intestinal parasites is the second most predominant cause of outpatient morbidity (Tesfa-Yohannes and Kloos, 1988). Of which, intestinal schistosomiasis is the major causes of morbidity in most parts of the country (Tedla, 1981; Tedla and Jemaneh, 1998) at an altitude of 1000 to 2000 m above sea level (Yomi et al., 2010).

Information on countrywide epidemiology of schistosomiasis in Ethiopia is limited. But numberous studies conducted in different localities showed that the prevalence of *S. mansoni* was high among school children. For instance, 73.7% in Bushulo village near Lake Hawassa (Terefe et al., 2011), 43.2% in Ziway town and 31.6% in Wodogenet (Legesse et al., 2009), 67.6% in Finchaa valley (Haile et al., 2012), 82.8% in Sanja town (Alebie et al., 2014) and 73.9% in Waja-Timuga (Abebe et al., 2014) are prevalence values in some of the areas with high prevalence of *S. mansoni* infection in Ethiopia.

Schistosomes need fresh and still water for their survival. They are therefore found in lake-side communities, around water development schemes like dams. People are in contact with infested fresh water during their normal daily activities of fishing, farming, swimming, washing, bathing, recreation, and irrigation (WHO, 1999; WHO, 2007; NIPD, 2013).

The first step in human infection involves penetration of the skin by cercariae which come out from a freshwater intermediate host. Inside the human host, cercariae develop into sexually mature egg-laying worms in the hepatic portal system. The eggs are passed out from the human body in feces or urine, depending on the species of the parasite. The eggs hatch on contact with water and release free swimming miracidia which infect the intermediate snail host. This chronic infection can persist for decades causing disease of intestinal or urinogenital system depending on species types (Hotez et al., 2007). Children are the most heavily infected, because they spend hours playing and swimming in water and lack partial immunity (WHO, 1999).

The pathological effects of schistosomes cause a range of morbidities which seems to be influenced to a large extent by the nature of the induced immune response and its effects on granuloma formation and associated pathologies in target organs (Dunne and Pearce, 1999).

Diagnosis is made by microscopic identification of *S. mansoni* or *S. japonicum* eggs in stool or *S. haematobium* eggs in urine (Katz et al., 1972). Serologic tests are useful to diagnose light infections (in travelers) and in non-endemic or low transmission areas. Antibody tests do not distinguish between past and current infection (WHO, 2007; CDC, 2007).

The drug of choice for the treatment of schistosomiasis is praziquantel. However, despite successful treatment will improve the health of the people, schistosomiasis treated persons can become re-infected. Prevention is still preferable to treatment since in the absence of obvious symptoms, irreversible damages may be done before the infection is detected (IAMAT, 2012). Improved sanitation and potable water, environmental modification, health education, mass treatment and limiting human-water contact offer long-term control of schistosomiasis. Proper health impact assessment of new irrigation schemes and other water resources projects will provide a solid basis for the incorporation of health safeguards at design and construction plans (WHO, 2001).

The present review presents current challenges in the prevention and control of schistosomiasis in Ethiopia. The derived models used in prevention approaches are not practically applicable in Ethiopia. There are several reasons why prevention methods are not applicable and become challenging in Ethiopia. For instance, inadequate budget and lack of awareness are the dominant factors.

WATER HARVESTING PROGRAMS

A major mean of rehabilitating and reconstructing the natural resource base is through comprehensive water harvesting development. The impact of prolonged available surface water in newly developed irrigation areas is a predisposing factor for water and vector-borne diseases. Areas that are periodically affected by schistosomiasis are exposed to continued year round attack due to suitable environment for snails' survival (Behailu and Haile, 2002).

A major factor associated with the rise of schistosomiasis is water development projects, particularly manmade lakes (hydroelectric power) and irrigation schemes (agriculture), which can lead to shifts in snail vector populations (Patz et al., 2000; WHO, 2002). On the other hand, water stagnation and weed growing due to inadequate water management sustain the life of the snails to complete the life cycle of schistosomes (Boeele and Madsen, 2006).

Many surface irrigation systems in Africa create favorable snail-breeding conditions that facilitate the transmission of schistosomiasis (WHO, 2004). For instance, there was an introduction of *S. mansoni* to Mauritania and Senegal after construction of huge Diama dam on the Senegal River (Gryseels et al., 2006), and Koka dam in Ethiopia (Kloos et al., 1988).

Irrigation schemes are dynamic agro-ecosystems that can transport snails a long way along the canals and where local events can either provide habitat-friendly conditions or inhibit snail populations (Dale and Polasky, 2007). Generally, the variability within irrigated areas, the canal type, the distance of sites from the canal, the composition and density of aquatic vegetation (Appleton, 1978; Khallaayoune, 1998a), the season (WHO, 2001), specific local conditions such as water stagnation, water depth and shading (Dale and Polasky, 2007) and water flow velocities and the location of breeding sites affect the presence and density of snails among sites. Low-flow velocities and locations found at the starting point of low-order canals are favorable to get aeration of the water and food availability for snails (Boelee and Madsen, 2006).

Environmental changes linked to water resource

development like water harvesting for irrigation and population growth which lead to sanitation problem in Ethiopia have facilitated the recent spread of schistosomiasis to areas where it was not endemic before (Gryseels et al., 2006; Li, 2007). For instance, there is an introduction of *S. mansoni* into the upper and middle Awash valley in Ethiopia, following the establishment of large-scale irrigation schemes of Wonji sugar factory and the prevalence steadily increased up to 20% in 1980 (Kloos et al., 1988) and 81.9% in 1988 (Simonsen, 1990). Similarly, the prevalence of *S. mansoni* of 10.7% was recorded after the introduction of new irrigation schemes in South and Central Tigray (Dejenie and Petros, 2009).

Currently, agriculture development especially in water harvesting for irrigation in Ethiopia have increased from time to time. Similarly, Ethiopia is now constructing a huge hydroelectric power in East Africa in different rivers. The existence of surface water for irrigation and power will sustain the snails in endemic areas of the countries. Therefore, water harvesting program have a negative impact on the health of community by sustaining the life of the snails unless critical plans are taken by both Ministry of Health and Agriculture for snail control and preventions.

Vector control

The aquatic snails (*Biomphalaria* and *Bulinus* species) cannot usually survive without water serving as intermediate hosts of *S. mansoni* and *S. haematobium*, respectively. All species of *Biomphalaria* and *Bulinus* are hermaphrodite, possessing both male and female organs and being capable of self- or cross-fertilization. A single specimen can invade and populate a new habitat. A snail lays up to 1000 eggs during its life, which may last more than a year. All the characteristics mentioned earlier make it difficult to eliminate snails in schistosomiasis endemic areas (NIPD, 2013).

Recently, a standard cement tanker which completes the green-house effect heating system is developed to control snails. It constituted movable dark canvas covers, which allowed the temperature to be controlled between 20 to 24°C especially during the coldest months of the year. The tanker facilitates the mortality of *Biomphalaria tenagophila* (Rosa et al., 2013).

Due to the relative high cost of molluscicides, no public health intervention measure is taking place in Ethiopia. However, since 1991 interest has rekindled in the use of local molluscicidal plant called Endod (*Phylolacca dodecondra*) for community based intervention of snails (Lambert et al., 1991).

Sanitation

Sanitation is one basic problem in Sub-Saharan countries. The ova of the parasites can easily access to the water when infected people urinate or defecate close to a water source. The contaminated water will be used for different purpose for the community. Schistosomiasis is easily spread in communities that do not have access to toilets or sanitation facilities and awareness about the means of transmission (Assefa et al., 2013). Despite of higher coverage of latrine reports in some places of Ethiopia, prevalence of intestinal parasites is significantly higher due to poor water supply (Mengistu et al., 2007). In addition, open air defecation, washing and bathing also favor transmission of schistosomiasis in Tikur Wuha area of Ethiopia (Mitiku et al., 2010).

The awareness of the people regarding how schistosomiasis infection occurs is limited in Ethiopia. Majority of the people practice open field defecation and urination. Most of the people wash their body in the streams and cross the streams during their daily activities. As a result, the possibility of schistosomiasis infection is high. Unless community awareness is created on the transmission, prevention and control of the infection, the impacts of the disease will live for unlimited time in the community.

Mass deworming

Current population-based schistosomiasis treatment programs are a first step to reduce the global burden of Schistosoma-related disease; however, they might not dramatically reduce parasite transmission in highly endemic areas (King et al., 2006). Prazquantel is given based on the WHO recommendations in a community level; schools with schistosomiasis high prevalence were mass treated with praziquantel annually (Montresor et al., 2013).

Although 42 schistosomiasis endemic countries in Africa have been determined, factors such as absence of disease mapping in endemic areas, limited availability of praziquantel, and lack of health infrastructure hampered programme implementation and medication distribution (NIPD, 2013). The proportion of people treated for schistosomiasis in the African Region in 2011 was only 9.8% of the people requiring treatment (WHO, 2013). Data for 2012 also showed that only 14.4% of people requiring treatments were reached (WHO, 2014).

Periodic treatment of the risk population groups will cure mild symptoms and prevent infected people from developing severe, late-stage chronic disease. However, a major limitation to schistosomiasis control has been the limited availability of praziquantel. Treatment of the infection brings immediately benefit and has long-lasting effect on morbidity and prevent irreversible sequelae in adulthood (WHO, 2002). On the other hand, there are some disadvantages of praziquantel treatment, which include the appearance of drug resistance in the treatment of *S. mansoni* along with allergic or hypersensitivity reactions against praziquantel treatment (Chai, 2013).

Though, 50% of the people suffering from schistosomiasis

are school-aged children, and chemotherapy is mostly targeted at school-aged children; the community-based approach has been well received and implemented in schistosomiasis control to address out-of-school children in Africa (NIPD, 2013).

In Ethiopia, countrywide epidemiology of schistosomes is limited but there were some prevalence reports which showed high infection rates in endemic areas of the country (Terefe et al., 2011; Haile et al., 2012; Legesse et al., 2010). *S. mansoni* has been recorded in all regions of Ethiopia and is rapidly spreading with water resource development and population movements (Erko et al., 2002). But limited availability of praziquantel and absence of community-based mass treatment approach of children based on the WHO guide line in endemic areas are the challenges to control schistosomiasis. Unless mass treatment strategy is not applicable in endemic areas of schistosomiasis, prevention and control of schistosomes may not be effective in Ethiopia.

Co-infection with HIV

Mass deworming could have beneficial effects on HIV-1 transmission dynamics. *S. mansoni* infections can increase HIV-1 replication, cell-to-cell transmission and HIV progression as measured by reduced CD4+ T lymphocytes counts in HIV infected individuals (Mazigo et al., 2013). On the other hand, HIV affects *S. mansoni* egg excretion which potentially affects parasitic diagnosis of schistosomiasis infection and releasing of circulating schistosome worm antigens which is important for detection of the worm (Mwanakasale et al., 2003).

Immunological studies have demonstrated that T-cells from HIV-1 positive individuals co-infected with *S. mansoni* responded to egg antigens by producing less interleukin (IL)-4 and IL-10 and a lower amount of interferon-gamma (IFN-γ) as compared to those from individuals infected with *S. mansoni* alone, indicating immune skewing from Th2 to Th1 (Mwinzi et al., 2001).

Prevalence of HIV and *S. mansoni* is high in sub-Saharan Africa like Ethiopia. The impact of *S. mansoni* may be high if HIV infected individuals are co-infected with *S. mansoni.* So, deworming of HIV positive individuals living in *S. mansoni* endemic areas may decrease HIV-1 viral loads and increases CD4+ T lymphocyte counts. Deworming of *S. mansoni* will prevent HIV cases to early developing of AIDS stage. Prevention of HIV cases from *S. mansoni* co-infection especially in endemic areas of Ethiopia is limited.

Vaccination

Vaccine represents an essential component for the future control of schistosomiasis as an adjunct to chemotherapy. The immune response to schistosome infection, both in animal models and in humans, suggests that development of a vaccine may be possible (McManus and Loukas, 2008). For instance, DNA-based vaccines protect against zoonotic schistosomiasis in water buffalo and schistosome vaccines capable of reducing water buffaloes' fecal egg output by 45% (Dadara et al., 2008). The vaccine against the schistosomes decreases the fecundity of the adult worm, the worm burden or egg production and viability. But difficulties in obtaining good expression levels and in scaling up production for the limited number of antigens selected have turned out to be another major obstacle (Bergquist et al., 2005). Therefore, absence of vaccine in the country is also another challenge in the prevention and control of schistosomiasis in Ethiopia.

CURRENT POLICY OF THE COUNTRY

The major challenges in schistosomiasis control remain the scale-up of treatment and to the need to advocate for increased resources for implementation of treatment programmes, provision of potable water, adequate sanitation, hygiene education and snail control. In some countries, schistosomiasis transmission may have been interrupted through active control programmes and/or changing the socio-economic conditions (WHO, 2013).

Still today, most sub-Saharan African countries endemic for schistosomiasis are in a position to establish a country-wide elimination program. While elimination should be considered at the national level, there may be low prevalence areas within each country where elimination could be achieved given adequate resources and political commitments (Rollinson et al., 2012). For instance, schistosomiasis was endemic in Egypt since the ancient times and the most important public health problem but the present findings revealed a decrease in the prevalence of schistosomiasis due to the current policy of schistosomiasis control in Egypt (Zaher et al., 2011).

In Ethiopia, schistosomiasis infection is recorded in all regions of the country and the prevalence is significantly high, but there is no specific policy launched to decrease the disease burden especially in endemic areas.

POPULATION MOVEMENT AND MIGRATION

Population movement, rural to urban migration, forced displacement migration of workers and the rise of ecotourism have all contributed to the increase in schistosomiasis (WHO, 2007). Environmental changes linked to water resource development and migrations have facilitated the recent spread of schistosomiasis to areas where it is not endemic before (Gryseels et al., 2006; Li et al., 2007).

Migration of workers from endemic areas to irrigation projects or major hydroelectric power plants, followed by contamination of reservoirs with human feces, and subsequent inevitable increasing contact of communities adjacent to the reservoir for the purpose of bathing, laundering in adults or swimming and playing in children will result in an expansion of infection in regions once free of infection (Steinmann et al., 2006).

The people of Africa move from one place to another for different reasons such as war, famine, food and water. Migration is commonly cited as a significant factor in the spread of other infectious diseases (Lurie et al., 2003). Immigrants and refugees originating from areas where infections persist can pose a significant challenge for national disease control and/or elimination strategies.

The prevalence levels of infectious diseases in a world of increasing travel and migration make national disease control or elimination almost impossible (Gushulak and MacPherson, 2004). For instance, high schistosomiasis seroprevalence was reported from Sudanese lost boys and girls (Posey et al., 2007) and a single *S. hematobium* positive case was also identified from Somalian immigrants in America (Neal, 2004). The transmission of schistosomiasis is also high around water bodies due to movement of people around the great lakes such as Lake Victoria in Uganda (Standle et al., 2009). There is also a report that confirmed displacement of population has introduced *S. mansoni* into Somalia and Djibouti (Gryseels et al., 2006). In addition, 80% of the tourists were positive for schistosomula tegument antigens (Sm-Teg) which helps in community-based schistosomiasis control programs from non-endemic areas (Grenfell et al., 2013).

Due to its geographical position, as well as environmental and geo-political developments in the region, Ethiopia is likely to continue to receive migrants from neighborhood countries including Eritrea, Somalia, South Sudan and Sudan in 2014 and 2015 (UNHCR, 2014). Control and prevention of *Schistosoma* may be very difficult in Ethiopia due to migrant people from endemic countries like Sudan (Ahmed et al., 2012), Somalia and Eritrea (IAMAT, 2014) and South Sudan (Deribe et al., 2011).

It is very difficult to totally avoid sanitation problems in refugee camps of Ethiopia. In addition, there is no screening policy of migrants' people in the refugee camps for schistosomiasis as soon as they arrive. This will have its own contribution on the prevalence of schistosomiasis infection in the country unless measure is taken.

RECOMMENDATION

Environmental modification preventing snail vectors and limiting human water contact offers long-term control of schistosomiasis. Health education is a fundamental component that ensures community participation in control interventions. In areas of high prevalence and intensity of infection, community-based mass treatment of high-risk groups with praziquantel offers the most efficient way to achieve the recommended strategy for morbidity control. Proper health impact assessment of new irrigation schemes and hydroelectric water resources projects will provide a solid basis for the incorporation of health safeguards at design and construction plans. Considering migration as a risk factor, screening of schitosomiasis should be adopted for migrants from endemic areas in East Africa to Ethiopia.

Conflict Interests

The authors declare that there is no conflict of interests regarding the publication of this article.

REFERENCES

Abebe N, Erko B, Medhin G, Nega B (2014). Clinico-epidemiological study of *Schistosomiasis mansoni* Waja-Timuga, District of Alamata, Northern Ethiopia. Parasit. Vectors 7:158.

Ahmed AM, Abbas H, Mansour FA, Gasim G I, Adam I (2012). Schistosoma haematobium infections among schoolchildren in central Sudan one year after treatment with praziquantel. Parasit. Vectors 5:108.

Alebie G, Erko B, Aemero M, Petros B (2014). Epidemiological study on Schistosoma mansoni infection in Sanja area, Amhara region, Ethiopia. Parasit. Vectors 7:15.

Appleton CC (1978). Review of literature on abiotic factors influencing the distribution and life cycles of bilharziasis intermediate host snails. Malacol. Rev. 11(1/2):1-25.

Assefa A, Dejenie T, Tomass Z (2013). Infection prevalence of *Schistosoma mansoni* and associated risk factors among schoolchildren in suburbs of Mekelle city, Tigray, Northern Ethiopia. Momona Ethiop. J. Sci. 5(1):174-188.

Behailu M, Haile M (2002). Water harvesting in northern Ethiopia: Environmental, health and socio-economic impacts. Integrated water and land management research and capacity building priorities for Ethiopia. pp. 185-191.

Bergquist NR, Leonardo LR, Mitchell GF (2005). Vaccine-linked chemotherapy: can schistosomiasis control benefit from an integrated approach? Trends Parasitol. 21:112-117.

Boelee E, Madsen H (2006). Irrigation and schistosomiasis in Africa: Ecological aspects. Colombo, Sri Lanka: International Water Management Institute. IWMI Research Report 99:39.

Centers for disease control and prevention (CDC) (2007). Parasites-Schistosomiasis. Resources for Health Professionals.

Chai JY (2013). Praziquantel Treatment in Trematode and Cestode Infections. Infect. Chemother. 45(1):32-43.

Cheesbrough M (2000). District Laboratory Practice in Tropical Countries, 2nd ed., Part 1. Cambridge University Press, UK.

Dadara AA, Li YS, Xiong T, Zhou J, Williams GM, McManus DP, Feng Z, Yu XL, Gray DJ, Harn DA (2008). DNA-based vaccines protect against zoonotic schistosomiasis in water buffalo. Vaccine 26(29-30):3617-25.

Dale VH, Polasky S (2007). Measures of the effects of agricultural practices on ecosystem services. Ecol. Econ. 64:286-296.

Davis A (2009). Schistosomiasis. In: Cook GC, Zumla AI (eds.), Manson's Tropical Diseases, 22nd edition. Saunders: Elsevier. pp. 1425-1460.

Dejenie T, Petros B (2009). Irrigation Practices and Intestinal Helminth Infections in Southern and Central Zones of Tigray. Ethiop. J. Health Dev. 23(1):48-56.

Deribe K, Eldaw A, Hadziabduli S, Kailie E, Omer MD, Mohammed AE , Jamshed T, Mohammed EA, Mergani A , Ali GA, Babikir K, Adem A, Hashim F (2011). High prevalence of urinary schistosomiasis in two communities in South Darfur: implication for interventions. Parasit. Vectors 4:14.

Dunne DW, Pearce EJ (1999). Immunology of hepatosplenic

schistosomiasis mansoni: a human perspective. Microbes Infect. 1:553-560.
Erko B, Abebe F, Berhe N, Medhin G, Gebre-Michael T (2002). Control of Schistosoma mansoni by the soapberry Endod (*Phytolacca dodecandera*) in Wollo, Northeastern Ethiopia post intervention prevalence. East Afr. Med. J. 79:198-201.
Grenfell RF, Martins W, Drummond SC, Antunes CM, Voietal, Otoni A, Oliveira AA, Silva-Moraes V, Oliveira ER, Oliveira E, Lambertucci JR, Fonseca CT, Coelho PM (2013). Acute schistosomiasis diagnosis: a new tool for the diagnosis of schistosomiasis in a group of travelers recently infected in a new focus of *Schistosoma mansoni*. Rev. Soc. Bras. Med. Trop. 46(2):208-213.
Gryseels B, Polman K, Clerinx J, Kestens L (2006). Human schistosomiasis. Lancet 368:1106-1118.
Gushulak BD, MacPherson DW (2004). Globalization of Infectious Diseases: The impact of migration: Clin. Infect. Dis. 38:1742-1748.
Haile S, Golassa L, Mek Z (2012). Prevalence of *Schistosoma mansoni* and effectiveness of Praziquantel in school children in Finchaa valley, Ethiopia. J. Parasitol. Vector Biol. 4(3):25-30.
Hotez PJ, Molyneux DH, Fenwick A, Kumaresan J, Sachs SE, Sachs JD (2007). Control of neglected tropical diseases. N. Engl. J. Med. 357:1018-1027.
International Association for Medical Assistance to Travelers (IAMAT) (2012). World schistosomiasis risk chart. IAMAT http://www.iamat.org/pdf/World_Schistosomiasis_Risk_Chart.pdf
Katz N, Chaves A, Pellegrino J (1972). A simple device for quantitative stool thicksmear technique in *Schistosomiasis mansoni*, São Paulo. Rev. Inst. Med. Trop. 14:397-440.
Khallaayoune K, Laamrani H, Madsen H (1998a). Distribution of *Bulinus truncatus*, the intermediate host of *Schistosoma haematobium* in an irrigation system in Morocco. J. Freshwater Ecol. 13(1):129-133.
King CH, Sturrock RF, Kariuki HC, Hamburger J (2006). Transmission control for schistosomiasis-why it matters now. Trends Parasitol. 22:12.
Kloos H, Lo CT, Birrie H, Ayele T, Tedla S, Tsegay F (1988). Schistosomiasis in Ethiopia. Soc. Sci. Med. 26:803-827.
Lambert JD, Temmink JH, Marquis J, Parkhurst RM, Lugt CB, Lemmich E, Wolde-Yohannes L, de Savigny DE (1991). Safety evaluation of a plant molluscicide. Regul. Toxicol. Pharmacol. 14:189-201.
Legesse L, Erko B, Hailu A (2010). Current status of intestinal Schistosomiasis and soiltransmitted helminthiasis among primary school children in Adwa Town, Northern Ethiopia. Ethiop. J. Health Dev. 24(3):191-197.
Legesse M, Jones CR, Singh SK, Erko B, Mekonnen Y (2009). Community's awareness about intestinal schistosomiasis and the prevalence of infection in two endemic localities of Ethiopia. Ethiop. J. Health Sci. 19:2.
Li YS, Raso G, Zhao Y, He HK, Ellis M, McManus DP (2007). Predicted impact of large water management projects on schistosomiasis transmission and control in the Dongting Lake Region, China. Emerg. Infect. Dis 13:973-979.
Lurie MN, Williams BG, Zuma K, Mkaya-Mwamburi D, Garnett GP, Sweat MD, Gittelsohn J, Karim SS (2003). Who infects whom? HIV-1 concordance and discordance among migrant and non-migrant couples in South Africa. AIDS 17(15):2245-52.
Mazigo H, Nuwaha F, Wilson S, Kinung'hi SM, Morona D, Waihenya R, Heukelbach J, Dunne DW (2013). Epidemiology and interactions of Human Immunodeficiency Virus-1 and *Schistosoma mansoni* in sub-Saharan Africa. Infect. Dis. Poverty 2:2.
McManus DP, Loukas Alex (2008). Current Status of Vaccines for Schistosomiasis. Clin. Microbiol. Rev. 21(1):225-242.
Mengistu A, Gebre-Selassie S, Kassa T (2007). Prevalence of intestinal parasitic infections among urban dwellers in southwest Ethiopia. Ethiop. J. Health Dev. 21(1):12-17.
Mitiku H, Legesse M, Teklemariam Z, Erko B (2010). Transmission of Schistosoma mansoni in Tikur Wuha area, Southern Ethiopia. Ethiop. J. Health Dev. 24(3):180-184.
Montresor A, Crompton DWT, Hall A, Bundy DAP, Savioli L (2013). Schistosomiasis and Intestinal Parasites Unit, Division of Control of Tropical, WHO, Geneva.
Mwanakasale V, Vounatsou P, Sukwa TY, Ziba M, Ernest A, Tanner M (2003). Interactions between *Schistosoma haematobium* and human immunodeficiency virus type 1: the effects of co-infection on treatment outcomes in rural Zambia. Am. J. Trop. Med. Hyg. 69:420-428.
Mwinzi PNM, Karanja DMS, Colley DG, Orago ASS, Secor WE (2001). Cellular Immune responses of schistosomiasis patients are altered by Human Immunodefiency Virus Type 1 co-infection. J. Infect. Dis. 184:488-496.
National Institute of Parasitic Diseases (NIPD) (2013). China CDC, Shanghai and WHO China Representative Office, Beijing.
Neal PM (2004). Schistosomiasis-An Unusual Cause of Ureteral Obstruction A Case History and Perspective. Clin. Med. Res. 2(4):216-227.
Patz J, Graczyk T, Geller N, Vitor AY (2000). Effects of environmental change on emerging parasite diseases. Int. J. Parasitol. 30:1395-405.
Posey DL, Blackburn BG, Weinberg M, Flagg EW, Ortega L, Wilson M, Secor WE, Sanders-Lewis K, Won K, Maguire JH (2007). High prevalence and presumptive treatment of schistosomiasis and strongyloidiasis among African refugees. Clin. Infect. Dis. 45(10):1310-1315.
Rollinson D, Knopp S, Levitz S, Stothard R, Tchuente L-AT, Garba A, Mohammed KA, Schur N, Person B, Colly DG, Utzinger J (2012). Time to set the agenda for Schistosomiasis elimination. Acta Trop. 4:13.
Rosa FM, Marques DP, Maciel E, Couto JM, Negrão-Corrêa DA, Teles HM, Santos JB, Coelho PM (2013) Breeding of Biomphalaria tenagophila in mass scale. Rev. Inst. Med. Trop. 55(1):39-44.
Siddiqui AA, Siddiquid BA, Ganley-Leal L (2011). Schistosomiasis vaccines. Hum. Vaccines 7(11):1192-1197.
Simonsen PE, Nega A, Furu P (1990). Intestinal schistosomiasis among children in a labour village of Wonji Sugar Estate, Ethiopia. East Afr. Med. J. 67:532-538.
Standle CJ, Adriko M, Alinaitwe M, Kazibwe F, Kabatereine NB, Stothard JR (2009). Intestinal schistosomiasis and soil-transmitted helminthiasis in Ugandan schoolchildren: a rapid mapping assessment. Geospatial Health 4(1):39-53.
Steinmann P, Keiser J, Bos R, Tanner M, Utzinger (2006). Schistosomiasis and water resources development and estimates of people at risk. J. Lancet Infect. Dis. 6(7):411-425.
Tedla S (1981). Intestinal helminthiasis in Ethiopia. Helminthologia 23:43-48.
Tedla S, Jemaneh L (1998). Schistosomiasis and its distribution in Ethiopia and Eritrea. In: Birrie H, Tedela S, Jemaneh L (Eds.), Schistosomiasis in Ethiopia and Eritrea, 2nd ed. Institute of Pathobiology Addis Ababa University, Addis Ababa, Ethiopia. pp. 1-18.
Terefe A, Shimelis M, Mengistu T, Hailu A, Erko B (2011). *Schistosomiasis mansoni* and soil-transmitted helminthiasis in Bushulo village, southern Ethiopia. Ethiop. J. Health Dev. 25(1):46-50.
Tesfa-Yohannes TM, Kloos H (1988). "Intestinal parasitism." In: Zein AZ, Kloos H (Eds.), Ecology of Health and Disease in Ethiopia. Ministry of Health, Addis Ababa, Ethiopia. pp. 214-230.
United nation high commissioner for refugees (2014). UNHCR country operations profile – Ethiopia.
World Health Organization (2006). World Health Organization Weekly Epidemiologic Record. pp. 145-164.
World Health Organization (WHO) (2007). Schistomiasis. WHO Fact Sheet No 115. Geneva.
World Health Organization (WHO) (1963). Soil transmitted helminthes. WHO Technical Report Series, Geneva. P 277.
World Health Organization (WHO) (1981). Scientific Group, Intestinal protozoa and helminths infection. WHO Technical Report Series, Geneva. pp. 7-43.
World Health Organization (WHO) (1999). World Health Report: Action against worms, WHO, Geneva.
World Health Organization (WHO) (2002). Prevention and control of schistosomiasis and soil-transmitted helminthiasis, WHO technical report series, Geneva.
World Health Organization (WHO) (2004). WHO: Prevention and Control of Schistosomiasis and Soil-Transmitted Helminthiasis. WHO Technical Report Series, Geneva.
World Health Organization (WHO) (2013). Epidemiological record,

Schistosomiasis: number of people treated in 2011. WHO No. 8. pp. 81-88.
World Health Organization (WHO) (2014). Schistosomiasis. WHO Fact Sheet No 115. Updated February 2014.
Yomi A, Kebede S, Mamo Y (2010). Impact assessment of Gilgel Gibe hydroelectric dam on schitosomiasis: a cross sectional study in Southwest Ethiopia. Ethiop. J. Health Sci. 20:2.
Zaher T, Abdul-Fattah M, Ibrahim A, Salah H, El-M, Mona AA-DW, Refaey MM (2011). Current Status of Schistosomiasis in Egypt: Parasitologic and Endoscopic Study in Sharqia Governorate. Afro-Egyptian J. Infect. Endemic Dis. 1(1):9-11.

5

Prevalence, intensity and associated risk factors for *Toxocara canis* infection in Nigerian dogs

Akeredolu A. B. and Sowemimo O. A.

Department of Zoology, Obafemi Awolowo University, Ile-Ife, Osun State, Nigeria.

The objective of this cross-sectional study was to determine the prevalence and risk factors associated with *Toxocara canis* infections in dogs. The study was conducted in three communities in Southwest Nigeria between January, 2011 and February, 2012. Faecal samples (n=474) were processed by modified Kato-Katz procedure. Multiple logistic regression analysis was used to assess risk factors for *T. canis*. An overall prevalence of 34.6% was recorded for *T. canis*. Intensity of infection, determined by the mean number of eggs per gram of faeces (±standard error of mean [SEM]) was 4301.2 ± 348.4. Age of dogs and the mode of life were identified as the significant risk factors for *T. canis*. The odds of being infected with *T. canis* decreased with age. Dogs aged 0 to 6 months were 7.9 times more likely to harbour *T. canis* than dogs aged 31 months and above. The odds of harbouring *T. canis* by stray dogs were 2.7 times more than the kennel dogs. This study shows that *T. canis* infection is high in puppies and stray dogs. There is the need for establishment of deworming program to improve the health status of the dogs and to reduce the risk of zoonotic infection in humans.

Key words: Parasitism, epidemiology, pets, toxocariasis, larva migrans.

INTRODUCTION

Toxocara canis is a widespread gastrointestinal parasite of dogs and a causative agent of zoonotic disease in humans (Barutzi and Schaper, 2003; Ramirez-Barries et al., 2004). Infected dogs can shed large number of eggs into the environment causing infection in other dogs and in paratenic hosts including small mammals and humans (Holland and Smith, 2006). High prevalences of this parasite had been reported by various researchers both in stray dogs and wild canids such as foxes (O'Lorcain, 1994; Dubinsky et al., 1995). Studies have also shown that well-cared for dogs are also infected with *T. canis* (Sowemimo and Asaolu, 2008).

Previous studies conducted in tropics and sub-tropics including Nigeria have reported *T. canis* as one of the most frequently encountered canine parasites. In a study investigated by Degefu et al. (2011) to determine the gastrointestinal parasites among 334 household dogs in Jimma town, Ethiopia. They reported *T. canis* as the second most frequently observed parasite with a prevalence of 27.3%. In another study, Swai et al. (2010) examined faecal samples collected from 241 non-descript healthy dogs in and around Arusha municipality, Tanzania for gastrointestinal helminth infections.

They reported an overall prevalence of 73.8% for

helminth parasites. Further, it was reported that *T. canis* was the second most frequently observed parasite with a prevalence of 13.7%.

In Nigeria, Kutdang et al. (2010) examined faecal samples collected from 1000 dogs from two Local Government Area (LGA), Jos North and Jos South LGAs, Plateau State, Nigeria for eggs of intestinal helminth parasites. They reported an overall prevalence of 66.1% for helminth parasites. In addition they reported a prevalence of 31.8% for *T. canis*. In another study, Biu et al. (2012) investigated gastrointestinal helminth infections among 138 dogs resident on the University of Maiduguri campus, Borno State, Nigeria. They reported *T. canis* as the second most frequently observed parasite with a prevalence of 27.3%.

T. canis infections have, because of their zoonotic significance, important public health consequences especially in developing countries and communities that are socioeconomically disadvantaged (Craig and Macpherson, 2000). There are factors which have been reported to play a role in the epidemiology of *Toxocara* infections in dogs and they include age of dogs, sex, breed type and dog breeding kennels (Habluetzel et al., 2003; Senlik et al., 2006; Pullola et al., 2006). The risk of *T. canis* infection has also been reported in people who suffer from pica, a compulsion to eat mainly non-nutritive items, and particularly soil (Holland et al., 1995; Mizgajska-Wiktor and Uga, 2006). In Nigerian dogs, age and sex of dogs have been reported as risk factors for *T. canis* transmission (Sowemimo and Asaolu, 2008).

There is the need to investigate whether there are other factors that can contribute to the prevalence of *T. canis* infections in Nigerian dogs. In attempt to fill this gap, a cross-sectional study was conducted to determine the prevalence, intensity and the associated risk factors for *T. canis* infection in dogs in three communities in Southwest, Nigeria.

MATERIALS AND METHODS

Study area

This study was carried out in Ile-Ife, Ede and Ondo communities, Ile-Ife has been described in earlier report (Sowemimo, 2007). Ede is in Ede South Local Government Area in Osun State. It has a population of 76,035 according to National Population Commission (NPC) census 2006 estimates. It is located on Latitude 7° 42'N and Longitude 4° 27'E. The climate is typically tropical with well-defined wet (April-October) and dry (November-March) seasons, a mean annual precipitation of over 1100 mm, mean annual temperature of 27°C and relative humidity of 85% (Ayoade, 1982).

Ondo (Latitude 7° 05'N; Longitude 4° 50'E) is located within the humid region of Nigeria. Ondo lies in the rainforest zone with an annual rainfall range between 1300 and 1600 mm and a temperature range between 21 and 29°C. The tropical climate is broadly of two seasons'; rainy season (April to October) and dry season (November to March). The vegetation is typical rainforest and sub-savannah forest (Ayoade, 1982).

The inhabitants of the three communities are a mixture of people from different ethnic groups in Nigeria, although the majority is the Yoruba-speaking people of the Southwest. They are mainly peasant farmers growing cocoa, vegetables, maize, yam, and cassava. Traders, civil servants, artisan workers fishermen (common among Ondo people) and transport workers are found in smaller numbers.

Faecal sampling

Individuals owning dogs in each of the community were visited between January 2011 and February 2012 and the purpose of the study was explained. Questionnaires were then distributed to owners of dogs for the collection of demographic and socio-economic information which include the approximate age of dog, sex, mode of life, type of breed (African and exotic) type of food, anthelminthic regimen, occupation of owners of dog, place of defecation and awareness and knowledge of canine parasite zoonoses. Thereafter, clean specimen bottles were distributed to owners of dogs for the collection of specified quantity of faecal sample. The specimen bottles containing the faecal specimen were collected from owners of dogs and transported immediately to the Department of Zoology, Obafemi Awolowo University for preservation. About 10 g of faeces from each dog was mixed thoroughly with 10% aqueous solution of formaldehyde for preservation. Samples were processed for egg concentration in the laboratory using modified Kato-Katz method (Forrester and Scott, 1990) and examined for *T. canis* eggs using a light microscope at a magnification of ×100. In addition to qualitative diagnosis, an indirect measure of parasite intensity was obtained by counting eggs, expressed as eggs per gram of faeces (epg).

Statistical analysis

Version 16.0 of the SPSS for Windows Software Package (SPSS Inc, Chicago, IL) was used for all the data analysis. Chi-squared analysis was used to determine whether the prevalence of *T. canis* varied significantly among communities, and between age and sex. A Mann-Whitney U test was used to test the difference in intensity (epg) between sexes, modes of life and breeds while Kruskal-Wallis tests were used to test the difference in epg among age groups and communities. Multivariate logistic regression was further carried out to assess the predictive effect of the various variables measured on the prevalence of *T. canis* infection in the communities. All the variables were initially entered into the analysis and stepwisely removed (Backward Method, Wald) or added (Forward Method, Wald) until only the significant variables remained in the analysis. P-value less than 0.05 were taken as statistical significant association between dependent and independent variables.

RESULTS

An overall prevalence of 34.6% was recorded for *T. canis* infection in the dogs examined (n = 474). The prevalence of *T. canis* in relation to different factors is shown in Table 1.

The highest prevalence (56.6%) of *T. canis* infection was recorded in dogs age 0 to 6 months old and significantly higher than those of older age groups ($P<0.05$). The prevalence showed a decreasing trend with increasing age with the lowest recorded in dogs aged 19 to 24 months.

Male dogs were more frequently infected (37.0%) than females (32.3%) ($P>0.05$). The prevalence of *T. canis* infections was significantly higher in stray (42.4%) than in domiciliated dogs (20.6%). Similarly, the prevalence of infection was significantly higher in African (African shepherd) breed (39.3%) than in exotic breeds (24.3%) (Labrador, Mongrel, Alsatians, Ridgeback, Doberman) ($P<0.05$).

Table 1. Prevalence (%) and intensity (I) of *T. canis* infection in relation to age, gender, breed types and mode of life in the three communities combined' Nigeria, 2014.

Variable	No. examined	No. (%) infected	I (Mean ± SEM)
0 – 6	205	116 (56.6)	7588.4 ± 638.9
7 – 12	118	24 (20.3)	2152.3 ± 489.2
13 – 18	51	11 (21.6)	2178.4 ± 654.8
19 – 24	33	2 (6.1)	601.5 ± 474.3
25 – 30	17	3 (17.6)	1140.7 ± 702.5
≥ 31	50	8 (16.0)	1576.3 ± 752.2
Total	474	164 (34.6)	4301.2 ± 348.4
P-value	-	< 0.05	< 0.05
Sex			
Male	243	90(37.0)	4415.2 ± 500.0
Female	231	74(32.3)	4181.2 ± 485.0
P-value	-	>0.05	>0.05
Mode of life			
Stray	304	129 (42.4)	5438.5 ± 497.6
Kennel	170	35 (20.6)	2267.3 ± 421.2
P-value	-	< 0.05	< 0.05
Breed type			
Local	326	128 (39.3)	4827.4 ± 429.7
Exotic	148	36 (24.3)	3142.1 ± 581.5
P-value	-	< 0.05	< 0.05

I: Intensity; SEM: Standard Error of Mean.

Intensity of infection

The overall mean intensity recorded for *T. canis* infection among the infected dogs was 4301.2 ± 348.4 (Table 1). The intensity pattern was similar to that of prevalence with intensity of infection being the highest in dogs aged 0 to 6 months old and significantly higher than in older age groups (P<0.05). The intensity of *T. canis* infection was higher among the males than female dogs, however, there was no significant difference between the intensities of the two sexes (P>0.05). The intensity of *T. canis* infections was significantly higher in stray than in domiciliated dogs (U=19920, df=1: P=0.000). Similarly, the intensity of infection was also significantly higher in African breed than in exotic breed (U=20580, df=1; P=0.003) (P<0.05).

Questionnaire analysis

The responses received to the questionnaire showed that 304 (64.1%) of the dogs defecate outside of which 42.4% were infected with *T. canis*, while 170 (35.9%) of the dogs defecate inside out of which 20.6% were infected (Table 2). 395 (83.3%) of the respondents were aware that dogs harbour worms, while 181 (38.2%) respondents know that dogs have zoonotic parasites. Among the respondents who have knowledge of canine zoonotic parasites, 23.2% of their dogs were infected with *T. canis*, while 41.6% of the dog's whose owners have no idea were infected.

Logistic regression analysis

Based on the results from the chi-squared analysis identifying potential risk factors, we specified a maximal logistic regression model comprising the following explanatory factors: mode of life, age of dogs, dog breed type, place of defecation, anthelminthic treatment, knowledge of dog harbouring worms and awareness of canine parasite zoonoses (Table 2). After adjusting for the effect of individual variables, *T. canis* infection in dogs can be explained by two of the variables entered into the model (Table 3). These are the age of dogs and mode of life. The risk factors analysis showed that the odds of being infected with *T. canis* decreased as the age of dogs increased. Dogs aged 0 to 6 months were 7.9 times more likely to harbour *T. canis* than dogs aged 31 months and above. Stray dogs were 2.7 times more likely to harbour *T. canis* eggs than kennel dogs.

DISCUSSION

The overall prevalence of *Toxocara canis* obtained in this study (34.6%) was comparable to 41.7% reported from Ilorin, Kwara State with a similar ecological and epidemiological settings in Nigeria (Ugbomoiko et al., 2008), 33.6% from Italy (Habluetzel et al., 2003), but higher than the prevalence of 14.54% reported from USA (Blagburn, 2011), 12.5% from Japan (Yamamoto et al., 2009), 8.7% from Fortaleza Brazil (Klimpel et al., 2010), 3.1% from Finland (Pullola et al., 2006) and 4.4% from Northern Belgium (Claerebout et al., 2009). This variation may be due to differences in management systems, health care and degree of environmental contamination with infective

Table 2. Chi-squared (χ^2) analysis for factors significantly associated ($p<0.05$) with *T. canis* infections among dogs.

Variable	Group	N	Prevalence	χ^2	P-value
Age (Months)	0 – 6	205	56.6	79.91	<0.001
	7 – 12	118	20.3		
	13 – 18	51	21.6		
	19 – 24	33	6.1		
	25 – 30	17	17.6		
	≥ 31	50	16.0		
Mode of life	Stray	304	42.4	22.99	<0.001
	Domiciliated	170	20.6		
Breed origin	African	326	39.3	10.04	0.002
	Exotic	148	24.3		
Place of defecation	Outside	304	42.4	22.99	<0.001
	Inside	170	20.6		
Anthelminthic treatment	Yes	215	28.8	5.77	0.02
	No	259	39.4		
Aware that dog harbor worm	Yes	395	31.1	12.54	<0.001
	No	79	51.9		
Aware of canine zoonoses	Yes	181	23.2	16.80	<0.001
	No	293	41.6		

[a]The prevalence of *T. canis* positive cases to show the direction of association.

stages and exposition to natural infection more than owned dogs. Studies have shown that dog's well cared for by their owners and given veterinary attention has lower prevalence of intestinal parasites than dogs lacking such priviledges (Hendrix, 2006; Sowemimo and Asaolu, 2008).

In this study, age was the most significant risk factor for *T. canis* infection in the logistic regression model. The prevalence pattern also shows that it was age-dependent; *T. canis* decreased with age of dog. The expected high prevalence of *T. canis* recorded in dogs aged 0 to 6 months is in accordance with the transmission pattern of the parasite, which is mainly by transplacental and transmammary routes; acquired age-dependent immunity may be caused by repeated exposure (Sprent, 1957, 1961). The high prevalence of *T. canis* (56.6%) in dogs aged 0 to 6 months compares well with studies under-taken in Netherlands (48%) (Overgaauw and Boersema, 1998), Ile-Ife Nigeria (51.4%) (Sowemimo, 2007) and UK (65.8%) (Jalaya, 1969).

Stray dog was significantly associated with the prevalence of *T. canis* and was retained as a significant risk factor in the logistic regression model. The high prevalence of *T. canis* infection in stray dogs is a reflection of the fact that stray dogs roam freely without proper care such as the regular administration of anthelminthic drugs. They are more susceptible to *T. canis* infection through the ingestion of paratenic hosts as food sources as a result of their scavenging habits. Several other studies have reported high prevalence of intestinal parasites among stray and uncared for dogs than in kennel dogs (Pandey et al., 1987; Minaar et al., 2002; Martinez-Moreno et al., 2007; Sowemimo, 2009). These findings are clear indications of the serious public health challenges attributable to stray dogs in Nigeria and many other third world nations where dog control laws are non-existing or loosely enforced by government agencies. Therefore, appropriate control laws and stricter enforcement are therefore inevitable components of any control initiatives in endemic societies.

Our questionnaire data on defecation sites revealed that over 40% of the dogs who defecate outside were infected with *T. canis*. The implication of this is that since dogs are left to roam the streets unhindered, there is the possibility of contaminating the environments with *T. canis* eggs which are voided with faeces. Many public places including children playing grounds are likely to be contaminated with these eggs, thereby exposing children

Table 3. Result of logistic regression analysis showing the effect of age and mode of life on the prevalence of *T. canis* in the three communities in Southwest, Nigeria.

Parameter[a]	No. examined	Odds ratio (95% CL)	P-value
Dog's age (months)			
0 – 6	205	7.87 (3.33 – 18.58)	0.00
7 – 12	118	1.45 (0.58 – 3.65)	0.43
13 – 18	51	1.48 (0.52 – 4.23)	0.46
19 – 24	33	0.32 (0.06 – 1.65)	0.17
25 – 30	17	1.21 (0.27 – 5.47)	0.81
≥31[b]	50		
Mode of life			
Stray	304	2.75 (1.30 – 5.84)	0.008
Kennel[b]	170		

95% CL, 95% Confidence Limit; [a]Variables are arranged in the order they were removed by the Backward Stepwise method (Wald); [b]Reference group.
χ^2 = Chi^2.

to accidental ingestion of the eggs through their habits of playing with soil.

Questionnaire results also showed that even though more than half of the individuals surveyed are aware that dog harbour worms, they lacked awareness and knowledge of canine parasitic zoonosis. The high prevalence of *T. canis* recorded among dogs whose owners have no knowledge of canine zoonosis calls for great concern. There is the need for greater awareness of the problem among the population in the area of study. Awareness of the possible infection routes need to be raised: for children and adults egg intake can be the consequence of soil related activities such as playing, gardening, sweeping or of contaminated vegetable consumption (Genchi et al., 1990)

From the results of this study, stray dogs and puppies (aged 0 to 6 months) have been identified as important source of *T. canis* infections in dogs. In addition to the health implications for dog with patent infections one might consider the health hazard to humans. In view of the fact that *T. canis* has been implicated as the causative agent of visceral larva migrans in humans, there is the need for intervention measures to reduce the risk of transmission of the parasite from dogs to humans. Interventions should focus on health education provided to dog owners such as using effective anthelminthic for deworming and establishment of a program based on zoonotic diseases.

Limitations

One of the limitations to this study includes a single stool sample collected from each dog examined; because only a single stool sample was collected on each dog in this study, using the Kato-Katz preparation to determine infection status, there is likely to be a significant underestimation of infection in individual dog.

ACKNOWLEDGEMENTS

The authors wish to thank dog-owning individuals from Ile-Ife, Ede and Ondo for help and cooperation during faecal sample collections. We also acknowledge the assistance of Dr. Ambrose Akinlo for the statistical analysis. This research received no specific grant from any funding agency, commercial or non-profit sectors.

Conflict Interests

The authors declare that there is no conflict of interests regarding the publication of this article.

REFERENCES

Ayoade SS (1982). Climate change. In: Baobour KM, Oguntoyinbo JS, Onyemelukwe JOC, Nwafor JC (eds.), Nigeria in maps. London: Holder & Stoughton. pp. 14-15.

Barutzki D, Schaper R (2003). Endoparasites in dogs and cats in Germany 1999-2002. Parasitol. Res. 90:148-150.

Biu AA, Aitiya G, Paul BT, Konto M (2012). Incidence of gastrointestinal helminthes ova of Nigerian local dogs (*Canis familiaris*) resident on University of Maiduguri Campus, Maiduguri, Nigeria. J. Agric. Vet. Sci. 4:63-68.

Blagburn BL (2001). Prevalence of canine and feline parasites in the United States of America. Compend. Contin. Educ. Pract. Vet. 23:5-10.

Claerebout E, Casaert S, Dalemans, AC, De Wilde N, Levecke B, Vercruysse J Geurden, T (2009). Giardia and other intestinal parasites in dog populations in Northern Belgium. Vet. Parasitol. 161:41-64.

Craig PS, Macpherson CNL (2000). Dogs and cestode zoonoses. In: Macpherson CNL, Meslin FX, Wandeler AI (Eds.), Dogs, Zoonoses and Public Health. Oxon, United Kingdom. CAB International, pp. 149-211.

Degefu H, Tefera A, Yohannes M (2011). Zoonotic helminth parasites in faecal samples of household dogs in Jimma town, Ethiopia. J. Public Health Epidemiol. 3(4):138-143.

Dubinsky P, Havasiova – Reiterova K, Petko B, Hovorka I, Tomasovicova O (1995). Role of small mammals in the epidemiology of toxocariasis. Parasitol. 110:187-193.

Forrester JE, Scott ME (1990). Measurement of *Ascaris lumbricoides* infection intensity and dynamic of expulsion following treatment with mebendazole. Parasitol. 110:303-308.

Genchi C, Di Sacco B, Gatti S, Sangalli G, Scaglia M (1990). Epidemiology of human toxocariasis in Northern Italy. Parassito 32:313-319.

Habluetzel A, Traldi G, Ruggieri S, Attili A R, Scuppa R, Marchetti R, Menghini G, Esposito F (2003). An estimation of *Toxocara canis* prevalence in dogs, environmental egg contamination and risk of human infection in the Marche region of Italy. Vet. Parasitol. 113:243-252.

Hendrix CM, Robinson ED (2006). Diagnostic Veterinary Technicians. 3rd Edition, Mosby, Inc., USA. P 285.

Holland CV, O'Lorcain P, Taylor MRH, Kelly A (1995). A seroepidemiology of toxocariasis in school children. Parasitology 110:535-545.

Holland CV, Smith HV (2006). *Toxocara*: The Enigmatic Parasite. CABI Publishing. P 301.

Jalaya T (1969). Studies on *Toxocara canis* and visceral larva migrans. Ph. D. Thesis, University of Liverpool, UK.

Klimpel S, Heukelbach J, Pothmann D, Ruckert S (2010). Gastrointestinal and ectoparasites from urban stray dogs in Fontaleza (Brazil): High infection risk for humans? Parasitol. Res. 107:713-719.

Kutdang ET, Bukbuk DN, Ajayi AA (2010). The prevalence of intestinal helminthes of dogs (*Canis familiaris*) in Jos, Plateau State, Nigeria. Researcher 2(8):51-56.

Martinez – Moreno F J, Hernandez S, Lopez – Cobos E, Becerra C, Acosta I, Martinez-Moreno A (2007). Estimation of canine intestinal parasites in Cordoba (Spain) and their risk to public health. Vet. Parasitol. 143:7-13.

Minaar WN, Krecek RC, Fourie LJ (2002). Helminths in dogs from a peri – urban resource – limited community in Free State Province, South Africa. Vet. Parasitol. 107:343-349.

Mizgajska-Wicktor H, Uga S (2006). Exposure and environmental contamination. In: Holland CV, Smith HV (Eds.), *Toxocara*: The Enigmatic Parasite. CABI International. pp. 211-227.

O'Lorcain P (1994). Epidemiology of *Toxocara* spp. in stray dogs and cats in Dublin, Ireland. J. Helminthol. 68:331-336.

Overgaauw PAM Boersema JH (1998). Nematode infections in dog Breeding kennels in the Netherlands, with special reference to *Toxocara*. Vet. Q. 20:12-15.

Pandey VS, Dakkak A, Elmamoune M (1987). Parasites of stray dogs in Rabat region, Morocco. Ann. Trop. Med. Parasitol. 81:53-55.

Pullola T, Vierimaa J, Saari SA, Virtala M, Nikander S, Sukura A (2006). Canine intestinal helminths in Finland: Prevalence, risk factors and endoparasites control practices. Vet. Parasitol. 140:321-326.

Ramirez-Barrios RA, Barboza-Mena G, Munoz J, Angulo-Cubillan F, Hernandez E, Gonzalez F, Escalona F (2004). Prevalence of intestinal parasites in dogs under veterinary care in Maracaibo, Venezuela. Vet. Parasitol. 30:1369-1377.

Senlik B, Cirak YY, Karabacak A (2006). Intestinal nematode infections in Turkish military dogs with special reference to *Toxocara canis*. J. Helminthol. 80(3):299-303.

Sowemimo OA (2007). Prevalence and intensity of *Toxocara canis* (Werner, 1782) in dogs and its potential public health significance in Ile-Ife, Nigeria. J. Helminthol. 83:27-31.

Sowemimo OA (2009). The prevalence and intensity of gastrointestinal parasites of dogs in Ile-Ife, Nigeria. J. Helminthol. 83:27-31.

Sowemimo OA, Asaolu SO (2008). Epidemiology of intestinal parasites of dogs in Ibadan, Nigeria. J. Helminthol. 82:89-93.

Sprent JFA (1957). Observations on the development of *Toxocara canis*. J. Parasitol. 48:184-193.

Sprent JFA (1961). Post-parturient infection of the bitches with *Toxocara canis*. J. Parasitol. 48:284.

Swai ES, Kaaya EJ, Mshanga DA, Mbise EW (2010). A survey of gastrointestinal parasites of non-descript- dogs in around Arusha Municipality, Tanzania. Int. J. Anim.Vet. Adv. 3(2):63-67.

Ugbomoiko SU, Ariza L, Heukelbach J (2008). Parasites of importance for human health in Nigerian dogs: high prevalence and limited knowledge of pet owners. BMC Vet. Res. 4:49.

Yamamoto NM, Kon T, Saito N, Maeno K, Koyama M (2009). Prevalence of intestinal canine and feline parasites in Saitama Prefecture, Japan. Kansenshogaku Zasshi 83: 223-228.

6

Prevalence of lungworm infection in small ruminants in North Gondar zone, Amhara National Regional State, Ethiopia

Yitagele Terefe[1], Ketema Tafess[2], Getasew Fekadie[1] and Nigatu Kebede[3]

[1]Faculty of Veterinary Medicine, Haramaya University, Ethiopia.
[2]Faculty of Veterinary Medicine, Gondar University, Gondar, Ethiopia.
[3]Aklilu Lemma Institute of Pathobiology, Addis Ababa University, Addis Ababa, Ethiopia.

The study was conducted to determine the prevalence, associated risk factors and identification of the species of lungworms of small ruminants in four districts of Northwestern Amhara National Regional State. A total of 632 small ruminants were included in the study using coprologic procedures (565 animals) and post-mortem examination (87 animals). The study showed that the overall prevalence of lungworm infection was 46.0 and 56.3% based on coprological and postmortem examination, respectively. Significant difference (P < 0.05) in prevalence of lungworm was found between animals under semi-intensive and extensive management systems. The prevalence of lungworm infection was significantly different (P < 0.05) between seasons, the highest being during the rainy season (57.1%) and the lowest in the dry season months (35.0%). The prevalence of *Muelleris capillaris* infection was highest (50.4%) during the study period based on postmortem examination. The prevalence of lungworm infection showed significant association (P < 0.05) with different age groups of animals, where *Dictiocaulus fillaria* was highly prevalent in young age group while *M. capillaris* and *Protstrongylus rufescens* were highly prevalent in adult age groups of study animals. In the current study, three respiratory helminthes of small ruminants were identified and management system, age and season are important risk factors associated with lungworm infection in the study area. Due to its impact on production, emphasis should be given for the prevention and control of lungworm infection in the study area.

Key words: *Dictiocaulus fillaria, Muelleris capillaries, Protostrongylus rufescens*, lungworm, small ruminant, helminthes.

INTRODUCTION

Sheep and goats are the most numerous of man's domestic livestock and are especially important in the more extreme climates. Of the worlds' 1,614 million sheep and 475 million goats, 65 and 95%, respectively, are located in developing countries. Small ruminants in Africa are noted for their ability to convert low cost feed into high value products, namely: meat, milk, fiber, manure and hides (Wilsmore, 2006; Food and Agriculture Organization (FAO), 1986).

The current sheep and goat populations of Ethiopia are approximately 23 and 17 million, respectively (ILRL, 2000). Small ruminants are important contributors to food production in Ethiopia, providing 33% of meat consumption and 14% of milk consumption (Asfaw, 1997). In the central highlands of Ethiopia where mixed crop-livestock production system is practiced, small ruminants account for 40% of cash income and 19% of the house hold meat consumption (Zelalem and Fletcher, 1993). Owing to their high fertility, short generation interval and adaptation even in harsh environments, sheep and goats are considered as investments and insurance to provide income to purchase food during seasons of crop failure and to meet seasonal purchases such as improved seed, fertilizer and medicine for rural households (Asfaw, 1997).

Even though small ruminants are important components of the Ethiopian farming system, their contribution to food production, rural income and export revenue are far below than their expected potential. This is because small ruminant production is constrained by the compound effects of disease, poor feeding and poor management (Getachew, 1995). Studies in the central highland of Ethiopia have shown that lungworm parasites are a major problem in small ruminants and cause disease, increase mortality and production losses (Bekele et al., 1992). Up to half of all sheep deaths and morbidity on farms in Ethiopian highlands are caused by parasitic pneumonia and endoparasites (ILCA, 1990).

The incidence of lungworm infection varies geographically. Only a few limited studies have been completed on respiratory helminthes of small ruminants in the Amhara National Regional State, northwestern Ethiopia. Therefore, the aims of the present study are (i) to determine the prevalence of lungworm infection in small ruminants; (ii) to identify the species of lungworms and (iii) to identify associated risk factors for lungworm infection in north Gondar zone, Amhara National Regional State, northwestern Ethiopia.

MATERIALS AND METHODS

Study area

The study was conducted in four districts of northwest Amhara region including Abaintonios, Anchew, Belagig and Gondar town districts. The study areas are about 710 km North West of Addis Ababa at an altitude ranging from 1500 to 3500 meters above sea level. Numerous mountains, plateaus, sloped areas, rivers, streams and lakes mark the topography of the area. The climate of the study area is divided into subhumid ("wurch and Dega"), semiarid ("Woynadega") and arid ("Kola"). The average maximum and minimum daily temperature of the area varies between 22 to 30.7°C and 12.3°C, respectively.

The region receives a bimodal rainfall, the average annual precipitation rate being 1000 mm. The short rains occur during the months of March to May while the long rains extend from June to September. The production system observed in the area combines cereal-based agriculture and livestock farming. The study area has a livestock population of 2.03 million cattle, 0.6 million sheep, 0.54 million goats, 0.25 million equine species of which 62.5% are donkeys, 5% are mules, and 32.5% are horses, and 1.9 million poultry (CSA, 2003).

Study animals and sampling

Small ruminants in the study area are kept mainly under an extensive traditional management system, involving small household flocks of mixed age. The sample size was determined according to Thrusfield (1995) for an infinite population. Since there has been no previous work done in the study area, an estimated prevalence of 50% was used for sample size determination. A total of 652 small ruminants (412 sheep and 153 goats from study districts, and 58 sheep and 29 goats from Gondar ELFORA abattoir) were used in this study. Four representative districts were purposely selected; Abaintoniose, Anchew, Blagig and Gondar town. Sampling animals was conducted at random from study districts and small ruminants slaughtered in ELFORA abattoir. Animals from Gondar town were managed under a semi-intensive system while the rest were under an extensive production system.

Study design

Coprological examination

Faecal samples were collected directly from the rectum of all selected animals and stored in vials containing 10% formalin and transported to the Gondar veterinary clinic for laboratory examination. In the laboratory, the collected fecal samples were processed according to conventional methods for lungworm larvae (Baermanization) (Hansen and Perry, 1994). During sample collection, the species of the animal, sex, age, date of sampling and the area were recorded. In the laboratory, 25 g of faeces was weighed from each sample for the extraction of larvae using modified Baermann technique (Hansen and Perry, 1994). The faeces were enclosed in a gauze fixed on the string rod and submerged in a clean glass tube filled with warm water. The whole apparatus was left for 2 to 3 h. The larva leaves the faeces and migrates through the gauze and settles at the bottom of the glass. After siphoning off the supernatant, the sediment was examined under the low power of the microscope (Fraser, 1991; Urquhart et al., 1996). All the area under the cover slip was thoroughly and uniformly searched for the presence of lungworm larvae (Hendrix, 1998).

Abattoir survey

Small ruminants slaughtered at Gondar ELFORA abattoir were examined during the study period. The lungs were palpated for the presence of protostrogylidea nodules. These are brood nodules and worm nodules. Brood nodules are cone shaped granuloma-like areas of the affected lung tissue varying in size and color, which contains active worm and caused by parenchyma-dwelling small lungworms (Protostrongylidae). The worm nodules are pin head shaped mostly subpleural, grey (*Mullerius*) cyst which contain adult worms. If the nodules are present, they are trimmed off and worms extracted from the tissue by gentle pressing of a small non-calcified nodule or part of large nodule between two glass slides and then carefully teasing the worm away from the tissue. The air passages were opened starting from the trachea down to the small bronchi with fine blunt pointed scissors, to detect parasites (Kassai, 1999; Schneider, 2000).

Data analysis

Prevalence was determined by the percentage (%) positive and chi-square (x^2). To measure association were

Table 1. Prevalence of lungworm infection in different districts of the study area.

District	Number examined	Number positive (%)
Abaintonios	176	77 (43.7)
Anchew	88	50 (56.8)
Belagig	56	24 (42.8)
Gondar town	245	109 (44.4)
Total	565	260 (46.0)

P > 0.05.

the statistical tools applied and determined using statistical package for social sciences (SPSS) V 17.0. In all the analysis, confidence level was held at 95% and 5% level of significance (Thursfield, 1995).

RESULT

Coprological examination

Out of 565 (412 sheep and 153 goats) fecal sample examined, 260 (46.0%) were positive for lungworm infection. No statistically significant difference was observed in the prevalence of the lungworm infection of small ruminants in different district (x^2 = 28.22 P > 0.05). The higher prevalence rate was encountered in Anchew (56.8%) while the lowest was in Belagig (42.8%) (Table1).

The prevalence rate of lungworm infection was found to be higher in extensive management system (51.8%) than in semi-intensive management system (24.5%). Comparison between sheep and goats showed statically significant difference between the two species (x^2 = 33.2, P < 0.05). Similarly, there was association between the prevalence rate of lungworm infection with management system, and months where x^2 = 36.71, P < 0.0 and x^2 = 47.55 P < 0.05, respectively. The prevalence rate of lungworm infection did not show statistically significant association with sex of the animals (x^2 = 4.55, P > 0.05). Lungworm infection in small ruminants and associated risk factors in the study area are shown in Table 2.

In comparison of age groups, significant difference (x^2 = 33.22, P < 0.05) in the prevalence rate of lungworm infection was observed. Among lungworms identified, high prevalence rate of *Muelleris capillaries* (50.4%) followed by *Dictiocaulus fillaria* (26.5%) was recorded. The prevalence of *M. capillaris* and *Protostrongylus rufescens* increases with age of the animal, whereas the prevalence of *D. filaria* was high in young age groups of the animals. Mixed infection was common in old age group of the animals (Table 3).

Abattoir survey

Of the 87 animals (58 sheep and 29 goats) examined in the ELFORA abattoir, 49 (56.3%) were positive for lungworm infection. There was a significant difference in prevalence of lung worn infection between the two species (x^2 = 3.95, P < 0.05), the prevalence being higher in sheep (63.8%) than goats (41.3%). The distribution of lungworm species detected in the lung during postmortem examination of positive animals is shown in Table 4. Higher prevalence of *M. capillaris* (51.0%) was observed than *P. rufescens* (28.6%) and *D. fillaria* (20.4%). The relative percentage of adult lungworms recovered during postmortem examination is described in Table 4.

DISCUSSION

This study revealed the overall prevalence of lungworm infection of small ruminants to be 46.0%, with 42.3% in sheep and 46.7% in goats. It also disclosed that animals are infected with three nematode species parasitizing the respiratory tract of small ruminants of which *Mullerius* and *Dictyocaulus* species are the most abundant. This finding is in agreement with Regassa et al. (2010) and Sissay (1996) who have reported 40.4 and 44% in small ruminants in Dessie and Kombolcha districts, north-eastern Ethiopia and Bahir Dar, respectively. The present finding is lower than reports of Alemu et al. (2006) in northeast Ethiopia and Bekele and Abu (2011) in Tiyo District, South-East Ethiopia, with an observed prevalence of 53.6 and 57.1%, respectively. However, it is higher than that of Addis et al. (2011) and Weldesenebet and Mohamed (2012), with 33.83% prevalence at Gondar and 26.7% at Jimma, respectively. These differences might be due to the method used for the detection of the larvae or difference in the study area of topography which has a conducive environment for the survival of larvae and intermediate hosts, slug or snails and also the nutritional detection.

In this study, small ruminants under extensive management system had more infection than those kept under semi-intensive management system. The reason for this could be increased cultivation of land which restricts animals on communal grazing land so that large numbers of the animals are kept together. This could increase the degree of pasture contamination leading to higher prevalence rate. Management practice such as provision of ample nutrition increases the resistance of the host under the semi-intensive system, contrary to this malnutrition which reduces the host-parasite response and favors the fecundity of the parasites that allows the animals for continuous larvae exposure under extensive system (Soulsby, 1986).

The reason for this high prevalence of *Mullerius* comparied with *Dictyocaulus* could be partly attributable to its wide range of intermediate host and the ability of larvae to over winter in the mulluscs. Additional factors which play a part in ensuring the endemicity of the worm are, first, the ability of L1 (First stage larva) to survive for months in faecal pellets and secondly, the persistence of

Table 2. Lungworm infection in small ruminants and associated risk factors in the study area.

Variable	Number examined	Number positive (%)
Species		
Ovine	412	164(39.8)
Caprine	153	96(62.7)
Sex		
Female	329	158(48.0)
Male	236	102(43.2)
Management system		
Extensive	444	230(51.8)
Semi-intensive	121	30(24.5)
Season		
Dry	285	100(35.0)
Rainy	280	160(57.1)

Table 3. Prevalence of the species of lungworm in small ruminants of different age groups.

Age (years)	Prevalence (%)	Positive							
		Single infection			Mixed infection				Total
		D (%)	M (%)	P (%)	DM (%)	DP (%)	MP (%)	DMP (%)	
≤1/2	29(11.6)	14(20.3)	12(9.2)	2(4.9)	-	-	1(14.3)	-	81
1/2-2	80(30.8)	24(34.9)	41(31.3)	10(24.4)	2(33.3)	1(25.0)	2(28.6)	-	193
2-4	69.(26.5)	12(17.4)	41(31.3)	11(26.8)	3(50.0)	1(25.0)	1(14.3)	-	146
>4	82(31.5)	19(27.4)	37(28.2)	18(43.9)	1(16.7)	2(50.0)	3(42.9)	2(100.0)	145
Total	260(46.01)	69(26.5)	69(26.5)	41(15.8)	6(2.3)	4(1.5)	7(2.7)	2(0.8)	565

D:*Dictyocalus filaria*, M: *Muellerius capillaries*, P: *Protostrongylus rufescens*; $x^2 = 33.72$, $P < 0.05$.

Table 4. Relative percentage of adult lungworm recovered from postmortem examination.

Species of lung worm	Number of positive	Percentage (%)
D. filarial	10	20.4
M. capilaris	25	51.0
P. rufescens	14	28.6
Total	49	100.0

$x^2 = 87.0$, $P < 0.05$.

the L3 (third stage larva) in the intermediate host for the life time of the mollusks (Taylor et al., 2007). On the other hand, the longevity and development of free larvae of *Dictyocaulus* are known to be dependent on humidity and temperature condition. Dry seasons are characterized by high mortality of larvae in the pasture (Gallia and Nunns, 1976), where dry, hot summer and cold winter is the climatic condition of the study area. The prevalence of *D. filaria* is higher in infants than adults. However, the prevalence of *M. capillaris* and *P. rufescens* is higher in adult animals than younger ones. This might be due to the long period of potency and the apparent inability of the final host to develop acquired immunity, so that adult sheep have the heaviest infection, highest infection and the highest prevalence (Taylor et al., 2007). The study further revealed that goats were found to be more susceptible to lungworm infection than sheep. This could be because most of the goats in this study were from lowland and mid altitude areas, which are thought to be suitable for survival of the larval stage of the parasite. Likewise, in the lowland areas of the country where goats are mostly reared, there is poor veterinary infrastructure and medication.

More importantly, the condition could be due to less or slow development of immunity in goats compared to sheep (Wilsmore, 2006).

The higher prevalence of lungworm infection during the rainy season of the months might be the presence of sufficient moisture during the rainy season which favored the survival of infective larvae and stimulated the activity of mollusks in the pasture resulting in higher probability of up take of the infective larvae, leading to higher prevalence rate. In northwestern part of Ethiopia, under local production system, the animals that are completely managed on pasture grazing throughout the year succumb to seasonal variation of availability of forgeable feed and then difference in plane of nutrition. Thus, the presence of sufficient feed in rainy season could in turn increase the nutritional status, and these well fed animals develop good immunity that suppressed the fecundity of the parasites. This situation was reported by Bisset et al. (1996) that increased plane of nutrition increased the immunity and reduced the fecundity of the worm, contrary to the higher prevalence rate observed during the rainy season.

The prevalence of lungworm infection obtained by postmortem examination was higher (56.3%) than the result obtained by coproscopic examination (46.0%). This difference might be related to worm nodes of Protostronglidae. Worm nodules contain immature parasite in general (Schneider, 2000). Higher infection rate observed on postmortem examination may be related to these worm nodules detected. In addition, in the prepatent or postpatent phase or during hypobiosis, it is impossible to detect this parasite by fecal examination (Fraser, 1991). Furthermore, egg laying adult female parasites might be inhibited by the immune reaction of the host (Hansen and Perry, 1994).

Conclusion

The study revealed that lungworm infection in small ruminant is highly prevalent in North Gondar zone. *D. filaria, M. capillaris* and *P. rufescens* were the lungworms identified during the study. Young animals are mostly affected by most pathogenic species of lungworm, *D. filarial*, than adults on the contrary which are highly affected by *M. capillaris* and *P. rufacens*. Animals under semi-intensive management system have been found less affected than those under extensive management system. Therefore, avoidance of overstocking and grazing in watery and damp areas of arid pastures, treatment of potential worm carriers and separate grazing of young stocks are recommended.

REFERENCES

Addis M, Fromsa A, Ebuy Y (2011). Study on the Prevalence of Lung worm Infection in Small ruminants in Gondar Town, Ethiopia. Vet. Res. 4(3):85-89.

Alemu S, Leykun EG, Ayelet G, Zeleke A (2006). Study on small ruminants lung worms in northeastern Ethiopia. Vet. Parasitol. 142(2-3):330-335.

Asfaw W (1997). Country report, Ethiopia. In: proceeding of a seminar on livestock Development policies in Eastern and Southern Africa, 28th July-1st August, 1997, Mbabany. Organized by CTA, OAU/IBAR, the Ministry of Agriculture Cooperative, Swaziland.

Bekele M, Abu A (2011). Ovine lungworms in Tiyo District, South-East Ethiopia: Prevalence, effect of altitude and major host related risk factors. Glob. Vet. 7(3):219-225.

Bekele T, Woldeab T, Lahlou-lasso A, Sherington J (1992). Factors affecting Morbidity on-farm and on-station in Ethiopian highland sheep. Acta. Trop. 52(2-3)99-109.

Bisset ST, Thomsborg SM, Mainqi N, Munyua WK (1996). Nematode burdens and immunological responses following natural challenges in Romney lambs selectively breed for low or high fecal worm egg count. Vet. Parasitol. 61(3-4):249-263.

Central Stastical Authority (CSA) (2003). Ethiopian agricultural sample Enumeration report on livestock and farm implement part IV. Addis Ababa, Ethiopia. pp. 29-136.

FAO (1986). Small ruminant production in the developing countries. In: proceedings of an Expert consultation held in Sofia, Bulgaria, 8-12 July 1985 (http://www.fao.org/docrep/009/ah221e/ah221e00.htm).

Fraser CM (1991). The Merk veterinary manual. A hand book of diagnosis, therapy and disease prevention and control for the veterinarians, 7th ed. Mark and Co. Inc., Rahway, Nit, USA. pp. 714-614.

Gallie GJ, Nunns VJ (1976). The bionomics of free living larvae and transmissionof *Dictyocaulus filarial* between lambs in North-East England. J. Helminthol. 50(2):79-89.

Getachew MR (1995). Parasite of small ruminants. In: Gray GD, Uilengerg G (eds), Parasitological Research in Africa

Hansen J, Perry B (1994). The Epidemiology, Diagnosis and Control of Helminth parasites of ruminants. ILRAD, Nairobi, Kenya. P 38.

Hendrix C (1998). Diagnostic of veterinary parasitology, 2nd ed. Mosby, London. p 49.

International Livestock Center for Africa (ILCA) (1990). Annual report, 1989. (International Livestock Center for Africa, Addis Ababa, Ethiopia. p 37.

International Livestock Research Institute (ILRI) (2000). Hand book of livestock statistics for developing counties. Socio-economics and policy research working paper No. 26. ILRI Nairobi, Kenya. p 299.

Kassai T (1999). Veterinary Helminthological. Butterworth helminthology, Budapest, Hungary. pp. 85-93.

Lebbie SHB, Rey B, Irungu EK (1994). Small ruminant research and development in Africa. Proceeding of the second Biennial conference of the African Small Ruminant Research Network. ILCA, Ethiopia. pp. 1-5.

Regassa A, Toyeb M, Abebe R, Megersa B, Mekibib B, Mekuria S, Debela E, Abunna F (2010). Lungworm infection in small ruminants: prevalence and associated risk factors in Dessie and Kombolcha districts, Northeastern Ethiopia. Vet. Parasitol. 169(1-2):144-148.

Schneider T (2000). Helminths of respiratory system. In: Veterinary Medicine Parasitology, Vol. 5. Voll standing auflage.parey Buch vet log. Berlin. pp. 193-198.

Sissay A (1996). Preliminary study on the prevalence of ovine lungworm infection in and around Bahir Dar. DVM thesis, Faculty of Veterinary Medicine, Addis Ababa University, Debre Zeit, Ethiopia. pp. 14-16.

Solusby EJL (1986). Helminthes, arthropods and protozoa of domesticated animals, 7th ed. Bailliere, Tindall, London. pp. 262-268.

Taylor MA, Coop RL, Wall RL (2007). Parasites of the respiratory system. In: Veterinary parasitology, 3rd ed. Black well publishing Ltd., Oxford, UK. pp. 195-199.

Thrusfield M (1995). Surveys. In: Veterinary Epidemiology, 2nd ed. Black well science Limited, Cambridge, USA. pp. 178-198.

Urquhart HM, Armour Duncan JL, Dunn AM, Jennings FW (1996). Veterinary parasitology, 2nd ed. Black Well science limited, London. pp. 301-309.

Weldesenebet D, Mohamed A (2012). Prevalence of small ruminant lung worm infection in Jimma town. Glob. Vet. 8(2):153-159.

Wilsmore T (2006). Disease of small ruminants in Ethiopia, the veterinary epidemiology and economics research unit of agriculture's policy and development the university of read, UK, pp: 602.

Zelalem A, Fletcher IC (1993). Small ruminant productivity in the central Ethiopia Mixed farming system. In: Proceeding of the 4th National Livestock Improvement Conference, 13-15 November, IAR, Addis Ababa, Ethiopia.

7

Development and evaluation of lateral flow test for the detection of trypanosomes in tsetse flies

Enyaru C. K. John[1], Njuguna James[9], Alibu Vincent Pius[1], Matovu Enock[2], Malele I. Imna[3], Chisi E. John[4], Mbongo Nicolas[5], Mansinsa Philemon[6], Intisar E. I. Rayah[7], Mohammed Yassir[7], Mubarak Mustafa[7], Ochi Erneo[8] and Nantulya Vinand[9]

[1]Department of Biochemistry and Sports Sciences, Makerere University, P. O. Box 7062, Kampala, Uganda.
[2]College of Veterinary Medicine and Biosafety, Makerere University, P. O. Box 7062, Kampala, Uganda.
[3]Vector and Vector Borne Diseases Institute, Tanga, Tanzania.
[4]College of Medicine, University of Malawi, Private bag 360, Blantyre, Malawi.
[5]Laboratoire national de Sante Publique, Brazzaville, Congo.
[6]National Program of Human African Trypanosomosis, Kinshasa, Congo.
[7]Tropical Medicine Research Institute, Khartoum, Sudan.
[8]Ministry of Animal Resources and Fisheries, Juba, South Sudan.
[9]Astel Diagnostics, Uganda.

Lateral flow test (LFT) which is based on antibodies raised against the *Trypanosoma brucei* peptide was evaluated for its analytical sensitivity, specificity, positive predictive value and negative predictive values in order to assess its utility for surveillance of trypanosome infections in tsetse flies. The diagnostic test agreement of the FLT and microscopy was 91.0% and kappa value of 80.5% at the confidence interval 0.72 to 0.80 which is a high level of test agreement. Furthermore, the relative diagnostic sensitivity and specificity of lateral flow test were 83.0 and 96.0%, respectively. Both the positive predictive and the negative predictive values were high at 92.7 and 90.0%, respectively. The LFT is, therefore, recommended for surveillance of trypanosomes in tsetse flies in order to: (i) indicate areas with tsetse infected with trypanosomes; (ii) indicate tsetse infected with potentially human pathogenic trypanosomes and (iii) guide in prioritizing control strategies for human African trypanosomiasis (HAT).

Key words: Lateral flow test, detection, xenomonitoring, trypanosomes, tsetse flies.

INTRODUCTION

Tsetse fly-transmitted african trypanosomes (*Trypanosoma brucei*) are major pathogens of humans and livestock, causing Human African Trypanosomiasis (HAT) and Nagana, respectively. In Eastern and Southern Africa, including Western and Central Africa, trypanosomes are transmitted by various species of tsetse flies, the most important being *Glossina fuscipes fuscipes*, *Glossina pallidipes* and *Glossina morsitans*

morsitans. Animal trypanosomiasis (Nagana) is one of the major constraints in livestock production in Sub-Saharan Africa. Current losses due to trypanosomosis in cattle are valued at 4.5 billion dollars (U.S.) per year, with about 3 million cattle deaths (Budd, 1999), while farmers spend about US $35 million annually on nagana treatment.

Human trypanosomiasis (HAT) also called sleeping sickness affects people in 37 of 52 countries in Africa. The most highly endemic/epidemic countries are: Democratic Republic of Congo, Congo Republic, Angola, Central African Republic, Southern Sudan, Uganda and Tanzania. About 60 million people in Sub-Saharan Africa are at risk of sleeping sickness of whom about 4 million people are under surveillance and about 300,000 sleeping sickness cases are reported yearly. It is estimated that about 40,000 people diagnosed and treated and the rest left undiagnosed/untreated and a case fatality rate in untreated cases is 100% (WHO Fact sheet No 259, 2014). *T. brucei gambiense* is found in West and Central Africa. This form currently accounts for over 98% of reported cases of sleeping sickness and causes a chronic infection. A person can be infected for months or even years without major signs or symptoms of the disease. The second type of HAT called *T. brucei rhodesiense* is found in Eastern and Southern Africa. Nowadays, this form represents under 2% of reported cases and causes an acute infection. First signs and symptoms are observed a few weeks or months after infection. The disease develops rapidly and invades the central nervous system. Only Uganda presents both forms of the disease (WHO, Fact sheet No 259 of March, 2014). However, other parasite species and sub-species of the *Trypanosoma* genus are pathogenic to animals and cause animal trypanosomiasis in wild and domestic animals. In cattle the disease is called *Nagana*.

The pathway taken by trypanosomes through the tsetse fly is complicated and the flies have low infection rates even in endemic areas (Okoth and Kapata, 1986). Furthermore, it is difficult to differentiate the types of trypanosomes pathogenic to humans from those non-infective ones that circulate in tsetse flies. Trypanosomes of different species undergo cycles of development of varying complexity within the tsetse fly, transforming from bloodstream forms to procyclic non-mammalian infective forms in the fly midgut. *T. brucei* undergoes complex cycles of differentiation and multiplication in the tsetse fly. Flies are refractory to infection and the bloodstream form trypanosomes in the infected feed differentiate into viable procyclics. Different life cycle stages of *T. brucei* are: (a) Procyclic form in fly midgut, (b) long trypomastigote in fly proventriculus, (c) asymmetric dividing epimastigote in fly proventriculus, (d) short epimastigote in fly proventriculus, (e) attached epimastigotes in fly salivary gland, (f) metacyclic trypanosome in salivary gland, prepared for host inoculation. The 3 subspecies *T. brucei brucei*, together with the causative organisms of human African trypanosomiasis, *T. b. rhodesiense and T. b. gambiense*, must migrate from the midgut of the tsetse fly and transform to infective metacyclic forms in the salivary glands. Only a proportion of midgut infections mature into salivary gland infections with male tsetse maturing significantly more midgut infections than females (Dale et al., 1995; Milligan et at., 1995). Trypanosome genotype also influences maturation; those resistant to human serum (*T. b. rhodesiense*) are less likely to produce mature infections than human serum sensitive (*T. b. brucei*) parasites (Welburn et al., 1995). The mature *T. brucei* called the metacyclic form in the tsetse salivary glands are the ones regurgitated during the bite of the fly to the mammalian hosts.

A crucial component in HAT vector control and monitoring programmes is the sensitive detection of human-infective trypanosome species in the tsetse fly vector. The infection status of a tsetse fly is usually assessed by the presence/absence of trypanosomes by dissection and microscopic examination to identify the trypanosomes in the mouthparts, mid-guts and salivary glands. This is tedious and laborious.

It has already been demonstrated that it is possible to extract and amplify trypanosome DNA from whole tsetse flies (Ferreira et al., 2008). Detection of trypanosomes that cause disease in human beings and livestock within their tsetse fly hosts is an essential component of vector and disease control programmes. Several molecular-based diagnostic tests have been developed for this purpose (Malele et al., 2013). Many of these tests, while sensitive, require analysis of trypanosome DNA extracted from single flies. Previous serological tests used in diagnosis of HAT are immune-fluorescent test (IFAT) by Wery et al. (1970); enzyme linked immunoabsorbent (ELISA) by Komba et al. (1992) and Nantulya et al. (1987); card agglutination test (CATT) by Magnus et al. (1978); card indirect agglutination test for trypanosomiasis (CIATT) by Asonganyi et al. (1998) which are cumbersome to perform and have various sensitivities and specificities. However, recently developed LFT is an antigen test which is simple to perform and detects circulating *T. brucei* antigens in tsetse flies. In LFT, a filter paper is coated with antibodies against *T. brucei* and when homogenized tsetse midgut sample is added to the filter, agglutination is produced if trypanosome antigens of *T. brucei* are present.

The aim of this study was to evaluate LFT for detection and xenomonitoring of trypanosomes in tsetse flies under field conditions.

MATERIALS AND METHODS

Tsetse fly collection for trypanosome screening

Cross sectional surveys were done between April to July, 2013 in selected sleeping sickness endemic areas in each of the 6 countries (Uganda in Alwa, Dokolo, Tororo and Omugu, Tanzania in Serengeti, and Urambo, Malawi in Nkhotakota and Liwonde, Democratic Republic of Congo in Kwamouth and Itubi, South

```
Clone2        MFVFICRDFVSFFPLLSLFIPFFLFLLFHFECPYVSKCFLLIEDCPWQFLSYHFIFLSLV 60
TB927-Target  MFVFICRDFVSFFPFLSLFIPFFLFLLFHFECPYVSKCFLLIEDCPWQFLSYHFIFLSLV 60
Clone4        MFVFICRDFVSFFPLLSLFIPFFLFLLFHFECPYVSKCFLLIEDCPWQFLSYHFIFLSLV 60
Clone3        MFVFICRDFVSFFPLLSLFIPFFLFLLFHFECPYVSKCFLLIEDCPWQFLSYHFIFLSLV 60
              **************:*********************************************

Clone2        RFILYHFSFRFILYHFSFRFILYHFSFRFIYYFILLLFLFLFR---------CSHTH--- 108
TB927-Target  LFVLYHFSFRFILYHFSFRFILYHFSFRFIYYFILLSFHLLFYFTFVSIFVSAAHTHT-H 119
Clone4        RFILYHFSFRFILYHFSFR---------FIYYFILLLFLFLFP--------LLTYTHTPC 103
Clone3        RFILYHFSFRFILYHFSFRFILYHFSFRFIYYFILLLFLFLFP--------LLTYTHTPC 112
              *:*****************     ******** * :**      ::**

Clone2        -------------
TB927-Target  THTLLFQLHDIITH 133
Clone4        SSSCMTSLHT---- 113
Clone3        SSSCMTSLHT---- 122
```

Figure 1. CLUSTAL 2.0.12 multiple sequence alignment of Tb927.1.4620 and orthologues Clones 2, 3, and 4.

Sudan and Sudan in Kajokeji, Yambio and Tambura). The rest of the work was done in the laboratories in each country.

Fly dissection to screen trypanosomes

All live flies were dissected as described in Msangi et al. (1998). Midguts found positive by microscopy were transferred individually into labelled vials for analysis by TBR-PCR, ITS-PCR, LAMP and lateral flow tests.

Development of lateral flow test

Trypanozoon markers were selected using the *T. brucei* genome (*Trypanosoma brucei* TREU927), the *T. gambiense* (Tbg 972) genome, *T. congo*, and *T .vivax* genomes all contained in Tritryp data base (tritrypdb.org/). The criteria used was: (a) mining the genes only peculiar to the *Trypanozoon* parasites and (b) further checking *in silico* for mRNA expression profiles as an indicator that the gene phenotype is present. One of these genes which was selected was Tb927.1.4620 and its orthologues and specific primers were designed to confirm its unique presence in *Trypanozoon.*

Uninfected tsetse midguts from teneral and non-teneral flies for assessing specificity of lateral flow test

Midguts from 45 teneral and 25 non-teneral tsetse were tested to assess specificity of the lateral flow test. The samples were tested singly by both microscopy and lateral flow tests.

Experimentally infected tsetse flies for assessing sensitivity of lateral flow test

Midguts from a total of 128 experimentally infected tsetse flies were analyzed by both microscopy and Lateral Flow tests.

Statistical analysis

Statistical analyses were performed using Prism version 5.0 GraphPad Software (GPW5, USA) and StatMate version 2.0 GraphPad Software (GSW5, USA). The sensitivity and specificity of lateral flow assays were calculated from data entered into contingency tables. Sensitivity was defined as the proportion of positive cases that are positive by the LFT test and specificity as the proportion of controls that are negative by the LFT. Differences in sensitivity and specificity between the different tests were estimated by the Mc Nemar test while differences between LFT and microscopy were estimated with the Fisher exact test, setting the level of significance at 0.05. Agreement between the two tests was determined using the kappa index. A kappa index ranges from 0 to 1 and the higher the value the stronger the agreement. All calculations were estimated at a 95% confidence interval (95% CI).

RESULTS

The Tb 927.1.4620(TB927-Target) gene was selected to prepare the *T. brucei* peptide Tb04.29m 18.750 used to immunize rabbits. The gene was specific for *Trypanozoon* as shown in Table 1 by Tb gene PCR. This is a family of related genes that shares a highly conserved N terminal and a variable C terminal. The fact that we are dealing with a family of genes with a conserved terminal N and variable C terminal may offer some interesting potential downstream application. The amino acid translated sequences of the genes are shown in Figure 1 and compared to the target sequence (TB927-Target).

Four of these genes (Tb927.1.4620 and orthologues clones 2, 3, and 4) were cloned and expressed. Rabbits were immunized with *T. brucei* peptide Tb04.29m18.750 and orthologues clones 2, 3, and 4 antigens. The antibodies produced in the rabbits were purified by chro-matography for assay. The antibodies were assem-bled into LFT pilot kits and tested against laboratory derived *T. brucei* procyclics (Figure 2). Antisera raised against the peptide were employed in assembling the lateral flow test kits. Best results were obtained with antiserum against the peptide Tb04.29m18.750. The test kit based on the antiserum was then evaluated against midguts from both

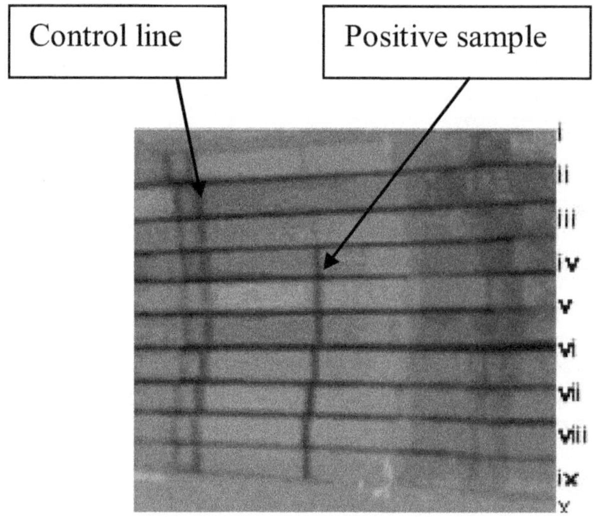

Figure 2. Homogenized tsetse midgut positive by microscopy is used as positive control sample , iv=positive control. Buffer reagents are used as negative samples seen in ii and iii=negative samples; all the rest are homogenized tsetse midguts which tested positive by microscopy in the field , I = weak positive sample and v-x=positive samples.

Table 1. Parasite Species/Subspecies results for the Tb927.1.4620 (TB927-Target) demonstrating the gene is specific to the *Trypanozoon.*

Isolate	Parasite species	Tb gene PCR
ILtat 1.1	*T.brucei*	+
ILtat 1.4	*T.brucei*	+
IL 1325	*T. b. gambiense*	+
IL2002	*T. b. rhodesiense*	+
IL1180	*T. congolense*	-
ILRAD 1895	*T. vivax*	-

Table 2. Evaluation of specificity of Lateral Flow Test using uninfected midguts from teneral and non-teneral tsetse flies.

Tsetse type	No. tested	No. Positive	No. Negative	Specificity %
Teneral tsetse midguts	45	0	45	100.0
Non-infected non-teneral tsetse midguts	25	0	25	100.0

The specificity of LFT using uninfected teneral and non-teneral tsetse flies was 100.0%.

from both experimentally infected tsetse and those from field collections.

Specificity of lateral flow test (LFT)

The LFT developed by Astel Diagnostics, Uganda, which is based on antibodies raised against the *T. brucei* peptide was evaluated for its analytical sensitivity, specificity, positive predictive value and negative predictive values in order to assess its utility for surveillance of trypanosome infections in tsetse flies. The evaluation of analytical specificity of lateral flow test using uninfected midguts from teneral and non-teneral tsetse flies was carried out and the results are summarized in Table 2.

Sensitivity of Lateral Flow Test (LFT)

The evaluation of analytical sensitivity of LFT using experimentally infected tsetse flies was carried out and the results are summarized in Table 3. From these results, the sensitivity of LFT is 97.3%; specificity 74.0% and the test was statistically significant at $P < 0.0001$.

Performance of lateral flow test in detecting trypanosomes in wild tsetse flies from 6 countries

The diagnostic sensitivity, specificity, positive predictive and negative predictive values of lateral flow test were evaluated using wild tsetse midguts from 6 different countries (Table 4) and the test was statistically significant at $P < 0.0001$.

Performance of lateral flow test using a combined data from the 6 countries against microscopy

The diagnostic test agreement or concordance of the lateral flow test and microscopy was 91.0% and kappa value of 80.5% at the confidence interval 0.72 to 0.80 which is a high level of test agreement. Furthermore, the relative diagnostic sensitivity and specificity of lateral flow test were 83.0 and 96.0%, respectively. The positive and negative predictive values were at 92.7 and 90.0%, respectively which were high (Table 5). The lateral flow test is therefore recommended for surveillance of trypanosomes in tsetse flies in order to: 1) Indicate areas with tsetse infected with trypanosomes; 2) Indicate tsetse infected with potentially human pathogenic trypanosomes; 3) Guide in prioritizing control strategies for HAT. The lateral flow test is a rapid serological test that detects *T. brucei* trypanosomes in tsetse flies which transmit trypanosomes to humans through its bite. Its ease of applicability, use of crude extracts of tsetse midguts with no specialized equipment requirement will make it popular for use in typically rural poor settings. The proof of the principle of lateral flow test has shown that it has a great potential for monitoring current trypanosome infections in tsetse flies, thus indicating areas of high tsetse infection rates and risk of outbreak and spread of HAT.

DISCUSSION

Serological tests based on antigen-detection enzyme-linked immunosorbent assay (ELISA) have been, to a limited extent, applied for detecting trypanosomes in tsetse flies. Based on trypanosome-specific monoclonal antibodies developed for revealing trypanosomes in mammalian hosts (Nantulya et al., 1987), Bosompem (1993) developed a monoclonal antibody-based dot-ELISA for detecting trypanosomes in tsetse flies. This method was found to be more useful in differentiating culture-derived insect stages of major pathogenic trypanosomes (Bosompem et al., 1996). However, large scale application of these antigen-detection tests has been limited because they required that species specific antibodies be developed and tested for cross-reactivity against all trypanosome species.

This work, therefore, provides an option for a vast terrain infected with tsetse for surveillance targeting the vector with limited resources. This project was conceived to come up with a tool to facilitate prioritization control areas with high tsetse infection rates. The linkage in research with private partnership for development, production and dissemination of developed technologies to the rural poor at low costs was coined at the beginning of the project development. Astel diagnostics, Uganda, a private sector was engaged to work in collaboration with other partners to design a simple LFT for detecting trypanosomes in tsetse flies. The LFT is a rapid test that detects *T. brucei* trypanosomes in tsetse flies. The diagnostic test agreement or concordance of the lateral flow test and microscopy was 91.0% which is a high level of test agreement.

Furthermore, the diagnostic sensitivity and specificity of LFT were 83.0 and 96.0%, respectively. Both the positive predictive value at 92.7% and the negative predictive value at 90.0% were high. The proof of the principle of LFT has shown that it has a great potential for monitoring current trypanosome infections in tsetse flies. It is a cheap, simple and rapid test which gives results within 20 min, read by naked eyes and can be performed by personnel with minimal training. The use of the LFT for control activities will be guided by the prevailing circumstances. The subspecies of *T. brucei* in positive samples in the test can be confirmed using other available molecular tests (TgsGP-PCR and SRA-PCR) since LFT does not discriminate between *T.b. brucei, T.b, rhodesiesne and T.b. gambiense* all of which belong to the subspecies *T. brucei*.

In Uganda, prevention of the merging of rhodesiense and gambiense foci requires intensive integrated interventions. Keeping the two foci separate in Uganda is the most logical approach to control and elimination of African trypanosomiasis. There is need to keep the two foci apart while integrated control strategies applied in each focus. It is equally important to note that elimination of *T.b. rhodesiense* is more complicated because of its zoonotic transmission which requires involvement of tsetse vector control. Thus the LFT has application in checking infections in the buffer zone that should be demarcated between the 2 endemic foci.

In Tanzania, strategic control of tsetse and African Trypanosomiasis in these areas will lead to significant contribution to improving human health, livestock sector and boost tourism through tsetse control in areas with tsetse pathogenic trypanosomes. *T.b. rhodesiense* sleeping sickness in Tanzania has to date been sporadic and of low endemicity in Tanzania. However, in the previous years, there were reports of epidemics that suggest that the situation of the disease has changed probably due

Table 3. Comparison of microscopy and LFT using experimentally infected tsetse.

Lateral flow test	Microscopy results on samples of *T. brucei* infected tsetse guts			Sensitivity (%)	Specificity (%)	PPV (%)	NPV (%)	Level of significance
	Positive	Negative	Total					
Positive	72	14	86					
Negative	2	40	42	97.3	74.0	83.7	95.2	P<0.0001
Total	74	54	128					

PPV= Positive predictive value, NPV= Negative predictive value.

Table 4. Performance of lateral flow test in detecting infections in wild tsetse flies

Country	Lateral flow test results	Microscopy results on samples of midguts			Sensitivity	Specificity	PPV	NPV	Level sign
		Positive	Negative	Total					
	Positive	21	1	22					
Tanzania	Negative	7	39	46	75.0	97.5	95.5	84.8	P<0.0001
	Total	28	40	68					
	Positive	20	0	20					
DRC	Negative	5	40	45	80.0	100.0	100.0	88.9	P<0.0001
	Total	25	40	65					
	Positive	36	4	40					
Malawi	Negative	4	36	40	90.0	90.0	90.0	90.0	P<0.0001
	Total	40	40	80					
	Positive	31	0	31					
Sudan	Negative	9	40	49	77.5	100.0	100.0	81.6	P<0.0001
	Total	40	40	80					
	Positive	34	6	40					
South Sudan	Negative	6	34	40	85.0	85.0	85.0	85.0	P<0.0001
	Total	40	40	80					
	Positive	24	2	26					
Uganda	Negative	3	118	121	88.9	98.3	92.3	97.5	P<0.0001
	Total	27	120	147					

Sens=Sensitivity, Spec= Specificity, PPV= Positive predictive value, NPV= Negative predictive value, Level sign. = Level of significance.

Table 5. Performance of Lateral Flow test using combined data from all countries against microscopy as a gold test

	Lateral flow test results	Microscopy test results		Total	Sensitivity (%)	Specificity (%)	PPV (%)	NPV (%)	Test agreement (%)
		Positive	Negative						
All countries	Positive	166.0	13.0	179.0					
	Negative	34.0	307.0	341.0	83.0	95.9	92.7	90.0	90.9
	Total	200.0	320.0	520.0					

Positive predictive value=PPV, Negative predictive value= NPV

changed probably due to ecological as well as socio-demographic changes favouring development of new active foci hence a need for instituting sustainable, user and environmental friendly techniques against the vectors of African Trypanosomiasis. Thus applying xenomonitoring tools such as LFT will be useful to monitor and predict the eminent outbreaks of HAT for timely intervention of control activities against tsetse to prevent and or minimize the outbreaks of sleeping sickness in the area would be a feasible option. The documented HAT cases of tourists together with local park residents infected with *T.b. rhodesiense* demonstrate the need to halt transmission by identifying areas with tsetse flies having pathogenic trypanosomes to humans and animals for targeted control.

In Malawi, tourists together with local population around the National Parks and the Game Reserves have been infected with *T.b. rhodesiense* and this has a negative effect on tourism industry as well as on the health of the population. There is, therefore, need to reduce the incidence of infection and prioritize control strategies in and around National Parks and Game Reserves by regular use of xenomonitoring tools to detect trypanosomes in tsetse flies. Besides, there is emerging evidence that tsetse flies are highly infected with *T. brucei* from this study where in some cases up to 85% of tsetse flies caught in Liwonde National Park were habouring *T. brucei* infections yet no recorded case of HAT has been reported in and around Liwonde National Park to our knowledge. This could be due to lack of active surveillance for HAT in human population in the areas.

Recently, foreign visitors to Kasungu National Park have reported to have been infected with HAT. Also the HAT cases caused by *T.b. rhodesiense* in Malawi are subacute as most people infected present themselves passively to district hospitals mostly in second stage of the disease (MacLean et al., 2010). The case fatality rates are very high due to this fact (Chisi et al., 2011). Furthermore, there are very few studies in Malawi that have concentrated on the role that tsetse flies play on the cycle of HAT transmission. The Congo River of Democratic Republic of Congo and Congo Brazzaville trans border is an active Gambiense HAT transmission area with Bandundu and Ngabé HAT foci. This socioeconomically active zone would benefit on surveillance strategy using xenomonitoring tools such as LFT. In addition, at Congo Brazzaville one could monitor spreading of trypanosomiasis from Congo River foci to neighboring trypanosomiasis free areas using LFT.

Most detection techniques for trypanosomes require technical knowhow in their use, hence restricted to developed labs, requires refrigeration which is non-reliable in rural areas where infected tsetse organs are collected, and some essential reagents are not readily and easily available to execute different trypanosome detection techniques. In this paper we have highlighted the need to come up with simple, user friendly, ultra-sensitive technique which can be used to detect trypanosomes especially the human infective ones as an early warning tool to predict the outbreak of tsetse borne diseases in human beings. Future work may include development of a LFT that discriminate between *T. brucei* subspecies and or *T. congolense* and *T. vivax*.

ACKNOWLEDGEMENTS

We are greatly indebted to Global Heath of the Bill and Melinda Gates Foundation for the financial assistance and to all its staff members who entirely assisted and encouraged us during the implementation of this project for five years. Secondly, we would like to thank the participating institutions: Tsetse and Trypanosomiasis Research Institute (TTRI), Tanzania, Tropical Medicine Research Institute (TMRI) Sudan, Ministry of Animal Resources and Fisheries South Sudan, College of Medicine, University of Malawi, The National Program of Control of Human African Trypanosomiasis (NPHAT), Congo

Kinshasa, Astel Diagnostics in Uganda especially their Technical and Administrative personnel in each of the mentioned institutions. Thirdly, we thank the Makerere University Administration and the Department of Biochemistry and Sports Science in particular for hosting the project so well for the last five years. Lastly, but not least, the Uganda National Council of Science and Technology, (UNCST), the Ministry of Health, the Ministry of Agriculture, Animal Industry and Fisheries for the ethical clearance and permission to carry out the research.

Conflict of interests

The authors declare that there is no conflict of interests

REFERENCES

Asonganyi T, Dono F, Kibona S N, Nyasulu Y M, Masake R, Kuzoe F (1998). A multi-centre evaluation of the card indirect agglutination test for trypanosomaisis (TryTect CIATT). Ann. Trop. Med. Parasitol. 92(8):837-844.

Bosompem KM (1993). Development of monoclonal antibody-based assays for detection and differentiation of trypanosome species in the tsetse fly (*Glossina* spp). PhD thesis, Department of Animal Science, University of Ghana, Legon.

Bosompem KM, Assoku RKG, Nantulya VM (1996). Differentiation between culture-derived insect stages of *T. brucei, T. vivax, T. congolense* and *T.simiae* using monoclonal antibody-based dot-ELISA. Parasitology 112:59-66.

Budd L (1999). DFID-Funded Tsetse and Trypanosomiasis Research and Development since 1980. Economic Analysis. Department of International Development, UK, London.

Chisi JE, Muula AS, Ngwira B, Kabuluzi S (2011). A retrospective study of Human African Trypanosomiasis in the three Malawi districts. Tanzania. J. Health Res. 13(1):79-86.

Dale C, Welburn SC, Maudlin I, Milligan PJ (1995). The kinetics of maturation of trypanosome infections in tsetse. Parasitol. 111(Pt 2):187-191.

Ferreira F, Cano J, Furtado A, Ndong-Mabale N, Ndong-Asumu P (2008). An alternative approach to detect *Trypanosoma* in *Glossina* (Diptera, Glossinidae) without dissection. J. Infect. Dev. Ctries. 2:63-67.

Komba E, Odiit, Mbulamberi DB, Chimfurembe EC, Nantulya VM (1992). Multi-centre evaluation of an antigen detection ELISA for the diagnosis of *Trypanosoma brucei rhodesiense* sleeping sickness. Bull. World Health Organ. 70(1):57-61.

MacLean LM, Odiit M, Chisi JE, Kennedy PGE, Sternberg JM (2010). Focus-Specific Clinical Profiles in Human African Trypanosomiasis caused by *Trypanosoma brucei rhodesiense*. PLOS Negl. Trop. Dis. 4 (12):e906.

Magnus E, Vervoort T, Van Meirvenne N (1978). A Card agglutination test with stained trypanosomes (CATT) for the serological diagnosis of *T.b.gambiense trypanosomiasis*. Ann. Soc. Belg. Med. Trop. 53(3):169-176.

Malele II, Ouma JO, Enyaru JC, Matovu E, Alibu V, Auma JE, Onyoyo SG, Bateta R, Changasi RE, Mukiria PW, Ndung'u K, Gitonga PK, Mwaniki LM, Nyingilili HS, Lyaruu EA, Kapange LA, Kamau PK, Masiga DK (2013). Comparative diagnostic and analytical performance of PCR and LAMP-based trypanosome detection methods estimated using pooled whole tsetse flies and midgets. Vet. Parasitol. 197:549-556

Masiga DK, Smyth AJ, Hayes P, Bromidge TJ, Gibson WC (1992). Sensitive Detection of Trypanosomes in Tsetse flies by DNA Amplification. Int. J. Parasitol. 22:909-918.

Msangi AR, Whitaker CJ, Lehane MJ (1998). Factors influencing the prevalence of trypanosome infection of *Glossina pallidipes* on the Ruvu flood plain of Eastern Tanzania. Acta Trop. 70:143-155.

Milligan PJ, Maudlin I, Welburn SC (1995). *Trypanozoon*: infectivity to humans is linked to reduced transmissibility in tsetse. II. Genetic mechanisms. Exp. Parasitol. 81:409-415.

Nantulya VM, Musoke AJ, Rurangirwa FR, Saigar N, Minja SH (1987). Monoclonal antibodies that distinguish *Trypanosoma congolense, T. vivax* and *T. brucei*. Parasite Immunol. 8:421-431.

Okoth JO, Kapaata R (1986). Trypanosome infection rates in *Glossina fuscipes fuscipes* Newst. in the Busoga sleeping sickness focus, Uganda. Ann. Trop. Med. Parasitol. 80:459-61.

Welburn SC, Maudlin I, Milligan PJ (1995). *Trypanozoon:* infectivity to humans is linked to reduced transmissibility in tsetse. I. Comparison of human serum-resistant and human serum-sensitive field isolates. Exp. Parasitol 81:404-408.

Wery M, Wery-Paskoff S, Van Wettere N (1970). The diagnosis of human African trypanosomiasis (*T. gambiesnes*) by the use of fluorescent antibody test. 1. Standardization of an easy technique to be used in mass surveys. Ann. Soc. Belg. Med. Trop. 50 (5):613-634.

World Health Organization (WHO) (2014). WHO fact sheet on Human African trypanosomiasis. Fact Sheet No 259.

8

Parasitic contamination on vegetables irrigated with Awash River in selected farms, Eastern Showa, Ethiopia

Girmaye Benti and Fekadu Gemechu

Department of Biology, Dilla University College of Natural and Computational Science, P. O. Box-419 Dilla, Ethiopia.

This study was conducted from February, 2013 to April, 2013 to evaluate parasitic contamination of vegetable at Melka Hida and Wonji Gefersa wastewater irrigated vegetable farms that are found in Adama Woreda. A total of 72 vegetables samples were collected from both farms using a random sampling technique. Among all vegetables examined, *Ascaris lumbricoides* eggs were detected in 22.22% (16/72) of spinach (*Spinacea oleracea*), 16.67% (12/72) of lettuce (*Lactuca sativa*) and 23.61% (17/72) of cabbage (*Brassica oleracea* Linn). *Giardia intestinalis* cysts were detected in 18.06% (13/72) of spinach, 19.44% (14/72) of lettuce and 15.28% (11/72) cabbage samples. Of all parasitic contaminants *Entamoeba histolytica* cyst was the least detected parasite; 8.33% (6/72) of spinach, 12.5% (9/72) of lettuce and 8.33% (6/72) of cabbage samples from both farms. The high parasite contamination rates associated with these vegetable samples indicated poor farming practice employed in the overall production in the study area.

Key words: Awash River, contamination, parasite, vegetables.

INTRODUCTION

Vegetables are essential for good hearth, and they form a major component of human diet in every family. They are vital energy contributors that are depended upon by all levels of human as food supplement or nutrient (Duckworth et al., 1996). Food safety regardless of the specific food product should be a paramount concern to everyone, with parasites from contaminated vegetables being a potential health risk. According to Speer (1997), vegetables can become contaminated with enteric bacterial, viral and parasitic pathogens throughout the process of planting to consumption. The extent of contamination depends on several factors that include, among others, use of untreated wastewater and water supplies contaminated with sewage for irrigation (Amoah et al., 2007; Beuchat, 2002; Simões et al., 2001).

Intestinal parasites are common in fresh vegetables. Vegetables are reported to harbor intestinal parasites such as *Ascaris lumbricoides, Taenia* species, *Fasciola hepatica*, *Hymenolepis nana*, *Echinococcus* species, *Trichuris* species, *Enterobius vermicularis*,

Trichostrongylus species, *Toxocara* species, *Strongyloides stercoralis*, *Giardia intestinalis*, *Entamoeba* species, *Iodamoeba butschlii*, *Blastocystis hominis* and *Cryptosporidium parvum* (Gharavi et al., 2002; Gupta et al., 2009; Abougrain et al., 2009; Uga et al., 2009). Intestinal parasitic infections are the top global health problems, whereas amoebiasis and ascariasis, are among the ten most common infections (WHO, 1987). Hence, consumption of raw vegetables plays an important role in the transmission of human parasitic infection (Amoah et al., 2005; Choi et al., 1982; Coelho et al., 2001; Daryani et al., 2008; Erdogrul and Sener, 2005).

Contamination of vegetables may occur through contacting with the soil, raw manure and sewage used as fertilizer on vegetable farms. In most cases, contamination of vegetables is associated with the water used for irrigation (Simoes et al., 2001). Use of sewage contaminated water for irrigation of vegetables is a common practice in developing countries. The access to clean water for irrigating vegetables is a major challenge. As an alternative, urban and peri-urban vegetable farmers in search of water for their crops have no other choice than to use water from these highly polluted sources. This raises public health concerns due to possible crop contamination with pathogens where vegetables are eaten uncooked (Amoah et al., 2006).

Unhygienic sewage disposal and absence of its treatment facilities pose potential health hazards through contaminating irrigated food crops with parasites in urban and suburban areas of African countries including Ethiopia (Damen et al., 2007; Nyarango et al., 2008; Srikanth and Naik, 2004; Weldesilassie et al., 2009). Many farm households in Eastern Showa that are irrigating their farmlands with wastewater are not aware of the risks or the potential harmful environmental consequences. This may be attributed to illiteracy, lack of adequate information and exposure to poor sanitary conditions for most of their lives. Several studies documented prevalence of intestinal parasites in different parts of Ethiopia including Oromia region through microscopic examination of stool samples collected from suspected human population (Legesse and Erko, 2004; Tadesse, 2005; Dejenie and Petros, 2009). However, research endeavors to detect parasites from environmental sources such as vegetables are scarce in Ethiopia (ROSA, 2009). Therefore, the aim of this study was to assess the degree of parasitic contaminations on selected vegetables irrigated with Awash River at Melka Hida and Wonji Gefersa farms.

MATERIALS AND METHODS

Description of the study area

This study was conducted at Melka Hida and Wonji Gefersa wastewater irrigated vegetable farms that are found in Adama Woreda. Wonji Gefersa is located in East Showa zone, Oromia region, 107 km away from Addis Ababa. It is located at a latitude of 8° 26' 59" North and longitude of 39° 16' 48" East. It has an elevation of 1588 masl and its temperature and rain fall is 23°C and 500 to 800 mm, respectively (Environmental Protection Authority, 2005). While Melka Hida is found in Adama town administrative zone, Oromia region, 99 km away from Addis Ababa. It is located at latitude of 8° 33′ 0″ North and longitude 39° 16′ 12″ East. It has an elevation of 1620 masl (Environmental Protection Authority, 2005) (Figure 1).

Study Design

Preliminary surveillance study was conducted before starting the test experiment at Melka Hida and Wonji Gefersa vegetable farms using Awash River for irrigation. Accordingly 160 sample respondents were purposively selected and provided with already prepared questionnaire. The questions were interpreted for few producer respondents who can't understand the matter. Out of the selected producer 135 were males and 25 were females, all of which are taken as literates. According to the answer obtained and analyzed from the respondents most of the problems that frequently faced them were recognized by the professional physicians and so related with the parasitic contaminations from the vegetables and fruits. Based on these the following three parasitic contaminants namely giant roundworm *(A. lumbricoides)*, Giardia *(Giardia intestinalis) and* parasitic protozoan *(Entamoeba histolytica*) have seen to be the most dominant and leading parasite diagnosed. Then, a cross sectional survey test experiment was conducted to assess contamination of specific parasites on the main leafy vegetables [lettuce (*Lactuca sativa*), cabbage (*Brassica oleracea* Linn), and spinach (*Spinacea oleracea*)] that were grown in Melka Hida and Wonji Gefersa vegetable farms. The samples were regularly collected three times at three week interval from February, 2013 to April, 2013.

Sample collection

A total of 72 samples comprising three types of fresh vegetables (cabbage, lettuce and spinach) were collected from Melka Hida and Wonji Gefersa vegetable farms using a random sampling technique. Recently, mature leaves of lettuce, cabbage and spinach were sampled at early maturity according to methods used by Itanna (1998). All samples were collected aseptically in a sterilized universal container and plastic bags and transported to Dilla University using a cooler box for laboratory processing. Analysis was conducted within 24 h of arrival at the Parasitology Laboratory of Dilla University.

Parasitological analyses of vegetables

In the laboratory, 100 g of each fresh vegetable sample was chopped into small pieces and put into a clean beaker containing enough physiological saline solution (0.85% NaCl), to wash the sample. After removing fragments of the vegetable sample from the washing saline using clean forceps, it was kept about 24 h for sedimentation to take place. After 24 h sedimentation, the top layer of the washing saline was carefully discarded leaving 5 ml of the sediment. This was finally centrifuged at 2000 rotations/min for 5 min by HERMLE Z200A centrifuge. After discarding the supernatant, the residue was mounted on slides, stained with Lugol's iodine solution and examined under the compound light microscope to examine the samples for intestinal parasites: *A. lumbricoides* eggs, *E. histolytica* cysts and *G. intestinalis* cysts (Abougrain et al., 2009; Uga et al., 2009). To increase the chance of parasite detection, three slides were prepared for each vegetable sample.

Data analysis

For data analyses, descriptive statistics in percentage was computed for the pre surveillance data and parasitic contamination of vegetable samples. Chi-square test was used for comparison between vegetable contamination rates using SPSS version 16.0. A P-value ≤0.05 was considered statistically significant.

RESULTS AND DISCUSSION

The study results of the pre surveillance and test experiment data on parasitological analysis of leafy vegetables were explained as shown in Tables 1 and 2, consecutively.

The aforementioned results (Table 1) show that 67.5% (108) of the respondents said that their income depends on the vegetable production. Producers also asked for why and which types of vegetables they produce and so, 50% (80) and 64.4% (103) explained that they were produced for the purpose of income generation and vegetable types like cabbage, lettuce and spinach, respectively. Similarly, 87.5% (140) of the producer said that the disease agents are caused by different parasitic worms and other pathogenic microbes. Among the parasites *A. lumbricoides, E. histolytica* and *G. intestinalis* are the predominantly mentioned by respondent.

Based on the information obtained from the surveillance, regular experiment was conducted to test for the presence of the aforementioned parasite. The results (Table 2) showed that cabbage samples collected from both Melka Hida and Wonji Gefersa farms, 17 (23.61%) were found to be positive for *A. lumbricoides* eggs. In developing countries, intestinal parasites are very common. Fresh vegetables are an important route of their transmission. Daryani et al. (2008) reported the detection of intestinal parasites in 29% (13/45) of native garden vegetables. The distribution of parasites on vegetables between Melka Hida and Wonji Gefersa was significantly different (χ^2=3.772, P=0.042). Variation may be due to different sections of the wastewater drainage canal in Wonji Gefersa farm which receive different load of faecal contamination.

A. lumbricoides eggs were the predominant intestinal parasite in this work. These include 22.22% (16/72) in spinach, 16.67% (12/72) in lettuce and 23.61% (17/72) in cabbage. The results showed that cabbage is relatively highly contaminated and lettuce is the least contaminated (Table 2). This might be due to cabbage nearby the soil as compared to the other vegetables that was directly contaminated with irrigation water or indirectly by contact with contaminated soil, may aggravate the rate of contamination. Other studies show that vegetables with dense foliage were most contaminated than those growing on surface (Idrissa et al., 2010). The dense foliage would protect the helminth eggs against unfavourable conditions to their survival and persistence, such as sunlight, drying, and wind (Dssouli 2001; Dssouli et al., 2006).

In relation to this finding, Abougrain et al. (2009) reported on other vegetables that helminth eggs and *G. intestinalis* cysts were detected in 58% (73/126) of vegetables examined. These include 14% (5/36) of tomato, 42% (15/36) of cucumber, 96% (26/27) of lettuce and 100% (27/27) of cress samples. This shows that the use of sewage water plays important role in the transmission of parasitic disease to human through consumption of such vegetables (Gupta et al., 2009). *A. lumbricoides* eggs were detected from lettuce with a value of 25% (9/36) from Melka Hida and 8.33% (3/36) from Wonji Gefersa. There was significant variation between the sites (χ^2=3.600, P=0.050), Melka Hida having the highest rate. This may be due to contamination during production and application of human excreta, animal manure as a fertilizer. The results of this study is comparable with the previous finding that showed Chinese cabbage (*Brassica pekinensis*) had the highest degree of contamination (91.1%) by *A. lumbricoides* eggs (Choi and Lee, 1992).

G. intestinalis cysts were detected in 18.06% (13/72) of spinach, 19.44% (14/72) of lettuce and 15.28% (11/72) of cabbage samples from both farms. There was also significant difference in the prevalence of parasites stages (χ^2 = 5.675, P = 0.017). This indicates that the prevalence of *G. intestinalis* cyst was more prevalent on lettuce collected from Wonji Gefersa farm than the other vegetables. This may be due to the production related sources of parasitic contaminants associated with practices of using untreated wastewater for irrigation.

On the other hand, the prevalence of *G. intestinalis* cyst examined on cabbage and spinach from both farms were not significantly different (χ^2 = 2.683, P = 0.101; χ^2 = 2.347, P = 0.126), respectively (Table 2). The finding of this study is in line with Daryani et al. (2008) who reported detection of intestinal parasites in 29% (13/45) of native garden vegetables consumed in Ardabil city, Iran. Similarly, Abougrain et al. (2009) examined 126 samples of four different types of fresh salad vegetables from wholesale and retail markets in Tripoli, Libya of which 58% were positive for helminth eggs and *G. intestinalis* cysts. The variation in the prevalence might be due to sample size difference.

Moreover, *E. histolytica* cyst was detected in 8.33% (6/72) of spinach, 12.5% (9/72) of lettuce and 8.33% (6/72) of cabbage samples from both farms. The results of the study indicated that the prevalence of *E. histolytica* cyst was the least. However, there was significant difference (χ^2 = 3.175, P = 0.045). The result of the study revealed that lettuce collected from Wonji Gefresa was more contaminated by *E. histolytica* cysts. As shown in Table 2, *E. histolytica* cysts was not significantly different (χ^2 = 2.683, P = 0.101; χ^2 = 2.347, P = 0.126) between cabbage and spinach, respectively.

In connection with this finding, Damen et al. (2007)

Table 1. The percentage and frequency of pre surveillance data collected.

Parameter	Frequency	%
Sex		
Male	135	84.4
Female	25	15.6
What is your source of income?		
Vegetables production	108	67.5
Other crops production	15	9.4
Both	37	23.1
Which type of vegetables do you cultivate?		
Cabbage	8	5.0
Lettuce	10	6.2
Spinach	12	7.5
Beetroot	4	2.5
Tomato	23	14.4
Cabbage, lettuce and spinach	103	64.4
Purpose of production		
Income generation	80	50
Home consumption	24	15.0
Both	56	35.0
Source of water for the production of vegetables		
Awash River	153	95.6
Under ground water	7	4.4
Would you think this water (Awash River) is contaminated?		
Yes	136	85
No	24	15
If you say yes for above question, what is the possible sources for contamination		
Wastewater from the municipality directly join Awash River	34	21.2
Community nearby Awash River defecated on open space	47	29.4
Both are the source of contamination of Awash River	79	49.4
Did you get sick in last one year?		
Yes	115	71.9
No	45	28.1
What is the possible source for disease occurrence?		
Consumption of improperly cooked vegetables	110	68.8
Personal hygiene	28	17.5
Utensil used	22	13.2
Which types of disease causing agents are identified in laboratory diagnosis?		
Parasite	140	87.5
Bacteria	20	12.5
If you say parasite for above question, which types of the parasites predominantly observed?		
Ascaris lumbricoides	40	25
Entomoeba histolytica	20	12.5

Table 1 cont'd

Giardia intestinalis	15	9.4
Fasciola hepatica	4	2.5
Strongyloides stercoralis	10	6.2
Cryptosporidium parvum	6	3.8
Ascaris lumbricoides, Entomoeba histolytica and *Giardia intestinalis*	65	40.6
What types of recommendation recommended by the physician about the quality of food consumptions?		
Not to eat improperly cooked vegetables	90	56.2
Wash the vegetables properly with clean water before consumption	50	31.2
Treat the water used for vegetables production	20	12.5

Table 2. Prevalence of three purposively selected parasites from cabbage, lettuce, and spinach samples of both farms (Melka Hida and Wonji Gefersa).

Vegetable type	No. of examined sample	Detected organisms	Site					χ^2 test	P-value
			MH	%	WG	%	Total (%)		
Cabbage	12	*A. lumbricoides* eggs	5	13.89	12	33.33	23.61	3.772	0.042
		G. intestinalis cyst	3	8.33	8	22.22	15.28	2.683	0.101
		E. histolytica cyst	1	2.78	5	13.89	8.33	2.909	0.088
Lettuce	12	*A. lumbricoides* eggs	9	25.0	3	19.44	16.67	3.600	0.050
		G. intestinalis cyst	3	8.33	11	30.56	19.44	5.675	0.017
		E. histolytica cyst	2	5.56	7	19.44	12.50	3.175	0.045
Spinach	12	*A. lumbricoides* eggs	6	16.67	10	27.78	22.22	1.286	0.257
		G. intestinalis cyst	4	11.1	9	25.00	18.06	2.347	0.126
		E. histolytica cyst	2	5.56	4	11.11	8.33	0.727	0.394

MH: Melka Hida, WG: Wonji Gefersa.

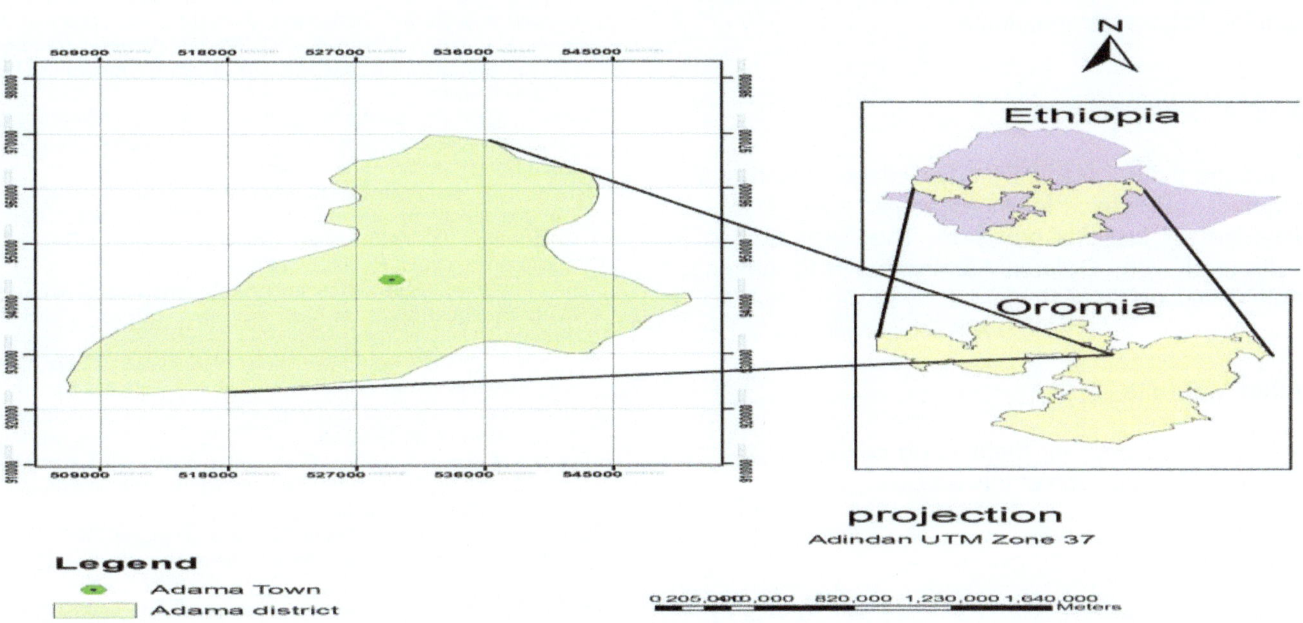

Figure 1. Location of the study area (Aynalem, 2010).

isolated ova of *A. lumbricoides* eggs (17.1%) and *E. histolytica* (14%) from different vegetables in Jos, Nigeria. In general, the study showed significant differences in the occurrence of pathogenic intestinal parasites detected from fresh vegetables. Several factors may include geographical location, type and number of samples examined, methods used for detection of the intestinal parasites, type of water used for irrigation, and per harvesting handling methods of such vegetables may contribute to such differences. According to Bethony et al. (2006) environmental factors play a great role in the incidence of intestinal parasitic infection as hot and humid tropical climate favour increased parasite prevalence. Based on variation in climatic and geographic zones in Ethiopia, it should be evident that there are macro and micro environmental factors contributing to the differences in prevalence of intestinal parasites (Jemaneh, 2000).

Conclusions

This study showed that *A. lumbricoides* eggs, *G. intestinalis* cysts and *E. histolytica* cyst were recorded on vegetable samples of both farms. Taking account of results obtained, it is clear that irrigation process of these vegetables fields has been performed using contaminated water sources (wastewater). Therefore, great attention should be given in using contaminated water for production of vegetables for the public health perspective. An adequate treatment of the sewage water and banning wastewater use for irrigation of plants intended for human consumption, among others, should be implemented. Health promotion and education on the mode of transmission of disease, environmental sanitation and eating habits will enhance the prospect for the control of parasitic infections.

ACKNOWLEDGEMENTS

The authors thank Dilla University College of Natural and Computational Science for the financial, material support and laboratory facilities provision. They also appreciate the efforts of Wonji Gefersa Administrative for helping in sample collection.

Conflict Interests

The authors declare that there is no conflict of interests regarding the publication of this article.

REFERENCES

Abougrain AK, Nahaisi MH, Madi NS, Saied MM, Ghenghesh KS (2009). Parasitological contamination in salad vegetables in Tripoli-Libya. Food Control 21(5):760-762.

Amoah P, Drechsel P, Abaidoo C (2005). Irrigated urban vegetables production in Ghana: Sources of pathogen contamination and health risk elimination. Irrigation Drainage 54(1):49-61.

Amoah P, Drechsel P, Abaidoo RC (2006). Pesticide and microbiological contamination of vegetables in Ghana's urban markets. Arch. Environ. Contam. Toxicol. 50(1):1-6.

Amoah P, Drechsel P, Abaidoo RC, Klutse A (2007). Effectiveness of common and improved sanitary washing methods in selected cities of West Africa for the reduction of coliform bacteria and helminth eggs on vegetables. Trop. Med. Int. Health 12(Suppl)2:40-50.

Bethony J, Brooker S, Albonico M, Geiger SM, Loukas A (2006). Soil transmitted helminth infections: Ascariasis, trichuriasis, and hookworm. Lancet 367(9521):1521-32.

Beuchat LR (2002). Ecological factors influencing survival and growth of human pathogens on raw fruits and vegetables. Microb. Infect. 4:413-423.

Choi DW, Lee S (1992). Incidence of parasites found on vegetables collected from markets and vegetables gardens in Taegu area. Korean J. Parasitol. 10:44-51.

Choi DW, Ock MS, Suh JW (1982). Recent demonstration of helminth eggs and larvae from vegetables cultivating soil. Korean J. Parasitol.20:83-92.

Coelho LM, Oliveira SM, Milman MH, Karasawa KA, Santos RD (2001). Detection of transmissible forms of enteroparasites in water and vegetables consumed at school in Sorocaba, Sao Paulo state, Brazil. Rev. Soc. Bras. Med. Trop. 34(5):479-482.

Damen JG, Banwat EB, Egah DZ, Allanana JA (2007). Parasitic contamination of vegetables in Jos, Nigeria. Ann. Afr. Med. 6:115-8.

Daryani A, Ettehad GH, Sharif M, Ghorbani L, Ziaei H (2008). Prevalence of intestinal parasites in vegetables consumed in Ardabil, Iran. Food Control 19(8):790-794.

Dejenie T, Petros B (2009). Irrigation Practices and Intestinal Helminth Infections in Southern and Central Zones of Tigray. Ethiop J. Health Dev. 23(1):48-56.

Dssouli K (2001). Traitement et réutilisation des eaux usées en agriculture au Maroc oriental: Etude des helminthes parasites. Thèse de doctorat en sciences, Faculté des Sciences, Oujda. French.

Dssouli K, El Halouani H, Berrichi A (2006). L'impact sanitaire de la réutilisation des eaux usées brutes de la ville d'Oujda en agriculture: Etude de la charge parasitaire en oeufs d'helminthes au niveau de quelques cultures maraîchères. Biol. Santé 6(1):51-58.

Duckworth RB (1996). Farming systems for the production of fruits and vegetables. Fruits and vegetables oxford: Pergama press. pp. 48-62.

Erdogrul OR, Sener H (2005). The contamination of various fruit and vegetable with *Enterobius vermicularis*, *Ascaris* eggs, *Entamoeba histolytica* cysts and *Giardia lamblia* cysts. Food Control 16:557-560.

Gharavi MJ, Jahani MR, Rokni MB (2002). Parasitic contamination of vegetables from farms and markets in Tehran. Iranian J. Publ. Health 31(3-4):83-86.

Gupta N, Khan DK, Santra SC (2009). Prevalence of intestinal helminth eggs on vegetables grown in wastewater irrigated areas of Titagarh, West Bengal, India. Food Control 20(10):942-945.

Idrissa S, Belghyti D, Kharrim KE, Yoro C (2010). Parasitic Contamination of the Mint and Turnip Irrigated by untreated wastewater in Sidi Yahia Gharb (Morocco). BALWOIS 2010 Conference, Ohrid, Republic of Macedonia.

Itanna F (1998). Metal concentrations of some vegetables irrigated with industrial liquid waste at Akaki, Ethiopia. Ethiop. J. Sci. 21(1):133-144.

Jemaneh L (2000).The Epidemiology of Schistosomiasis mansoni and soil transmitted helminths in elementary school children from the South Gonder zone of the Amhara National Regional State, Ethiopia. Ethiop. Med. J. 38:105-118.

Legesse M, Erko B (2004). Prevalence of intestinal parasites among schoolchildren in a rural area close to the southeast of Lake Langano. Ethiop. J. Health Dev. 18(2):116-120.

Nyarango RM, Aloo PA, Kabiru EW, Nyanchongi BO (2008). The risk of pathogenic intestinal parasite infections in Kisii Municipality, Kenya. BMC Public Health 8:237.

ROSA (2009). A preliminary survey of the microbiological and parasitological quality of some locally produced and marketed vegetables done as part of the ROSA research activity in Arba Minch,

Ethiopia. pp. 43-44.
Simoes M, Pisani B, Margues EGL (2001). Hygienic sanitary conditions of vegetables and irrigation water from kitchen gardens in the municipality of Campinas, SP. Braz. J. Microbiol. 32:331-333.
Speer CA (1997). Protozoan parasites acquired from food and water. Food Microbiology: Fundamentals and Frontiers, Washington DC, American Society for Microbiology. pp. 478-493.
Srikanth R Naik D (2004). Prevalence of Giardiasis due to wastewater reuse for agriculture in the suburbs of Asmara City, Eritrea. Int. Environ. Health Res. 14:43-52.
Tadesse G (2005). The prevalence of intestinal helminthic infections and associated risk factorsamong school children in Babile town, Eastern Ethiopia. Ethiop. J. Health Dev. 19:140-147.
Uga S, Hoa NTV, Noda S, Moji K, Cong L, Yaoki SK, Rai Fujimaki Y (2009). Parasite egg contamination of vegetables from a suburban market in Hanoi, Vietnam. Nepal Med. Coll. J. 11:75-78.
Weldesilassie AB, Frör O, Boelee E, Dabbert S (2009). The economic value of improved wastewater irrigation: A contingent valuation study in Addis Ababa, Ethiopia. J. Agric. Resour. Econ. 34:428-44.
World Health organization (WHO) (1987). Prevention and control of intestinal parasitic infections. Technical Report Series 749.WHO, Geneva.

Viability loss and ultrastructural changes on protoscolices of human hydatid cysts induced by retinoic acid

Doaa Abdelhafeez Yones[1] , Enas Abdelhameed Mahmoud[1], Ragaa Ali Othman[1], Mohamed Korany[2] and Tarek Hassan El-Metwally[3]

[1]Department of Medical Parasitology, Faculty of Medicine, Assiut University, Assiut, Egypt.
[2]Department of General Surgery, Faculty of Medicine, Assiut University, Assiut, Egypt.
[3]Department of Medical Biochemistry, Faculty of Medicine, Assiut University, Assiut, Egypt.

Surgical removal of intact hydatid cyst is the most effective treatment for hydatid disease. Recurrence of hydatid cyst is mainly due to dissemination of protoscolices (PSCs) rich fluid during the surgical operation. Therefore, preoperative instillation of a scolicidal agent into the cyst is a common practice with adverse side effects of the used drugs. All-trans retinoic acid (atRA) is the physiological mediator of most of the functions of vitamln A, particularly as a cellular differentiation and apoptosis regulating factor. We hypothesized that instillation of atRA could provide an alternative safe scolicidal approach. We tested the scolicidal effects and ultrastructural changes imposed by atRA on human hydatid cyst PSCs *in vitro*. Freshly isolated hydatid cyst PSCs were subjected to atRA (at 16.7, 1.67, 0.167 µM and 16.7 nM/L). Changes in protoscolices viability (0.1% eosin exclusion) and morphology (scanning and transmission electron microscopy; SEM and TEM) were investigated. Dose-dependent PSCs death within few minutes to 7 days of exposure to atRA was observed. SEM demonstrated ultrastructural damages including rosteller disorganization, loss of hooks and distortion of hooks morphology. TEM revealed loss of the integrity of the internal tissues of PSCs, an increased vacuolization, formation of large lipid droplets in the distal cytoplasm and aberrant, rounded abnormally large sized mitochondria. atRA is a promising alternative to the available synthetic and chemical scolicidal agents. However, *in vivo* scolicidal activities of atRA and the possible side effects necessitate further studies.

Key words: Hydatid cyst, protoscolices, retinoic acid, ultrastructure, scolicidal activity.

INTRODUCTION

Cystic echinococcosis (CE) is a chronic zoonosis caused by the larval stage of the cosmopolitan parasitic cestode *Echinococcus granulosus* (Lv et al., 2013). It is a serious disease to human and domestic livestock having as significantly high fatality rate (Himsworth et al., 2010; Grosso et al., 2012). In all North African countries, including Egypt, CE is highly endemic and is re-emerging as a major public health issue (Sadjjadi, 2006; Elshazly et al., 2009; Dakkak, 2010; Cardona and Carmena, 2013).

Although surgery is still the preferred method for symptomatic cases of CE, there is an increasing risk of intra-operative spillage of protoscolices rich fluid leading to local reoccurrence (10% in post operative cases) or secondary echinococcosis (Kilicoglu et al., 2006; Moro and Schantz, 2009). Intracystic instillation of scolicidal agents is a common practice to prevent such spillage along with percutaneous aspiration (Brunetti et al., 2010). However, no ideal (both effective and safe) scolicidal agent has been deviced yet. The unacceptable side-effects of the scolicidal agents cause limitation to their use (Naguleswaran et al., 2006).

Surgery (radical or conservative) is usually complemented by pre and/or post-surgical chemotherapy. In massive infestation and in inoperable cases, chemotherapy is the sole intervention. Several drugs, notably mebendazole, albendazole, praziquantel, mitomycinic, isoprinosine and other antihelminthic drugs showed promising reduction and induction of ultrastructural damage of the larval cystic mass (Satoh et al., 2005; Taran et al., 2009). However, their adverse side effects included abnormalities in liver function, leucopenia and alopecia, along with teratogenesis (albendazole). It should therefore be avoided during pregnancy and lactation (Kern, 2003; Nunnari et al., 2012; Verma et al., 2013). Moreover, drug resistance and high recurrence rates after interruption of therapy added further complications (Reuter et al., 2004; Sadjjadi et al., 2008). This mandates the development of new thera-peutic agents for such disease (Moazeni et al., 2012).

Retinoids are a class of lipophilic chemicals that include active natural metabolites of vitamin A (retinol), as well as a diverse array of synthetic derivatives. The major physiological retinoid, all-trans retinoic acid (atRA) modulates a wide variety of cellular processes including; cell proliferation, differentiation and apoptosis, cell-matrix interaction, angiogenesis, homeostasis, polyamine metabolism and immune performance through modulation of hormonal/growth factor signal transduction and expression hundreds of genes. It is a promising chemopreventive agent for bladder cancer (Al Tanoury et al., 2013), and the first line therapeutic agent for acute promyelocytic leukemia and is a potent inducer of liver and other tissue regeneration (Degos and Wang, 2001; Gudas, 2012; Hx et al., 2014). Anti-inflammatory (Thielitz et al., 2006; Andrea and Zaenglein, 2008), antifungal and antimicrobial effects of RA has also been observed against several fungi and bacteria (Mikhlin et al., 1983). RA antiparasitic activity was rarely investigated (Hurst and Else, 2012). For instance, RA was correlated with successful clearing of *Trichinella. spiralis* (Carman et al., 1992) and vitamin A supplementation reduced re-infection of *Ascaris lumbricoides* in children (Long et al., 2006; Payne et al., 2007). Hall et al. (2011) demonstrate that vitamin A insufficient animals experienced lack of Th1 responses to *Toxoplasma gondii* infection leading to increased parasite burdens.

Hypothesizing that atRA could prove scolicidal, for the first time, the present work planned to investigate changes in viability and ultrastructure of hydatid cyst PSCs treated with different concentrations of atRA *in vitro* as an initial screening of the anti-parasitic potential of retinoids against hydatid cysts.

MATERIALS AND METHODS

Parasitological procedures

For this *in vitro* study, intact hydatid cysts were removed aseptically from seven in-patients with hepatic hydatidosis underwent pericystectomy following laparotomy in the General Surgery Department, Faculty of Medicine, Assiut University Hospitals during May, 2012 to April, 2013. The intact cysts were immediately placed in an ice box and transported within 10 min to the Parasitology Laboratory, Medical Parasitology Department, Faculty of Medicine, Assiut University, Assiut, Egypt.

Collection of hydatid fluid

Cysts were washed three times in sterile phosphate-buffered saline (PBS; pH 7.2) and cyst fertility was determined by the presence of free PSCs in hydatid fluid by wet mount drop. Sterile cysts were discarded.

PSCs preparation

Fertile cyst surfaces were disinfected with 70% ethyl alcohol to guard against surface contaminations and hydatid fluid was evacuated completely and transferred into 15 ml falcon tubes by aseptic techniques. The fluid was left to precipitate at room temperature for 15 min to obtain hydatid sand without centrifugation. PSCs with enough volume were obtained and the supernatant was removed into another 15 ml tubes. The viability of PSCs was confirmed prior to the experiments by 0.1% eosin staining exclusion test and flame cells motility according to Esfandiari et al. (2010). Only PSCs with an initial viability of at least 95% or more were used.

Protoscolices preservation fluid

Preservative solution consisted of hydatid cyst fluid and normal physiological saline (0.85% NaCl) 4:1 mixture, v/v, without antibiotic or antifungal drugs.

All-trans retinoic acid solution

atRA was dissolved in endotoxin-free dimethysulfoxide (DMSO) for biological uses (Sigma-Aldrich Chemical Co., Saint Louis, MA,

USA) at 100, 10, 1 µM, and 100 nM/L stock concentrations under dim dark yellow light (the condition under which all experiments were conducted) and maintained in dark brown vials.

Treatment

The effect of different concentrations of atRA on the viability of PSCs was compared to albendazole sulfoxide (ABZSO) 800 µg/ml as a positive standard anti-hydatidosis control drug (Polat et al., 2009). Treatment of PSCs was initiated within 15 min of isolation. Five ml of the preservative solution was added to 1 ml of viable PSCs (approximately 2,000 PSCs) in separate tubes and 1.0 ml of each stock concentrations of atRA (100, 10, 1.0 µM and 100 nM/L) was added into three tubes for each dose to give rise to final atRA concentrations of 16.7, 1.67, 0.167 µM and 16.7 nM/L in the experiment mix. All tubes were mixed gently and incubated at 30°C without changes of preservative solution. Percentage of viable PSCs was determined in triplicate samples withdrawn from each tube after 5, 10, 15, 30, 60 and 120 min and every 6 h thereafter, until all of the PSCs were dead. ABZSO tubes were processed the same. One ml volume of each of the sterile preservative solution and the DMSO vehicle were used as negative controls in their respective tubes.

Viability assay

The viability of PSCs was assessed by microscopic observation of biological feature represented by muscular movements, flame cell activity in 0.1% aqueous eosin stain. The corresponding numbers of viable/non-viable PSCs were determined in 10 randomly chosen fields by phase contrast microscopy under sterilized condition using sterilized materials.

Ultrastructural investigations

Topographic studies were performed to visualize the ultrastructure changes in PSCs imposed by atRA and controls. PSCs were processed for scanning and transmission electron microscopy (SEM and TEM) by thorough washing in distilled water, fixation in 2.5% glutaraldehyde (in 100 mM sodium cacodylate buffer; pH 7.2) for 2 h at room temperature, followed by post-fixation in 100 mM sodium cacodylate buffer (pH 7.2) for 2 h at room temperature. Samples were then washed in distilled water and treated with 1% uranyl acetate for 30 min. Subsequently, the specimens were extensively washed in distilled water and dehydrated by sequential incubations in increasing concentrations of ethanol (Elissondo et al., 2009). For SEM, dehydrated specimens were immersed in hexamethyl disilazane and air dried under a fume hood, sputter coated with gold and inspected on a JEOL 840 SEM operating at 25 kV (Hayat, 1981). For TEM, the specimens were embedded in Epon 812 resin that was polymerized overnight at 65°C. Ultramicrotome sections were loaded onto 300-mesh copper grids (Plano GmbH, Marburg, Germany). Ultrathin 80 to 100 nm sections were stained with uranyl acetate and lead citrate as described previously (Hemphill and Croft, 1997) and inspected on a JEOL-JSM-5400 LV (Electron Microscope Unit, Assiut University).

Statistical analysis

Statistical calculations were carried out by one way analysis of variance (ANOVA), using the software statistical package for social sciences (SPSS 16) for windows. P-values of < 0.05 were considered significant.

RESULTS

Intact hepatic hydatid cysts were dissected by pericystectomy following laparotomy (Figure 1A for example). Cyst fertility was examined by wet mount drop detection of free PSCs in cystic fluid (Figure 1B). PSCs viability was determined using eosin staining exclusion test; where dead PSCs absorbed eosin and stained red and alive PSCs remained colorless and showed characteristic motility (Figure 1C). The viability of PSCs was also confirmed by flame cell activity and muscular movements.

Changes in the viability of hydatid cyst PSCs induced by atRA

In preservative solution and dimethyl sulfoxide (DMSO) vehicle (negative and vehicle controls, respectively), PSCs viability decreased from more than 95% at the beginning to total death in an average of 13 days incubation at 30°C (Figure 1D). Microscopically, viable PSCs exhibited distinct movements, membrane integrity and ordered hooks. By time, most of these PSCs attained evaginated form (scolices) with clearly visible suckers. In preservative solution and DMSO, vehicle controls were non-significantly different.

On survivability assessment, it was apparent that atRA at 16.7 µM/L possessed profound scolicidal activities, as it reduced the number of viable PSCs by 90% within a period of 5 min. All PSCs were dead within 15 min ($p < 0.05$). At 1.67 µM/L, atRA reduced PSCs viability to 40% at 2 days time-point and reached total death within 3 days ($p < 0.05$). Although it was significantly scolicidal, atRA at 0.167 µM/L was clearly less efficient than the higher concentrations. It increased the percentages of dead PSCs with time where a small fraction of PSCs could be seen viable at day 5 and total death occurred at 7 days time-point of incubation with 0.167 µM/L atRA ($p < 0.05$). The viability of PSCs treated with 16.7 nM/L atRA did not show any changes till the 13th day time-point similar to the performance of the preservative and vehicle controls without significant difference amongst them. ABZSO (800 µg/mL) caused a significant reduction in the viability of PSCs at day 6 incubation time-point, where 70% of the incubated PSCs died. All of the remaining PSCs died by the end of 10th day post treatment ($p < 0.05$). The effect of different concentration of atRA on PSCs of hydatid cyst viability in comparison with preservative solution, DMSO vehicle and ABZSO is illustrated in Figure 2.

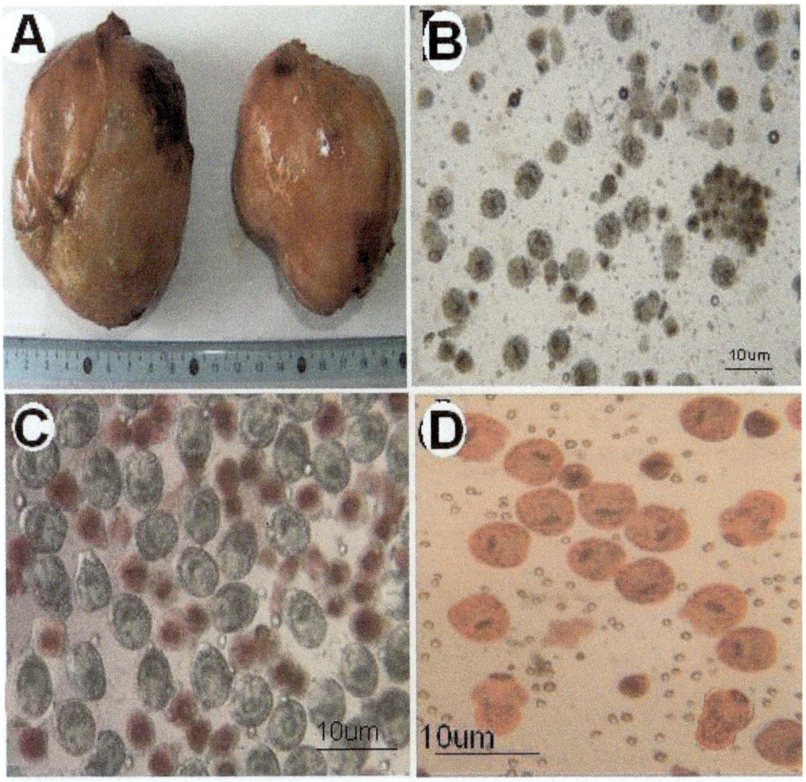

Figure 1. (A) Examples of dissected intact hepatic hydatid cysts (×40), (B) Free PSCs in cystic fluid examined by wet mount drop (×40), (C) Eosin staining exclusion test of living (unstained) and dead (stained) PSCs (×100), (D) Total death of PSCs (×200).

Ultrastructural changes in hydatid cyst PSCs induced by atRA

Morphological changes were observed in tegument and parenchyma of PSCs treated with 1.67 and 0.167 μM/L atRA at the end of the experiments. These changes were more evident in evaginated than invaginated PSCs. Similar changes were observed in PSCs treated with ABZSO. No similar tegumental or parenchymal ultrastructural alterations were observed in control PSCs. SEM showed that the primary site of damage was the rostellum of the scolices. Compared to the normal arrangement of hooks on the rostellum (Figure 3A), damages varied from disorganization of rostellar hooks (Figure 3B) to total loss of rostellar hooks (Figure 3C). Distortion of hooks morphology was also observed especially in the guard and handle regions of the hook (Figure 4A). The tegumental scolex (especially the sucker regions) appeared to be altered or contracted (Figure 4B). Several blebs were detectable in the surface of PSCs (Figure 4C). There was also aggregation and adherence of the PSCs together (Figure 4D).

Transition electron microscope (TEM) showed that the internal tissues of PSCs were severely affected in response to treatment with atRA compared to the normal tissues (Figure 5A). Loss of its integrity, an increased vacuolization and significant reductions in the relative numbers and lengths of the microtriches of the distal portion of the tegument were observed (Figure 5B). There were formation of large lipid droplets in the distal cytoplasm and many undifferentiated cells, the cytoplasm of which contained large vacuoles filled with electron-dense bodies, granular particles, and small vesicles (Figure 5C). These vacuoles occupied a large portion of the cytoplasm of these cells. The treatment with atRA also resulted in the formation of aberrant, rounded mitochondria which had abnormally increased in size, and the cellular disintegration process had advanced, rendering the tissue largely necrotic (Figure 5D).

DISCUSSION

Chemotherapy, surgery and puncture with aspiration are the three traditional treatments for hepatic hydatid cysts. Treatment option is still subject of controversy (Frider and

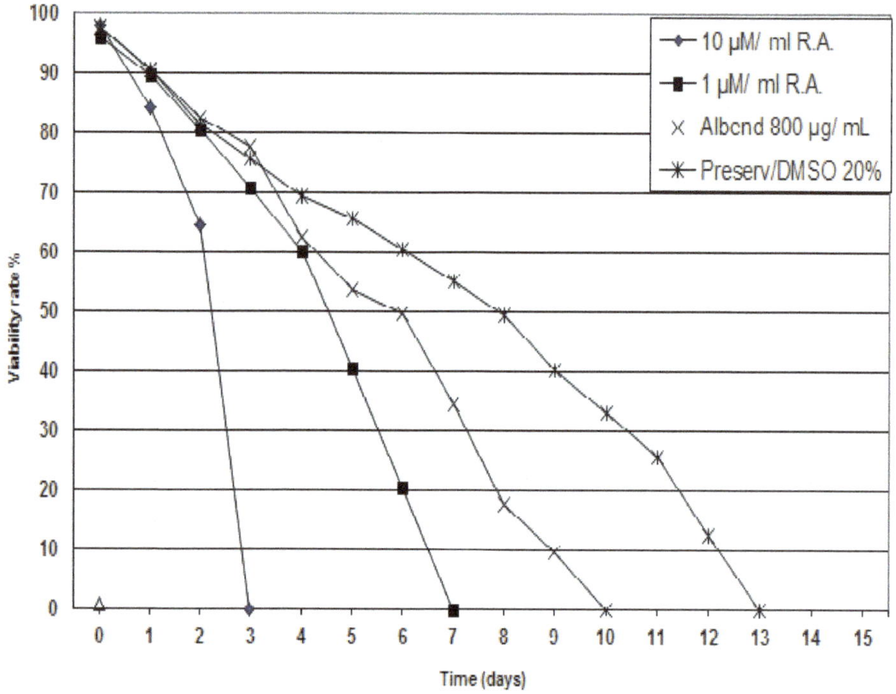

Figure 2. Loss of viability of human hydatid protoscolices during *in vitro* atRA -treatment. Viability was determined through eosin staining exclusion test.
Note the dose-dependent effect of atRA in comparison with preservative solution, DMSO vehicle and ABZSO.

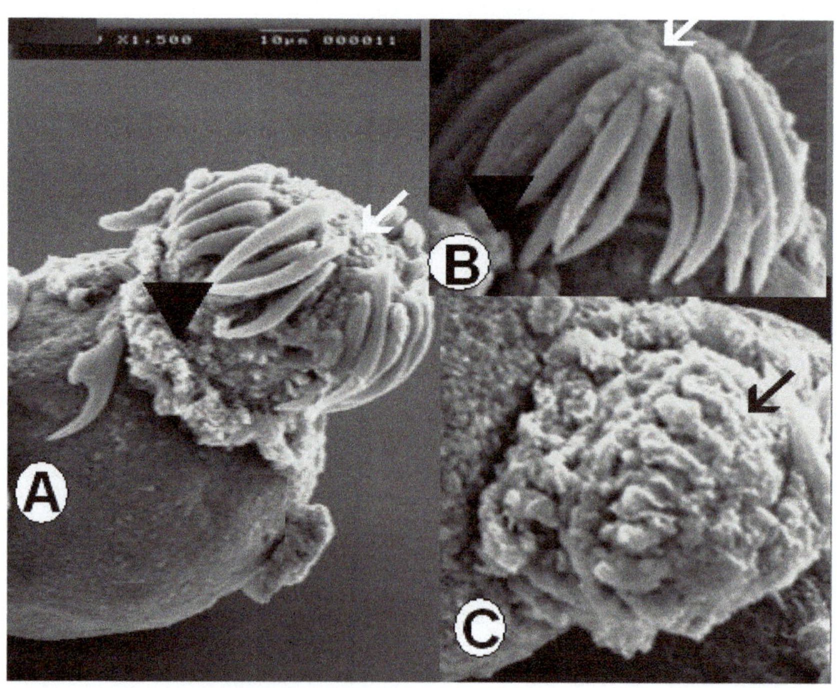

Figure 3. SEM of scolices in presence or absence of retinoic acid (A) the normal arrangement of hooks on the rostellum (arrow), (B) disorganization of rostellar hooks (x), (C) total loss of hooks from the rostellum. Arrow-heads pointed to the hooks and arrows to the rostellum.

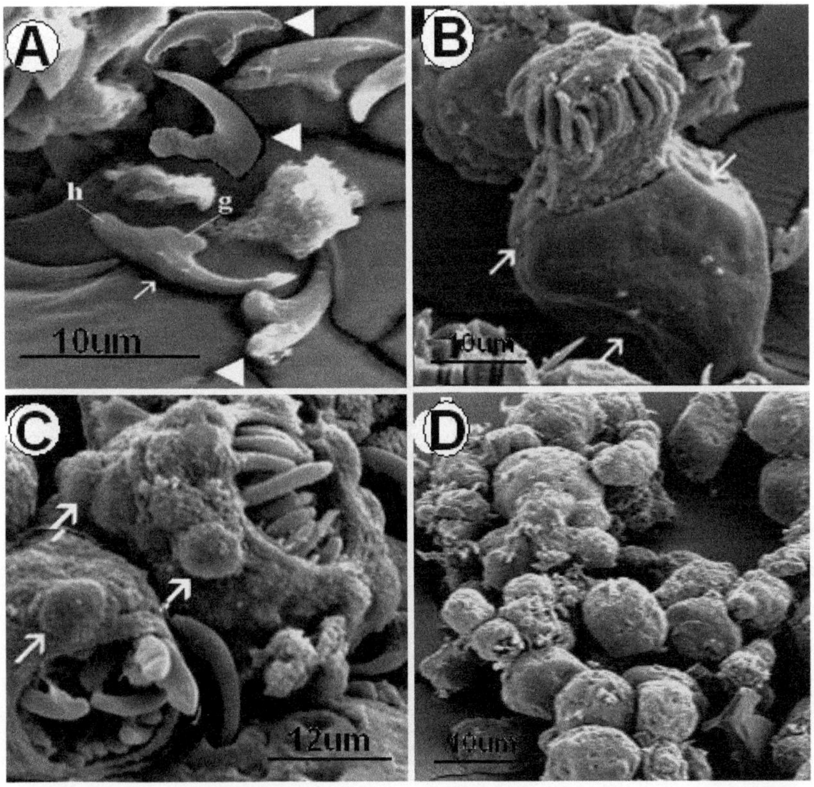

Figure 4. SEM of the effects of retinoic acid on different components of hydatid cyst (A) distorted hooks (arrow-heads) morphology especially in the guard (g) and handle (h) regions of the hook (arrow pointed to normal hook) (×2000), (B) altered or contracted tegumental scolex especially the sucker regions (arrows) (×1000), (C) Several blebs (arrows) in the surface of PSCs (×1500), (D) aggregation and adherence of the PSCs together (×200).

Larrieu, 2010). For inoperable patients and those refusing surgery, chemotherapy and puncture are recommended as alternative medication (Lv et al., 2013). Inactivation of the scolex with a protoscolicidal agent before surgery is intensely recommended because the spillage of the cyst contents is a major cause of return (Zhang et al., 2003). A number of drugs are being used and various degrees of success have been claimed. However, the metabolites of certain drugs including benzimidazole mebendazole, albendazole and albendazole sulfoxide are potentially toxic in some subjects (Polat et al., 2005; Mandal and Mandal, 2012).

Manterola et al. (2006) reported that the prevalence of fertile human hydatid cyst was low and fertility is associated with the type of cyst and the presence of biliary communications. The viability assessment in CE is important for *in vitro* and *in vivo* studies and also drug experiments. Several vital stains like neutral red, methylene blue, eosin, giemsa, Ziehl-Neelsen and trypan blue have been used to assess viability of protoscolices in fertile cysts (Yildirim et al., 2007). In the current study, cyst fertility was determined by the presence of free PSCs in cystic fluid by wet mount drop. This was important to exclude sterile hydatid cyst which is common in human. In the present study, 0.1% aqueous eosin stain exclusion was used for initial PSCs viability assessment and its time- and treatment-induced changes *in vitro*.

Hydatid PSCs remain viable for long periods in different media at different temperatures. In the absence of nutritive factors, the PSCs of *Echinococcus granulosus* do not survive *in vitro* for more than 4 day at 37°C. The survival time depends on temperature, pH and ionic concentration (Al-Sudani, 2007). In the present study hydatid PSCs remain viable for 13 days after being incubated in the preservative solution (HCF and NS; 4:1 v/v) that provided PSCs with the minimal requirement to stay alive for such period of time at 30 °C. Moreover, evagination of some PSCs may indicate further growth in such conditions. Similar PSCs morphological changes in culture media were discussed to indicate morphological

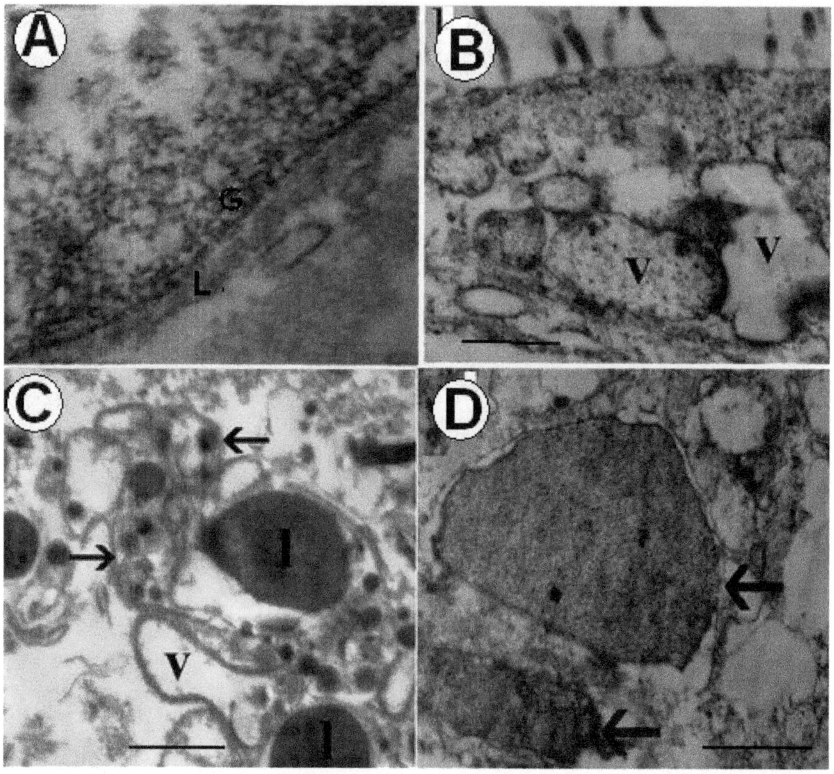

Figure 5. TEM of scolices in presence or absence of retinoic acid (A) Normal internal tissues of PSCs; G pointed to germinal layer and L to laminated layer (×7200), (B) Significant reductions in the relative numbers and lengths of the microtriches of the distal portion of the tegument and increased vacuolization (v) (×14000), (C) Formation of large lipid droplets (l) in the distal cytoplasm, many undifferentiated cells (arrows) (×19000), (D) Formation of aberrant, rounded, abnormally increased in size mitochondria (arrows) and advanced cellular disintegration (×29000). Bars in A = 3.5 mm, B = 2.9 mm, C = 5.8 mm, D = 1.6 mm.

evolution, whereas, presence of free hooks in culture medium indicated their detachment and destruction of the tegument (Al-Sudani, 2007).

The present study demonstrated the dose- and time-dependent protoscolicidal effect of atRA. The biologically apoptotic high atRA concentration (16.7 µM/L) caused such effects within standardized very short time of 6 min, whereas, the retinoid receptor saturating concentration of atRA (1.67 µM/L) induced such protoscolicidal effect within five days. The lowest concentrations atRA used (0.167 and 16.7 nM/L) induced no effect on PSCs compared to negative and vehicle controls. This was confirmed by morphological and ultrastructural deformities induced by the two effective atRA doses at different incubation periods. Such atRA-induced morphological and ultrastructural changes that included contraction in soma region, formation of blebs on the tegument, rosteller disorganization, loss of hooks, and destruction of microtriches were consistent with that observed with several established protoscolicidal compounds including praziquantel (Urrea-Parıis et al., 1999, 2000), benzimidazole (Pe´rez-Serrano et al., 2001), albendazole (Ceballos et al., 2011) and ivermectin (Casado et al., 1989; Elissondo et al., 2009).

Rosteller disorganization and destruction of microtriches might be major reason for loss of viability of PSCs as they are the means for nutrient absorption, physiological homeostasis and defense (Satoh et al., 2005; Afifi and Harba, 2012). Similarly, the formation of numerous blebs on the tegument which became detached, leaving debris was described for the effects of pure albendazole and its sulphoxide combination therapy on *E. granulosus*. These blebs could had enlarged with time and burst out leading to the disturbance in osmoregulatory system of PSCs (Lv et al., 2013).

Our TEM findings of increasing vacuolization, small

lamellated bodies, and presence of lipid droplets in atRA treated PSCs reflect general tissue stress for subsequent apoptosis/necrosis. These vacuolations might result in major leakage of the cytoplasm contents (Kim et al., 2011). The presence of lipid droplets in the cytoplasm of tegumentary cells of the germinal layers indicates metabolic disruption of the cyst (Verma et al., 2013). Very similar tissue degeneration was observed, when treated with netobimin (García-Llamazares et al., 2002) and isoprinosine (Sarciron et al., 1993). Disruption of the muscle and tegumental layers, which is the fundamental interface of the helminth body with its environment, and responsible for selective absorption of nutrients, secretory activities and sensory perception renders it specifically susceptible to anthelmintic agents (Roy et al., 2007; Shalaby et al., 2012).

Biologically, high concentrations of ~10 μM/L atRA reduce cell-cell and cell-matrix interactions, affect cell differentiation, and mitochondrial mass and membrane potential as tools towards its apoptotic effect (El-Metwally et al., 2005). Such retinoid receptor mediated-effects utilize modulation of rate of expression of hundreds of genes (Brun et al., 2013; Hx et al., 2014). This study as the first worldwide attempt to screen the retinoic acid parasiticidal activity against *E. granulosus* was based on such abilities of atRA to regulate cellular differentiation, proliferation, metabolism and apoptosis as very dynamic tools used by the ever evolving parasite. Our results are quite encouraging since atRA concentrations around the effective 1.0 μM/L is therapeutically systemically achievable rather than intra-cyst instillation in human patients. This low dose showed a profound effect on the viability and ultrastructure of PSCs very similar to some existing scolicidal drugs. These results pave the road for future studies implicating the cestodicidal capabilities of atRA and other natural and synthetic retinoids as alternative therapeutic agents - particularly against hydatidosis with massive evolutionary cellular dynamics of proliferation, differentiation and apoptosis sustained with highly active metabolism. It is worth mentioning that reviewing the existing literatures, no previous publications were found about the protoscolicidal effect of retinoic acid either *in vitro* or *in vivo*.

Conclusion

The *in vitro* scolicidal activity of all-trans retinoic acid was described for the first time. Retinoic acid that is a major physiological player with low drug toxicity in a proof-of-the concept study showed promising scolicidal activity - indicated by massive reduction in PSCs viability and unfavorable ultrastructural deformations. Further *in vitro* studies are required to dissect the molecular machinery of such atRA actions along with *in vivo* experimentation using *Echinococcus* infection models.

Conflict of interest

The authors declare that there is no conflict of interests.

REFERENCES

Afifi FA, Harba MN (2012). Chemoprophylaxis and chemotherapeutic effect of Flubendazole on *Echinococcus granulosus* experimentally infected dogs. Res. J. Parasitol. 7:1-16.

Al Tanoury Z, Piskunov A, Rochette-Egly C (2013). Vitamin A and retinoid signaling: genomic and nongenomic effects. J. Lipid Res. 54: 1761-1775.

Al-Sudani NM (2007). Effect of extract of *Ziziphus spina-christi* L. and *Olea europaea L.* on the larval stage of *Echinococcus granulosus.* Int. J. Parasitic Dis. 2 (1):17-22.

Andrea L, Zaenglein MD. (2008). Topical retinoids in the treatment of acne vulgaris. Semin. Cutan. Med. Surg. 27:177-182.

Brun PJ, Yang KJ, Lee SA, Yuen JJ, Blaner WS (2013). Retinoids: Potent regulators of metabolism. Biofactors 39(2):151-63.

Brunetti E, Kern P, Vuitton DA (2010). Expert consensus for the diagnosis and treatment of cystic and alveolar echinococcosis in humans. Acta Trop. 114:1-16.

Cardona GA, Carmena D (2013). A review of the global prevalence, molecular epidemiology and economics of cystic echinococcosis in production animals. Vet. Parasitol. 192(1-3):10-32.

Carman JA, Pond L, Nashold F, Wassom DL, Hayes CE (1992). Immunity to *Trichinella spiralis* infection in vitamin A-deficient mice. J. Exp. Med. 175:111-120.

Casado N, Rodrı´guez-Caabeiro F, Jime´nez A, Criado A, de Armas C (1989). In vitro effects of levamisol and ivermectin against *Echinococcus granulosus* protoscoleces. Int. J. Parasitol. 19:945-947.

Ceballos L, Elissondo C, Bruni SS, Denegri G, Lanusse C, Alvarez L (2011). Comparative Performances of Flubendazole and Albendazole in Cystic Echinococcosis: Ex Vivo Activity, Plasma/Cyst Disposition, and Efficacy in Infected Mice. Antimicrob. Agents Chemother. 55(12):5861-5867.

Dakkak A (2010). Echinococcosis/hydatidosis: a severe threat in Mediterranean countries. Vet. Parasitol. 174:2-11.

Degos L, Wang YZ (2001). All-trans retinoic acid in acute promyelocytic leukemia. Oncogene 20:7140-7145.

Elissondo MC, Ceballos LL, Bruni SS, Lanusse C, Denegri G (2009). Flubendazole and ivermectin in vitro combination therapy produces a marked effect on *Echinococcus granulosus* protoscoleces and metacestodes. Parasitol. Res. 105:835-842.

El-Metwally TH, Hussein MR, Pour PM, Kuszynski CA, Adrian TE (2005). High concentrations of retinoids induce differentiation and late apoptosis in pancreatic cancer cells in vitro. Cancer Biol. Ther. 4(5):602-611.

Elshazly MA, Manar AS, El Beshbishi NS, Elsheikha MH (2009). Hepatic hydatid disease: Four case reports. Cases J. 2(58):1-4.

Esfandiari B. Youssefi MR, Ziapour SP, Fard SM, Abouhossein M (2010). Evaluation of Hydatid Cyst Surgeries in Northern Iran (Mazandaran Province) During 2001-2007. J. Anim. Vet. Adv. 9 (7):1128-1130.

Frider B, Larrieu E (2010). Treatment of liver hydatidosis: How to treat an asymptomatic carrier? World J. Gastroenterol. 16(33):4123–4129.

García-Llamazares G, Merino-Pelaez G, Prietio-Fernande JG (2002). Effects of netobimin treatment on the glucose and glycogen contents of *Echinococcus granulosus* cysts from gerbils. Vet. Res. Commun. 26(1):55-59.

Grosso G, Gruttadauria S, Biondi A, Marventano S, Mistretta A (2012). Worldwide epidemiology of liver hydatidosis including the

Mediterranean area. World J. Gastroenterol. 18(13):1425-1437.
Gudas LJ (2012). Emerging roles for retinoids in regeneration and differentiation in normal and disease states. Biochim. Biophys. Acta 1821:213-221.
Jason AH, Jennifer LC, John RG, Liliane MDS, Timothy WH, Shruti N, Elizabeth AW, David BC, Guillaume O, Melody R, Michael EG, Robin K, Pamela LS, Yasmine B (2011). Essential role for retinoic acid in the promotion of CD4(+) T cell effector responses via retinoic acid receptor alpha. Immunity 34:435-447.
Hayat MA (1981). Fixation for Electron Microscopy. Academic Press, NY.
Hemphill A, Croft SL (1997). Electron microscopy in parasitology. In: Analytical parasitology. Springer Berlin Heidelberg. pp. 227-268.
Himsworth CG, Jenkins E, Hill JE, Nsungu M, Ndao M et al (2010). Short report: emergence of sylvatic *Echinococcus granulosus* as a parasitic zoonosis of public health concern in an indigenous community in Canada. Am. J. Trop. Med. Hyg. 82(4):643-645.
Hurst RJ, Else KJ (2012). Retinoic acid signaling in gastrointestinal parasite infections: lessons from mouse models. Parasite Immunol. 34:351-359.
Hx L, Ly I, Hu Y, Wan YJ (2014). Retinoic acid regulates cell cycle genes and accelerates normal mouse liver regeneration. Biochem. Pharmacol. 31(14):390-396.
Kern P (2003). *Echinococcus granulosus* infection: clinical presentation, medical treatment and outcome. Langenbecks Arch. Surgery 388:413–420.
Kilicoglu B, Kismet K, Koru O, Tanyuksel M, Oruc MT, Sorkun K, Akkus MA (2006). The scolicidal effects of honey. Adv. Ther. 23(6):1077-1083.
Kim HS, Lee SH, Ryu SD, Choi SJ, Lee DS (2011). Antibacterial Activity of Silver-nanoparticles Against *Staphylococcus aureus* and *Escherichia coli*. Korean J. Microbiol. Biotechnol. 39(1):77–85.
Long KZ, Estrada T, Santos JI, Haas M, Rosado JL, Firestone M, Bhagwat J, Young C, DuPont HL, Hertzmark E, Nanthakumar N (2006a). The impact of vitamin A supplementation on the intestinal immune response in Mexican children is modified by pathogen infections and diarrhea. J. Nutr. 136:1365–1370.
Lv H, Li S, Zhang J, Liang W, Mu X, Jiang Y (2013). In vitro effects of SB202190 on *Echinococcus granulosus*. Korean J. Parasitol. 51(2): 255-258.
Mandal S, Mandal MD (2012). Human cystic echinococcosis: epidemiologic, zoonotic, clinical, diagnostic and therapeutic aspects. Asia Pac. J. Trop. Med. 5(4):253-260.
Manterola C, Vial M, Melo A, Oberg C, Fonseca F (2006). Viability and fertility of human hepatic hydatid cysts. World J. Surg. 30(2):227-232.
Mikhlin ED, Radina VP, Dmitrovskiĭ AA, Blinkova LP, Butova LG (1983). Antifungal and antimicrobial activity of beta-ionone and vitamin A derivatives. Prikl. Biokhim. Mikrobiol. 19(6):795-803.
Moazeni M, Saharkhiz MJ, Hosseini AA (2012). *In vitro* lethal effect of ajowan (*Trachyspermum ammi L.*) essential oil on hydatid cyst protoscoleces. Vet. Parasitol. 187:203-208.
Moro P, Schantz PM (2009). Echinococcosis: A review. Int. J. Infect. Dis. 13:125-133.
Naguleswaran A, Spicher M, Vonlaufen N, Ortega-Mora LM, Torgerson P, Gottstein, Bruno G, Hemphill A (2006). *In vitro* metacestodicidal activities of genistein and other isoflavones against *Echinococcus multilocularis* and *Echinococcus granulosus*. Antimicrob. Agents Chemother. 3770-3778.
Nunnari G, Pinzone MR, Gruttadauria S, Celesia BM, Madeddu G, Malaguarnera PP, Alessandro C, Bruno C (2012). Hepatic echinococcosis: Clinical and therapeutic aspects. World J. Gastroenterol. 18(13):1448-1458.
Payne LG, Koski KG, Ortega-Barria E, Scott ME (2007). Benefit of vitamin A supplementation on *Ascaris* reinfection is less evident in stunted children. J. Nutr. 137:1455-1459.
Nunnari G, Pinzone MR, Gruttadauria S, Celesia BM, Madeddu G, Malaguarnera G, Pavone P, Cappellani A, Cacopardo B (2001). In vitro effect of albendazole and albendazole sulphoxide on 70 and 60 kDa heat shock proteins in *Echinococcus granulosus* protoscolices. Res. Rev. Parasitol. 61:67-71.
Polat C, Dervisoglu A, Hokelek M, Buyukkarabacak Y, Ozkutuk Y, Erzurumlu K (2005). Dual treatment of albendazole in hepatic hydatidosis: new therapeutic modality in 52 cases. J. Gastroenterol. Hepatol. 20:421-425.
Polat E, Aslan M, Cakan H, Saribas S, Ipek T, Kocazeybe k (2009). The effects of albendazole and povidone iodine for hydatid cysts protoscolices, *in-vitro* and *in vivo*. Afr. J. Microbiol. Res. 3(11):743-746.
Reuter S, Andreas B, Burkhard M, Wolfgang K, Hanns M S, Kassa D, Sven NR, Peter K (2004). Structured treatment interruption in patients with alveolar echinococcosis. Hepatology 39:509-5017.
Roy B, Lalchhandama K, Dutta K B (2007). Anticestodal efficacy of *Acacia oxyphylla* on *Raillietina echinobothrida*: A light and electron microscopic studies. Pharmacologyonline 1:279-287.
Sadjjadi SM (2006). Present situation of echinococcosis in the Middle East and Arabic North Africa. Parasitol. Int. 55:197-202.
Sadjjadi SM, Zoharizadeh MR, Panjeshahin MR (2008). In vitro screening of different *Allium sativum* extracts on hydatid cysts protoscolices. J. Invest. Surg. 21(6):318-322.
Sarciron ME, Walbaum S, Arsac C, Raynaud G, Petavy AF (1993). *Echinococcus granulosus:* isoprinosine treatment of the metacestode stage. Am. J. Trop.Med. Hyg. 48(5):658-665.
Satoh M, Nakaya K, Nakao M, Xiao N, Yamasaki (2005). *Echinococcus* confirmed on Kunashiri Island. Am. J. Trop. Med. Hyg. 72:284-288.
Shalaby HA, Namaky AE, Khalil FA, Kandil OM (2012). Efficacy of Methanolic extract of *Balanites aegyptiaca* fruits on *Toxocera vitulorum*. Vet. Parasitol. 183:386-392.
Taran M, Azizi E, Shikhvaisi A, Asadi N (2009) The anthelmintic effect of *Pistacia khinjuk* against Protoscoleces of *Echinococcus granulosus*. World J. Zool. 4(4):291-295.
Thielitz A, Krautheim A, Gollnick H (2006). Update in retinoid therapy of acne. Dermatol. Ther. 19:272-279.
Urrea-Paris M, Moreno MJ, Casado N, Rodriguez-Caabeiro F (1999). *Echinococcus granulosus*: Praziquantel treatment against the metacestode stage. Parasitol. Res. 85:999-1006.
Urrea-Paris M, Moreno MJ, Casado N, Rodriguez-Caabeiro F (2000). *In vitro* effect of praziquantel and albendazole combination therapy on the larval stage of *Echinococcus granulosus*. Parasitol. Res. 86:957-964.
Verma VC, Gangwar M, Yashpal M, Nath G (2013). Anticestodal activity of endophytic Pestalotiopsis sp. on protoscoleces of hydatid cyst *Echinococcus granulosus*. Biomed. Res. Int. 308515:1-11.
Yildirim A, Iça A, Düzlü O, Inci A (2007). Statistical analyses of different staining methods for viability assessment in cystic echinococcosis. Turk. . Derg. 31:105-108.
Zhang W, Li J, McManus DP (2003). Concepts in immunology and diagnosis of hydatid disease. Clin. Microbiol. Rev. 16:18-36.

10

The prevalence and risk factors associated with ovine lungworm infestation in the Asella province, Central Ethiopia

E. Eyob and L. Matios

Department of Parasitology, Yabello Regional Veterinary Laboratory, Ethiopia.

A cross sectional study design was conducted in Asella province, Central Ethiopia from November 2009 to May 2010. The study was carried out for the purpose of determining the prevalence of lungworm infection in ovine through coproscopic and postmortem examination in the Asella province. Moreover, the study assesses the association of some bionomic factors on the different risk factors and the correlation between adult worm burden and fecal larval output during the study time. Faecal samples (381) were collected from sheep of different age groups, husbandry system and health care, of both sexes; and additionally from 71 sheep slaughtered at Asella town restaurant and hotels, with both intact lung and fecal samples were randomly selected. The investigation also consists of questionnaire survey. Out of 381 fecal samples coproscopically examined, 276 (72.44%) were found to be infected with one or more species of lungworm. Sheep showing respiratory clinical signs, non-dewormed with anthelmintics and raised in extensive husbandry system when compared with those sheep that apparently appeared healthy, dewormed ones and of semi-intensive husbandry system shows statistically significant variation (P<0.05). The difference in the prevalence of infection, both among age groups and between sexes did not show statistically significant variation (P>0.05). The postmortem examination carried out on 71 sheep revealed that coproscopic examination had limited value in terms of estimating the prevalence of lungworm infection in live sheep and implies as there was no direct correlation between the adult worm burden and faecal larval output. The prevalence of lungworm infection observed through postmortem coproscopic examination and postmortem intact lung incision was 39.4 and 26.8%, respectively.

Key words: Asella province, Central Ethiopia, lungworm, ovine, prevalence, risk factors.

INTRODUCTION

Ethiopia has estimated 12 million sheep together with its great variation in agro climatic zones represents a good reservoir of sheep genotypes. Despite the low input extensive management system an estimated, 132,000 tons of sheep meat is produced annually, providing more than 30% of all domestic meat consumption. Sheep also generate cash income to the farming community and significantly contribute to the foreign exchange earnings of about 43 and 5% economy from exports and meat, respectively. Inspite of the large population and importance of sheep however, morbidity and mortality are high in the traditional agro-pastoral production system. In highland area including Arsi Asella, respiratory lungworm parasites are the most common cause of high mortality and morbidity rates of sheep (FAO, 2006). Lungworms of small ruminants are limited to two super families such as Dictyocaulidae and Metastrongylidae. Although these parasites are widely prevalent, the clinical signs they showed in infected sheep can be less obvious than signs of other ovine diseases (Hansen and Perry, 1996). Infestation to lungworms in sheep is characteristics by dyspnea, sneezing and cough due to pneumonia. Lungworms infection produces nodules and brownish spots in lung. This signs are diagnosed at postmortem examination during slaughter (Etminani, 1980).

Primary finding of lungworm infection in Ethiopia (Netsanet, 1992; Teffera, 1993; Ayalew et al., 2011; Bekele and Aman, 2011; Hasen et al., 2013) indicated high prevalence and importance of the infection in certain area of the country. In order to investigate a sound lungworm control strategy at local and regional level, further and detailed investigation on epidemiology and importance of lungworm infections with respect to its temporal distribution is necessary. Therefore, the objective of this research is to determine the prevalence of lungworm infection in ovine through coproscopic and postmortem examination, to assess the association of some bionomic factors on the different risk factors during the study period and to determine the correlation between adult worm burden and fecal larval output during the study in the Asella province.

MATERIALS AND METHODS

Description of study area

Asella is situated at 6'59' to 8'49'N latitudes and 38'41' to 40'44' E longitude in Central Ethiopia, 175 km South East of Addis Ababa. The area covers 23674.72 km^2 representing only 2% of the total land surface of the country. Topographically, Asella province has high land escarpment and low land areas. Vegetation of the area changes with altitude and rainfall ranging from scattered trees and bushes to dense shrubs and bushes. Livestock are the major agricultural resources in the area and has livestock population of 82190 cattle, 51292 sheep, 11479 goats, 162015 poultry, 11,716 camels and 22055 equines (APEDO, 1999).

Study animal

The study population includes indigenous sheep raised in the Asella province and includes sheep grazed in extensive system, reared in in-door and those brought to Asella Veterinary Clinic. Sheep of 1 to 7 years of age and of both sexes, with extensive and semi-intensive husbandry system, dewormed (oral administration of Albendazole at 7.5 g/kg) and non-dewormed sheep; and additionally sheep with no clinical respiratory signs and those that appeared apparently healthy are included. For convenience, the age of every sampled sheep was categorized into young and adult (ESGPIP, 2000) using dentition and information from the owners.

Sample size and sampling methods

The desired sample size for the study was calculated using the formula given by Thrusfield (1995) with 95% confidence interval and at 5% precision. As pulmonary helminthosis prevalence of 54.6% (previous average prevalence of sheep lungworm infection in the areas recorded by the Asella Veterinary Regional Laboratory) was taken in to consideration, a sample size of 381 sheep was considered. A stratified sampling method was used to sample sheep. For postmortem examination, from sheep which was presented for slaughter at Asella hotels and/or restaurants, 71 sheep intact lungs and fecal samples were randomly selected. The desired sample size for the study was calculated using the same formula as for the coproscopic examination (except that the estimated previous average prevalence of pulmonary helminthosis in the area by postmortem examination is 4.86%).

Experimental procedures

For coproscopic examination purpose, about 10 to 15 g of fresh fecal pellets taken from the rectum of sheep were transported to Asella Regional Veterinary Laboratory within 4 h. Individual samples were processed by the modified Baermann technique described by Glovirina (1984) and Hanse et al. (1994). All larvae were identified morphologically as described by Soulsby (1982) and Urquhart et al. (1992). For postmortem intact lung incision, the lungs were removed with trachea immediately after slaughter, then put on clean metallic frailly and examination of adult lungworm was done as per the method described by William (1998). Hence, it was described as: after intact lung removed incision was made with fine blunt-pointed scissors and scalpel blade along trachea, bronchi and then bronchioles. Description was given by Soulsby (1982) used for identification of each adult lungworm morphology.

Closed type questionnaire survey was also carried out to interview individual owners to obtain general information about sheep age, husbandry practices, usage of previous anthelmintics administration and appearance of respiratory clinical signs.

Statistical analysis

A computer based statistical software package interceded (STATA-7) was used to carry out different statistical analysis. The prevalence variation of species within parameters as well as statistical significance of the association between risk factors and the infection was assessed using descriptive statistics and Chi-square (χ^2). Additionally, to make comparison between post-mortem (PM)-intact lung incision and PM-coproscopic examination, sensitivity and negative predictive value were used. For the purpose of this study, a 95%, significant level was considered.

RESULTS

Overall prevalence

Of the total 381 sheep examined, 72.44% (confidence interval (CI)=95%; χ^2=3.1484) were found to be infected with either one or more of the lungworm species. The lungworm species encounters during the study period were *Dictyocaulus filaria*, *Protostrongylus rufescens* and *Muellerius capillaries* which accounted for a prevalence of 12.3, 33.7 and 5.4%, respectively.

Prevalence of lungworm infestation by age and sex

The overall lungworm infection prevalence in young and adult sheep was accounted to be 70.7 and 73.2%, respectively; revealing no statistically significant variation (χ^2=0.2562; P>0.05) between the two age groups. Lungworm infection in relation to the sex of sheep

Table 1. Prevalence of lungworm infection in ovine hosts in relation to hypothesized risk factors (variables).

Variable	No. of sheep examined	No. of sheep infected				
		D. filaria (%)	*P. rufescens* (%)	*M. capillaries* (%)	Mixed infection* (%)	Overall infection (%)
Age						
Young	116	3 (2.6)	34 (29.3)	7 (6.0)	38 (32.8)	82 (70.7)
Adult	265	31 (11.7)	59 (22.3)	8 (3.0)	96 (36.2)	194 (73.2)
						χ^2=0.26
Sex						
Male	168	16 (9.5)	42 (25.2)	4 (2.4)	59 (35.1)	121 (72.2)
Female	213	18 (8.5)	51 (23.9)	11 (5.2)	75 (35.2)	155 (72.8)
						χ^2=0.03
Husbandry system						
Extensive	244	25 (10.2)	69 (28.3)	12 (4.9)	82 (33.6)	188 (77.0)
Semi-intensive	137	9 (26.6)	24 (17.5)	3 (2.2)	52 (37.9)	88 (64.2)
						χ^2=7.23
Anthelmintics usage						
Dewormed	130	8 (6.2)	18 (13.8)	5 (3.8)	39 (30.0)	70 (53.8)
Non-dewormed	251	26 (10.0)	75 (29.0)	10 (3.9)	95 (37.8)	206 (82.1)
						χ^2=34.18
Resp. clinical sign						
Showing respiratory clinical sign	197	23 (11.7)	59 (29.9)	11 (5.6)	78 (39.6)	171 (86.8)
Apparently healthy	184	11 (5.9)	34 (18.5)	4 (2.2)	56 (30.4)	105 (57.1)
						χ^2=42.14

*Infestation of sheep with more than one lungworm species.

accounted for 72.8 and 72.2% prevalence in female and male, respectively. Statistically, significant variation (χ^2= 0.0262; P>0.05) was not found between the two sex groups (Table 1).

Prevalence of lungworm infection by husbandry system and anthelmintics usage

The result of questionnaire survey revealed the effect of husbandry system on the prevalence rate of lungworm infection and it has statistically significant association (χ^2=7.2180; P<0.05). The analysis of questionnaire survey from sheep owner also revealed the effect of anthelmintics usage on the prevalence rate of lungworm infection and the variation was found to be statistically significant (χ^2=34.1766; P<0.05) (Table 2).

Prevalence variation with clinical respiratory sign

Of the total sheep sampled for the study; 184 of them were apparently found clinically healthy while the remaining 197 sheep presented with clinical respiratory signs (like coughing and nasal discharge). Thus, there was a statically significant (χ^2=42.1404; P<0.05) difference in lungworm infection prevalence between sheep those showing respiratory clinical signs and apparently healthy (Table 1).

Postmortem examination

With postmortem intact lung incision, 39.4% (28) sheep were found to contain lungworm infection. However, only 26.8% (19) sheep were detected to have the infection with the coproscopic examination and presented in lung incision to the coproscopic examination was 3:2.04. The estimated probability of missing sheep with lungworm infection using coproscopic examination was 47.4%.

DISCUSSION

In the present study, coproscopic results disclosed an overall lungworm infection prevalence of 72.44%

Table 2. Prevalence of lungworm infection by postmortem examination of slaughtered sheep.

Lung worm species	Larval encounter during coproscopic examination		Adult worm encountered during PM intact lung incision	
	Male	**Female**	**Male**	**Female**
D. filarial	1	3	1	4
P. rufescens	1	4	2	6
M. capillaries	1	3	2	3
Mixed infection	2	4	3	7
Total (%)	5 (20)	14 (30.43)	8 (32)	20 (43.48)

(276/381) in sheep of Asella province. Of the total infected, 12.3, 33.7, 5.4 and 48.6% prevalence was due to *D. filaria*, *P. rufescens*, *M. capillaries* and carried mixed infection with two or all the three species of lungworm, respectively. The overall prevalence result almost coincides with the previous Ethiopians researchers report, 70.7% prevalence in DebreTabor Awraja (Yohannes, 1989), 73.75% in Debre Birhan (Netsanet, 1992) and 71.3% in six district of Wollo at high altitude (Sefinew, 1999). But, it was lower than the observation made from the same area by Bekele et al. (1981) 59.4%, Wondwossen (1992) 58.8%, Paulos (2000) 52.54%, Mireteab and Aman (2011) 57.1% and Hasen et al. (2013) 55.10%. The studies analysis observed in other area of the country also disagrees very highly than with this study conducted in different parts of Ethiopia; 50% in Dessie and Kombolcha (Teffera, 1993), 13.24% in and around Mekele (Frewengel, 1995), 44.7% in BahirDar (Sisay, 1996), 53.6% in Northwestern Ethiopia (Amenu et al., 2006), 40.4% in Dessie and Kombolcha districts, Northern Ethopia (Regassa et al., 2010), 32.7% at Gonder (Mekonnen et al., 2011) and 13.4% in Mekele town (Ibrahim and Degefa, 2012), while 83 to 99.5% in highland of Shoa province (Jovanovic et al., 1962). The possible explanation for such infection rate variation could be attributed variation in agro-ecology, altitude, rainfall, humidity and temperature difference and season of examination on the respective study areas, which favor or disfavor the survival of parasite larvae (Blood, 1976; Soulsby, 1982; Bradford, 2002).

An attempt was furthermore made to know the influence of age and sex on the overall prevalence of lungworm infection and there is no statistically significant difference (P>0.05) in the infestation rate of all lungworm species. However, a high infection prevalence of overall lungworm was recorded in youngest than adults which is in agreement with the observations of Geottray (1962), Blood (1979), Soulsby (1982) and Mireteab and Aman (2011). This has been partly explained by the analysis given as though, infection with *M. capilleries* and *P. rufescens* did not show development of acquired immunity. The acquired immunity developed older sheep due to previous exposure of *D. filaria* and that recovered from the infection have better immunity against re-infection (Soulsby, 1982; Craig, 1998). In the present study, sex dependent variation was not encountered, hence both sexes showed almost equal susceptibility to infection with lungworms. This was coinciding with research study reported by Addis et al. (2011), Nibret et al. (2011), Dawit and Abdu (2012) and Hasen et al. (2013), but disagree with report of Alemu et al. (2006) and Mireteab and Aman (2011). These may be due to the fact that improper distribution of sample selection between the two sexes (Paulos, 2000); or else most of the sampled females are not in preparturient period during the study time (Urquhart et al., 1996).

In addition, husbandry system revealed statistical significant variation on the infection prevalence of sheep. Higher prevalence of infection was noted out where the husbandry of sheep was extensive type than in the semi-intensive. This result coincides with that of the observation recorded by Sisay (1996) and Dhar (1982). The reason why sheep which have extensive management type has the highest infection prevalence can be explained due to the sheep in these managemental system that have higher chance to ingest the intermediate host (Snail and slugs) for *P. rufescens* and *M. capillaries* in large number; or are they possibly infested with larvae as well as easily obtained *D. filaria* from the herbage (Radostitis, 1994; Souls, 1982). Another possible explanation for the massive infection of sheep in extensive managemental type is that they were not supplied with appropriate nourishment which provides high computation, getting wide of lungworm infection (Kimberting, 1988).The variation with anthelmentic usage is clearly indicating as the non-dewormed sheep have high infection prevalence than dewormed counter parts. When the infection prevalence on antihelmentic usage base was subjected to analysis, the difference is statistically significant (P<0.05). The observation noted on the dewormed sheep in this study was in agreement with the work of Yohannes (1989), Netsanet (1992) and Sefinew (1999). Even though the dewormed sheep revealed low infection prevalence compared to non-dewormed groups, about 53.8% of them are infected with lungworm. The reason behind this result probably, is that sheep which have only cough and/or tachypnoeic are usually in the prepatent stage of the disease or have small adult worm burden and the anthelmentic used for the treatment of these sheep may be only temporarily suppress egg production of the adult worms (Blood et al., 1997; Urquhart et al.,1996).

The higher infection prevalence was recorded in sheep showing clinical respiratory signs than those apparently healthy and significantly higher ($P<0.05$) prevalence infection was revealed on comparison of coproscopic larvae burden. The result coincides with the observation of Paulos (2000) and Hasen et al. (2013). *D. filaria* appeared with the highest prevalence and coproscopic larvae output in sheep with respiratory sign. This may indicate that *D. filaria* is responsible for clinical respiratory sign of sheep in Asella. However, *P. rufescens* infection may weaken the lung and assists in reducing the appearance of respiratory clinical signs (Geottray, 1956; Radostitis, 1994).

The observation of intact lung from slaughtered sheep revealed 39.4% infection, whereas 26.8% infection prevalence was discovered in the examination of intestinal pellet. Wondwossen (1992) and Teffera (1993) reported the prevalence of 60.5% in Asella area and 67.83% in Dessie and Kombolcha, respectively. Infection prevalence of 40.09% (Paulos, 2000) and 53.6% (Sefinew, 1999) were observed by coproscopic examination of intestinal pellets from sheep slaughtered in Chilalo area and six district of Wollo, respectively. The prevalence of coproscopic examination reported by the aforementioned two investigator have significant variation with their respective prevalence revealed by intact lung incision which accounted 72.8% in Chilalo area and 62% in the six district of Wollo. Thus, the statistical variation of infection prevalence between the two techniques observed by Paulos (2000) and Sefinew (1999) almost coincides with the present study.

In an attempt done to assess the value of coproscopic examination in estimating the worm larvae in infected sheep, it was found that, of the confirmed adult worm infection, only 67.86% infection prevalence was detected. It was also shown that there was no statistically significant correlation ($P>0.05$) between adult worm recovery and faecal larvae outputs for any of the three lungworm species as well as mixed lungworm species. One of the probable reasons attributed for such difference could be that those larvae which reach the lung of sheep remain in the parenchyma and became encysted in fibrous nodules. Because, such nodules may not contain adult worm of both sexes; fertile egg may not be deposited in air passage (Radostitis, 1994; Blood et al., 1997). Therefore, the finding of the present study strongly supports that coproscopic examination had limited value in estimating the burden of lungworm infection in sheep and interpretation of coproscopic value. Hence, copro-scopic examination requires serious cautions. In addition, it was more appropriate to use coproscopic examination for clinical purpose than in scientific investigations. The other reason may be that the larvae are found (50 to 1000 g) only in the faces/pellet of patent cases, otherwise, in the rest phase it may not be dispersed throughout the pellet (Urquhart et al., 1996; Rose, 1965). Likewise, through PM-intact lung incision, the present study revealed that about 60.6% of sheep were found to be apparently healthy, but it may not mean that only 28 sheep were infected with lungworm. This is because of the fact that, although sheep are infected with lungworm at the penetration, prepatent and post-patent phases, the adult worm may not be observed with the aid of necked eye or else there be the case of completion of sufficient duration and self-cure, at these time no adult worm was visible and careful microscopic examination of bronchial mucus is to find out the larvae (Urquhart et al., 1996; Soulsby, 1982; Blood et al., 1997).

In conclusion, the infection prevalence of lungworm has no statistical significant association with the age and sex of sheep. Husbandry system and anthelmintics usage has great contribution on the prevalence of lungworm infection. Although most sheep infected with lungworm are clinically characterized by respiratory signs; some sheep appeared apparently healthy. Both coproscopic and postmortem examination results justify high prevalence of lung worm infection during the study period. The postmortem examination also implies as the coproscopic examination has limited value in terms of estimating the prevalence of lungworm infection in live sheep.

ACKNOWLEDGEMENTS

The authors are indebted to the Asella Regional Veterinary Laboratory for material and logistic supports, and cooperation of the staff members to bring this research to completion.

REFERENCES

Addis M, Fromsa A, Ebuy Y (2011). Study on the prevalence of Lungworm Infection in Small Ruminants in Gondar Town, Ethiopia. J. Anim. Vet. Adv. 10:1683-1687.

Alemu S, Gelay E, Ayele G, Zeleke A (2006). Study on small ruminant lungworm in North East Ethiopian. Egyptian J. Vet. Parasitol. 14(2):330-335.

Arsi Plan and Economic Development Office (APEDO) (1999). Socio Economic on Arsizone; Asella, Arsi. pp. 1-105.

Ayalew A, Tewodros D, Alemayehu W (2011). Prevalence and risk factors of intestinal parasites among Dergi school children, North Gonder, Ethiopia. J. Parasitol. Vector Biol. 3(5):75-81.

Ayalew L, Frechette JL, Malo R, Beaukegard G (1973). Studies on the incidence of *D. filaria* in sheep of Ramoski region, Can. Vet. J. 14(12):301-303.

Bekele M, Feseha G, Shibru T (1981). Observation of D.filaria in Wollo and Arsi Administrative Region of Ethiopia, Eth. J. Agric. Sci. 3(21):75.

Bekele M, Aman A (2011). Ovine lungworms in Tiyo district, South-east Ethiopia: Prevalence, effect of altitude and major host related risk factors. Studies 2(15):16.

Blood DC (1979). Veterinary Medicine, A text Book of the disease of cattle, sheep, goat and horse; lungworm, 3rd Ed.; London.

Blood DC, Henders JA, Radostitis OM (1997). Veterinary Medicine, A text Book of the disease of cattle, sheep, goat and horse; lungworm, 3rd Ed.; London.

Bradford P (2002). Large Animal Internal Medicine: disease of horses, cattle, sheep and goats, 3 ed. Mosby Inc., pp. 514-515, 1452-1455.

Craig TM (1998). Epidemiology of internal parasites: Effect of climate and host reproductive cycles on parasite survival. Proceedings of the Small Ruminants for the Mixed Animal Practitioner, Western Veterinary Conference, Las Vegas.

Dawit W, Abdu M (2012). Prevalence of Small Ruminant Lung Worm Infection in Jimma Town. Global Vet. 8(2):153-159.

Dhar DN (1982). A note on the immunological response of goats to vaccination with the radiated attuaneted *D. filaria* vaccine, Ind. J. Anim. Sci. 48(10):762.

Dunn AM (1978). Veterinary Helminthology, 2nd Ed., William Helmann Medical Books Ltd., London. p. 203.

Etminani A (1980).Veterinary Respiratory Diseases. Publication center of Tehran University. p. 470-472.

Food and Agricultural Organization (2006). International scheme for the coordination of dairy development and International meat development scheme, Report of a mission to Ethiopia; May 27- June 28; Working paper 23; Rome, Italy.

Frewengel S (1995). Prevalence of ovine *Dictyocoulus* in and around Mekele, DVM thesis, Faculty of Veterinary Medicine, Addis Ababa University Debreziet, Ethiopia.

Gallie GJ, Nunns VJ (1976). The bionomics of the free living larvae and the transmission of d. filarial between lambs in North eat England. J. Helmth. Engl. 50(2):79-89.

Geottray L (1962). Veterinary Helmenthology 9th Ed., Baille, Jindail, London. pp 68.

Glovirina LP (1984). Search for and testing of some anthelmentic against D. filaria with inhibited developments in sheep; Bulletin, Institute Gelmintologic, Skriabina.

Hansen J (1994). The epidemiology, diagnosis and control of Helment parasite of ruminant, ILRAD, Kenya. p. 83.

Hansen J, Perry B (1996). The epidemiology, Diagnosis and control of Helminths parasites of ruminants, ILRAD, Kenya. p. 29-31.

Hasen A, Takele S, Simenew K (2013). Ovine lungworm infestation rate on fecal larvae recovery basis. Acta Parasitol. Globalis 4(1):29-33.

Ibrahim N, Degefa Y (2012). Prevalence of Ovine Lung Worm Infection in Mekelle Town, North Ethiopia. Int. J. Vet. Med. 9(1):1-15.

Institute of Agricultural Research (IAR) (1974). Ethiopian National, Sheep searching development policy report by a technical committe, Addis Ababa, Ethiopia.

Jovanovic M (1962). Observation of D. filaria infection in and around Shoa, a report to the ministry of Agriculture, Addis Ababa, Ethiopia. p. 12.

Kimberling CV (1988). Disease of sheep, 3rd ED., Leaftbiger, Philadelphia. p. 99.

Mekonnen A, Abebe F, Yohannes E (2011). Study on the Prevalence of Lungworm Infection in Small Ruminants in Gondar Town, Ethiopia. Vet. Res. 4(3):85-89.

Mihreteab B, Aman A (2011). Ovine Lungworms in Tiyo District, South-East Ethiopia: Prevalence, Effect of Altitude and Major Host Related Risk Factors. Global Vet. 7(3):219-225.

Natsanet B (1992). Study on prevalence and control of lungworm in local Ethiopian highland sheep in and around Debere Birhan, DVM thesis, Faculty of Veterinary Medicine Addis Ababa University Debreziet. pp 42.

Nibret M, Basazne B, Mersha C (2011). *Dictyocaulus filaria* and *Muellerius capillaris* are Important Lungworm Parasites of Sheep in Wogera District, Northern Ethiopia. Int. J. Anim. Vet. Adv. 3(6):465-468.

Paulos A (2000). Importance and seasonal dynamics of lungworm infection of small ruminants in Chilalo area, Arsizone, DVM thesis, Faculty of Veterinary Medicine, Addis Ababa University Debreziet.

Radostitis OM (1994). Veterinary Medicine a Text Book of the Disease of cattle, sheep, pigs, goats and horses, 8th Ed., Baillae Jindali, England. p. 916.

Regassa A, Toyeb M, Abebe R, Megersa B, Mekibib B, Mekuria S, Debela E, Abunna F (2010). Lungworm infection in small ruminants: Prevaence and associated risk factors in Dessie and Kombolcha districts, northeastern Ethiopia. Vet. Parasitol. 169(1-2):144-148.

Rose JH (1965). Some observation on the transmission of lungworm infestation in a flock of sheep at pasture, Res. Vet. Sci. 6(189):189.

Sefinew A (1999). Survey of small ruminant lungworm in six district of Wollo, DVM thesis, Faculty of Veterinary Medicine, Addis Ababa University Debreziet. p. 69.

Sisay A (1996). Preliminary study on the prevalence of Ovine lungworm infection in and around BahirDar, DVM thesis, Faculty of Veterinary Medicine, Addis Ababa University Debreziet. p. 36.

Soulsby EJL (1982). Helmenths, Arthropods and Protozoa of domesticated animals, 6th Ed., Baillare Jindall, London. pp 492.

Teffera S (1993). The prevalence of ovin4e lungworm around Dessie and Kombolcha, DVM thesis, Faculty of Veterinary Medicine, Addis Ababa University Debreziet. p. 65.

Thriest Field M (1995). Department of Vet. Clinical Studies, Royal (Disk) School of Vet. Studies, University of Edinburgh 2nd Ed. London.

Urquhart GM, Armous J, Duncan JH, Jenings FW (1992). Veterinary parasitologica, Glasgow, Scotland. p. 395.

Urquhart GM, Armous J, Duncan JH, Jenings FW (1987). Veterinary parasitologica, Glasgow, Scotland. p. 395.

William WC (1998). Special Veterinary Pathology, Postmortem examination of respiratory tract, New Bolton Center, 2nd Ed., America. p. 122.

Wondowessen T (1992). Prevalence of lungworm in and around Asella, DVM thesis, Faculty of Veterinary Medicine, Addis Ababa University Debreziet.

Yohannes G (1989). Epidemiological study and anthelmentic trail of ovine *Dictyocoulosisi* in Debretabor Awraja, DVM thesis, Faculty of Veterinary Medicine, Addis Ababa University Debreziet. p. 81.

11

Prevalence and intensity of nodular oesophagostomosis in West African Long Legged goats in Northern Benin

Sahidou Salifou[1], Dadjo Florian Daga[2] and Sabbas Attindehou[1]*

[1]Laboratory of Applied Biology Research, University of Abomey-Calavi, 01 BP 2009 Cotonou, Benin.
[2]Department of Animal Production, Faculty of Agricultural Sciences, University of Abomey, 01 BP 526 Cotonou, Benin.

This study was conducted between January and December, 2011 to assess the prevalence and intensity of nodular lesions caused by *Oesophagostomum* species into the intestine of Benin goats. Intestines of 384 West African Long Legged (WALL) goats, from Northern Benin, were examined during slaughter, and 241 intestines (62.8%) were found to be positive for nodular lesions. The infection rate was significantly ($p < 0.05$) higher during the rainy season in comparison with dry season. Animals above one year of ages were mostly affected (82.5%). Extensive lesions were observed in 95 large intestines (24.9%) and calcified nodules were observed in 130 intestines (34%).

Key words: *Oesophagostomum* species, nodules, goats, prevalence, Northern Benin.

INTRODUCTION

The climatic conditions of tropical Africa are conducive for development of parasites, notably the exogenous stage of their life cycle. Therefore, helminthosis is an absolute and important constraint in livestock production in Tropical Africa. The prevalence of gastrointestinal strongyles and their impact on yields are described by many authors (Sykes, 1994; Mortensen et al., 2003) and several reports are also available regarding haemonchosis, most detrimental of which is strongylosis which is synonymous to small ruminants production in Sub-Saharan Africa (Ouattara and Dorchies, 2001; Achi et al., 2003; Osakwe and Anyigor, 2007; Attindéhou et al., 2012). However, information is available regarding other pathogens like *Oesophagostomum,* which also causes important economic losses (Olivares et al., 2001) by decreasing animal productivity and market value of the viscera due to the presence of nodules in the intestines. The assessment of nodular oesophagostomosis importance in Benin is very timely and relevant, especially in the Northern region where farming is a major activity. This study aimed to determine the prevalence and intensity of nodules due to *Oesophagostomum* spp. in West African Long legged Goats in Benin.

MATERIALS AND METHODS

Period and study area

The study was undertaken between January and December, 2011 in the region located between latitudes 9.5° and 12.5° North and longitudes 1° and 4° West, covering four agro-ecological zones of Northern Benin. During the study period, rainfall was 921.5 mm and temperature varied between 16.2 and 40.6°C. Wet season was from May to November.

Animal

A total of 384 West African Long-Legged (WALL) goats (196 female and 188 males) were selected randomly for the study. On average, 30 animals were screened each month.

Parasitological examination

The parasitological examination consisting of usual slaughterhouse

*Corresponding author. E-mail: sabbastino@yahoo.fr.

Table 1. Prevalence of nodular oesophagostomosis relative to the seasons and the animal's age and sex.

Variable	Category	Number of examined animals	Number of infected animals	Infection rate (%)	p value
Season*	Dry	159	91	57.2	0.020
	Wet	225	150	66.7	
Age*	Below one year	144	49	34.0	0.000
	Above one year	240	192	80.0	
Sex	Female	193	123	63.7	0.278
	Male	191	118	61.8	
Total	-	384	241	62.8	

*Significant variation factor.

inspection has been used to assess the prevalence and the intensity of nodular lesions in animals. After slaughter, the intestines were isolated, emptied and washed with pressured tap water injection. The external and internal mucosa of both intestines was inspected carefully in search of nodules. Some nodules were incised with a knife in order to characterize the nature (calcified or not calcified). Nodular lesions affecting more than 50% of the length of the small intestine or large intestine were called generalized and considered as a result of heavy infection. When nodules affect less than 25% of the tract, the infection was declared localized. The other cases are referred to as fairly extensive lesions. A logistic regression was performed in Stata 11 (Nicoletti, 2011) to analyze and evaluate age, sex and season influence upon the prevalence and intensity of nodular lesions.

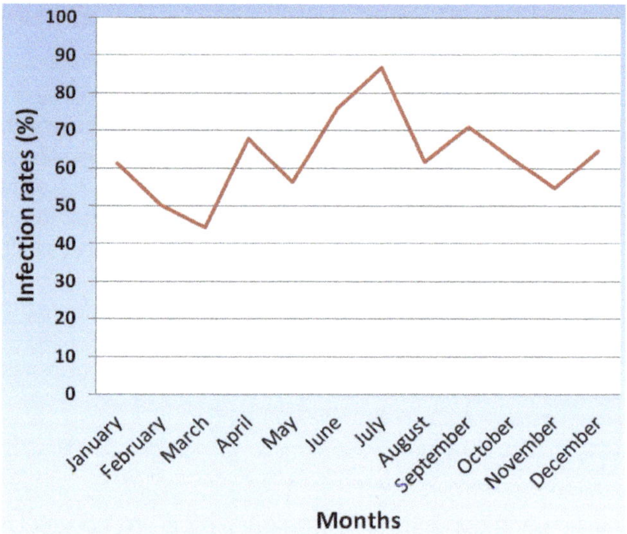

Figure 1. Nodular oesophagostomosis prevalence.

RESULTS

Prevalence of nodular oesophagostomosis and variation factors

Out of 384 examined animals, 241 (62.8%) were positive to nodular oesophagostomosis. Figure 1 shows the infection rates recorded monthly. It highlights the endemic nature of the parasitosis whereas Table 1 clearly shows a higher prevalence in the wet season ($p < 0.05$). Animals aged one and older were significantly ($p < 0.001$) more infested than younger ones. No influence of the animal sex could be proven.

Location and intensity of nodular lesions

Parasitic nodules due to *Oesophagostomum* spp. were found on the terminal portion of the abomasum, on the small intestine and on the large intestine. The mucosa of the large intestine was the most dominant location of nodules (Figure 2). Respectively, 17.8, 25.7, and 56.4% of infected viscera had generalized, fairly extensive and localized lesions.

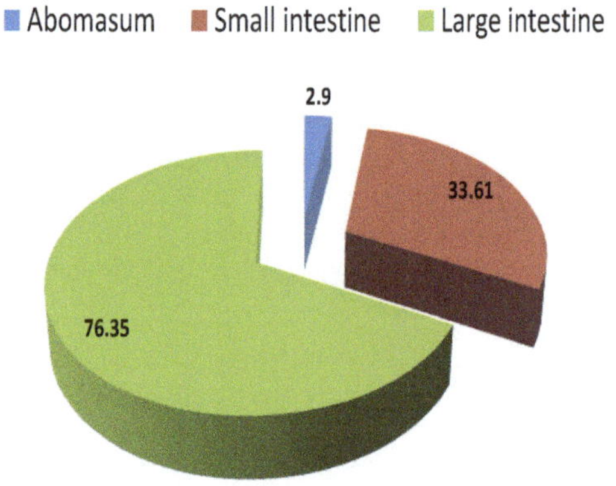

Figure 2. Nodular lesions localisation (Frequency of infection; %).

DISCUSSION

The prevalence of nodular oesophagostomosis was important (almost 63%) even if lower values have been recorded in neighboring regions. This is the case in Northern Nigeria where Nwosu et al. (2011) reported a prevalence of 71% in a slaughterhouse investigation. This difference is probably due to the fact that slaughterhouses surveys are likely to overestimate pathology's prevalence. Indeed, the candidates for the slaughter often look bad, health wise. However, it is clear that oesophagostomosis is an important strongylosis in West African Long Legged goats in West Africa.

O. columbianum have been reported in 88.3% of goats in the Central Plateau of Burkina (Belem et al., 2005). The nodular oesophagostomosis was reported in 62% of goats in the Middle Guinea by Barry et al. (2002). Contrary to this, a low rate of 43% was reported by Achi et al. (2003) in savannah region in north of Ivory Coast. The results of the study revealed a fairly clear seasonal variation, with high levels of infection during wet, which was consistent with the observations of Nwosu et al. (2011) but somewhat contradicted the results recorded in Burkina Faso and Ivory Coast (Belem et al., 2005; Achi et al., 2003). A highly significant effect of the age was observed. The elderly were most affected. This is understandable considering the chronicity of nodules, especially in the calcification phase. The intensity of the lesions has been similar to reported findings (Kulo and Seme, 2007; Nwosu et al., 2011).

Conclusion

Nodular oesophagostomosis must be considered as the second main strongyle of high sanitary importance for livestock in Northern Benin after *Haemonchus contortus*. The control of this *Oesophagostomum* spp is very necessary and need to be conducted based on climatic calendar, in association with haemonchosis management.

ACKNOWLEDGEMENTS

The study was funded by Association Doguici NGO through its West African Zoopole Program and with the support of Benin Ministry of High Education and Scientific Research. The authors express their gratitude to these partners.

REFERENCES

Attindéhou S, Salifou S, Biaou CF, Gbati OB, Adamou-N'diaye M, Pangui LJ (2012). Epidemiology of haemonchosis in sheep and goats in Benin. J. Parasitol. Vector Biol. 4(2):20-24.

Osakwe II, Anyigor SI (2007). Prevalence of gastrointestinal helminths in West African Dwarf (WAD) goats in an agrarian agro-ecosystem. Anim. Res. Int. 4(3):728-732.

Belem AMG, Kaboré A, Bessin R (2005). Variations saisonnières des helminthes gastro-intestinaux chez la chèvre du plateau central du Burkina Faso. Rev. Elev. Med. Vet. Pays Trop. 58(1-2):37-43.

Ouattara L, Dorchies PH (2001). Helminthes gastro-intestinaux des moutons et chèvres en zones sub-humide et sahélienne du Burkina Faso. Rev. Méd. Vét. 152(2):165-170.

Nicoletti NP (2011). Introduction to stata 11 - Getting started with stata programming. University at Buffalo (Suny), Department of Political Science. p 56.

Barry AM, Pandey VS, Bah S, Dorny P (2002). Etude épidémiologique des helminthes gastro-intestinaux des caprins en Moyenne Guinée. Rev. Élev. Méd. Vét. Pays Trop. 55(2):99-104.

Achi YL, Zinsstag J, Yèo N, Dea V, Dorchies PH (2003). Épidémiologie des helminthoses des moutons et des chèvres dans la région des savanes du Nord de la Côte d'Ivoire. Rev. Méd. Vét. 154(3):179-188.

Sykes AR (1994). Parasitism and production in farm animals. Anim. Prod. 59:155-172.

Mortensen LL, Williamson LH, Terrill TH, Kircher R, Larsen M, Kaplan RM (2003). Evaluation of prevalence and clinical implications of anthelmintic resistance in gastro-intestinal nematodes of goats. J. Am. Vet. Med. Assoc. 223(4):495-500.

12

Prevalence of helminthic fauna in wild pigs in comparison with domestic pigs: A study in the adjoining areas of Mudumalai, Anamalai and Sathyamangalam tiger reserves, Tamil Nadu South India

Boon Allwin[1] , M.G. Jayathangaraj[1], M. Palanivelrajan[1], S. Gomathinayagam[2], M. Raman[3] and S.T. Bino sundar[2]

[1]Department of Wildlife Science, Madras Veterinary College, Chennai, India.
[2]Department of Veterinary Parasitology, Madras Veterinary College, Chennai, India.
[3]Programme director, TRVPB, TANUVAS, Chennai, India.

Study on endoparasites in wild pigs (*Sus scrofa*) interfering with agriculture, was carried out in areas adjoining the Western Ghats (Mudumalai tiger reserve, Anamalai tiger reserve) and Eastern Ghats (Sathyamangalam region) of Tamil Nadu state In India during November, 2013 to May, 2014. Ninety faecal samples in total (n=30 of Wild pigs, n=30 of Desi pigs and n=30 Cross bred pigs each of the study areas) were subjected to the parasitological examination using standard methods in wild pigs as well as in domestic pigs. Domestic pigs were involved to find the difference in the prevalence level between wild pigs and domestic pigs as there is a wide variation in their habitat. Prevalence of endoparasitic infections revealed the evidences of *Ascaris suum, Trichuris suis,* Strongyles, *Strongyloides* sp. and mixed parasitic infections comprising of *Ascaris* sp. with *Trichuris* sp., in addition to Strongyles with *Strongyloides* sp. The overall positivity of internal parasitic prevalence in wild pigs, desi pigs and cross bred pigs pertaining to Mudumalai, Anamalai and Sathyamangalam were documented, and differences among their prevalence data indicated an over-dispersed helminth distribution. These results indicate that populations of wild pigs although living under optimal conditions, are heavily affected by a burden of parasitic disease and some parasites are likely to limit population growth via a high mortality of piglets and infections throughout the lifespan of adults.

Key words: Wild pigs, domestic pig, endoparasites, prevalence.

INTRODUCTION

Although, the helminth parasites of domestic pigs are well documented there is paucity of information with regard to wild pigs. In wild populations there is usually a balanced host-parasite relationship, but pathogenicity. However, anthropogenic changes of the environment, the increase of human populations and the introduction of other animal species, may introduce unknown factors that can disrupt the natural balance and induce pathological conditions. This study is a contribution to the knowledge of wild pig helminthic fauna as these animals have been co-inhabiting with human beings and domestic pigs sharing the same resources such as land, water and also air. The epidemiology of parasitic diseases is very important as they have a zoonotic potential that can lead to various deleterious effects. In this study, a comparison of the prevalence rate of various parasitic infections in wild pigs, desi pigs and cross bred pigs have been recorded by standard techniques like floatation and centrifugation and their Egg Per Gram (EPG) was constituted to know the intensity of the infections.

MATERIALS AND METHODS

Study area

The study on endoparasites in wild pigs (*Sus scrofa*) interfering with agriculture was carried out in areas adjoining the Western Ghats (Mudumalai tiger reserve, Anamalai tiger reserve) and Eastern Ghats (Sathyamangalam region) of Tamil Nadu state in India during November, 2013 to May, 2014.

Collection and preservation of coprological samples for endoparasitic examination

Throughout this study programme, 30 fresh faecal samples of wild pigs from described sampling areas were collected in small containers with 10% formalin for parasitic examination, and were properly labelled and sealed with parafilm subsequently. The wild pigs were tracked by foot and their locations and resting nests were identified by their field signs (Boon et al., 2015). Similarly, 30 faecal samples from desi-pigs that are semi-free ranging and 30 faecal samples from cross-bred pigs that were maintained in organised farms were collected and processed.

Examination of samples

The faecal samples were processed by both centrifugal sedimentation technique and floatation technique as described by Soulsby (1982).

Centrifugal sedimentation technique

Approximately 2 g of faeces was taken in a 100 ml beaker and was thoroughly mixed with about 10 to 15 ml of tap water. The mixture was strained through a tea strainer into a cup and then, it was transferred into a centrifuge tube. The centrifuge tubes were placed in a balanced state and were subsequently centrifuged for 2 to 4 min, at 1500 rpm. Then, the supernatant was discarded leaving 1 to 2 ml of supernatant, without disturbing the sediment at the bottom and finally, small drop from thoroughly homogenized sediment was taken on clean glass slide and was observed under both low and high power objectives of microscope (Soulsby, 1982).

Floatation technique

Faecal samples were taken in a 100bml beaker and were thoroughly emulsified with about 10 to 15 ml of saturated solution of sodium chloride with a specific gravity of 1.18 to 1.20. The mixture was strained into a cup and then, it was transferred into a floatation tube till the mixture reaches the brim of the tube and forms a positive meniscus and was left undisturbed for 15 to 20 minutes. The tip of the positive meniscus was gently touched with a clean cover slip and then the cover slip was placed on a slide, and was examined microscopically under both low and high power objectives (Soulsby, 1982).

Quantitative analysis

One gram of faecal sample was mixed with 4 to 5 ml of saturated solution of sodium chloride and was strained trough tea-strainer into floatation tube and volume was adjusted up to 12 ml with saturated salt solution and kept for 20 min undisturbed. From this, 0.3 ml of suspension was added to McMaster slide chamber and the eggs were counted and multiplied by 40 and thus, the number of eggs per gram of faeces was calculated.

STATISTICAL ANALYSIS

The statistical analysis of the data was carried out as per the guidelines, using one way ANOVA, wherever applicable.

RESULTS

Evidence of endoparasitic fauna was recorded in free ranging wild pigs as well as domestic pigs, comprising of desi pigs and cross bred pigs. *Ascaris suum, Trichuris suis*, Strongyles, *Strongyloides* sp. and mixed parasitic prevalence comprising of *Ascaris* sp. with *Trichuris* sp. in addition to Strongyles with *Strongyloides* sp. were documented during the study programme with all these pigs with a varying level of intensity that was high in wild pigs, moderate in desi pigs and negligible in cross bred pigs (Figure 1).

Prevalence of endoparasites with regard to different internal helminthic fauna of wild pigs, desi pigs and cross bred pigs pertaining to adjoining areas of Mudumalai, Anamalai, and Sathyamangalam wildlife regions were presented in Table 1 to 3. Positivity for internal parasitic prevalence with regard to wild pigs, desi pigs and cross bred pigs were 62, 29 and 9% respectively (Figure 2).

The EPG values pertaining to the parasitic prevalence comprising of *A. suum, T. suis*, Strongyles, *Strongyloides* sp. and mixed parasitic prevalence comprising of *Ascaris* sp. with *Trichuris* sp. in addition to Strongyles with *Strongyloides* sp. in wild pigs, desi pigs and cross bred pigs were presented in Table 4. The mean ±S.E. values of EPG among wild pigs, desi pigs and cross bred pigs with regard to individual endoparasitic species during this study in different adjoining areas of different wildlife regions were documented in Table 4. Highly significant variations ($P \leq 0.01$) were encountered with regard to *A. suum, T. suis,* Strongyles in Mudumalai regions, and with

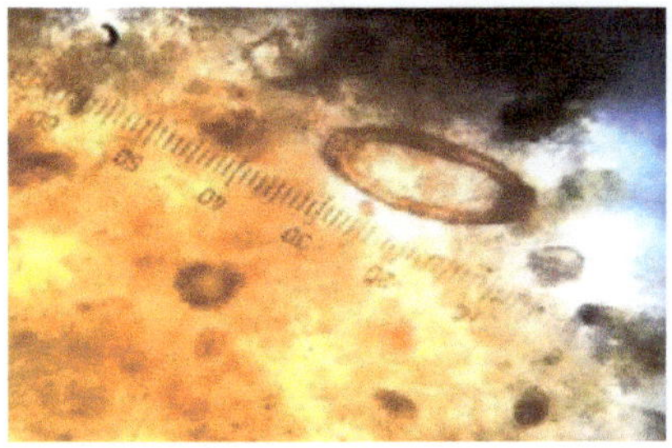

(A) Egg of *Trichuris suis* 40X

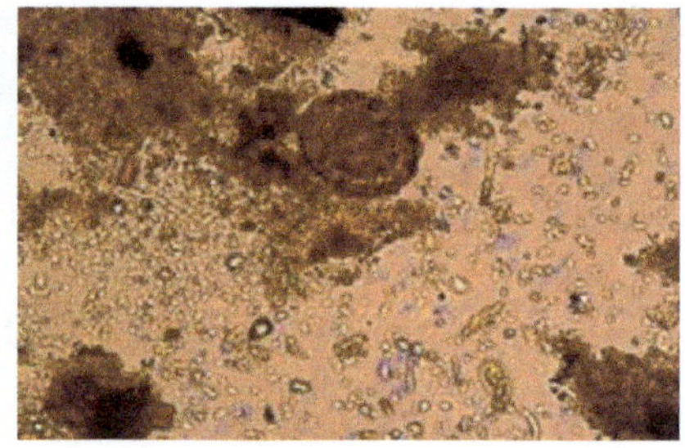

(B) Egg of *Ascaris suum* 40X

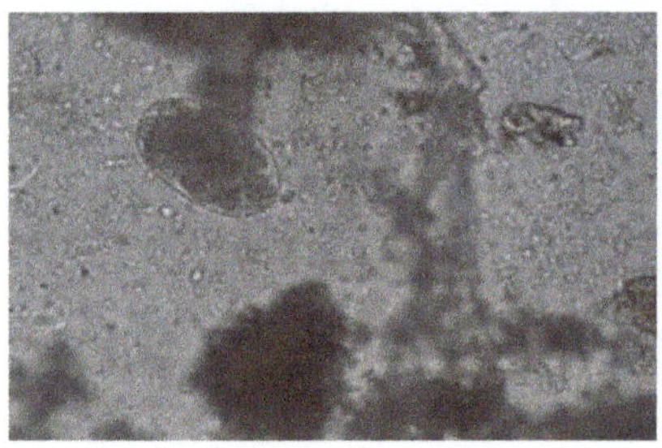

(C) Egg of Strongyles 10X

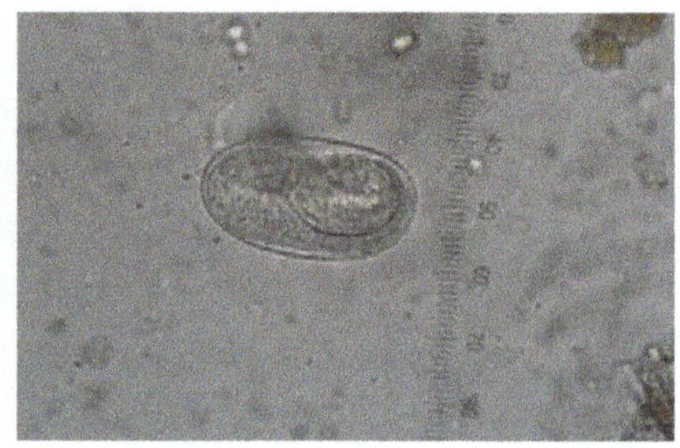

(D) Egg of *Strongyloides* sp. 40X

Figure 1. Endoparasites of wild pigs.

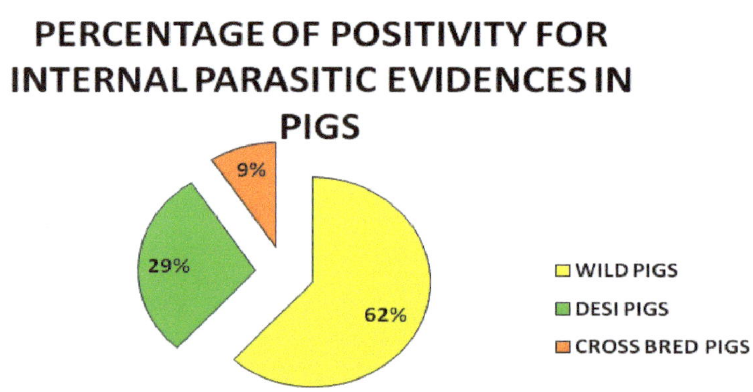

Figure 2. Parasitic evidence in pigs.

regard to *A. suum* in Sathyamangalam region. Similarly, significant variations (P≤0.05) were encountered among these pigs with *A. suum* in Anamalai region, and also with regard to mixed parasitic infection comprising of *Ascaris* sp. with *Trichuris* sp. among the pigs in Mudumalai region.

DISCUSSION

Parasitic prevalence

Overall parasitism in wild pigs and other pigs

The overall parasitic prevalence pertaining to the internal parasites in wild pigs (Figure 2) was found to be 62% wild pigs, 29% in desi pigs and 9% in cross bred pigs. Even though literatures pertaining to the occurrence of internal parasites in domestic swine are more, there is paucity of information in helminthic fauna of wild pigs in general. The parasitic eggs in this study were identified based on the morphological keys furnished by Encountering the increased percentage of parasitic prevalence in the samples from wild pigs which was supported by the report furnished by Jarvis et al. (2007) who quoted that none of the 25 examined carcasses of wild boars from Central Spain and those imported from France was free of helminths. Similarly, Eslami and Hamdi (1992) opined that the majority of wild boars examined (74%) had at least one species of helminth in the internal organs, and

Table 1. Prevalence of endoparasitic infections of wild pigs (n=30).

S/N	Parasites	Places under study		
		Mudumalai (%)	Anamalai (%)	Sathyamangalam (%)
1	*Ascaris suum*	8 (80)	8 (80)	7 (70)
2	*Trichuris suis*	6 (60)	7 (70)	8 (80)
3	Strongyles	7 (70)	6 (60)	7 (70)
4	*Strongyloides* sp	7 (70)	6 (60)	5 (50)
5	Mixed infections (*Ascaris +Trichuris)*	8 (80)	7 (70)	7 (70)
	Strongyle+Strongyloides	4 (40)	3 (30)	2 (20)
Total number of faecal samples examined		10	10	10

Table 2. Prevalence of endoparasitic infections of desi pigs (n=30).

S/N	Parasites	Places under study		
		Mudumalai (%)	Anamalai (%)	Sathyamangalam (%)
1	Ascaris suum	5 (50)	4 (40)	4 (40)
2	Trichuris suis	5 (50)	4 (40)	4 (40)
3	Strongyles	4 (40)	3 (30)	2 (20)
4	Strongyloides sp	3 (30)	4 (40)	1 (10)
5	Mixed infections *(Ascaris +Trichuri)*	4 (40)	2 (20)	1 (10)
	Strongyle+Strongyloides	2 (20)	1 (10)	NIL
Total Number of faecal samples examined		10	10	10

Table 3. Prevalence of endoparasitic infections of cross bred pigs (n=30).

S/N	Parasites	Places under study		
		Mudumalai (%)	Anamalai (%)	Sathyamangalam (%)
1	Ascaris suum	2 (20)	1 (10)	2 (20)
2	Trichuris suis	2 (20)	1 (10)	1 (10)
3	Strongyles	2 (20)	1 (10)	0
4	Strongyloides sp	1 (10)	1 (10)	2 (20)
5	Mixed infections *(Ascaris +Trichuris)*	NIL	1 (10)	NIL
	Strongyle+Strongyloides	NIL	NIL	NIL
Total Number of faecal samples examined		10	10	10

parasitic infections with several species were common in the wild boars. Further, encountering the increased overall positivity of parasitic prevalence in the wild pigs under study was in argument with the reports furnished by Bhat and Manickam (1998) who opined that high faecal egg counts were common in free ranging animals' state than those living in captivity.

Similarly, when compared to the overall parasitic prevalence between desi pigs and the cross bred pigs of the adjoining areas studied; the percent prevalence of helminthic fauna was more in the case of desi pigs. The reasons for such an intensive prevalence of helminthic fauna in case of desi pigs might be assigned to the lesser veterinary care, straying of desi pigs outside consuming different kinds of intermediate hosts like earthworms, small sized reptiles or other creatures in or around the drainage areas noticed in the adjoining areas of wildlife regions. Comparatively, due to a better management

Table 4. Mean±standard error value of egg per gram of faeces in the different pigs.

S/N	Endoparasitic infection		Mudumalai	Sathyamangalam	Anaimalai
1	*Ascaris suum*	Wild pig	756.25±57.82^{b}	575±31.34^{a}	664.29±84.34
		Desi	650±74.36^{b}	575±59.51^{ab}	412.5±82.60
		Cross bred	125±25.00^{a}	100±0.00^{b}	125±25.00
		F value	12.845**	10.680**	6.558*
2	*Trichuris suis*	Wild pig	225±21.41^{b}	221.43±18.45	206.25±19.91
		Desi	290±18.76^{b}	250±20.4	225±12.5
		Cross bred	125±25.00^{a}	100±0.00	100±0.00
		F value	9.073**	4.201^{NS}	2.126^{NS}
3	Strongyles	Wild pig	214.29±14.29^{a}	166.66±21.09	164.29±21.03
		Desi	77.5±30.38^{b}	100±28.87	125±25.00
		Cross bred	75±25.00^{b}	50±0.00	50±0.00
		F value	14.643**	3.230^{NS}	2.195^{NS}
4	*Strongyloides* sp	Wild pig	150±22.01	133.33±18.26	160±12.95
		Desi	100±25.00	137.5±21.35	100±0.00
		Cross bred	100±0.00	50±0.00	75±25.00
		F value	1.018^{NS}	0.875^{NS}	1.469^{NS}
5	*Ascaris* sp +*Trichuris* sp	Wild pig	137.5±15.67	142.86±17.01	118.75±16.02
		Desi	150±20.41	125±53.03	150±0.00
		Cross bred	Nil	Nil	Nil
		F value	5.051*	2.672^{NS}	3.411^{NS}
6	*Strongyle* sp+*Strongyloides* sp	Wild pig	75±14.43	133.33±33.33	125±25.00
		Desi	125±25.00	150±0.00	Nil
		Cross bred	Nil	Nil	Nil
		F value	5.619^{NS}	2.300^{NS}	6.250^{NS}

Means bearing different superscript differ significantly (NS-Non Significant).

system that includes provision of required health care measures, maintenance of good feeding regime, feeding based enhanced immune status of the animal etc. while the overall prevalence in the case of cross bred pigs was found to be less than 9%.

The reasons for encountering higher percentage of overall positivity of parasitism in wild pigs might be assigned to the reasons like diversified feeding activities of the animals, straying into the peripheral-areas of the wild regions, consumption of different types of intermediate host, absolute lack of health care related measures, consumption of feed materials contaminated by excreta of the co-existing wild animals like gaur, nilgiri tahr, spotted deer and sambar deer that co-exist in the wild environment etc. Different types of management patterns, variations in feed offered, extent of health care measures like deworming activities, varying immune status of the swines might be the reasons that could be attributed to the variation in the positivity of parasitism in desi as well as the cross bred pig of the adjoining areas studied. However, since wild pigs entering the agricultural fields adjoining those regions in the study are basically wild in nature with more or less similar type of feeding activities, wild pigs of all these regions were found to have higher percentage of overall parasite prevalence with values of 90% and above, and this could be a potential source of transmitting infection to the domestic pigs. Further, the encountering of increased overall parasitic prevalence in wild pigs under study was in agreement with the report given by Jarvis et al. (2007) who opined that the natural peculiarities of the area, including the sufficient availability of intermediate hosts of helminths were the important factors that affected wild pigs with helminths.

Parasitic prevalence in different adjoining regions and the intensity of infection

The percentage of positivity pertaining to different parasitic fauna encountered in wild pigs,desi pigs and cross bred pigs are presented in Tables 1 to 3. *A. suum* and *T. suis* were the frequently encountered parasitic fauna, with wild pigs in general. The EPG values of *A. suum,* in case of wild pigs as well as desi pigs differed highly significantly when compared to the value encountered in case of cross bred pigs in Mudumalai as well as Sathyamangalam area. Similarities in certain feeding related activities, varying immune status of pigs etc. might be assigned as the etiological factors pertaining to such highly significant variations in them.

In this regard, it was noteworthy to mention the report furnished by Tiwari et al. (2009) who quoted that in Botswana, *A. suum* was the most prevalent helminth encountered in the case of pigs. Further variations in the number of wild pigs affected with *A. suum* was encountered in this study was supported by Foata et al. (2005) who opined that among the numerous most often cited in the literature, *A. suum* was found to be one of the only three parasitic species that were index at the time of their study in case of wild boars. Variations in EPG values pertaining to the *A. suum* as encountered in the wild pigs as well as in others was further in agreement with the findings revealed by Popiolek et al. (2010) who further stated that though the *A. suum* was common in pigs and wild boars worldwide, the level of wild boar infection was not very high. In this regard, Tomass et al. (2013) quoted that *A. suum* was the most common helminth in all age categories of pigs and *Ascaris suum* was a natural parasite of pigs as it could also infect human, and the potential of the *A. suum* to infect human might be due to the fact that it stored similar protein molecular with *A. lumbricoides* for which man is the natural host. Urquart et al. (1996), also quoted that the eggs pertaining to *A. suum* were very resistant to extreme temperature and the eggs of *A. suum* were found to be viable for more than four years.

Encountering *A. suum* in domestic pigs as noticed with desi as well as cross bred pigs under study, was in agreement with the findings presented by Radostitis et al. (2007) and Urquart et al. (1996). Encountering *T. suis* in the wild pigs in this study was in agreement with the report furnished by Jarvis et al. (2007) who revealed that though, *T. suis* was noticed as one of the seven helminth species in wild boars. During this study, the EPG value pertaining to wild pigs as well as the desi pigs were found to be significantly different, when compared with the values encountered with cross bred pigs in Mudumalai region alone. The different types of management systems that are maintained with the cross bred pigs including the reasonably intensive health care related measures lack of opportunities for the consumption of diversified types of feeds including different types of intermediate hosts etc. might be assigned as the etiological factors for the encountering of highly significant variations in the EPG values of *T. suis* in the case of cross bred pigs when compared to that of wild pigs as well as desi pigs. Encountering the *T. suis* as noticed in the studyis also in agreement with the report presented by Tomass et al. (2013). It becomes noteworthy to mention the report presented by Nansen and Roepstorff (1999) who opined that often moderate numbers of adult *T. suis* were present in the caecum and colon, and if there was high prevalence of *T. suis* infections in the swines, it might lead to unthriftiness and death, and it was demonstrated that severe clinical disease might be associated with *T. suis* induced suppression of mucosal immunity to resident bacteria.

The EPG value of strongyles in wild pigs was found to be 214.29±14.29, where it was 77.50±30.38 in desi pigs and 75.00±25.00 in cross bred pigs. Findings on the prevalence of strongyles in wildlife under study was in agreement with the report presented by Souse et al. (2004) who encountered presence of gastrointestinal strongyles eggs in all the faecal samples of wild boars examined and further quoted about the average EPG of 2142 in the case of gastrointestinal strongyles in wild pigs when compared to the EPG values of desi pigs, in addition to the cross bred pigs which might be attributed to the existing different kinds of biotic as well as the abiotic environmental factors. The encountering of prevalence of strongyles was further in agreement with the report furnished by Magi et al. (2005)

The EPG values of *Strongyloides* sp. were presented in Table 4 and however, there was no statistical significance among the pigs comprising of wild pigs, desi pigs and cross bred pigs. Encountering the *Stronglyloides* sp. with pigs under this study was in agreement with the findings reported by Varadhrajan and Pythal who however, encountered mixed infection associating with *Fasciola* sp., Strongyles and *Strongyloides* sp. in wild boars. In this regard, Tiwari et al. (2009) quoted that parasitic fauna like *Strongyloides* sp. were found to be related to the occurrence of clinical signs like diarrhoea as well as emaciation in pigs. With regard to encountering the prevalence of *Strongyloides* sp. in this study, Coombs and Springer (1974) opined that since wild pigs were associated with feeding of earthworms, beetles, bugs and numerous larvae which functioned intermediate or paratenic hosts for various helminthic fauna, more helminth fauna were encountered including the *Strongyloides ransomi*. Urquart et al. (1996), reported the experimental demonstration of prenatal infection associated with *Stronglyloides* sp. in pigs.

Conclusion

Even though increased parasitic prevalence was encountered in the faecal samples of wild pigs entering the agriculture fields near wildlife regions, immediate conclusion about the existence of a clinical disease condition could not be drawn. Many factors may be involved and lack of extensive studies pertaining to the parasitic fauna in both core and buffer zone areas of wildlife regions. However, it might be understood that wild pigs revealed existence of different helminthic fauna comprising of *A. suum, T. suis,* Strongyles, *Strongyloides* sp. and mixed parasitic infections associated with *Ascaris* sp. with *Trichuris* sp., in addition to Stongyles with *Strongyloides* sp. Availability and feeding of diversified feed materials which comprises of earthworms, beetle larvae, insects, small rodents, egg and chicks of birds nesting on the ground as well as in short grass, including the feeding of carrion sometimes might be the significant factors that lead to the encountering of different parasitic fauna in the wild pigs studied. The encountering of significant variations in the mixed parasitic prevalence associated with *Ascaris* sp. and *Trichuris* sp. might be attributed to differences in feed materials consumed, variations in the immune level, management practices etc. The finding of the presence of endoparasites in wild pigs will be helpful in designing strategy management practices and curbing the disease at the initial levels and this can be done with ease.

Conflicts of interest

The authors declare that they have no conflicts of interest.

REFERENCES

Boon A, Jayathangaraj MG, Palanivelrajan M, Raman M (2015). Wild pig field signs - identity and confirmation. Int. J. Life Sci. Res. 3(1):205-213.

Bhat MN, Manickam (1998). Coproculture and Study of seasonal variation in parasitic burden in spotted deer (*Axis axis*) in different sanctuaries in Tamil Nadu. Indian Vet. J. 75:469-471.

Coombs DW, Springer DM (1974). Parasites of Feral pigs and European wild boar hybrids in South Texas. J. Wildl. Dis. 10:436-441.

Eslami A, S Farsad-Hamdi (1992). Helminth parasites of wild boar, *Sus scrofa*, in Iran. J. Wildl. Dis. 25(2): 316-318.

Foata J, Culioli JL, Marchand B (2005). Helminth fauna of Wild boar in Corsica. Acta Parasitol. 50(2):168-170.

Jarvis H, Kapel C, Moks E, Talvik H, Magi E (2007). Helminths of wild boar in the isolated population close to the northern border of its habitat area. Vet. Parasitol. 150:366-369.

Nansen P, Roeptstorff A (1999). Parasitic helminths of the pig: factors influencing transmission and infection levels. Int. J. Parasitiol. 29:877-891.

Magi M, M Bertani, M Dell'omodarme, MC Prati, G. Poglayen (2005). Seasonal Egg output of gastro-intestinal parasites in wild ungulates in a Mediterranean area (Central Italy). Hystrix 16(2): 169-177.

Popiolek MD, Knecht JS, Staskiwicz A, Czerwinska R (2010). Helminths of the wild boar (*Sus scrofa* L.) in natural and breeding conditions. Bull. Vet. Inst. Pulawy 54(1):161-166.

Soulsby EJ (1982). Helminths, Arthropods and Protozoa of domestic animals, 10th Edn., ELBS, Bareilly Tindall, London p. 809.

Tiwari KP, Chikweto A, Belot G, Vanpee G, Deallie C, Stratton G, (2009). Prevalence of intestinal parasites in pigs in Grenada, West Indies. West Indies Vet. J. 9(1):22-27.

Tomass Z, Inam E, Kifleyhannes T, Tekle Y, Weldu K (2013). Prevalence of gastrointestinal parasites and Cryptosporidium species in extensively managed pigs in Mekelle and urban areas of southern zone of Tigray region, Northern Ethiopia. Vet. World 6(7):433-439.

The epidemiology of malaria among pregnant women in Garoua, Northern Cameroon

Thomas Hile[1], J. P. Assam Assam[1,2,4], E. Amali[3] and E. Amuta[3]*

[1]Laboratory of Parasitology, Department of Biomedical Science, Cosendai Adventist University, Eboko, Cameroon.
[2]Biotechnology Center, University of Yaoundé I, Cameroon.
[3]Department of Biological Science, University of Makurdi, Benue State, Nigeria.
[4]Department of Biochemistry, Faculty of Science, University of Douala.

Epidemiological studies of malaria in pregnant women who live in Garoua, malaria endemic areas are scarce. The present study was undertaken to investigate the prevalence and distribution of malaria in the area. A total of two hundred and forty (240) pregnant women participated in the survey from January to March, 2007. Out of the 240 pregnant women examined, 97 were found to have malarial infections during the study, with *Plasmodium falciparum*. There was significant difference between prevalence of the infection at different stages of pregnancy (χ^2 = 0.088, df = 2, P < 0.05). Further investigations showed that 101 (44.44%) of the pregnant women had no bed nets for various reasons. Implication of malaria in pregnancy cannot be over emphasized. The need to vigorously pursue acceptance of control methods like the bed nets are discussed.

Key words: Malaria, prevalence, pregnancy, women.

INTRODUCTION

Malaria is endemic throughout most of the tropics. Of the approximately three billion people living in 108 countries who are exposed, approximately 243 million will develop symptomatic malaria annually World Health organisation (WHO, 2008). Malaria deaths peaked at 182 million in 2004 and felt to 124 million in 2010. Over 80% of the deaths occur in Sub-Saharan Africa (WHO, 2011).

Malaria during pregnancy is a major public health problem; it increases the risk of low birth weight, infant mortality and morbidity during the first year of life by inducing intra-uterine growth retardation, prematurity, and infant anaemia (Nosten et al., 1994). In malaria-endemic countries, placental malaria is associated with a 2-fold higher risk of still birth and is responsible for up to 35% of preventable low birth weight (Van et al., 2004), the prevalence of malaria is higher during pregnancy when compared with the non-pregnant state (Menedez et al., 1995). Garoua is the capital of the Northern Cameroon; there are very scarce reports and information on the incidence or prevalence of the disease or methods of control in the area. The study of malaria epidemiology in such area cannot be over emphasized.

MATERIALS AND METHODS

Study area and study population

The study was conducted in Garoua, the North region of Cameroon. The region is bounded by the far north region to the north, the Adamawa region to the south, Nigeria to the west, Chad to the east, and Central African Republic to the southeast. The study was conducted from January to March 2007. Before commencement of the study, permission was sought from the government health. Meetings with pregnant women were held in each community (health care centers), to explain the objectives of the study and to obtain their consent. Structured questionnaires were used to collect information on socio demographic data, malaria mode of prevention and control, complaints, scientific knowledge about malaria, and laboratory examination.

*Corresponding author. E-mail: lizamuta@yahoo.com.

Table 1. Distribution of malaria parasite among pregnant women living in different density in Garoua.

Location	No. examined	No. positive (%)
Low density	13	1 (1.03)
High density	163	77 (79.38)
Intermediate	64	19 (19.58)
Total	240	97

$\chi^2 = 5.99$, df = 2, $P < 0.05$.

Table 2. Distribution of malaria parasite among pregnant women using various control and preventive methods.

Preventive method	No. examined	No. positive (%)
None	101	40 (46.5)
Bed net	82	27 (32.9)
Drug	28	8 (28.6)
Insecticide	29	15 (51.7)
Total	240	97 (40.4)

$\chi^2 = 6.65$, df = 3, $P < 0.05$.

Table 3. Distribution of malaria among pregnant women using none or more than one methods of prevention and control.

Methods of prevention	No. examined	No. positive (%)
With control		
None	101	40 (43.9)
Bed nets + insecticide	5	0 (0.0)
With drug intake		
Bed nets + drug intake	45	4 (4.3)
Bed nets + insecticide	17	7 (7.6)
Insecticide + drug	72	40 (43.9)
Total	240	91

$\chi^2 = 0.74$ df = 4, $P < 0.05$.

Table 4. Distribution of malaria parasite among pregnant women of different occupation.

Occupation	No. examined	No. positive (%)
House wife (complete)	123	53 (54.63)
Trader/Business	76	28 (28.86)
Farmer	34	16 (16.49)
Student	5	0 (0. 0)
Civil service	2	0 (0.0)
Total	240	97

$\chi^2 = 0.087$, df = 4, $P < 0.05$.

Laboratory investigations

Thick and thin blood films were made on clean slides properly labelled for each subject. After patient's information has been recorded in the questionnaire, palm upwards, generally the third finger is selected and clean with cotton wool lightly soaked in the alcohol. With a sterile lancet, the ball of the finger is punctured using a quick rolling action. Pressure is applied to express blood about two to three drops on the slide (about 1 cm) for thick smear. The thick film was labelled with a marker at the edge of the slide. Slides were allowed to dry in a protected place from flies, dust and extreme heat in a slide space. Thick films slides were placed face downwards in a shallow tray on the slides and allowed to stain for 10 to 15 min. Afterward, the stain was gently flushed off the slides by adding drops of water, and the slides were placed in the film rack downwards to drain off water and dry (Monica, 1998).

Data collection and analysis

A computer program, Statistics Program of Social Sciences (SPSS) was used for data analysis. Frequency distribution tables, percentage prevalence and intensity of malaria infection were estimated using standard formulae; chi-square test was used as appropriate to test the possible effect of malaria parasite, as socio-demographic behaviour. 5% level of significance was used.

RESULTS

Table 1 describes the prevalence of malaria parasite among pregnant women in relation to population density in Garoua. Malarial infection rate was observed higher among pregnant women, living in high density area 79.38% (77/163). Women of low density area were relatively less infected 1.03% (1/13), especially as compared to pregnant women of intermediate area 19.58% (19/64). There was significant difference found among different density ($\chi^2 = 0.088$; $P < 0.05$). Preventive method against malaria infection was surveyed as shown in Table 2. Pregnant women using insecticide were less infected 16.66% (15/29) than those that were not using preventive method 44.44 (40/101) and bed net 30% (27/82). With regards to the use of prophylactic drugs, among the pregnant women, those that used prophylactic drugs had the lowest number of malarial infection with 8.88% (8/28). There was significant difference between the preventive methods used by pregnant women ($\chi^2 = 0.98$; $p < 005$). Table 3 shows the prevalence of malaria among pregnant women using none or more than one method of prevention and control. Malaria infection was found among women using insecticide and drug 40.81% (40/72), while those using bed nets, insecticide and drug together were not infected at all. Pregnant women that were using bed nets and drug intake were relatively less infected 4.08% (4/45) as compared to those with bed nets and insecticide 7.14% (7/17). Pregnant women that were not having preventive method were infected 47.9% (47/101). A significant difference was found between the methods of prevention ($\chi^2 = 0.036\%$; $P < 0.05$). The prevalence of malaria parasite among pregnant women in relation to their occupation is as shown in Table 4. The results show that farmer were highly infected 16.49% (16/34) than complete housewife 54.63% (53/123) and trader/business

Table 5. Distribution of malaria parasite among pregnant women at different stages of pregnancy.

Period of pregnancy (Months)	No. examined	No. positive (%)
0 – 3	30	13 (13.4)
3 – 6	132	48 (49.48)
6 – 9	78	36 (37.11)
Total	240	97

$\chi^2 = 0.088$, df = 2, $P < 0.05$.

28.86% (28/76). Civil servant and student had no infection. Table 5 shows prevalence of malarial in-fection among different stage of pregnancy. Women at their first trimester had 13.40% (13/30) infected, while those at their third and second trimester recorded 37.11% (36/78) and 49.48% (48/132), respectively. A significant difference was observed between malarial infection and pregnancy period ($\chi^2 = 5.99$; $P < 0.05$).

DISCUSSION

In low density settlement areas, the prevalence of malaria was relatively low among pregnant women. The infection is 1.03% conversely to those living in high density areas where the infection was high to 79.38%. However, given the large at-risk population in this malaria-endemic region of Garoua, low density settlement area is not a guaranty against malaria. High density settlement areas are characterised by the presence of dustbins, grasses, congestion, polluted gutter, stagnant waters, and dirty surroundings covered with weeds in the rivers that are suitable breeding sites for the presence of mosquitoes in the areas. This is consistent with the reports of Shr-jie et al. (2005) in Ouagadougou, Burkina Faso who found that higher prevalence rates of malaria occurred in areas where larvae breeding sites were semi-permanent. The study found that control and preventive methods commonly used were bed nets, drug's intake (sulphadoxine pyrimethamine (sp)), intermittent preventive treatment (IPT) in pregnancy, and insecticide in homes among pregnant women. Those with no prevention and control method (44.44%) were more infected than those that were having bed net (30%) as method of control. This showed that bed net is a good mode of controlling mosquito's bites. This corroborate with Mouhamadou et al. (2006) that found that bet net efficacy help in reducing man-vector contact, malaria morbidity and mortality. The research proved that insecticide treated bed nets are amongst the effective tools at our disposal for reducing pregnant women mortality caused by malarial parasite. At dusk, some family will light flammable insecticide (coiled insecticide) that burns gradually during the night, repelling or killing mosquitoes. However, the smoke produced by such insecticide is sometime unbearable. The efficacy of insecticide has been proved in South Africa by Hargreaves et al. (2000) where dichlorodiphenyltrichloroethane (DTT) was used for house spraying to restore malaria control. Insecticide treated bed nets in pregnancy were associated with the first four months of pregnancies in Kenya where women receive as many as 45 to 230 infective bites on the average during 40 weeks gestation, and the result showed a significant reduction in parasitaemia during pregnancy, in the risk of low birth weight (D'alessandro and Olaye, 1997). The fact that the infection rate among those using bet nets and those using insecticide is almost equal explain that there may not be proper use of bed nets generally, they use the bed net after they have been exposed to mosquitoes outdoor. Series of trials in Africa have shown that proper use of mosquito net reduced malaria incidence among children from 63 to 14% in Ugandan (Frederick and Arinaitwe, 2003). Further evidence came from a recent study in a highly malarious area of Kenya. During the first four months of pregnancies, women who were protected by insecticide-treated bed nets at night gave birth to 25% fewer premature or small for gestational age babies than women who did not sleep under insecticide-treated bed nets; However, during the past decades, potentially more effective strategies for prevention and control of malaria in pregnancy have been developed and shown to have remarkable impact on the health of the mothers (Kuile et al., 2003). In Cameroon, the government via the public health centre protects pregnant women by giving bed nets to them (Plan Cameroon, 2005) with respect to WHO recommendations. The various reasons why some women were not sleeping under bed nets are: discomfort, sleepless, and difficulty in breathing. These days, antimalarial combination therapy is widely advocated and the use of artemesinin containing regimens is encouraged (WHO, 2000). In Thailand, when other treatments were failing, artesunate + mefloquine combination as first line treatment was successful and remains so many years later. In tropical Africa, WHO have pushed for artemether + lumefantrine and artesunate + amodiaquine combinations. Theoretically, artemisinin containing combinations might not only improve cure rates, but also reduce the speed at which resistance develop (Peter, 2004). In Cameroon, sulphadoxine-pyrimethamine (sp) is given at a therapeutic dose to pregnant women attending prenatal centre. It is a single antimalarial dose with the best overall effectiveness for prevention of malaria in pregnancy in areas with high transmission and low resistance to sulphadoxine-pyrimethamine (WHO, 2000). The combine preventive methods and prevention in our research showed that people using bed nets and prophylactic drugs or insecticides were less infected than those using just bed nets. The lowest rate of infection was found among pregnant women using drugs as prophylactic treatment. However, 8.88% (8/28) were found infected. Interview with some women reveal that not all were having theirs drugs taken. They were flinging away the drugs just after they had left the medical center. Here again, reason like vomiting, discomfort, and dizziness was given. Though, in recommended dosage, drug should be generally well tolerated. Women are either stubborn or simply ignorant of the risk they are taking. The survey showed that infection was found at the three stages of pregnancy which was rather unfortunate, because of the adverse effect of malaria in pregnancy. Pregnant women at their third trimester were more infected than those at their first semester. The low infection rate recorded among pregnant women at their first trimester may be explained by the fact that some women are receptive and excited during the first trimester of their pregnancy. It also shows that proper and reinforced information could protect pregnant women in a better way and at different period of the pregnancy. Women at their third trimester are more infected than others; this may explain the fact that prevention against mosquitoes for many women is temporary and after some time, fail to protect themselves; this is clearly explain by the infection rate recorded by pregnant women at their second trimester, who are gradually neglecting malaria's prevention and increasing with infection rate.

The use of combined methods of control is good, because those that were using three control methods were not as infected as those that used only two prevention methods. This shows that the more the numbers of the combined preventive methods used the more the confidence about freedom from malaria. According to WHO (2001), new regiment should be incorporated into strategies aimed at controlling the spread of the disease, such as insecticides, spraying, and use of impregnated bed nets, sustained chemoprophylaxis and intermittent prophylactic treatment. Since none of these is used routinely in most parts of Africa, effective chemotherapy for malaria has a pivotal role in reducing morbidity and mortality (Schultz et al., 1994). Malaria infection was related to the occupation; civil servants and students were less infected than house wives who were majority Moslems and farmer in partial purdah, most of these pregnant women being illiterates are quite ignorant about malaria transmission and treatment. The considerable number of infection observed among traders/business women could be due to the fact that they are always travelling and do not have enough time to protect themselves against mosquitoes bite.

Conclusion

A combination of up to three methods of control/prevention is most active against acquisition of disease. Malaria is commonly found in pregnant women though many are asymptomatic, because of earlier contacts and development of immunity.

This study has evidenced the abundance of infection among pregnant women. Pregnant women at their third trimester were more infected than others. Pregnant women living in high density settlement area were more infected than those of low and intermediate areas. This study has shown that the infections were related to occupations and also to the poor knowledge of malaria mode of transmission. The infection is related to the level of education and the lack of improved continual education. Pregnant women are not participating in the fight against malaria. Bed nets are effective for reducing malaria transmission and mortality. The usage of bed nets, insecticide and therapeutic drugs gave considerable protection and should be promoted. These findings show that determining local endemicity and the rate of clinical malaria cases are urgently required in order to target control activities and avoid over-treatment with antimalarial.

ACKNOWLEDGEMENTS

We thank the staff of all the district and regional hospitals of the North region of Cameroon for accepting the work to be done in their hospitals and for their technical assistance during sample and data collections, microscopy.

REFERENCES

D'alessandro U, Olaye B (1997). the Gambian National impregnated bed net programme: evaluation of effectiveness by means of case-control studies. Trans. Royal Soc. Trop. Med. Hyg. 91:638-642

Frederick M, Arinaitwe J (2003). Sleeping arrangements and mosquito net use among under five: results from the Uganda demogr. Health Surv. Malar. J. 2:40.

Hargreaves K, Koekemoer LL, Brooke BD, Hunt RH, Mthembu J, Coetzee M (2000). Anopheles funestus resistant to pyrethroid insecticides in South Africa. Med. Vet. Entomol. 14:181-189.

Kuile FOT, Terlouw DJ, Kariuki SK, Phillips-Howard PA, Mirel LB, Hawley WA, Friedman JF, Shi YP, Kolczak MS, Lal AA, Vulule JM, Nahlen BL (2003). Permethrin - treated bed nets reduce malaria in pregnancy in area of intense perennial malaria transmission in western Kenya. Am. J. Trop. Med. Hyg. 68(4 Suppl):50-60.

Menedez C, D'Alessandro U, ter Kuile F (2007). Reducing the burden of malaria in pregnancy by preventive strategies. Lancet infect. Dis. 7:126-135.

Monica C (1998). Examination of blood for malaria parasites. District Laboratory practice in tropical countries part 1:454 pp.

Mouhamadou C, Fabrice C, Josiane E (2006). Efficacy of bifenthrin-impregnated bednets against anopheles funestus and pyrethroid-resistant anopheles gambiae in North Cameroon. Malar. J. 5:77.

Nosten F, Terkuile F, Maelankiri L (1994). Mefloqine prophylaxis prevents malaria during pregnancy: a double-blind, placebo-controlled study. J. infect. Dis. 169:595-603.

Plan Cameroon (2005). Renforcement de la capacité des femmes à

reduire la morbidité liée au paludisme chez les enfants et les femmes enceintes du district de santé de Akonolinga. Projet de lutte contre le paludisme au cameroun 2005-2008.
Peter GK (2004). Antimalarial combinations. Lancet 364:285-294.
Schultz LJ, Steketee RW, Macheso A, Kazembe P, Chitsulo L, Wirima JJ (1994). The efficacy of antimalarial regimens containing Sulphadoxine -pyrimethamine and/or chloroquine in preventing peripheral and placental *Plasmodium falciparum* infection among pregnant women in Malawi. Am. J. Trop. Med. Hyg. 51(5):378-385.
Shr-jie W, Christian L, Thomas AS, Penelope V, Diallo AD, Xavier P, Natalie C, Mathieu K, Marcel T (2005). Rapid urban malaria appraisal epidemiology of urban. Malaria J. 4:43.
Van G, Thomas F (2004). The contribution of malaria in pregnancy to perinatal mortality. Am. J. Trop. Med. Hyg. 71:35-40
WHO (2000). Twentieth Report of the WHO Expert Committee on Malaria. Technical Report Series 892, Geneva. World Health Organization.
WHO (2001). WHO recommended strategies for prevention and control of communicable disease WHO/CDLS/CPE/SMT/ 13:107-110.
WHO (2008). World malaria report. Geneva Switzerland.
WHO (2011). World malaria report. WHO.int/malaria/world-malaria-report.

14

Two avian cestodes parasitic to *Corvus* species of Kashmir, India

Javid Ahmad Dar and Syed Tanveer

Department of Zoology, University of Kashmir, Srinagar-190 006, India.

Three species of birds belonging to the genus *Corvus* (*Corvus monedula*, *Corvus splendens* and *Corvus macrorhynchos*) were collected from nine different localities of Kashmir valley and investigated for the presence of helminthes. *Dilepis undula* (Schrank, 1788) was recovered from all the three host species. While, *Mayhewia kavini* Chishti and Khan, 1982 was recovered only from *C. monedula* and no specimen of this cestode was obtained from *C. splendens* and *C. macrorhynchos* during the present study. The specimens were identified and redescribed on the basis of various morphological and morphometric characters when compared to the known species of genera as *Dilepis* and *Mayhewia*, respectively. However, some intraspecific variations were observed when the present measurements were compared with those of previous authors. The prevalence, mean intensity and abundance of the parasites were determined.

Key words: Cestode, crows, *Corvus*, *Dilepis*, *Mayhewia,* prevalence, Kashmir.

INTRODUCTION

The present study was a part of helminthological investigation carried out on three *Corvus* species (*Corvus monedula*, *Corvus splendens* and *Corvus macrorhynchos*) of Kashmir from November, 2007 to May, 2009. *C. monedula* Linnaeus, 1758 (Jackdaw) is a black-plumaged passerine bird with distinctive white irises; sexes and ages are alike (Goodwin, 1983). It is omnivorous and feeds on plant material and invertebrates (Lockie, 1956). *C. splendens* Vieillot, 1817 (House crow) is about 40 cm in length with lighter grey-brown neck and breast. The wings, tail and legs are black. It appears to be associated with humans and no populations are known to exist independently of humans (Nyari et al., 2006). *C. macrorhynchos* Wagler, 1827 (Jungle crow) is a Large-billed crow with black glossy wings, tail, face and cause diseases include viruses, bacteria, protozoans and throat.The groups of parasites which infect birds and helminthes besides some arthropod ectoparasites. Sexually mature tapeworms live in the intestine or the diverticula of all classes of vertebrates (Roberts and Janovy, 2005). These parasites are found more frequently in warmer seasons, when intermediate hosts are abundant. Birds become infected by eating an intermediate host (infested with larval stage of a cestde), which may be an insect, crustacean, earthworm, slug, snail or leech depending upon the species of tapeworm (Calnek, 1997). Over 4,000 species of cestodes have been described from animals (Schmidt, 1986). Recently many workers have contributed to the knowledge of avian cestodes such as evaluation of genetic basis of host specificity of cestodes (Benesh, 2010; Henrich et al., 2013), description of ultrastructural characters of the spermiogenesis and mature spermatozoon of *Notopentorchis* sp. by means of transmission electron microscopy (Yoneva et al., 2012), description of some new cestode species like *Spiniglans sharpiloi* from *Pica pica* (Kornyushin et al. 2009), *Cotugnia orientalis* from *Gallus gallus domesticus* (Nanware et al., 2011).

Table 1. Prevalence, mean intensity and abundance of *Dilepis undula* in *Corvus* species of Kashmir.

Host spp.	NE	NI	NP	P%	MI	AB
C. monedula	30	6	18	20	3	0.6
C. splendens	23	3	11	13.04	3.66	o.47
C. macrorhynchos	12	3	11	25	3.66	0.92
Total	65	12	40	18.46	3.33	0.61

NE = number examined; NI = number infected; NP = number of parasites; P = prevalence; MI = mean intensity; Ab = Abundance.

Table 2. Prevalence, mean intensity and abundance of *Mayhewia kavini* in *Corvus* species of Kashmir.

Host spp.	NE	NI	NP	P%	MI	AB
C. monedula	30	4	10	13.3	2.5	0.33
C. splendens	23	0	0	0	0	0
C. macrorhynchos	12	0	0	0	0	0
Total	65	4	10	6.15	2.5	0.15

NE = number examined; NI = number infected; NP = number of parasites; P = prevalence; MI = mean intensity; Ab = abundance.

However, systematic knowledge of cestodes from birds of Kashmir including *Corvus*, is still represented through a few references as is obvious by tracing the historical review of cestodes from aves of Kashmir (Gupta, 1967; Fotedar et al., 1970; Chishti, 1973, 1980a, b, c; 1981; Fotedar and Chishti, 1973, 1974, 1976a, b; 1977; Chishti and Khan, 1978; 1979; 1982; Chishti et al., 1986). *Choanotaenia infundibulum*, *Choanotaenia micracantha*, *Anomotaenia galbulae* and *Mayhewia kavini* are the only cestodes reported from *Corvus* species in Kashmir so far (Foterdar and Chishti, 1974; 1976; Chishti and Khan, 1982; Chishti et al., 1986). Thus a thorough investigation of helminth parasitism of *Corvus* species of Kashmir was imperative. The present paper gives redescription of two cestodes (*Dilepis undula* and *M. kavini*) collected from the intestines of three species of crows (*C. monedula, C. splendens and C. macrorhynchos*) caught from different localities in Kashmir.

MATERIALS AND METHODS

Study area

Kashmir valley is a temperate, North-west Himalayan region of Jammu and Kashmir states in India. It lies between 33° 20' and 34° 54' N latitudes and 73° 55' and 75° 35' E longitudes, covering an area of about 15,948 km^2. It is a deep bowl shaped valley bounded by lofty mountains of the Pir Panjal and the great Himalayan ranges. The floristic and faunal diversity of the valley is considerably rich owing to its unique topography, temperate climate and geographical isolation from the surrounding plains (Dar et al., 2002). The birds were collected from different localities of Kashmir valley like Barzulla, Bugam, Chadoora, Khansahib, Naseembagh, Rajbagh, Rambagh, Wathora, and Yousmarg. These collection sites lie within the radius of five to fifty Kilometers (km) from the center of Srinagar city, the summer capital of Jammu and Kashmir States in India and are easily accessible by road transport.

Collection and processing of cestodes

During the present study, 65 birds belonging to three species of *Corvus* (*C. monedula* Linnaeus, 1758; *C. splendens* Vieillot, 1817 and *C. macrorhynchos* Wagler, 1827) were caught alive with the help of nylon net traps, locally known as "*Walwash*" using suitable baits. The hosts were slaughtered and dissected for parasitological investigation, and the cestode parasites thus collected were fixed in Carnoy's fixative, stained in acetoalum carmine and transferred to xylene for clearing before mounting them in dextrine plasticised xylene (DPX) (Meyer and Olsen, 1975). The drawings of the specimens were made with the help of prism type camera Lucida. Measurements were taken with objective and stage micrometers and expressed in mm (unless otherwise stated). The specimens were identified on the basis of various taxonomic characters using Yamaguti (1961), Schmidt (1986) and Chishti and Khan (1982). Photomicrography was conducted with the help of Digital Olympus Camera.

RESULTS

Two cestode species (*D. undula* and *Mayhewia kavini*) recovered from the intestines of three species of crows (*C. monedula, C. splendens and C. macrorhynchos*) are

Table 3. Comparative measurements of *Dilepis undula* (Schrank, 1788) with present form (measurements in microns = μ).

Characteristic	Davies (1935)	Mattrick (1958)	Chishti (1974)	Present author
No. of rostellar hooks	45-60	48-62	64-70	46-58
Size of outer rosterllar hooks (μ)	84	91-116	72-80	74-85
Size of inner rostellar hooks (μ)	72	70-88	60-64	67-78
No. of testes	28-35	28-36	24-26	24-32

redescribed based on their detailed morphological and morphometric studies, revealing some minor intraspecific variations. The prevalence, mean intensity and abundance of each parasite were recorded (Tables 1 and 2).

Dilepis undula (Schrank, 1788)

Hosts: *C. monedula*, *C. splendens* and *C. macrorhynchos*;
Locality: Naseembagh, Rajbagh, Wathora and Yousmarg;
Location: Intestine.

The following redescription is based on twelve cestode specimens, taking four specimens from each *Corvus* species. The cestodes measure 38 to 43 mm in length and 1.36 mm in maximum breadth. All the proglottids are broader and longer, and the posterior margin is broader than anterior margin. The mature proglottids measure 0.18 to 0.27 mm in length and 1.16 to 1.28 mm in breadth. The gravid proglottids measure 0.16 to 0.23 × 0.8 to 1.25 mm in length and 1.12 to 1.25 mm in breadth (Table 3 and Figure 1: A to E).

The scolex is globular and measures 0.44 to 0.54 mm in length and 0.60 to 0.64 mm in breadth across the suckers. The four suckers are rounded, muscular and measure 0.16 to 0.20 mm in diameter. The rostellum with its sac extends below the posterior margin of suckers and measures 0.44 to 0.58 mm in length and 0.14 to 0.20 mm in maximum breadth; it bears 46 to 58 hooks arranged in a double crown; each crown alternating with the other. The blade of hooks is smaller than handle, with guard as knob-like structure. The hooks of anterior and posterior crown measure 74 to 85 μ and 67 to 78 μ in length, respectively. The scolex is followed by a short neck that measures 0.24 to 0.33 mm in length and 0.50 to 0.55 mm in breadth. The dorsal longitudinal excretory duct is 14 μ in diameter and the ventral duct is 17 μ in diameter. The testes lie posterior to female reproductive organs, numbering 24 to 32 in each proglottid. They are rounded and measure 48 to 53 μ in diameter. The vas-deferens forms many coils in the anterior region of the proglottid. The vesicula seminalis are absent and the ductus ejaculatorius is straight. The cirrus pouch is cortical, extending up to the longitudinal excretory ducts of poral side; measuring 0.12 to 0.15 mm in length and 0.04 to 0.07 mm in width. The cirrus is slender and small 82 to 88 μ in length, armed with small spines. The genital pores are unilateral and present just near the anterior margin of proglottids, laterally.

The ovary is bilobed, with the poral lobe being smaller than the aporal one; the two lobes show further lobulations. The two lobes are connected in the middle by a narrow U-shaped isthmus. The ovary extends 0.65 to 0.72 mm across. The vitelline gland is compact, horse-shoe shaped and lies posterior to the ovary. It measures 0.12 to 0.17 × 0.04 to 0.06 mm. An oval receptaculum seminis, 0.08 to 0.10 × 0.03 to 0.05 mm in size, lies dorsal and anterior to the ovary. The uterus occupies the whole of the gravid proglottid filled with eggs, which measure 23 to 26 × 14 to 17 μ.

Mayhewia kavini (Chishti and Khan, 1982)

Hosts: *Corvus monedula*;
Locality: Barzulla, Naseembagh, Rambagh and Wathora;
Location: Intestine.

The following redescription is based on ten cestode specimens, collected from the intestines of four, out of thirty jackdaws, *C. monedula.* The cestodes measures 55 to 84 mm in length and 0.77 to 1.0 mm in their maximum breadth. The strobila consists of many proglottids which are broader than being long. The mature proglottids measure 0.20 to 0.26 mm in length and 0.8 to 0.92 mm in breadth. The gravid proglottids are 0.30 to 0.36 mm in length and 0.77 to 1.0 mm in breadth (Table 4 and Figure 2: A to D).

The scolex is small and globular, measuring 0.15 to 0.18 mm, with everted rostellum up to the base of suckers, and 0.12 to 0.14 mm without rostellum. It measures 0.22 to 0.23 mm across suckers. The rostellum with its sac measures 0.17 to 0.21 mm in length and extends below the posterior margin of suckers. The rostellum bears a single crown of 10 wrench-shaped hooks which measure 26 to 28 μ in length. The handle of hook is larger than the blade and is parallel to the equally long blade. The suckers are rounded and muscular, measuring 0.06 to 0.08 mm in diameter. The scolex is followed by a short neck that measures 0.17 to 0.22 mm in length and 0.10 to 0.15 mm in breadth. The dorsal longitudinal excretory duct is 10 to 13 μ and ventral excretory duct is 16 to 20 μ in diameter.

Testes are three in number, disposed in a triangle, two being aporal and one poral. They measure 92 to 100 μ in diameter. Cirrus puch extends into the medullary

Table 4. Comparative measurements of *Mayhewia kavini* Chishti and Khan, 1982 with present form. (measurements in mm unless stated otherwise).

Particulars	Chishti and Khan, 1982 *Mayhewia kavini*	Present author
Max. length	60-90	55-84
Max. breadth	0.72-0.785	0.77-1.0
Mature proglottids	0. 22-0.25 × 0.65-0.72	0.20-0.26 × 0.8-0.92
Gravid proglottids	0.26-0.35 × 0.72-0.785	0.3-0.36 × 0.77-1.0
Scolex length upto rostellum	0.19-0.198	0.15-0.18
Scolex length without rostellum	0.138-0.146	0.12-0.14
Width across suckers	0.242-0.249	0.22-0.23
Rosetellum length	0.222-0.228	0.17-0.21
Number of rostellar hooks	10	10
Hook size	25-30 µ	26-28 µ
Sucker size	0.09-0.1 × 0.075-0.082	0.06-0.08 (Dia.)
Dorsal excretory duct	10-12 µ	10-13 µ
Ventral excretory duct	18-22 µ	16-20 µ
Testis diameter	72-84 µ	93-100 µ
Cirrus pouch extent	Medullary	Medullary
Cirrus pouch size	0.14-0.155 × 0.052-0.057	0.16-0.19 × 0.050-0.053
External vesicula seminalic	0.16-0.18 × 0.06-0.072	0.18-0.25 × 0.07-0.072
Internal vesicula seminalis	0.06-0.072 × 0.048-0.052	0.05-0.065 × 0.42-0.50
Genital pore position	Anterior 1/3 rd.	Anterior 1/3 rd.
Ovary extend across	0.172-0.23 × 0.052-0.09	0.25-0.3 × 0.06-0.08
Vitelline gland	50-60 µ × 40-50 µ	65-72 µ × 48-53 µ
Receptaculum seminis	0.121-0.15 × 0.088-0.11	0.14-0.16 × 0.07-0.08
Egg size	42-54 µ × 30-41 µ	38-44 µ × 28-33 µ
Embryo size	28-32 µ × 18-22 µ	22-26 µ × 16-20 µ
Embryonic hook size	12-15 µ	9-12 µ

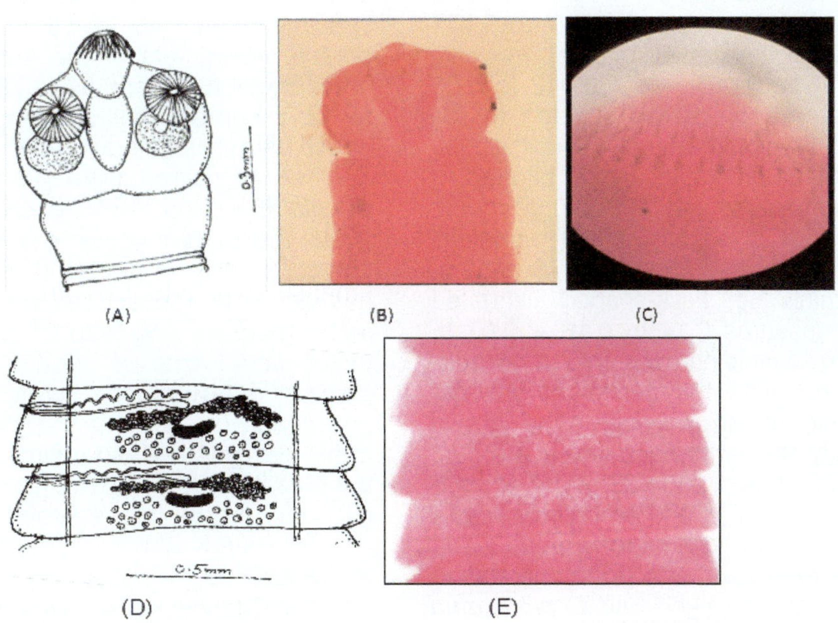

Figure 1. *Dilepis undula* Schrank, 1788. (A, B) Scolex showing suckers and rostellum with hooks; (C) rostellum showing hooks; (D, E) mature proglottids showing reproductive organs.

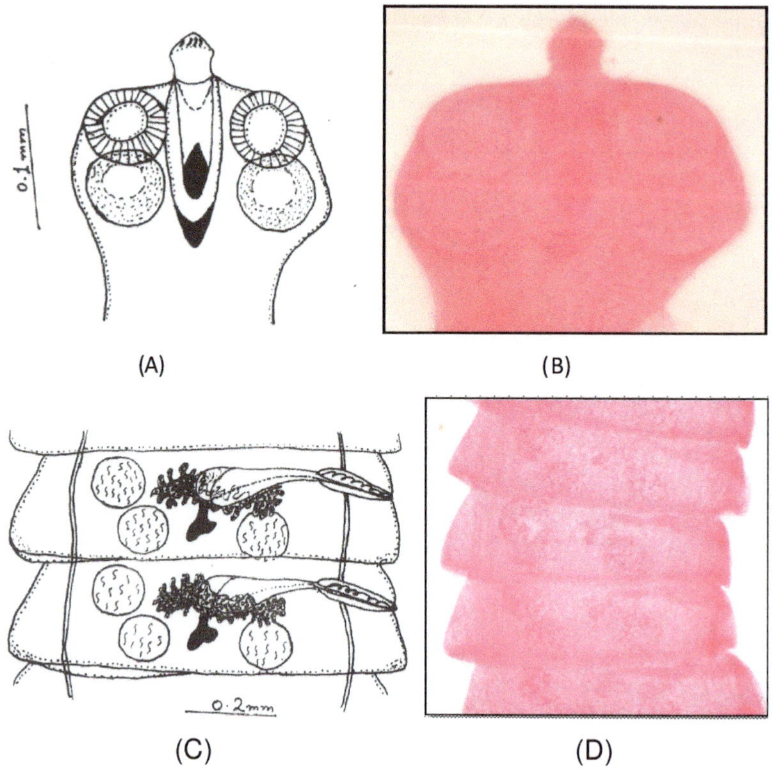

Figure 2. *Mayhewia Kavini* Chishti and Khan, 1982. (A, B) Scolex showing suckers and rostellum with hooks; (C, D) mature proglottids showing reproductive organs.

region and measures 0.16 to 0.19 mm in length and 0.05 to 0.053 mm in width. The external vesicula seminalis is large, extending to the middle of progrottids and measures 0.18 to 0.25 mm in length and 0.07 to 0.072 mm in width. The small and oval internal vesicula seminalis measures 0.05 to 0.065 mm × 0.042 to 0.05 mm. The genital pores are unilateral, opening in the anterior third margin of proglottid laterally. The follicular ovary lies in the middle of proglottid and is bilobed. It measures 0.25 to 0.3 mm across and 0.06 to 0.08 mm in thickness. Vitelline gland is slightly lobed and lies posterior to the ovary. It measures 65 to 72 × 48 to 53 μ. The receptaculum-seminis is flask-shaped, situated anterior to ovary and measures 0.14 to 0.16 × 0.07 to 0.08 mm. Receptaculum seminis narrows into a slender vaginal tube which opens into the genital atrium just posterior to the opening of cirrus pouch. The gravid proglottids are compactly filled with eggs which are oval in shape and measure 38 to 44 × 28 to 33 μ.The embryo measures 22 to 26 × 16 to 20 μ, and embryonic hooks measure 9 to 12 μ in length. In addition to *D. undula* and *M. kavini* redescribed in this paper, concurrent infestation with other helminth parasites such as *Echinostoma revolutum, Capillaria anatis, Anomotaenia galbulae* and *Choanotaenia micracantha* was also observed in some of the bird hosts during the present investigation.

DISCUSSION

The distinguishing characteristics of the present *Dilepis* specimens are: 46 to 58 rostellar hooks arranged in double crown; that of the anterior and posterior crown measure 74 to 85 and 67 to 78 μ, respectively; testes 24 to 32 in number, postovarian; cortical position of cirrus pouch; large bilobed ovary and unilateral genital pores.

When compared with the known species of genus *Dilepis* Weinland, 1858 described from different avian hosts, the present specimens shows a large similarity of characters with *D. undula* (Schrank, 1788). This species has also been redescribed several times, giving variation in the number and size of rostellar hooks, number of testes and length of strobila (Table 1). However, the present measurements come in the range of those described by various authors. In view of these measurements and other similarities as described above, the present cestode specimens are assigned to *D. undula* (Schrank, 1788). It is recorded for the first time in *Corvus* species from Kashmir. The earliest record of this species in Kashmir is from *Sturnus vulgaris humii* and *Turdus unicolor* (Chishti, 1974). The present cestode specimens of genus *Mayhewia* collected from *C. monedula* possess characteristics like rostellum with single crown of 10 small hooks; strobila comprising of numerous transversely elongated progrottids; two pair of excretory stems; three testes arranged in a triangle; cirrus pouch extending into medullary region; external and internal seminal vesicles present; ovary bilobed and uterus succular. All these characters are in conformity with the description of genus *Mayhewia* Yamuguti, 1956.

While comparing with the known species of genus *Mayhewia* Yamaguti, 1956, the present form shows a similarity of characters with those of *Mayhewia kavini* Chishti and Khan, 1982 as regards the number (10) and size of rostellar hooks, number of testes, extent of cirrus pouch, presence of seminal vesicles etc. However, some intraspecific variations were recorded in different structures as indicated in Table 2, the reason for which could be the age of parasite, host species, intensity of infection (higher intensity, smaller parasites), methodology (fixation), environmental factors of the study area, body conditions of the host, etc. (Ternopolskaya, 1984; Kuchai et al., 2012). Since these variations are of minor significance, therefore the present specimens in view of above similarities, are assigned to *Mayhewia kavini* Chishti and Khan, 1982.

Conclusion

The present paper redescribes only two species of cestode parasites (*D. undula* and *M. kavini*) recovered from three species of the genus *Corvus*. This work adds to the parasite species diversity infecting these avian hosts and may prove helpful for the future research on helminthes of birds. In addition, the present study showed that *M. kavini* was recovered only from *C. monedula*, hence the parasite may be host specific in nature.

ACKNOWLEDGEMENTS

The authors would like to thank all the members of the Department of Zoology, University of Kashmir who helped during the study. The authors express their deep sense of gratitude and sincere thanks to Prof. M. Z. Chishti, Professor Emeritus, Centre of Research for Development (CORD), University of Kashmir for helping in the identification of the cestode parasites.

REFERENCES

Benesh DP (2010). Developmental inflexibility of larval tapeworms in response to resource variation. Int J. Parasitol. 40(4):487-497.

Calnek BW (1997). Diseases of Poultry. 10th Ed. Iowa State University Press, Ames, Iowa 50014. pp. 850-851.

Chishti MZ (1973). On a new species of the cestode genus *Choanotaenia* Railliet, 1896 from *Acridotheres tristis* in Kashmir. J. Sci. Univ. Kashmir 1(1-2):51-54.

Chishti MZ (1980a). A new record of *Choanotaenia gondwana* Inamdar, 1934 (Choanotaenidae: Cestoda) from *Passer domestica* in Kashmir. Proceedings of the 5th All India Congress on Zoology, Bhopal, India.

Chishti MZ (1980b). *Dilepsis fotedari* n. sp. (Dilepididae Fuhrmann 1907: Cestoda) from *Anas Platyrhynchos* in Kashmir. Indian J. Helminthol. 32(1):1-3.

Chishti MZ (1980c). On the infection of Cestode genus *Sobolevicanthus* Spassky et Spasskeja, 1954 from Aves in Kashmir. Proceedings of the 3rd National Congress on Parasites, Hissar. pp. 13-14.

Chishti MZ (1981). On a new species of the genus *Choanotaenia* Railliet, 1826 from *Corvus monedula* in Kashmir. Proceedings of the 68th Session of the Indian Science Congress.

Chishti MZ, Khan AR (1978). Epidemology of Cestode infection in Snipe, *Tringa hypoleuca* from Kashmir, with description of a new species of the genus *Amoebotaenia*. Proceedings of the 65th Session of the Indian Science Congress. p 316.

Chishti MZ, Khan AR (1979). A new record of *Dilepsis undula* (Schrank, 1788) from some avian hosts in Kashmir. Proceedings of the 66th Session Indian Science Congress. p 107.

Chishti MZ, Khan AR (1982). *Mayhewia kavini* sp. nov. (Hymenolepididae Railliet et Henry, 1909: Cestoda) from *Corvus monedula* in Kashmir. Indian J. Helminthol. 34(2):139-142.

Chishti MZ, Mir AA, Rasool A (1986). *Choanotaenia micracantha* sp. nov. (Dilepoidea: Cestoda) from *Corvus monedula* in Kashmir. Indian J. Helminthol. 38(2):107-111.

Dar GH, Bhagat RC, Khan MA (2002). Biodiversity of the Kashmir Himalaya, 1st Ed. Valley Book House, Kashmir University Road, Srinagar-190006, India.

Davies TI (1935). The anatomy of *Dilepis undula* (Schrank, 1788). Proc. Zool. Soc. London. pp. 717-722.

Fotedar DN, Chishti MZ (1973). On a new species, *Anomotaenia kashmirensis* (Choanotaeniidae, Methevossian 1953) from *Sturnus vulgaris* in Kashmir. J. Sci. Univ. Kashmir 1(1-2):48-50.

Fotedar DN, Chishti MZ (1976). *Anomotaenia acrocephali* n. sp. and first record of *A. galbulae* (Gmelin, 1709) Fuhr; 1932 from some birds of Kashmir. Riv. Parassitol. 37(2):247-252.

Fotedar DN, Chishti MZ (1976). *Pseudoschistotaenia* n. gen. (Amabillidae, Fuhrman, 1908: Cestoda) from *Podiceps ruficollis capensis* in Kashmir. Proceedings of 63rd Session Indian Science Congress, Part III (Abt). p 21

Fotedar DN, Chishti MZ (1977). On a new species of the genus *Pseudoschistotaenia* Fotedar and Chishti, 1976. Proceedings, All Indian Symposim of Helmintology held at Srinagar (Abt). p 37.

Fotedar DN, Mahajan, R, Dhar RL, Chishti MZ (1970). New variety of *Raillietina* from common Blue Rock Pigeon in Kashmir. Kashmir Sci. 7(1-2):103-106.

Foterdar DN, Chishti MZ (1974). Redescription of *Choanotaenia oriole* Joyeux et Baer 1955 and *C. infundibulum* (Bloch, 1979) with a note on the synonymy of *C. dutii* Mukherjii 1964. J. Sci. Univ. Kashmir 2(1-2):73-78.

Goodwin D (1983). Crows of the World. Queensland University Press, St Lucia, Qld.

Gupta SP (1967). Helminthic-fauna of Kashmir. Kashmir Sci. 4(12):56-61.

Henrich T, Daniel PB, Martin K (2013). Hybridization between two cestode species and its consequences for intermediate host range. Parasit. Vectors 6:33.

Khan AR, Chishti MZ (1982). On *Echinostoma revolutum* (Froelich) Looss, 1899 and synonymy of *Neoechinostoma spinosa* Agarwal, 1963. Kashmir Univ. Res. J. 2:22-24.

Kuchai JA, Fayaz A, Chishti MZ, Tak H, Javid AD, Dar SA, Muzaffar R (2012). A study on morphology and morphometry of *Haemonchus contortus*. Pak. J. Zool. 44(6):1737-1741.

Kornyushin VV, Salamatin RV, Greben OB, Georgiev BB (2009). *Spiniglans sharpiloi* sp. n. (Cestoda, Dilepididae), a parasite of the common magpie, *Pica Pica,* in the palaearctic. Vestn. zool. 23:85-93.

Lockie JD (1956). The Food and Feeding Behavior of the Jackdaw, Rook and Carrion Crow. J. Anim. Ecol. 25(2):421-428.

Meyer CM, Olsen WO (1975). Essentials of Parasitology. WM. C. Brown Company Publishers, Dubuque, Iowa (USA).

Mettrick DF (1958). Helminth parasites of Hertfordshire Birds-II Cestoda. J. Helminthol. 32:157-194.

Nanware SS, Dhondge RM, Bhure DB (2011). Biosystematic studies on *Cotugnia orientalis* sp. nov. (Cestoda: Davaineidae, Fuhrmann 1907) from *Gallus gallus domesticus*. Bioscan (Ranchi) 6(1):71-75.

Nyari A, Ryall C, Peterson AT (2006). Global invasive potential of the house crow *Corvus splendens* based on ecological niche modeling. J. Avian Biol. 37:306-311.

Roberts LS, Janovy JJ (2005). Gerald D. Schmidt and Larry S. Roberts' Foundation of Parasitology. 7th Ed., Mc Graw-Hill Companies. p 311.
Schmidt GD (1986). Handbook of Tapeworm Identification. CRC Press, Boca Raton, FL.
Ternopolskaya LD (1984). Variability of *Fasciola hepatica* L., 1785 in different hosts. Bulletin Vsesoyuznogo Instituta Gel. Mintologil Im. K. I. Skrybin. 38: 47-51.
Yamaguti S (1961). Systema Helminthum. Vol. 2. The Cestodes of Vertebrates. Interscience Publisher, John Wiley Sons, New York.
Yoneva A, Céline L, Pavel NN, Yana M, Jean M, Boyko BG (2012). Spermiogenesis and spermatozoon ultrastructure of the paruterinid cestode Notopentorchis sp. (Cyclophyllidea). Parasitol. Res. 111(1):135-142.

15

Factors associated with ascariasis among pre-school aged children in Ile-Ife, Osun state, Nigeria

Ogunkanbi, A. E. and Sowemimo, O. A.

Department of Zoology, Obafemi Awolowo University, Ile – Ife, Osun State, Nigeria.

The risk factors predisposing children to ascariasis transmission in a peri-urban community of Osun state, Nigeria were investigated from November, 2012 to June, 2013. Preschool children aged 24 to 60 months were tested by faecal examination for *Ascaris lumbricoides* infection using Kato thick smears after information on personal biodata, access to water supply, sanitation, socio-economic status of their parents, hygiene and deworming history was collected using a questionnaire. Out of 416 children examined, 117(28.1%) were infected with an overall mean intensity of 1846.33±235.23 eggs per gram (epg). Infection patterns were gender comparable and age dependent with peak prevalence (33.1%) and mean intensity (2264.14±493.58 epg) occurring in children aged 48 to 59 months. Parental educational background, sanitation, washing hands with soap and water and deworming history were identified as the significant risk factors which correlated with *Ascaris* infection. These findings suggest that socio-economic risk factors which play a role in disease transmission need to be taken into account when formulating sustainable control strategies for ascariasis and other soil-transmitted helminths in Nigeria.

Key words: Prevalence, ascariasis, risk factors, preschool aged children, Nigeria.

INTRODUCTION

Faecal contamination is one of the most environmental health problems in poor countries, where an estimated three million children die of enteric diseases annually and even more suffer from debilitating diseases due to soil-transmitted helminths (STHs) (Carneiro et al., 2002). Ascariasis is one of the important soil-transmitted diseases with enormous health and social implications for pre-school and school aged children in impoverished and endemic communities of tropical Africa, Asia and South America. Children are infected after ingesting emryonated eggs from contaminated soil or food. Although, the infection is often asymptomatic, its effect may contribute substantially to child mortality when associated with malnutrition, pneumonia, enteric diseases and vitamin A deficiency (Montressor et al., 2003). In developing countries the burden of helminthic diseases had been reported to be influenced by socio-economic status, poor environmental sanitation, lack of personal hygiene and use of unsafe water (Soares Magalhaes et al., 2011). Although, recent control strategies of targeted chemotherapy (that is, periodic administration of antihelminthic drugs to vulnerable groups) have achieved significant reduction in morbidity, they have failed to significantly break transmission in areas where socio-environmental risk factors have continued to play vital roles in the daily life of vulnerable population. Crompton (1989) reported that there is a direct relationship between some risk factors (such as water supply, sanitation and poverty) and patterns of ascariasis transmission in most endemic communities.

Other studies have reported the likely role of the age of the child, father's occupation, dog ownership, (Kirwan et al., 2009), defecation practices and mother's occupation (Nmorsi et al., 2009). In spite of the health risks associated with soil-transmitted helminthiasis, there is still paucity of epidemiological data regarding the endemicity of soil-transmitted helminth infection among pre-school children who are also vulnerable. Therefore, the present study was aimed to investigate the effect of water, sanitation, hygiene and other socio-economic parameters on the prevalence of ascariasis in pre-school children aged 24 to 60 months of age in Ile-Ife, Osun state, Nigeria. It is hoped that our findings will contribute to sustainable ascariasis control in this and other endemic communities in Nigeria and other countries with similar environmental conditions.

MATERIALS AND METHODS

Study area and population

The study was conducted in peri-urban community, Ile-Ife in Ife Central Local Government area of Osun state, Nigeria located between latitudes 7°26´´N and 7°35´´N and longitudes 4°24´´E and 4°39´´E. The climate is typically tropical with well defined wet (April to October) and dry (November to March) seasons with mean annual precipitation of 1220mm and mean annual temperature of 27°C (Ayoade, 1982). The estimated population of people in the peri-urban city according to 2006 census is 167,254. The inhabitants are a mixture of people from different ethnic groups in Nigeria, although the majority are the Yoruba speaking people of the South-west. They are mainly peasant farmers growing cocoa, vegetables, maize and cassava. Traders, civil servants, artisan workers and transport workers are also found in smaller numbers. The three major sources of water supply are tap water, shallow wells and streams. Toilet facilities included water closets, pit latrines or indiscriminate defecation in the bush.

Study design

A cross-sectional study was conducted among randomly selected pre-school aged children from Ife Central Local Government Area (LGA) between November, 2012 and June, 2013. Nine nursery and primary schools comprising four government-owned and five privately-owned schools were selected randomly from a total of 62 schools in Ife Central LGA. Introductory meetings were held in each selected school with the school authority where the purpose of the study was explained. Thereafter, the school authority called a Parents Teachers Association (PTA) meeting where the parents/guardians of the pre-school children aged 2 to 5 years were informed on the purpose of the study and their co-operation sought for the success of the study. The parents/guardians were informed that only children whose parents returned the signed fingerprint consent forms will be allowed to participate in the study. Prior to the commencement of the study, permission was sought and obtained from the Local Educational Authority of the Ife Central Local Government Area. Ethical clearance for the study was given by the Ethical Committee of the Obafemi Awolowo University Teaching Hospital Complex (OAUTHC) Ile-Ife, Osun State, Nigeria. The inclusion criteria for the study were (1) children aged 24 to 60 months; (2) parents or guardians gave written informed consent; (3) children lived in the study area.

Sample size determination and sampling techniques

The sample size for this study was calculated by using proportion formula at 95% confidence interval (CI) level (Z (1-½α) = 1.96), an expected prevalence of 10% study has been conducted before on this topic in the area of study and 5% marginal error. Then the sample size was calculated as n = (Z . 1- ½α) 2P (1 – P)/d², where n = sample size, P = proportion problem in the study area, Z 1- ½α = CI of 95%, d = Marginal error to be tolerated by adding 10% of contingency. This led to a total sample size of 416 children included in our study. The sample size which was determined earlier was used to calculate the proportion of pupil to be selected from each school. Using registration list, simple random sampling was employed to select pupils from each school using a table of random numbers.

Data and faecal sample collection

A pre-tested questionnaire was sent to the parents/guardian to be filled out to provide information on personal biodata (name, age, sex) of each child, socioeconomic status of parent (Educational background and Occupation) access to different facilities for water supply (boreholes, well and stream), sanitation (pit latrines, water closet, bush), hygiene (washing hands before eating and after using the toilet, washing hands with soap and water, washing hands with water only), deworming history (whether the child had taken anthelminthic medication within the last 12 months prior to faecal collection). A total of 416 children were randomly selected from a total of 1080 children in the list of the nine selected schools. Labelled plastic bottles (containing the name, age and sex of the selected children) with a clean wooden spatula for the collection of faeces were distributed to the children for a small portion of their early morning faeces. Faecal samples were collected from each selected child on the same day or within 24hrs of being passed and examined for presence and number of *A. lumbricoides* eggs.

Parasitological procedures

Each faecal sample was collected in a clean 30mL universal plastic bottle and stored in the laboratory refrigerator at 4°C. The Kato thick smear method (WHO, 1994) was used for the analysis of the presence and number of *A. lumbricoides* egg. Samples found to be positive for the *A. lumbricoides* eggs underwent egg burden counts to determine eggs per gram (epg) of faeces using standard WHO protocol (WHO, 2006). Those pupils who were positive for *A. lumbricoides* eggs were treated accordingly by appropriate antihelminthic drugs. Table 1.

Statistical analyses

SPSS version 17.0 (SPSS Inc., Chicago, Illinois, USA) was used for statistical analysis. The differences in the prevalence of ascariasis

Table 1. Prevalence and Intensity of ascariasis in preschool-aged children in relation to age, Gender, parental characteristic, water supply, sanitation, hygiene and deworming history in Ile-Ife, Osun State, Nigeria.

Variable	No. examined	No. infected	Prevalence (%)	Intensity (eggs per gram mean±SEM)
Age (months)	-	-	-	-
23-35	56	10	17.9	1459.71±595.48
36-47	103	32	31.1	1362.17±264.79
48-59	118	39	33.1	2264.14±493.58
≥60	139	36	25.9	2006.16±473.66
P-value	-	> 0.05	0.84	-
Gender	-	-	-	-
Male	219	64	29.2	2010.63±344.64
Female	197	53	26.9	1663.68±316.55
P-value	-	>0.05	0.00	-
Mother's educational background	-	-	-	-
No formal education	63	31	49.2	3284.59±816.08
Primary	63	29	46.0	3561.51±834.67
Secondary	116	40	34.5	2158.35±455.86
Tertiary	174	17	9.8	496.55±164.41
P-value	-	<0.05	0.00	-
Father's educational background	-	-	-	-
No formal education	45	23	51.1	3116.82±841.71
Primary	55	27	49.1	3248.71±927.53
Secondary	121	42	34.7	2663.60±522.26
Tertiary	195	25	12.8	650.46±170.02
P-value	-	<0.05	0.00	-
Sanitation	-	-	-	-
Bush	43	30	69.8	4653.21±1107.66
Latrine	150	55	36.7	2528.96±466.39
Water closet	223	32	14.3	845.91±192.95
P-value	-	<0.05	0.00	-
Water supply	-	-	-	-
Borehole	94	10	10.6	505.02±242.93
Well	309	95	30.7	2131.88±294.97
Stream	13	12	92.3	4747.54±1700.69
P-value	-	<0.05	0.00	-
Hygiene	-	-	-	-
Washing hands after toilet	-	-	-	-
Yes	305	51	16.7	982.58±189.76
No	111	66	59.5	4219.68±662.60
P-value	-	<0.05	0.00	-
Washing hands with soap and water	-	-	-	-
Yes	194	15	7.7	515.51±185.63
No	222	102	45.9	3009.30±394.06
Washing hands with water only				
Yes	111	36	32.4	1798.91±397.83
No	305	81	26.6	1863.58±286.67
P-value	-	>0.05	0.238	-

Table 1. Contd.

Deworming history	-	-	-	-
<3 months	78	2	2.6	2.46±1.94
6 month	26	1	3.8	76.62±76.62
1 year	96	10	10.4	547.25±220.93
Never dewormed	216	104	48.1	3302.56±418.58
P-value	-	<0.05	0.00	-

EPG: eggs per gram; SEM: standard error of mean.

between groups were determined using chi-squared (χ^2) tests. Mann-Whitney U test was used to test for differences in the intensity of infection between dichotomous groups and Kruskal-Wallis tests for groups with more than two levels. Multivariate logistic regression was further applied to assess the predictive effect of the various variables measured on the prevalence of ascariasis in the schools. P- value less than 0.05 were taken as statistical significant association dependent and independent variables.

RESULTS

Overall infection patterns

The cross-sectional study showed that out of the 416 children examined, 117(28.1%) were positive for *A. lumbricoides* eggs at an overall mean intensity of 1846.33 ± 235.23 eggs per gram (epg). In terms of distribution of infection among sub-groups, the prevalence increased from youngest age group, 24 to 35 months (17.9%) to a peak (33.1%) in the 48 to 59 months age group, while the intensity increased from 1362.17 ± 264.79 epg in children aged 36 to 47 months until it peaked at 2264.14±493.58 epg in children aged 48 to 59 month.. Neither prevalence nor intensity was significantly gender-dependent though each was higher in males than females. Prevalence and intensity of *Ascaris* infection were significantly lower among children with both mother (9.8%; 496.55±164.41 epg) and father (12.8%; 650.46±170.02 epg) having tertiary education than mother (34.5 to 49.2%; 2158.35 to 3561.51 epg) and father (34.7 to 51.1%; 2663.60 to 3248.71 epg) having at least a primary education or no formal education. Prevalence and intensity were significantly lower among children with access to boreholes (10.6%; 505.02±242.93 epg) and water closets (14.3%; 845.91±192.95 epg) than their counterparts utilizing other sources of water supply (30.7 to 92.3%; 2131.88 to 4747.54 epg) and/or lacking access to any toilet facility (36.7 to 69.8%; 2528.96 to 4653.21 epg). Both prevalence and intensity were significantly lower among children who wash their hands after using the toilet or before eating (16.7%; 982.58±189.76) compared to those that do not wash their hands (59.5%; 4219.68±662.60). Considering hand washing with water only and hand washing with soap and water, it was found that both prevalence and intensity were significantly lower in children who reported washing hands with soap and water (7.7% 515.51±185.63) than those who reported negatively (45.9%; 3009.30±394.06), while infection pattern was independent on whether children reported washing hands with water only or not.

Logistic regression analysis

Table 2 summarised the results of logistic analysis carried out to determine the relative effect of explanatory variables on the prevalence of ascariasis in the study area. After adjusting for the effect of individual variables, ascariasis can be explained by five variables which are considered to be the risk factors. These are parent's educational background, sanitation, hand washing with soap and water and deworming history. Risk of ascariasis decreases from children whose fathers had no formal education to a minimum among those whose fathers had secondary education. The odds of harbouring *Ascaris* for children whose mothers had a secondary education were 4.2 times than children whose mothers had a tertiary education. The risk of *Ascaris* infection is 1.7 times higher among children that used the bushes indiscriminately than those who had access to water closet. The risk of infection is lower among children that washed their hands with water and soap than those who did not. Children who had received anthelminthic treatment prior to faecal collections were less likely to be infected than those who had never received treatment.

DISCUSSION

The present study has revealed that an overall prevalence of 28.1% for *A. lumbricoides* infection was obtained among the pre-school aged children investigated in Ile-Ife, Osun State, Nigeria. The prevalence value obtained is higher than the national average of 18.6% (Crompton, 1989). The prevalence of infection is higher than prevalence of 13.1% reported among the same age group in the same region over a decade ago (Asaolu et al., 2002). The possible reason for this might be due to re-infection resulting from unhygienic habit of this age

Table 2. Logistic Regression model of the risk factors for *Ascaris lumbricoides* among the preschool-aged children.

Risk factor	Number examined	Prevalence (%)	Odds ratio (95% CI)	P-value
Fathers' educational level	-	-	-	0.035*
No formal education	45	51.1	16.687(1.733-160.682)	0.015
Primary	55	49.1	6.136 (1.041-36.158)	0.045
Secondary	121	34.7	0.941(0.311-2.845)	0.914
Tertiary[a]	195	12.8	-	-
Mothers' educational level	-	-	-	0.003*
No formal education	63	49.2	0.211(0.025-1.760)	0.151
Primary	63	46.0	1.881(0.433-8.174)	0.399
Secondary	116	34.5	4.270(1.339-13.614)	0.014
Tertiary	174	9.8	-	-
Sanitation	-	--	-	0.029*
Bush	43	69.8	1.691 (0.374-7.658)	0.495
Latrine	150	36.7	0.421 (0.151-1.174)	0.495
Water closet[a]	223	14.3	-	-
Water supply	-	-	-	0.204
Borehole	94	10.6	0.077 (0.005-1.291)	0.075
Well	309	30.7	0.112 (0.008-1.501)	0.098
Stream[a]	13	92.3	-	-
Hand-washing with water only	-	-	-	-
Yes	111	32.4	0.449 (0.171-1.178)	0.104
No[a]	305	26.6	-	-
Hand-washing with soap and water	-	-	-	-
Yes	194	7.7	0.107 (0.034-0.338)	0.000*
No[a]	222	45.9	-	-
Time since last anthelminthic treatment	-	-	-	0.000*
<3 months	58	2.6	0.011 (0.002-0.068)	0.000
6 months	42	3.8	0.008 (0.001-0.123)	0.001
1 year	40	10.4	0.153 (0.061-0.383)	0.000
Never[a]	172	48.1	-	-

[a]=Reference category
95% CL, 95% confidence limit

group playing with contaminated soil and irregular deworming exercise. The prevalence of *A. lumbricoides* infection recorded in this study is significantly lower than the prevalence of 85.7% reported from preschool aged children in Ibilo, Akoko-Edo LGA, Edo State, Nigeria (Nmorsi et al. 2009), and higher than prevalence of 12.5% reported from the same age group in three villages in Ife North LGA, Osun State (Kirwan et al., 2009). However, it is comparable to a prevalence of 30.9% reported among the same age group of children in selected Local Government units in the Philippines (Belizario et al., 2013). Both the prevalence and intensity of *A. lumbricoides* infection were higher in males than in females, although the gender of the children was not found to be significantly associated with infection. Since there was no gender significant difference in gender related prevalence pattern, the propensity to ingest contaminated soil is similar in the two sexes (Culha et al., 2007, Faiza et al., 2009). Our findings also showed a direct relationship between patterns of infection and sanitation.

The risk of *A. lumbricoides* infection is 1.7 times higher among children that practise indiscriminate defecation in the bush than those that have access to water closet, while those who used latrines are 0.4 times less likely to be infected. Though these findings may be explained by the varying degrees of environmental contamination associated with the various facilities for sanitation, it is also conceivable that where access to improved facilities for sanitation is a direct indicator of socio-economic stratification, differences in socio-economic status will play vital roles in explaining observed patterns of infection. For instance, it was observed in this study that there is a link between access to improved facilities and improved educational background. Thus, though it is well established that facilities like bucket latrines and defecation in the bush spread faecal matter in the environment, it is inconceivable that people are infected just because they use such facilities. Explaining the observed infection patterns should therefore also take into consideration earlier reports suggesting that the effect of any facility to reduce environmental contamination and disease transmission depends not just on access but also on educational background, awareness and ability to use improved facilities well. This conclusion is strongly supported by our findings and that of Quihui et al. (2006) which found that though both prevalence and intensity decreased significantly in children with access to water closets, the infection pattern was progressively better as parents' educational background improved.

Conclusion

This study also found a strong association between infection pattern in children and hand washing with soap and water. Children who washed their hands with water and soap were less likely to be infected than those who did not. Hand washing with soap might remove the parasite eggs, reduce the egg load and because of the adhesive nature of the eggs washing hands with water only may not totally remove the *A. lumbricoides* eggs. This finding corroborates the observation of Awasthi et al. (2008) who reported that the exclusive use of soap and water after defecation was protective against geohelminth infection. Mumtaz et al. (2009) also reported that majority of the infected pre-school aged children at the tertiary care hospital in Karachi did not wash their hands with soap and water regularly. Our study found a strong association between deworming history and the prevalence of *A. lumbricoides* infection. The prevalence of infection was lowest among children who dewormed three months before they were examined while the highest was recorded among children who had never received any antihelminthic treatment.

The multivariate analysis also showed that children who received antihelminthic treatment prior to the study were less likely to be infected than those who had never received any anthelminthic treatment. Similar result had been reported among primary school children in Ebenebe town, Anambra state, Nigeria that children who had never taken antihelminthic drugs in their life had high prevalence of geohelminth infection (Chukwuma et al., 2009). Damen et al. (2010) reported a lower prevalence of helminth infection among children who had received anthelminthic drugs prior to the time their faecal samples were collected while a high prevalence of infection was recorded among children who did not receive antihelminthic medication before their stools were collected for analysis. This study has been able to identify risk factors which facilitate the transmission of *Ascaris* infection and the findings from this study will help in planning a sustainable ascariasis control programme in the area of study. It is worthy of note that among children who had received antihelminthic treatment prior to the study, some were still infected with ascariasis. This could be attributed to the fact that the antihelminthic used might not be effective and could also be due to reinfection because of the contaminated environment.

ACKNOWLEDGEMENTS

We thank the parents of the children examined at Ile-Ife for their assistance in the collection of faecal samples from their children.

Conflict Interests

The authors declare that there is no conflict of interests regarding the publication of this article.

REFERENCES

Asaolu SO, Ofoezie IE, Odumuyiwa PA, Sowemimo OA, Ogunniyi TAB (2002). Effect of water supply and santation on the prevalence and intensity of *Ascaris lumbricoides* among preschool-aged children in Ajebandele and Ifewara, Osun State, Nigeria. Trans. Roy. Soc. Trop. Med. Hyg. 96:600-604.

Awasthi S, Verma T, Kotecha PV, Venkatesh V, Joshi V, Roy S (2008). Prevalence and risk factors associated with worm infestation in preschool-aged children (6-23 months) in selected blocks of Uttar Pradesh and Jharkand India. India J. Med. Sci. 62(12):484-491.

Ayoade SS (1982). Climate change .I. ITD movements and winds. In: Nigeria in maps, Barbour KM, Oguntoyinbo JS, Onyemelukwe JOC, Nwafor JC (editors). London: hodder and Stoughton. pp.14-15.

Belizario VY, Totanes FIG, De Leon WU, Ciro RNT, Lumampao YF (2013). Sentinel Surveillance of Soil-Transmitted Helminthiasis in preschool-aged and school-aged children in selected Local Government Units in the Philippines: Followup Assessment. Asia-Pacific J. Public Health 20(10):1-12.

Carneiro FF, Cifuentes E, Tellez-Rofo MM, Romieu I (2002). The risk of *Ascaris lumbricoides* infection in children as environmental health indicator to guide preventive activities in Caparao and Alto caparao, Brazil: Bull. World Health Organ. 80:40-46.

Chukwuma MC, Ekejindu IM, Agbakoba NR, Ezeagwuna DA, Anaghalu IC, Nwosu DC (2009). The prevalence and risk factors of geohelminth infections among primary school in Ebenebe Town, Anambra state, Nigeria. Middle-East J. Sci. Res. 4(3):211-215.

Crompton DWT (1989). Prevalence of *ascariasis*. In: Crompton DWT, Nesheim MC, Pawlowski ZS (eds.), *Ascariasis* and its prevention

and control. Taylor and Francis, London. pp. 45-69.
Culha G, Kemal M, Ozer C (2007). The distribution of intestinal parasites among Turkish children living in rural areas. Middle East J. Fam. Med. 3(1):20-25.
Damen JG, Lav P, Mershak P, Mbaawuga EM, Nyavy BW (2010). A comparative study on the prevalence of intestinal helminths in dewormed and non-dewormed students in a rural area of North-Central Nigeria. Lab. Med. 41(10):585-589.
Faiza AA, Rasha AS, Nabila ST (2009). Predictors of the intestinal parasitic infection among pre-school children in rural lower Egypt. Egypt J. Comm. Med. 27(1):17-34.
Kirwan P, Asaolu SO, Abiona TC, Jackson AL, Smith HV, Holland CV (2009). Soil transmitted helminth infections in Nigeria children aged 0-25 months. J. Helminthol. 83:261-266.
Montressor A, Awasthi S, Crompton DWT (2003). Use of benzimidazoles in children younger than 24 months for the treatment of soil-transmitted helminthiasis. Acta Trop. 86:223-232.
Mumtaz S, Siddiqui H, Ashfaq T (2009). Frequency and risk factors for intestinal parasitic infection in children under five years at a tertiary care hospital in Karachi. J. Pak. Med. Assoc. 59(4):216-219.
Nmorsi OPG, Isaac C, Aashikpelokhai JS, Ukwandu NCD (2009). Geohelminthiasis among Nigeria preschool age children. Int. J. Med. Med. Sci. 1(10):407-411.
Quihui L, Valencia ME, Crompton DWT, Phillips S, Hagan P, Morales G, Diaz-camacho SP (2006). Role of the employment status and education of mothers in the prevalence of intestinal parasitic infections in Mexican rural school children. BMC Public Health 6:225.
Soares Magalhaes RJ, Barnett AG, Clements ACA (2011). Geographical analysis of the role of water supply and sanitation in the risk of helminth infections of children in West Africa. Proc. Nat. Acad. Sci. 108(50):20085-20089.
World Health Organization (WHO) (1994). Bench aids for the diagnosis of intestinal parasites: laboratory manual. Geneva: World Health Organization. 14pp.
World Health Organization (WHO) (2006). Schistosomiasis and Soil-transmitted helminth infections: preliminary estimates of the number of children treated with albendazole or mebendazole. Wkly. Epidemiol. Rec. 16:145-164.

Entomological baseline data on malaria transmission and susceptibility of *Anopheles gambiae* to insecticides in preparation for Indoor Residual Spraying (IRS) in Atacora, (Benin)

Rock Aïkpon[1,2] , Razaki Ossè[1,2], Renaud Govoetchan[1,2], Arthur Sovi[1,2], Frédéric Oké-Agbo[1] and Martin C. Akogbéto[1,2]

[1]Centre de Recherche Entomologique de Cotonou (CREC), Cotonou, Benin.
[2]Faculte des Sciences et Techniques de l'Université d'Abomey Calavi, Benin.

To implement indoor residual spraying (IRS), the department of Atacora was selected in Benin. Entomological surveys were performed before IRS implementation. Mosquitoes were sampled by Pyrethrum spray catch and were identified morphologically and by molecular methods. The *Plasmodium falciparum* circumsporozoite indices were measured by enzyme linked immunosorbent assay (ELISA). Molecular detection of pyrethroid knock down resistance and that of insensitive acetylcholinesterase were performed. Susceptibility status of *Anopheles gambiae* was determined using World Health Organization (WHO) bioassay tests to various insecticides. *A. gambiae s.l.* was the main species harvested in houses (81.71%) and *A. gambiae s.s* is practically the only member that was found. Both M and S forms were in sympatry, but the molecular S form was predominant (94.42%). *A. gambiae s.l* were susceptible to bendiocarb but fully resistant to organochlorine (DDT), permethrin and deltamethrin. Entomological inoculation rate vectors (EIR) was 6 infectious bites per man per month on average during the study period. The average of *kdr* and *Ace-1* allelic frequency were 78 and 3%, respectively. *A. gambiae s.l* is characterized by a high endophilic behavior in Atacora, which is a good criterion for IRS implementation. The susceptibility to bendiocarb add to the low *Ace-1* mutation frequency found in *A. gambiae* populations could lead to the use of bendiocarb for IRS.

Key words: Entomological baseline data, IRS implementation, *Anopheles gambiae,* Atacora, Bénin.

INTRODUCTION

Malaria remains a major cause of morbidity and mortality in sub-Saharan Africa and represents one of the most critical public health challenges for Africa. More than two billion people around the world, particularly people living in South America, south-eastern Asia and sub-Saharan Africa, are at risk of contracting malaria. Besides, one million deaths are recorded yearly of which, 91% occur in sub-Saharan Africa (WHO, 2011). In Benin, in 2011, malaria was responsible for more than 1,753 deaths (MS, 2011). However, it incidence in Atacora in 2011 was

18.5%, which is higher than the national average (15.7%). The main malaria parasite was *Plasmodium falciparum* and was mainly transmitted by *Anopheles gambiae* s.*l*. This disease representing the primary cause of mortality and morbidity in health centres caused an enormous burden to health and economy in developing countries (WHO, 2011). Insecticide Treated Nets (ITNs) use represent the main approaches of malaria control (WHO, 2009). Household ITNs ownership reached more than 50% in several high burden African countries (WHO, 2009). Pyrethroids are the only insecticides used for net impregnation because of their strong efficacy, their fast acting effect at low doses and their low toxicity for mammals (Zaim et al., 2000). Unfortunately, resistance to pyrethroids in malaria vectors has spread across Africa and is now present in most of the countries where national malaria control programmes (NMCP) are implementing large scale distribution of long lasting insecticidal nets to populations at risk, that is children under five and pregnant women (Santolamazza et al., 2008).

There are numerous reports of resistance to pyrethroid throughout Africa (Elissa et al., 1993; Vulule et al., 1999; Koekemoer et al., 2002; Etang et al., 2006; Abdalla et al., 2008; Nwane et al., 2009; Bigoga et al., 2012). In Benin, the resistance of malaria vectors to pyrethroids observed first in Cotonou spread not only to central and southern regions of the country, but also in the northern localities (Akogbéto et al., 1999; Chandre et al., 1999; Corbel et al., 2007; Yadouleton et al., 2010).

The high level of resistance to pyrethroid in *A. gambiae* in Africa, particularly in Benin (Corbel et al., 2007; Yadouleton et al., 2010) is one of the major challenges being faced by malaria vector control programmes. It is then urgent to find ways to manage this resistance. In this context, an alternative insecticide to pyrethroids in IRS strategy supported by the President Malaria Initiative launched since 2008 in 14 African countries including Benin could be on the way to resolve the problem of resistance to pyrethroids in *A. gambiae*. After experimental hut evaluation (phase II), the non-pyrethroid that is effective seems to be bendiocarb (carbamate) (Akogbéto et al., 2010).

A study conducted in Ouémé, in southern Benin showed that the first and second rounds of IRS using bendiocarb were successful with a drastic decrease in malaria transmission in areas under IRS (Akogbéto et al., 2011). In spite of this success, NMCP decided to move IRS from Ouémé to Atacora (northern Benin). As matter of fact, Ouémé is characterized by two periods of transmission which last all the year. Implementation of IRS in such area needs two rounds of intervention. Contrary to Ouémé is characterized by a single transmission period. This means that just one spray round per year would be sufficient to cover the period of transmission. However, no entomological data from Atacora was available. In order to collect baseline data relating to diversity and abundance of mosquitoes, malaria transmission and the prevalence of insecticide-resistance alleles in malaria vectors, entomological surveys were performed between September and October, 2010 in all the districts of Atacora before IRS implementation. The above mentioned baseline data are reported in this paper.

METHODOLOGY

Study area

The study was carried out in Atacora, a department located in north-west of Benin (Figure 1). This department covers an area of 31,665 km^2 and counts a total of 735,845 inhabitants including 146,309 children under 5 years old in 2011 (INSAE, 2009). It is located at 10° 18' 46'' N and 1° 23' 19'' E. There are fifty health facilities for nine districts. This department is characterized by a sub-equatorial climate, with only one dry season (December to May) and only one rainy season (July to November). The annual mean rainfall is 1,300 mm and the mean monthly temperature varies between 23 and 31°C. The department is irrigated by three major rivers: the Mekrou, the Pendjari and the Alibori. The major economic activity is agriculture and it is characterized by the production of cotton and millet, where various classes of pesticides are used for pest control.

Mosquito sampling

Larvae collection

Mosquito larvae were collected from September to October, 2010 at the end of the rainy season. All instars of larvae were collected using dipping method from a wide range of breeding sites (puddles, shallow wells, gutters and rice fields). Then all larvae were brought back to the laboratory of Centre de Recherche Entomologique de Cotonou (CREC) for rearing. After that, emerging adult female mosquitoes were used for insecticide susceptibility tests and finally a susceptible strain of *A. gambiae* (Kisumu) was used as reference strain for bioassays.

Pyrethrum spray catch (PSC)

Adult mosquito collections were carried out from September to November, 2010 in all the nine districts in Atacora. In addition to larvae collection, mosquitoes resting in the house were collected through morning spray catch (MSC) from 7 a.m. to 9 a.m. In all the nine districts, we collected mosquitoes resting indoors. In each district, two areas were selected: a central area more or less urbanized and rural areas. In each area, some ten homes were randomly selected for mosquito collection. Three sessions were performed per month. Morning pyrethrum spray catches were performed using pyrethrum spray Rambo® and white canvas spread on the floor to collect knocked down mosquitoes. Knocked down mosquitoes falling on white bed sheets were kept separately and later, kept in labeled tubes containing silica gel and frozen at -20°C for further laboratory analysis. This sampling method led to an accurate estimation of the total density of mosquito species in the houses and the proportion of female mosquitoes resting from the

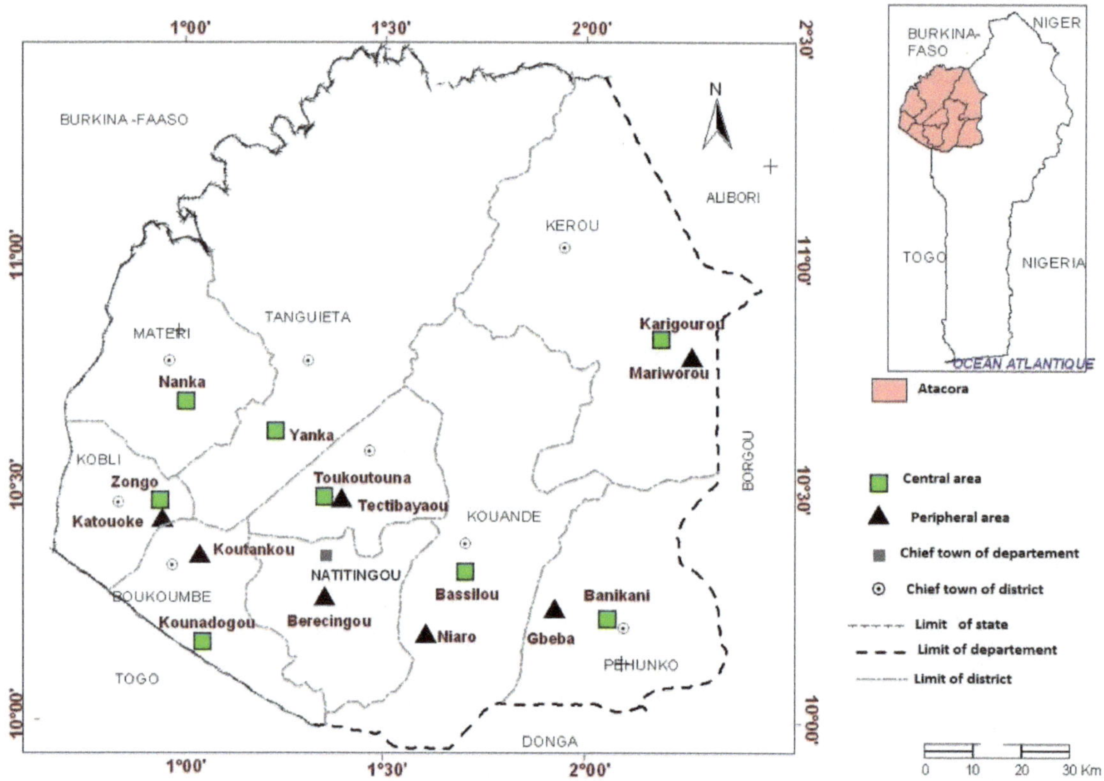

Figure 1. Map of study area.

houses. The recorded data were used to assess the human biting index (ma) and the entomological inoculation rate vectors (EIR).

Insecticide susceptibility tests

WHO insecticide susceptibility test-kits and standard procedures (WHO, 1998) were used to monitor the susceptibility of wild *A. gambiae* populations to the chemical groups of insecticides commonly used in public health and agriculture. Batches of 25 non-blood fed, 3 to 5 days old adult females were exposed to filter papers impregnated with 4% organochlorine (DDT), 0.1% bendiocarb (carbamate), 0.75% permethrin and 0.05% deltamethrin (pyrethroids). The number of mosquitoes knocked down was recorded every 5 min during exposure. After exposure, mosquitoes were supplied with glucose solution as food, and mortality was recorded 24 h post-exposure. Tests with untreated papers were systematically run as controls. WHO criteria were respected to classify populations as 'resistant' if less than 80% mortality was observed, as "suspected resistant" if mortality rates were between 80 and 97% and as 'susceptible' for mortality > 97% (WHO, 2001).

Laboratory processing

After each indoor pyrethrum spray, mosquitoes were sorted by genus as anopheline and culicine and counted. Anophelines were morphologically identified to species using taxonomic keys of Gillies and De Meillon (1968) and Gillies and Coetzee (1987). Ovaries from randomly selected female *A. gambiae s.l.* specimens captured by PSC were dissected to determine parity rate, by observing the coiling degree of ovarian tracheoles (Detinova and Gillies, 1964). Then, mosquito infectivity rates were determined from head and thorax of all female anopheline specimens by enzyme-linked immunosorbent assay (ELISA) using monoclonal antibodies against *Plasmodium falciparum* circumsporozoite protein (CSP) as described by Wirtz et al. (1987). The carcasses of these females (abdomens, wing and legs) were stored in individual tubes with silicagel and preserved at −20°C in the laboratory for identification of species and characterization of molecular forms within the *A. gambiae* complex as previously described (Scott et al., 1993; Favia et al., 2001).

Data analysis

The human biting index (ma) or *Anopheles* density (number of *Anopheles* per person and per night) and the circum sporozoite protein positive rate were calculated. The human biting rate was determined by dividing the number of mosquitoes fed and half gravid collected within a room by the number of sleepers in this room the night before morning spray catching. The circum sporozoite protein positive rate (% CS+) was calculated as the proportion of mosquitoes found to be positive for CSP. The entomological inoculation rate (EIR) was defined as *Anopheles* density by the CSP and estimated as the number of infectious bites per human per month. A Chi-square test with the MINITAB statistical software (Version 12.2) was used to compare the mortality rates among the

Table 1. Mosquito species caught by PSC from September to October, 2010 in the study area.

Species	Nb	%
Aedes aegypti	11	1.04
Aedes gr. Palpalis	5	0.47
Aedes longipalpis	2	0.19
Aedes vittatus	5	0.47
Anopheles broheri	6	0.57
Anopheles funestus	13	1.23
Anopheles gambiae	862	81.71
Anopheles pharoensis	7	0.66
Culex annulioris	3	0.28
Culex fatigans	45	4.27
Culex gr decens	9	0.85
Culex nebulosus	3	0.28
Culex quinquefasciatus	72	6,82
Mansonia africana	12	1,14
Total	1055	

localities. The genotypic differentiation of *kdr* and *Ace1* loci was tested using the Fischer exact test implemented in GenePop software (Raymond and Rousset, 1995), and the Fisher test was used to compare these frequencies. An analysis of variance (ANOVA) was performed to compare the entomological estimates (ma, EIR, CSP) among the sites.

Ethical consideration

Permission was sought from inhabitants to perform collections in their rooms. In addition, community consent had been obtained beforehand in all the villages. This study was approved by the Ethical Committee of the Ministry of Health in Benin.

RESULTS

Culicidae diversity and endophily

A total of 1,055 mosquitoes belonging to 14 different species were caught. They included 84.71% (888) *A. gambiae s.l.*, 2.18% (23) *Aedes* spp., 12.51% (132) *Culex* spp., and 1.14% (12) *Mansonia africana* (Table 1). During the period of study, anopheline species were the most important species collected in all the districts except Tanguiéta where *Culex* spp. was rather very important. The ovary physiological state of *A. gambiae s.l* collected from PSC were mostly fed, half-gravid and gravid, reflecting a strong endophily of this species (Table 2). The percentage of female fed, half-gravid and gravid, was higher than 80% in most of the districts ($p > 0.05$). Vector density by human habitation was on average 10 females in the nine districts.

Vectors infection to CSP and malaria transmission risk

Table 3 shows the circumsporozoite protein positive rate (%CS+) and the EIR at the study sites. Results from this section showed that the risk of malaria transmission was very high in the nine districts. In fact, the average of the sporozoite rate in the nine districts is 6.63% (53 positive thoraces on 800). Among the nine districts, the highest rate was obtained in Natitingou (55.6%) ($p < 0.05$). The average of EIR was 6 infectious bites per person per month (6 bi/p/m). The highest EIR was found in the districts of Cobly and Matéri, with 10 bi/p/m, which means that in these localities every inhabitant receives on infected bite every three days during the study period.

Resistance status

Figure 2 shows the resistance status of *A. gambiae s.l* populations collected in the various districts at Atacora. From this study, it appears that results are similar in the nine districts. *A. gambiae s.l* populations were susceptible to bendiocarb, with mortality rates ranging between 95 and 100%. For this carbamate, no significant difference was noticed from a district to the other ($p > 0.05$). On the other hand, *A. gambiae* was resistant to deltamethrin with mortality rate ranging between 27 and 54%. Moreover, *A. gambiae* was fully resistant to DDT and permethrin with a mortality rate which did not exceed 18%.

Species and molecular forms of *Anopheles gambiae*

Mosquitoes from PSC were analysed by polymerase chain reaction (PCR) for identification of sibling species among *A. gambiae* complex and molecular M and S forms of *A. gambiae s.s*. In all districts, *A. gambiae s.s* was predominant (99.49%). Only one specimen of *A. arabiensis* was found in Kérou (Table 4). Both M and S forms were found in sympatry, but the molecular S form was predominant (94.42%).

Kdr resistance gene status in *A. gambiae*

The average of *kdr* allelic frequency from September to October, 2010 was 78%. *kdr* allelic frequency was very high (98%) in Cobly and Kouandé.

Insensitive acetylcholinesterase gene status in *An. gambiae s.l*

In all the nine districts, the allelic frequency of this gene

Table 2. Vectors distribution and the physiological state of their abdomen.

Districts	Villages	Nb of houses sprayed	*A. gambiae*						*A. funestus*				
			Unfed	Fed	Half Gr	Gr	Total	Density/House	Unfed	Fed	Half Gr	Gr	Total
Pehunco	Gbéba	7	8	64	3	7	82	11.71	-	1	-	-	1
	Banikani	6	4	11	3	5	23	3.83	-	-	-	-	-
Kouandé	Niaro	5		70		1	71	14.20	-	3	-	-	3
	Bassilou	5		8	1		9	1.80	-	-	-	-	-
Cobly	Zongo	7	10	122	30	19	181	25.86	-	2	-	-	2
	katouokè	6	20	120	28	5	173	28.83	-	-	-	-	-
Boukoumbé	Coutankou	5	5	111	30	29	175	35	-	1	1	-	2
	Kounadogou	2	1	4	2	5	12	6	-	1	-	2	3
Tanguiéta	Yanka	7	3	21	1		25	3.57	-	-	-	-	-
	Thanwassaka	5	2	17		1	20	4	-	1	-	-	1
Toukountouna	Sanga	6	8	22		3	33	5.50	-	-	-	-	-
	Tectibayaou	4	13	15			28	7	-	1	-	-	1
Natitingou	Bérécingou	5	1	8		2	11	2.20	-	-	-	-	-
Matéri	Nanka	10		17		2	19	1.90	-	-	-	-	-
Kérou	Karigourou	4		1			1	0.25	-	-	-	-	-
	Maréworou	5		16			16	3.20	-	-	-	-	-
Total		89	75	610	98	79	862	9.69	0	10	1	2	13

was low. The average of *Ace-1* allelic frequency from September to October, 2010 was 3% and no homozygous individuals RR were found (Table 4 and Figure 3).

DISCUSSION

The present study provides entomological baseline data in Atacora for the first time. 14 different species of mosquito were collected during our investigation. A similar result was obtained by Huttel (1950) in southern Benin. As a matter of fact, he got the same species by using the same

Table 3. *A. gambiae* infectivity to circumsporozoïtic antigen (CS) and Entomological Inoculation Rate (EIR) in Atacora.

Locality	N tested	CS+	% CS+	Nb of Sleepers	ma	EIR (bi /p/m)
Pehunco	93	2	2.15ª [00.26-07.55]	43	2.16ᵇ [01.75-02.65]	1.4ª [01.06-01.80]
Kouandé	79	9	11.39ᵇ [05.34-20.53]	45	1.76ᵇ [01.49-02.19]	6ᶜ [05.30-06.76]
Cobly	330	15	4.55ª [02.57-07.39]	44	7.5ᶜ [06.71-08.36]	10.23ᵈ [09.30-11.22]
Boukoumbé	153	3	1.96ª [00.41-05.62]	22	6.95ᶜ [05.89-08.15]	4.09ᵇ [03.29-05.03]
Tanguiéta	44	4	9.09ᵇ [02.53-21.67]	20	2.2ᵇ [01.60-02.95]	6ᶜ [05.35-07.18]
Toukountouna	58	5	8.62ᵇ [02.86-18.98]	24	2.42ᵇ [01.83-03.12]	6.25ᶜ [05.29-07.34]
Natitingou	9	5	55.56ᵈ [21.20-86.30]	18	0.5ª [00.23-00.95]	8.33ᵈ [07.46-09.78]
Matéri	17	5	29.41ᶜ [10.31-55.96]	15	1.13ª [00.66-01.32]	10ᵈ [08.46-11.74]
Kérou	17	5	29.41ᶜ [10.31-55.96]	34	0.5ª [00.29-00.80]	4.41ᵇ [03.73-05.18]
Total	800	53	6.63	265	3.02	6

%CS+: the circum sporozoite protein positive rate; bi/p/m: infective bites per person per month; ma: Number of bites per person per night.

method. However, more species were supposed to be collected if the PSC method was associated with other catch methods such as Human Landing Catch (HLC), Center for Disease Control (CDC) and BG sentinel traps. In fact, a study conducted in southern Benin the same year by Lingenfelser et al. (2010), using HLC with conventional aspirators, showed the presence of 24 species belonging to 6 genera of mosquito. Therefore, the choice of sampling method influenced the entomological data recorded. A report conducted in Senegal by Ndiath et al. (2011) on the sampling method concluded that the sampling technique has to be chosen according to the vector studied and the aim of the study.

A. gambiae, the main malaria vector in sub-Saharan Africa, is the main species harvested in houses (81.71%). This predominance of *A. gambiae* is probably due to the low urbanization level of the districts. Indeed, unlike urban areas, in rural areas, larval habitats are less polluted and therefore, favorable for the production of anophelines; hence the higher production of anopheline fauna in rural areas. *A. funestus* was also present in the department but in small proportion (1.23%), probably due to the presence of dams which surface was covered by vegetation suitable to *A. funestus* development.

Anopheles arabiensis was also very poorly represented in our samples; one specimen was collected in Kérou whereas fifteen years before Akogbeto and Di Deco (1995) reported its presence in sympatry with *A. gambiae* s.s in northern Benin. The absence of *A. arabiensis* during the study period may appear strange. Indeed, the long drought that characterizes Atacora (Sudanian climate), and the presence of a large dry savannah area, satisfy the conditions for the development of *A. arabiensis*. In an article, Djogbénou et al. (2010) reported *A. arabiensis* disappearance in Atacora, especially in Tanguiéta and Natitingou between 2006 and 2007, without explaining the causes of this disappearance. In reality, it was not a problem of disappearance, but a problem of sampling and duration of the study instead.

Regarding our study, the catches were performed inside human dwellings while *A. arabiensis* endophilic trends is low. However, it is possible that the action of man on the environment be unfavorable to the development of *A. arabiensis* in this area. A study on the dynamics of *A. arabiensis* and its distribution showed that within 25 years, *A. arabiensis* moved from latitude 10° 10' N (northern Benin) to latitude 6° 40' N (southern Benin). This change of environment is made possible through deforestation, thus creating favorable conditions for the development of *A. arabiensis* in the central and southern Benin. As a matter of fact, shelters for its development are threatened in the north by a severe drought. During this period, the absence of pasturage in the north led to the movement of this species to the center and the south of Benin through transhumance. Climate change, therefore, favored the movement of *A. arabiensis* to the south. This new distribution of *A. arabiensis* should help the NMCP to adapt to its vector control programme. Molecular analyses performed on the same samples showed that the S form is more represented. This confirms that the S form of *A. gambiae* depends on dry savannah areas as pointed out by studies carried out in West Africa and which reported similar results in similar bioecological zones in Nigeria, Cameroon and Burkina Faso (Wondji et al., 2002; Onyabe et al., 2003; Dabiré et al., 2009). IRS implementation as vector control strategy is based on endophilic behavior of vector populations and their preference to rest inside houses. This is why the high vectors population collected indoors is an asset for the IRS implementation in Atacora. Indeed, the endophilic behavior of vectors facilitates their contact with the insecticide. The second argument for the need to strengthen

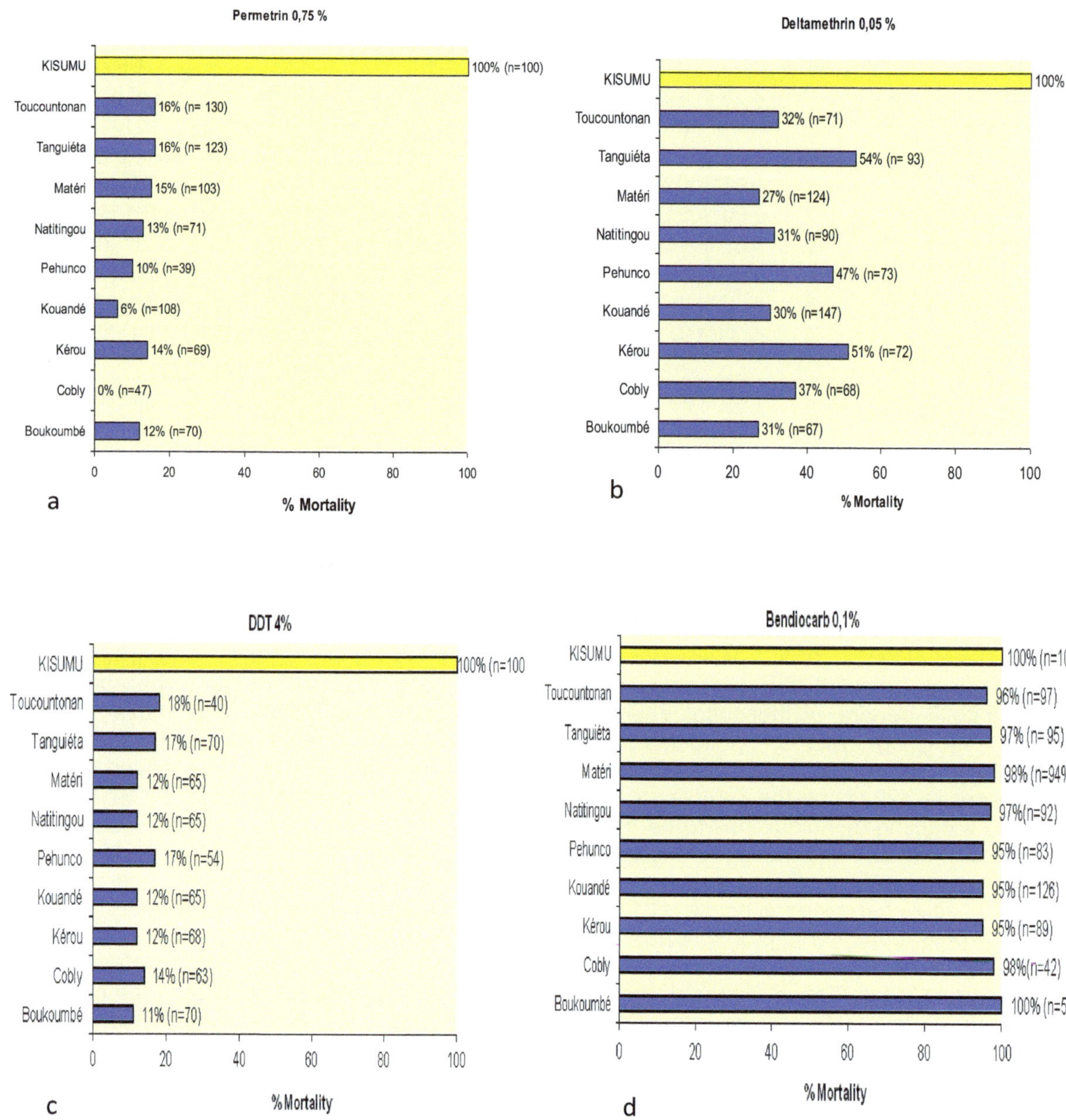

Figure 2. Insecticide susceptibility status of *An. gambiae s.l.* in the nine districts in Atacora.

the fight against malaria in Atacora by a rapid intervention effect on the vectorial capacity can be justified by the very high circumsporozoitic index in Atacora. The entomological inoculation rate recorded during the study (6 infected bites per man per month) was higher than that obtained in southern Benin (Djènontin et al., 2010) and in north-east of Benin (Yadouleton et al., 2010). The high level of this EIR could be explained by the low coverage of LLINs. In fact, a survey conducted in Atacora in 2010 on the use of bednets by the populations showed that very few people had a bednet (1 net for 5 people) and those who had it, would not use it frequently because of the strong heat (Aïkpon, personal communication).

Data recorded from the susceptibility test performed on *A. gambiae* showed a high level of resistance to permethrin, deltamethrin and DDT but susceptibility to bendiocarb. The emergence of resistance to pyrethroid in *A. gambiae* has become a serious concern for the success of malaria control in the last decades. In fact, in Benin, pyrethroids have been extensively introduced in

Table 4. Species and molecular forms of *An. gambiae* and *kdr* and *Ace.1* allelic frequency.

Locality	Species		Molecular forms		*kdr mutation*				*Ace-1* mutation			
	Aa	Ag	M	S	RR	RS	SS	F(*kdr*)	RR	RS	SS	F(*Ace-1*)
Pehunco	0	28	1	27	10	18	0	0.68	0	2	26	0.04
Kouandé	0	27	0	27	26	1	0	0.98	0	1	26	0.02
Cobly	0	21	2	19	20	1	0	0.98	0	1	20	0.02
Boukoumbé	0	27	0	27	10	17	0	0.69	0	0	27	0.00
Tanguiéta	0	26	6	20	20	4	2	0.85	0	0	26	0.00
Toukountounan	0	26	0	26	13	11	2	0.71	0	1	25	0.02
Natitingou	0	9	0	9	3	3	3	0.50	0	1	8	0.06
Matéri	0	17	0	17	10	5	2	0.74	0	1	16	0.03
Kérou	1	16	2	14	11	2	3	0.75	0	3	13	0.09

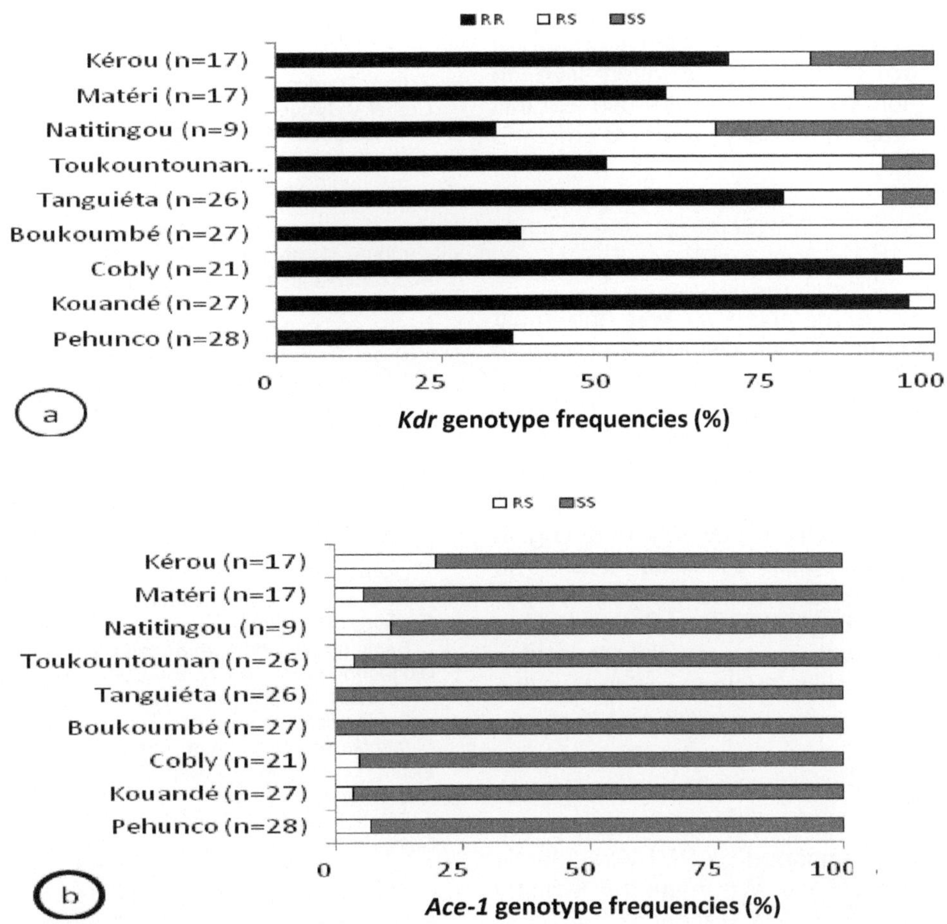

Figure 3. *kdr* (a) and *Ace-1* (b) allelic frequency.

agriculture since 1980s (Prudent et al., 2006). This factor is probably one of the causes of the selection of high resistance in *A. gambiae* to pyrethroids, with a high allelic frequency of the knock down resistance (*kdr*) mutation level. The *kdr* which seems to be the main mechanism conferring resistance in *A. gambiae* to pyrethroids (Leu-Phe *kdr* mutation) in West Africa was found in mosquito samples collected in different sites. The allelic frequency found in our study area was higher than what was reported in southern Benin by Djènontin et al. (2010), and similar to those reported in Parakou, a town in the north-east of Benin by Yadouleton et al. (2011). This would certainly explain the fact that most of the localities in northern Benin are cotton growing and use many pesticides. This massive use of insecticide selects the resistance of malaria vectors (Akogbéto et al., 2005; Dabiré et al., 2008; Yadouleton et al., 2009).

The *Ace-1* mutation was also found but at a very low frequency (< 0.1). Several reports in southern Benin (Yadouleton et al., 2009; Djènontin et al., 2010; Padonou et al., 2012) confirmed the low allelic frequency of this mutation. Basing on the information above related to the susceptibility tests, and for better management of resistance to insecticide in *A. gambiae* populations, the low frequency of *Ace-1* recorded in all populations encourages the use of bendiocarb as an alternative insecticide to pyrethroids for IRS in the study areas. A previous study conducted in southern Benin (Akogbéto et al., 2010) had shown bendiocarb as alternative to pyrethroids. The author showed that the use of this insecticide for IRS had drastically reduced malaria transmission.

Conclusion

This study shows that *A. gambiae s.l* was the most abundant species found in the nine districts. *A. gambiae s.s* was practically the only member of the complex found responsible for malaria transmission in Atacora. Besides, the high antigen CSP positivity rate observed proves the vulnerability of populations to malaria transmission and constitutes challenge not only for the reinforcement of existing control strategies (LLINs), but also resorting to other complementary strategies such as IRS. The susceptibility to bendiocarb added to the low frequency of *Ace-1* mutation found in *A. gambiae* populations could lead to the use of bendiocarb as a potential alternative against resistance of *A. gambiae* to pyrethroid in the nine districts for IRS.

ACKNOWLEDGEMENTS

This work was financially supported by PMI (President's Malaria Initiative) through USAID. We thank the Ministry of Higher Education and Scientific Research (MESRS) and the team of CREC for their technical assistance during field work and the laboratory. We also thank the people of Atacora for their collaboration.

REFERENCES

Abdalla H, Matambo TS, Koekemoer LL, Mnzava AP, Hunt RH, Coetzee M (2008). Insecticide susceptibility and vector status of natural populations of *Anopheles arabiensis* from Sudan. Trans. R. Soc. Trop. Med. Hyg. 102: 263-271.

Akogbeto M, Di Deco MA (1995). Répartition des membres du complexe *Anopheles gambiae* et de leurs variants chromosomiques au Bénin et au Togo, Afrique occidentale. J. Afr. Zool., 109: 443–454.

Akogbéto M, Yakoubou S (1999). Résistance des vecteurs du paludisme vis-à-vis des pyréthrinoïdes utilisés pour l'imprégnation des moustiquaires au Bénin, Afrique de l'Ouest. Bull. Soc. Path. Exo. 92: 123–130.

Akogbeto MC, Djouaka R, Noukpo H (2005). Utilisation des insecticides agricoles au Bénin. Bull. Soc. Pathol. Exot. 5: 400-405.

Akogbéto M, Padonou G, Gbénou D, Irish S, Yadouleton A (2010). Bendiocarb, a potential alternative against pyrethroid resistant *Anopheles gambiae* in Benin West Africa. Malar. J., 9:204.

Akogbéto M, Padonou G, Bankolé H, Kindé Gazard D, Gbédjissi G (2011). Dramatic decline of malaria transmission after implementation of large-scale Indoor Residual Spraying using bendiocarb in Bénin, West Africa, an area of high *Anopheles gambiae* resistance to pyrethroids. Am. J. Trop. Med. Hyg. 85(4): 586–593.

Chandre F, Manguin S, Brengues C, Dossou Yovo J, Darriet F, Diabate A, Carnevale P, Guillet P (1999). Current distribution of a pyrethroid resistance gene (kdr) in *Anopheles gambiae* complex from West Africa and further evidence for reproductive isolation of the Mopti form. Parassitologia., 41: 319–322.

Corbel V, N'guessan R, Brengues C, Chandre F, Djogbenou L, Martin T, Akogbéto M, Hougard JM, Rowland M (2007). Multiple insecticide resistance mechanisms in *Anopheles gambiae* and *Culex quinquefasciatus* from Benin, West Africa. Acta. Trop. 101(3): 207-16.

Dabiré KR, Diabaté A, Agostinho F, Alves F, Manga L, Faye O, Baldet T (2008). Distribution of the members of *Anopheles gambiae* and pyrethroid knock-down resistance gene (kdr) in Guinea-Bissau, West Africa. Bull. Soc. Pathol. Exot. 101(2): 119-23.

Dabiré KR, Diabaté A, Namountougou M, Toé KH, Ouari A, Kengne P, Bass C, Baldet T (2009). Distribution of pyrethroid and DDT resistance and the L1014F *kdr* mutation in *Anopheles gambiae s.l.* from Burkina Faso (West Africa). Trans. R. Soc. Trop. Med. Hyg. 103: 1113–1120.

Detinova TS, Gillies MT (1964). Observations on the Determination of the Age Composition and Epidemiological Importance of Populations of *Anopheles gambiae* Giles and *Anopheles funestus* Giles in Tanganyika. Bull. World. Health. Organ. 30: 23–28.

Djònontin A, Bio-Bangana S, Moiroux N, Henry MC, Bousari O, Chabi J, Ossè R, Koudénoukpo S, Corbel V, Akogbéto M, Chandre F (2010). Culicidae diversity, malaria transmission and insecticide resistance alleles in malaria vectors in Ouidah-Kpomasse-Tori district from Benin (West Africa): A pre-intervention study. Parasit. Vectors.3: 83.

Djogbénou L, Pasteur N, Bio-Bangana S, Baldet T, Irish SR, Akogbeto M, Weill M, Chandre F (2010). Malaria vectors in the Republic of Benin: distribution of species and molecular forms of the *Anopheles gambiae* complex. Acta. Trop.114(2): 116-22.

Elissa N, Mouchet J, Rivière F, Meunier JY, Yao K (1993). Resistance of *Anopheles gambiae s.s.* to pyrethroids in Côte d'Ivoire. Ann. Soc. Belge. Med. Trop. 73: 291–294.

Etang J, Fondjo E, Chandre F, Morlais I, Brengues C, Nwane P, Chouaibou M, Ndjemai H, Simard F (2006). First report of knockdown mutations in the malaria vector *Anopheles gambiae* from Cameroon. Am. J. Trop. Med. Hyg. 74: 795–797.

Favia G, Lanfrancotti A, Spanos L, Siden Kiamos I, Louis C (2001). Molecular characterization of ribosomal DNA polymorphisms discriminating among chromosomal forms of *Anopheles gambiae* s.s. Insect. Mol. Biol. 10: 19–23.

Gillies MT, De Meillon D (1968). The Anophelinae of Africa South of the Sahara. Publ. South. Afri. Inst. Med. Res., 54:343.

Gillies MT, Coetzee MA (1987). Supplement to the Anophelinae of Africa South of the Sahara, 2nd edn. Publ. South. Afri. Inst. Med. Res. 55: 143.

Huttel J: Note sur la répartition des moustiques dans le Bas-Dahomey.

Bull. Sot. Path. Exot., 43: 563-566.
INSAE (2004). Projection démographique révisée. Atacora. 29.
Koekemoer LL, Kamau L, Hunt RH, Coetzee M (2002). A cocktail polymerase chain reaction assay to identify members of the Anopheles funestus (Diptera: Culicidae) group. Am. J. Trop. Med. Hyg. 66(6):804-811.
reaction assay to identify members of the *Anopheles funestus* (Diptera: Culicidae) group. Am. J. Trop. Med. Hyg. 66: 804-11.
Lingenfelser A, Rydzanicz K, Kaiser A, Becker N (2010). Mosquito fauna and perspectives for integrated control of urban vector-mosquito populations in Southern Benin (West Africa). Ann. Agric. Environ. Med. 17(1): 49-57.
Ministry of Health (2011). Health Statistics Yearbook 2011. Cotonou: Department of Programming and Prospective.
Ndiath OM, Mazenot C, Gaye A, Konate L, Bouganali C, Faye O, Sokhna C, Trape JF (2011). Methods to collect Anopheles mosquitoes and evaluate malaria transmission: A comparative study in two villages in Senegal. Malar. J. 10: 270.
Nwane P, Etang J, Chouaibou M, Toto J C, Kerah-hinzoumbe C, Mimpfoundi R, Awono-Ambene HP, Simard F (2009). Trends in DDT and pyrethroid resistance in *Anopheles gambiae* s.s. populations from urban and agro-industrial settings in southern Cameroon. BMC. Infect. Dis. 9: 163.
Onyabe DY, Vajime CG, Noc IH, Ndams IS, Akpa AU, Alaribe AA, Conn JE (2003). The distribution of M and S molecular forms of *Anopheles gambiae* in Nigeria. Trans. R. Soc. Trop. Med. Hyg. 97: 605–608.
Padonou GG, Sèzonlin M, Ossè R, Aizoun N, Oké-Agbo F, Oussou O, Gbédjissi G, Akogbéto M (2012). Impact of three years of large scale Indoor Residual Spraying (IRS) and Insecticide Treated Nets (ITNs) interventions on insecticide resistance in *Anopheles gambiae s.l.* in Benin. Parasit. Vectors. 10: 72.
Prudent P, Loko S, Deybe D, Vaissayre M (2006). Factors limiting the adoption of IPM pratices by cotton farmers in Benin: a participatory approach. Cambridge. University. Press. 10: 10-17.
Raymond M, Rousset F (1995). GENEPOP (version 1.2): population genetics software for exact tests and ecumenicism. J. Heredity. 86: 248-249.
Santolamazza F, Calzetta M, Etang J, Barrese E, Dia I, Caccone A, Donnelly MJ, Petrarca V, Simard F, Pinto J, della Torre A (2008). Distribution of knock-down resistance mutations in *Anopheles gambiae* molecular forms in west and west-central Africa. Malar. J. 7: 74.
Scott JA, Brogdon WG, Collins FH (1993). Identification of single specimens of the *Anopheles gambiae* complex by the polymerase chain reaction. Am. J. Trop. Med. Hyg. 49: 520–529.
Vulule JM, Beach RF, Atieli FK, McAllister JC, Brogdon WG, Roberts JM, Mwangi RW, Hawley WA (1999). Elevated oxidase and esterase levels associated with permethrin tolerance in *Anopheles gambiae* from Kenyan villages using permethrin-impregnated nets. Med. Vet. Entomol. 13: 239-244.
Wirtz RA, Ballou WR, Schneider I, Chedid L, Gross MJ, Young JF, Hollingdale M,Diggs CL, Hockmeyer WT (1987). *Plasmodium falciparum*: immunogenicity of circumsporozoite protein constructs produced in *Escherichia coli*. Exp. Parasitol. 63: 166–172.
Wondji C, Simard F, Fontenille D (2002). Evidence for genetic differentiation between the molecular forms M and S within the Forest chromosomal form of *Anopheles gambiae* in an area of sympatry. Insect. Mol. Biol. 11: 11–19.
World Health Organization (WHO) (1998). Tests procedures for insecticide resistance monitoring in malaria vector, bio- efficacy and persistence of insecticides on treated surfaces. Document WHO/CDS/CPC/MAL/98.12 Geneva, Switzerland 1998 [http://whqlibdoc.who.int/hq/1998/ WHO_CDS_CPC_MAL_98.12.pdf].
World Health Organization (WHO) (2001). Supplies for monitoring insecticide resistance in disease vectors. WHO/CDS/CPE/PVC/20012 2001.
World Health Organization (WHO) (2009). World Malaria Report. Geneva.
World Health Organization (WHO) (2011). World Malaria Report. Geneva, Switzerland.
Yadouleton AW, Asidi A, Djouaka RF, Braïma J, Agossou CD, Akogbeto MC (2009). Development of vegetable farming: a cause of the emergence of insecticide resistance in populations of *Anopheles gambiae* in urban areas of Benin. Malar. J. 14: 103.
Yadouleton AW, Padonou G, Asidi A, Moiroux N, Banganna S, Corbel V, N'guessan R, Gbenou D, Yacoubou I, Gazard K, Akogbeto MC (2010). Insecticide resistance status in *Anopheles gambiae* in southern Benin. Malar. J. 9: 83.
Yadouleton A, Martin T, Padonou G, Chandre F, Asidi A, Djogbenou L, Dabiré R, Aïkpon R, Boko M, Glitho I, Akogbeto M (2011). Cotton pest management practices and the selection of pyrethroid resistance in *Anopheles gambiae* population in northern Benin. Parasit. Vectors. 4: 60.
Zaim M, Aitio A, Nakashima N (2000). Safety of pyrethroid-treated mosquito nets. Med. Vet. Entomol. 14:1-5.

Seasonal prevalence of *Oestrus ovis* L. (Diptera: Oestridae) larvae in infested sheep in Jazan Region, Saudi Arabia

Bosly A. Hanan

Biology Department, Faculty of Science, Jazan University, Jazan, Saudia Arabia.

Oestrus ovis **L. (*O. ovis*) (Diptera: Oestridae) is a cosmopolitan agent of myiasis in sheep and goats. Sneezing and nasal discharges are the major clinical signs in infested animals. The present study was carried out to estimate the annual abundance of *O. ovis* in randomly selected male sheep heads in Jazan region. Survey conducted weekly for *O. ovis* during the period from November, 2010 to October, 2011, from Slaughterhouses in Jazan and the ecological data including, temperature, relative humidity, were obtained from the Department of Meteorology, Jazan. The study investigated 480 sheep heads. The collected larval specimens were identified by British Museum (Natural History), London. Results indicated that 257 (53.54%) of the total examined sheep heads showed positive infestation. Infestation recorded four peaks at January, May, July and October. The highest infestation was detected at January (82.5%), while the lowest was at August (25.00%). Also, the first, second and third larval instars represented 3.02, 20.35, and 76.63% from the total collected larvae, respectively. The data represented mean larval monthly number per sheep head. The study suggested the cold seasonal related *O. ovis* infestation in Jazan region.**

Key words: Jazan, *Oestrus ovis*, sheep, seasonal prevalence, abundance.

INTRODUCTION

The present study continues the planned series of studies that aim to detect the prevalence and abundance of the dipterous insects associated with sheep in Jazan region. These studies were started by Bosly (2010), who recorded twelve dipterous species within eight genera, belonging to seven families: Calliphoridae, Sarcophagidae, Muscidae, Ceratopogonidae, Ulidiidae, Sphaeroceridae, Chloropidae on sheep in Abu Arish, Alquayiyah, (Eastern Jazan); Duhaygah and Mihliyah (Northern Jazan); Mizhirah and Industrial City (Southern Jazan). The present study focuses on the nasal botfly *Oestrus ovis*, Linnaeus 1761 (Diptera: Oestridae).

O. ovis is the almost cosmopolitan and well known myiasis causing dipterous insect species which affects the health of sheep and goats in many parts of the world (Zumpt, 1965; Nacapunchai et al., 1998; Gracía et al., 2006). The parasitic phase of *O. ovis* begins after adult females deposit first larval stage larvae into the nostrils of hosts; these larvae develop into second and third larval stages in the nasopharyngeal cavities and frontal sinuses (Zumpt, 1965). Sneezing and nasal discharges are the major clinical signs in infected animals (Angulo-Valadez et al., 2011). Sometimes, the larvae are present in human eyes, classified as ophthalmomyiasis externa if it is on

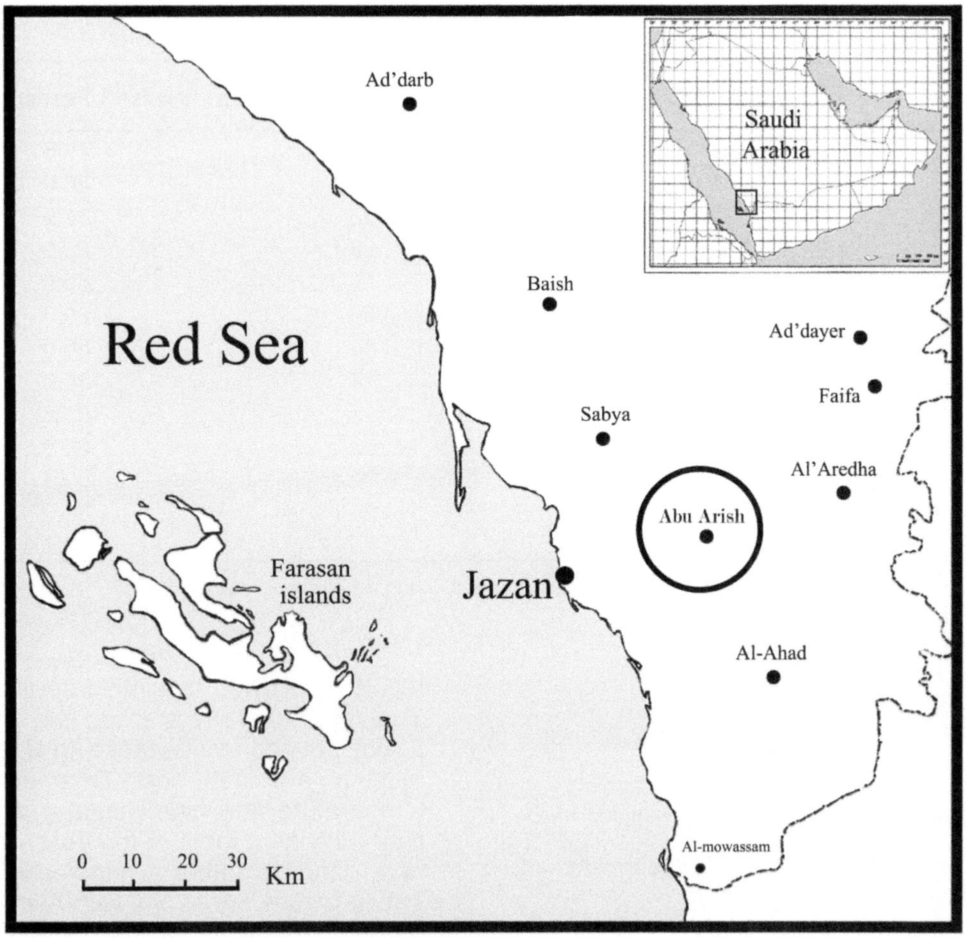

Figure 1. Map showing the location of the study area.

the conjunctiva and ophthalmomyiasis interna when there is intraocular penetration of larvae (Gregory et al., 2004; Dunbar et al., 2008; Khurana et al., 2010; Parikh et al., 2011). McGarry et al. (2012) observed infestation of dogs with first-stage larvae deposited by an adult fly the previous autumn in a farm in the Cotswolds, UK. Abundance of *O. ovis* larvae were done in many localities in central region of Saudi Arabia by Alahmed (2000). Despite the veterinary and medical importance of the nasal botfly in Saudi Arabia, little is known about its distribution and abundance in Jazan region. Accordingly, the present work aimed to detect the annual abundance of *O. ovis* on sheep in Jazan during the period from November, 2010 to October, 2011.

MATERIALS AND METHODS

The study area

The present study was carried out in Abu Arish area (Eastern Jazan), southern Saudi Arabi, (16° 58'N to 42° 47'E) (Figure 1).

Animals

The survey of *O. ovis* was conducted weekly during the period from November, 2010 to October, 2011 from Slaughterhouses in Jazan. A total of 480 randomly selected male sheep heads were examined for infestation.

Examination procedures

Heads were detached and opened from nose to the base by an electric saw. Larvae of the different instars were found alive inside heads; these larvae were picked up by forceps and preserved in 70% ethanol. Identification of specimens was performed by using keys of Zumpt (1965), Ferrar (1987) and Smith (1989), moreover some specimens were sent to British Museum (Natural History), London to confirm identification.

Weather data

Ecological data including the mean annual temperature, relative humidity and the received rainfall per annum were obtained from Meteorological Agency, Jazan.

Table 1. Monthly temperature, relative humidity and number of infested sheep head by *Oestrus ovis* in Jazan, Saudi Arabia.

Month	Temperature		Relative humidity (%)	No. of infested heads*	Percentage of infestation
	Max. Mean	Mean	Mean		
November, 2010	34.0	24.8	67	19	47.5
December	30.9	22.9	73	28	70.0
January, 2011	29.9	22.2	75	33	82.5
February	30.9	22.8	72	32	80.0
March	32.1	24.1	70	26	65.0
April	34.8	26.8	66	20	50.0
May	37.2	28.8	65	20	50.0
June	37.6	31.2	64	14	35.0
July	38.4	32.1	60	15	37.5
August	38.2	30.6	64	10	25.0
September	37.7	30.0	65	20	50.0
October	36.4	26.9	68	20	50.0
Total*	-	-	-	257	53.5

*Monthly examined 40 sheep head. *Total examined 480 sheep head during the study period.

Statistical analysis

The percentage of sheep head infestation (%) was calculated by the equation:

(Number of Infested heads / Total number of examined heads) × 100

The percentage of each *O. ovis* larval instar per infested sheep head was calculated by the equation:

(Number of larval instar / Total number of larvae) × 100

The mean larvae number per sheep head was calculated by the equation:

Number of monthly instar larval / Total monthly number of Infested sheep heads.

RESULTS AND DISCUSSION

Data in Table 1 and Figure 2 represent the monthly temperature, relative humidity and number of infested sheep heads by *O. ovis* larvae during the study period started in November, 2010 in Gizan region. Generally, the examination of 480 sheep heads revealed that 257 (53.5%) heads were infested with different larval instars of *O. ovis* (Table 1). Alahmed (2000) reported that during a period 12 months, 544 sheep heads were examined for infestation with *O. ovis* larvae in Riyadh abattoir, with recorded prevalence of infestation at 5.9%. The data reported that the lowest infestation percentage of sheep heads in August, 2011 was 25%, where temperature and relative humidity were 38°C and 64%, respectively. However, the percentage of infested heads in the present study increased to a peak abundance in January, 2011 (82.5%), followed by simple decline in February (80.0%). The results clearly showed the weather relation to the insect abundance because in Jazan region, the winter months are the best time of the year, with low temperature and high humidity while the decline in the flies' activity in most of the year was probably due to the heat and weather (http://en.wikipedia.org/wiki/Jizan). These results are in line with Alahmed (2000) who recorded the peak of infestation with *O. ovis* larvae in Riyadh as 23.3% in March (low temperature and humidity). Also, Dorchies et al. (2000) reported that the prevalence rate of *O. ovis* larvae in sheep and goats (in France) varied from 14.3% in February to 65% in October and Shoorijeh et al. (2009) found that prevalence of *O. ovis* larvae at the fairs abattoris in (Iran) ranged from 23.3% in spring to 80% in Winter.Datas in Table 2 and Figure 3 represent a total of 860 larvae of all three instars of *O. ovis* which were collected. The infestation percentages were 3, 20.4 and 76.6% of the first (L1), second (L2) and third instars (L3) larvae, respectively. The data in Table 3 showed the mean monthly number of larvae per sheep head (L/S). The L1 expressed two peaks, one in April (0.4 L/S) and the second in May (0.7 L/S) and L2 peaked in May (1.9 L/S) and in June (1.6 L/S) while L3 peaked only in August (8.4 L/S). In general, L2 and L3 instar larvae behaved in a similar trend throughout the experimental period as the fly had only one generation per year which started in November. Also data in Table 3 indicated that the mean monthly total number of all three instars had two peaks of abundance in August (8.7 L/S) and in May (6.3L/S). As regard to the predicted lower mean of the first instar larvae, Tabouret et al. (2001) previously reported that rapid larval development occurs in summer whereas L1

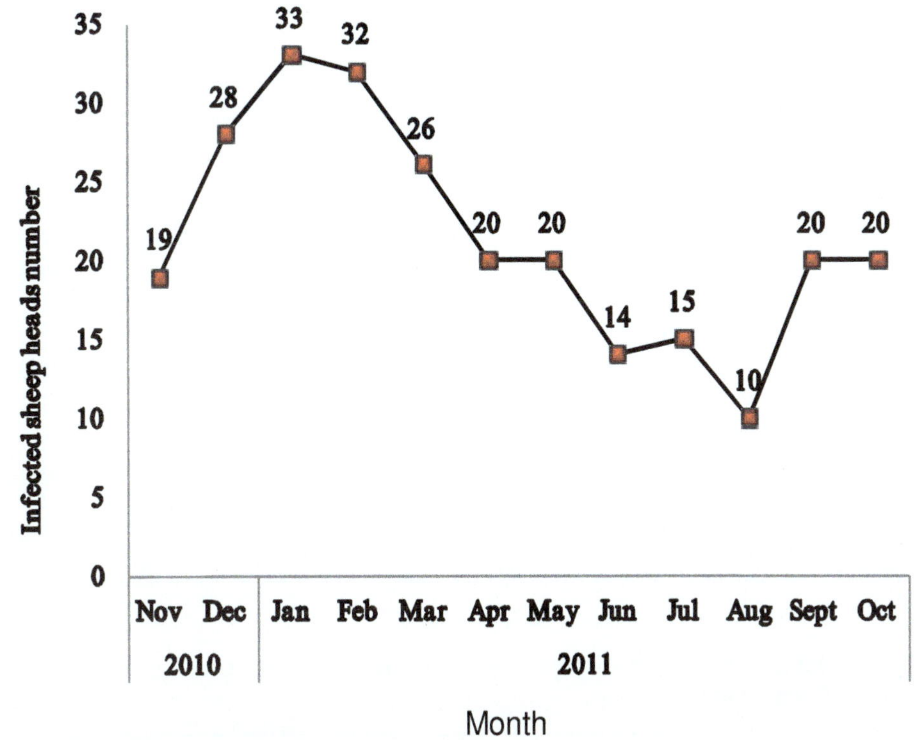

Figure 2. Monthly number of Infected Sheep Heads of *Oestrus ovis* Abu Arish area (Eastern Jazan), south Saudi Arabia.

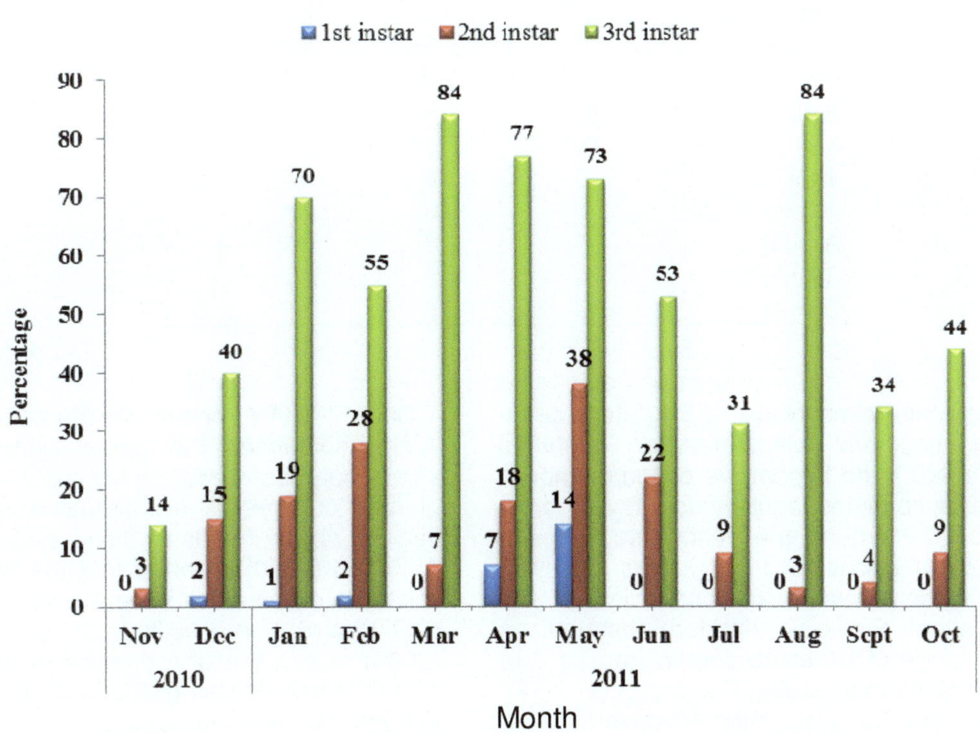

Figure 3. Monthly number intensity of *O. ovis* infection in sheep of different instars.

Table 2. Number of the three instars of *Oestrus ovis* larvae and number of each larval instar.

Month	Total No. of larvae	Larval instar					
		L1	%	L2	%	L3	%
November, 2010	17	0	0	3	17.7	14	82.35
December	57	2	3.51	15	26.3	40	70.18
January, 2011	90	1	1.11	19	21.11	70	77.78
February	85	2	2.35	28	32.9	55	64.71
March	91	0	0	7	7.7	84	92.31
April	102	7	6.86	18	17.7	77	75.49
May	125	14	11.20	38	30.4	73	58.40
June	75	0	0	22	29.3	53	70.67
July	40	0	0	9	22.5	31	77.50
August	87	0	0	3	3.5	84	96.55
September	38	0	0	4	10.53	34	89.47
October	53	0	0	9	16.98	44	83.002
Total	860	26		175		659	
Percentage		3.0		20.4		76.6	

Table 3. Monthly mean of first, second and third instars larvae *O. ovis* per sheep head (L/S).

Month	Mean larvae per sheep head (L/S)			
	1st instar	2nd instar	3rd instar	Mean total
November, 2010	0	0.2	0.7	0.9
December	0.1	0.5	1.4	2.0
January, 2011	0.03	0.6	2.1	2.7
February	0.1	0.9	1.7	2.7
March	0	0.3	3.2	3.5
April	0.4	0.9	3.9	5.2
May	0.7	1.9	3.7	6.3
June	0	1.6	3.8	5.4
July	0	0.6	2.1	2.7
August	0	0.3	8.4	8.7
September	0	0.2	1.7	1.9
October	0	0.5	2.2	2.7

whereas L1 hypobiosis takes place in the late autumn and winter. In southern Mediterranean countries (Morocco, Tunisia, etc.), the hypobiosis period is shorter and in humid tropical countries, adult fly activity and larval development occur all around the year. Also, Angulo-Valadez et al. (2010) discussed that *O. ovis* life cycle development is closely related to local climate and geographical location. Previous reported observations indicate that the range of L1 establishment rate is 0 to 48% in sheep (Dorchies et al., 1998; Frugére et al., 2000) and the L1 stage lasts from less than 10 days to more than 25 days under favorable temperatures (Cepeda-Palacios and Scholl, 2000).

Badawi (1994) previously mentioned that under Saudi Arabia conditions, the larvae spend about 9 months in sheep head before falling for pupation. Amin et al. (1997) pointed out that *O. ovis* larvae were found all over the year in sheep heads in Cairo and has one generation. Alahmed (2000) reported that the highest mean number of larvae per infested sheep head in L1, L2 and L3 were 5, 7 and 5.3, respectively in spring months. Alahmed (2004) stated that the myiasis incidences were highest during March to May (60%) and September to November (31.5%), where temperature and relative humidity are optimum. Arslan et al. (2009) found that the prevalence of sheep nasal myiasis in Turkey was 54.3% in spring, 41%

in summer and 38.9% in winter. Shoorijeh et al. (2009) indicated that prevalence of *O. ovis* in sheep head in Iran ranged from 23.3% in spring to 80% in winter. Also, Gracia et al. (2010) determined the prevalence of *O. ovis* larvae in sheep heads monthly for one year in Spain and reported that *O. ovis* recorded slower larval maturation in winter. Paredes-Esquivel et al. (2012) found that 46.03% of animals were infested in a 13-month period in the island of Majorca (Spain) and significant differences in oestrosis prevalences in winter and autumn. Silva et al. (2012) suggested that evolution and development of *O. ovis* practically occurs throughout the entire year, with larval infestation especially frequent during the spring and summer months.

This study suggested the risk of *O. ovis* larvae according to the abundance in Jazan region and the impact of climate on the infestation. Hence, the importance of controlling the sheep infestation which should be in the beginning of the winter season and for complete prevention, a seasonal treatment in April, is suggested, and every effort should be made to control them by sanitary measures and tools for the pest management. The results of this survey proved that the high infestation of sheep with *O. ovis* was monitored on January, 2011, and also revealed that this fly is becoming a serious pest in Gizan region. The prevention of myiasis is mainly achieved by treating the infested hosts. All efforts should be made to control this fly and the Ministry of Agriculture efforts should be intensified in the winter season to restrict and control the infection level.

REFERENCES

Alahmed AM (2000). Seasonal infestation of *Oestrus ovis* larvae in sheep heads in central region of Saudi Arabia. J. Egypt. Soc. Parasitol. 30(3):895-901.

Alahmed AM (2004). Myiasis in sheep farms in Riyadh Region, Saudi Arabia. J. Egypt. Soc. Parasitol. 34(1):153-160.

Amin AM, Morsy TA, Shoukry A, Mazyad SAM (1997). *Oestrus ovis of* maggots in head of slaughtered sheep in Cairo abattoir. J. Egypt. Soc. Parasitol. 27 (3): 855-861.

Angulo-Valadez CE, Scholl PJ, Cepeda-Palacios R, Jacquiet P, Dorchies P (2010). Nasal bots...a fascinating world! Vet. Parasitol. 174(1-2):19-25.

Angulo-Valadez CE, Ascencio F, Jacquiet P, Dorchies P, Cepeda-Palacios R (2011). Sheep and goat immune responses to nose bot infestation: a review. Med. Vet. Entomol. 25(2):117-125.

Arslan MO, Kara M, Gicik Y (2009). Epidemiology of Oestrus ovis infestations in sheep in Kars province of north-eastern Turkey. Trop. Anim. Health Prod. 41(3):299-305.

Badawi A (1994). Arthropods of medical and veterinary important in Saudi Arabia. King Saud University Press. Saudi Arabia.

Bosly, A.H. 2010. Prevalence of dipterous flies with veterinary importance in selected sheep's farms and slaughter houses in Jazan, Saudi Arabia. Egypt. Acad. J. Biol. Sci. 3 (2):81-91.

Cepeda-Palacios R, Scholl PJ (2000). Intra-puparial development in *Oestrus ovis* (Diptera: Oestridae). J. Med. Entomol. 37:239–245.

Dorchies P, Bergeaud JP, Tabouret G, Duranton C, Prevot F, Jacquiet P (2000). Prevalence and larval burden of *Oestrus ovis* (Linné 1761) in sheep and goats in Northern Mediterranean region of France. Vet. Parasitol. 88(3-4):269-273.

Dorchies P, Duranton C, Jaquiet P (1998). Pathophysiology of Oestrus ovis infection in sheep and goats: a review. Vet. Rec. 142:487-489.

Dunbar J, Cooper B, Hodgetts T, Yskandar H, van Thiel P, Whela, S, Taylor J, Woods DR (2008). An outbreak of human external ophthalmomyiasis due to *Oestrus ovis* in Southern Afghanistan. Clin Infect Dis. 46(11):e124-126.

Ferrar P (1987). A guide to the breeding habits and immature stages of Diptera (Cyclorrhapha) E.J.Brill/Scandinavian Science Press. Leiden, Copenhagen.

Frugére S, Cota Leon A, Prevot F, Cepeda Palacios R, Tabouret G, Bergeaud JP, Duranton C, Dorchies P, Jacquiet P (2000). Immunisation of lambs with excretory secretory products of Oestrus ovis third instar larvae and subsequent experimental challenge. Vet. Res. 31:527-535.

Gracía MJ, Lucientes J, Peribáñez MA, Calvete C, Ferrer LM, Castillo JA (2006). Kinetics of *Oestrus ovis* infection and activity of adult flies. Parasite 13(4):311-313.

Gracia MJ, Lucientes J. Peribáñez MA, Castillo JA, Calvete C, Ferrer LM (2010). Epidemiology of *Oestrus ovis* infection of sheep in Northeast Spain (Mid-Ebro Valley). Trop. Anim. Health Prod. 42(5):811-813.

Gregory AR, Schatz S, Laubach H (2004). Ophthalmomyiasis caused by the sheep bot fly Oestrus ovis in northern Iraq. Optom Vis Sci. 81(8):586-590.

Khurana S, Biswal M, Bhatti HS, Pandav SS, Gupta A, Chatterjee SS, Lyngdoh WV, Malla N (2010). Ophthalmomyiasis: Three cases from North India. Ind. J. Med. Microbiol. 28(3):257-261.

McGarry J, Penrose F, Collins C (2012). Oestrus ovis infestation of a dog in the UK. J. Small Anim. Pract. 53(3):192-193.

Nacapunchai D, Lamon C, Sukprasert N (1998). A first record from Thailand of human external ophthalmomyiasis due to *Oestrus ovis*. Southeast Asian J. Trop. Med. Pub.Health. 29(1):133-136.

Paredes-Esquivel C, del Rio R, Monerris M., Borràs D, Laglera LM, Miranda MÁ (2012). The influence of sheep age group on the seasonal prevalence of oestrosis in the island of Majorca. Vet Parasitol. 186(3-4):538-541.

Parikh V, Biswas J, Vaijayanthi K, Das D, Raval V (2011). Bilateral ocular myiasis interna caused by botfly (*Oestrus ovis*): a case report. Ocul. Immunol. Inflamm. 19(6):444-447.

Shoorijeh SJ, Negahban S, Tamadon A, Behzadi MA (2009). Prevalence and intensity of Oestrus ovis in sheep of Shiraz, southern Iran.Trop Anim Health Prod. ,41(7):1259-1262.

Silva BF, Bassetto CC, Amarante AF (2012). Epidemiology of Oestrus ovis (Diptera: Oestridae) in sheep in Botucatu, State of São Paulo. Rev. Bras Parasitol. Vet. 21(4):386-390.

Smith KGF (1989). An introduction to the immature stages of British flies: Diptera larvae, with notes on eggs, puparia and pupae. Handbook for Identification of British Insects. Royal Society of Entomology, London.

Tabouret G, Jacquiet P, Scholl P, Dorchies P (2001). *Oestrus ovis* in sheep: relative third-instar populations, risks of infection and parasitic control. Vet. Res. 32: 525-531.

Zumpt F (1965). "Myiasis in man and animals in the old world." In: F. Zumpt (Ed), A Textbook For Physicians, Veterinarians and Zoologists. Butterworth, London, UK. P 255.

18

Efficacy trial on susceptibility of *Amblyomma* ticks for commonly utilized acaricides in North Gondar

Achenef Melaku

Department of Veterinary Pharmacy and Biomedical Sciences, Faculty of Veterinary Medicine, University of Gondar, P. O. Box: 196, Gondar, Ethiopia.

Ticks are blood feeding ectoparasites that induce huge production losses in livestock industry and creating serious public health problems in the world. This study was conducted to assess the efficacy of commonly used acaricides against *Amblyomma variegatum* ticks in North Gondar. The assessment was conducted in laboratory and in the field. For laboratory experiment, adult ticks were collected and exposed to Amitraz or Diazinon. For live animal experiment, sheep infested naturally with *A. variegatum* ticks were kept in door and grouped into three. The first group was treated with Diazinon, the second group with Amitraz, and the last was left untreated as control. In the laboratory, significant difference ($P > 0.05$) was not observed between two acaricides on the killing effect at different concentrations. On live animals, there was a significant ($P < 0.05$) difference between two drugs. More ticks were died in Amitraz treated group than sheep treated with Diazinon. In conclusion, the two acaricides were performing almost equally in the laboratory but Amitraz was better on live animals. To confirm the presence of resistance on live animals, further studies using standard products are needed.

Key words: Acaricides, *Amblyomma* ticks, efficacy, North Gondar, susceptibility.

INTRODUCTION

Tick and tick born diseases (TBD) are widely distributed throughout the world particularly in tropical and subtropical countries, which cause tremendous economic losses in livestock production (Kettle, 1995). The economic losses caused by tick and TBD in cattle alone are estimated at 13.9 to 18.7 billion United States dollar annually worldwide (de Castro, 1997).

The problem is severe in developing countries where the resource for control and eradication is very limited (FAO, 1984). In most parts of Africa, including Ethiopia tick and TBD are economically very important. In Ethiopia, ticks occupy the first place among the external parasites and the economic loss incurred when they infest livestock, particularly cattle is enormous (Solomon et al., 2001).

Ticks are obligate ectoparasites of most types of terrestrial vertebrates. They are large mites and thus are arachnids, members of the subclass Acari. Tick bites, in addition to causing irritation and infestation, have been implicated in the transmission of serious diseases of livestock such as cowdriosis, babesiosis, anaplasmosis and others. These diseases are important causes of morbidity and mortality in livestock. By creating different grade of lesions on the skin, ticks down grade the quality of hides and skins up to 20 to 30%. They also predispose animals to secondary attacks from other parasites such as screw worm flies and infection by pathogens like *Dermatophilus congolensis* and other bac-terial diseases (Seyoum, 2001; Ghosh et al., 2007). Over 60 tick species are reported in Ethiopia. Especially, *Amblyomma* species are widely distributed and the most economically important ticks since they are known to transmit a fatal disease of domestic and wild ruminants in sub-Saharan Africa (heartwater) (Mekonnen, 1998). *Amblyomma* ticks are large; with three host parasites and long mouth parts which cause serious wounds (Fraser and Mays, 1991).Ticks are controlled by the application of acaricides to the body of the animal. Major chemicals used for this purpose are organophosphates, amidines, and synthetic pyrithroids. Some other compounds (chlorinated hydrocarbons and arsenicals) were used but phased out of the market mainly due to the development of tick resistance (Mekonnen, 1998).

In Ethiopia, acaricides are well utilized for the control of ticks, majority of these chemicals are utilized for two or three decades and it is also common to use acaricidal solutions too frequently at a lower concentration than recommended. All these promote the development of tick resistance against acaricides (Ethiopia Sheep and Goat Productivity Improvement Program (ESGPIP), 2010). Different acaricides are imported at the present without proper efficacy test, control and registration (Federal Environmental Protection Agency (FEPA), 2004). The purchase of acaricides by hard currency causes a major drain on the limited foreign exchange available in the country. Farmers and customers utilizing acaricides have complained about their effectiveness to kill or remove all ticks after application.

Moreover, the information about the efficacy of commonly utilized acaricides (Amitraz and Diazinon) on *Amblyomma* ticks in North Gondar zone is limited. Therefore, the objective of this study was to assess the efficacy of commonly used acaricides (Amitraz and Diazinon) against *Amblyomma* species of ticks in the area.

MATERIALS AND METHODS

The study area

The study was conducted from November, 2011 to June, 2012 in North Gondar, at latitude, longitude and altitude of 12.3 to 13.8 North, 35.3-35.7° East and 2,200 meter above sea level, respectively. The rainfall varies from 880 to 1772 mm Hg. The annual mean minimum and maximum temperature of the area varies between 12.3 to 17.7 and 22 to 30°C, respectively. The area has two seasons, the wet season from June to September in which the area gets its majority of rainfall, and the dry season from October to May which receives small and erratic rainfall (central statistical Authority (CSA), 2011).

METHODOLOGY

For laboratory experiment, unattached adult *Amblyomma* ticks were collected on the field and exposed to acaricides (Amitraz and Diazinon). The recommended (1:1000 for Diazinon, 1:625 for Amitraz), double (2:1000 for Diazinon, 2:625 for Amitraz) and half doses were prepared.

One ml of each liquor was added on Petri dish with a filter paper fit at its bottom. Then, the acaricide was evenly distributed and 10 ticks of equal size were placed on each Petri dish and it was closed.

Distilled water was used as a control. The number of ticks live or dead was counted after 24 h of exposure. The experiment was repeated three times for precision. For field trial, sheep infested naturally with *Amblyomma* ticks were kept in door and grouped into three containing five animals in each group. The first group was treated with Diazinon, the second group was treated with Amitraz, and the last group was left untreated as control. Acaricides were sprayed manually by giving more emphasis on tick infested areas. The acaricides' concentration was based on the manufacturer's recommendation for hand spray (1:1000 for Diazinon, 1:625 for Amitraz). The effects of acaricides on ticks were observed after 24, 48 and 72 h of exposure.

Data analysis

The data were recorded in excel spreadsheet. Descriptive statistics (mean, percentage and graphs) were used to express the results. Independent t-test was used to assess the difference between treatment groups. Statistical analysis was performed by using statistical package for social sciences (SPSS) version 19 and P value less than 0.05 was considered significant.

RESULTS

In laboratory

Figure 1 shows that a number of tick died after exposure with acaricides in laboratory at double, recommended and half doses. There was no significant difference ($P > 0.05$) between two acaricides on the killing effect in three different concentrations. The average number of ticks died after distilled water exposure was not more than one.

On live animals

The recommended doses of acaricides were applied on infested live animals. There was a significant ($P < 0.05$) difference between the two drugs. More ticks died in Amitraz treated group than sheep treated with Diazinon (Table 1). In addition to this, it was also observed that more dead ticks were not detached and more engorged ticks were observed in sheep treated with Diazinon.

DISCUSSION

The laboratory result showed absence of significant

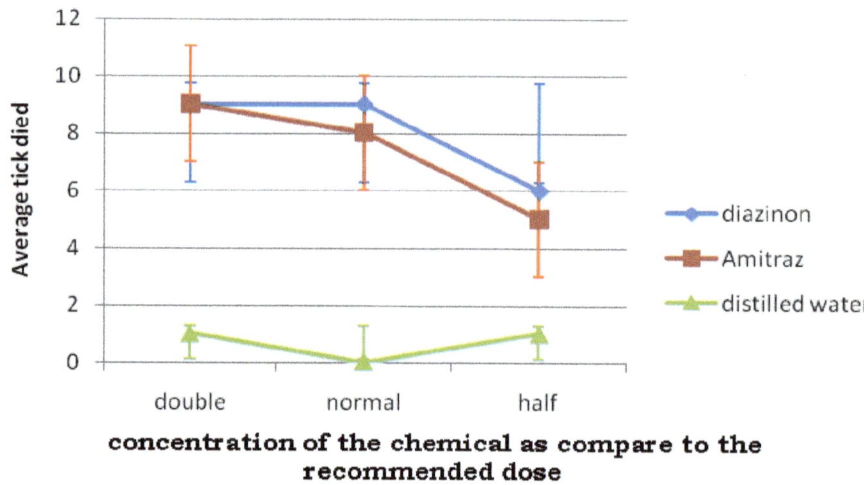

Figure 1. Killing effect of acaricides in the laboratory at double, recommended and half doses.

Table 1. Average tick count before and after treatment in different time interval.

S/n	Group	Average counts			
		Before treatment	After treatment		
			24 h	48 h	72 h
1	Group I (Diazinon)	39	27	14	12
2	Group II (Amitraz)	33	2	1	0
3	control	41	41	40	38

difference between two drugs on killing effect. However, on live animal, Amitraz was more effective than Diazinon. The difference in the result of the laboratory and on live animal may be related with different confounding factors on live animals. Significant proportion of ticks were not dead rather a lot of engorged ticks were observed in Diazinon treated group of sheep. The presence of engorged ticks in Diazinon treated group showed inability of the acaricide to inhibit growth and reproductive capacity of the ticks. This may be related to the decrease in the susceptibility of the tick for Diazinon. Such phenomenon was also complained of by livestock owners in the area. This finding is in line with report of Turkson and Botchey (1999) who reported that field strain of *Amblyomma* is resistant to organophosphates like Diazinon. Emergence of resistant tick against organophosphates acaricides was also stated by Tessema and Gashaw (2010). Eshetu et al. (2013) compared the efficacy of Amitraz and Diazinon on *Amblyomma* and other ticks and found that Amitraz at recommended concentration provides better efficient oviposition inhibition than Diazinon which agrees with the finding of this study.

The frequency of application of acaricides, dilution rate, storage as well as the quality of the product will affect the effectiveness of an acaricide (Turkson and Botchey, 1999).

The efficacy of Amitraz was compared with cyfluthrin, fipronil and permethrin by Burridge et al. (2003) on *Amblyomma* ticks in United States and found that Amitraz was less effective than others. However, Amitraz is an important acaricide because it does have some valuable properties for tick control. Amitraz has been shown to be an excellent detaching agent, inducing the rapid detachment of live ticks from infested animals (Mekonnen, 2001; Natala et al. 2005). However, some studies have shown that Amitraz can take several days to kill ticks and that some surviving ticks can complete engorgement and lay viable eggs. It is for these reasons that Amitraz is an acaricide useful for tick control but not for tick eradication (Burridge et al., 2003). However, still Amitraz is preferable than Diazinon in inhibiting oviposition (Eshetu et al., 2003). Cyfluthrin, fipronil and permethrin were not utilized in North Gondar for the control of ticks (Melaku, 2013) and can be used alternatively with other acaricides to reduce the emerging of drug resistance.

Conclusion

In the laboratory, two acaricides were equally performing but on live animals, Amitraz was preferable than Diazinon. Other types or group of acaricide should also be used rather than depending on limited type of acaricides.

REFERENCES

Burridge MJ, Simmons L, Allan SA (2003). Efficacy of Acaricides for Control of Four Tick Species of Agricultural and Public Health Significance in the United States. J. Agric. Urban Entomol. 20: 207-219.

CSA, 2011. Central statistical authority, Ethiopian statistical abstract. Addis Ababa, Ethiopia.

ESGPIP, 2010. Ethiopian Sheep and Goat Productivity Improvement Program, Control of External Parasites of Sheep and Goats Technical Bulletin No. 41, Addis Ababa, Ethiopia.

Eshetu E, Dinede G, Lakew M, Tolosa T (2013). *In-vitro* efficacy evaluation of Amitraz 0.025% and Diazinon 0.06% against *Rhipicephalus pulchellus* and *Amblyomma gemma* in Borena pastoral area, Southern rangeland of Ethiopia. J. Parasitol. Vector Biol. 5(6):72-76.

FAO (1984). Ticks and tick borne disease control. A practical field manual. Vol.1. Tick control, FAO, Rome pp, 1-299.

FEPA, 2004. Federal Environmental Protection Authority, Draft Assessment Guideline, On Pesticides, May 2004, Addis Ababa, Ethiopia.

Fraser GM, Mays A (1991). The Merck Veterinary Manual, Handbook of Diagnosis, Therapy and Disease Prevention and Control for Veterinarians, (Merck and Co. Inc., Rahway, NJ, USA, 604.

Ghosh S, Azhahianambi P, Yadav M P (2007). Upcoming and future strategies of tick control: a review. J. Vector Borne Dis. 44: 79-89.

Kettle DS (1995). Medical and veterinary Entomology. 2nd ed., CAB International, Wallingford, Oxon, UK. pp. 440-485.

Mekonnen S (1998). Ticks and tick born diseases and control strategies in Ethiopia Agricultural Research Council. Hoechst (Germany) OIE regional collaborating centre pp. 441-446.

Melaku A (2013). Ethnoveterinary practices and Potential Herbal Materials for the Treatment of Ticks in North Gondar. J. Intercul. Ethnopharm. 2(2):85-90. doi:10.5455/jice.20130311081716.

Miruts G (2010). The Distribution and Identification of Hard Tick Species on Cattle in and Around Gondar. DVM thesis, Faculty of Veterinary Medicine, University of Gondar, Ethiopia.

Natala AJ, Agyei AD, Awumbila B (2005). Susceptibility of *Amblyomma variegatum* ticks to acaricides in Ghana. Exp. Appl. Acarol. 35: 259-268.

Seyoum Z (2001). Study of ticks and tick born diseases on cattle at Girana valley in North Wollo zone proceeding of the Ethiopian veterinary association. Vol.15.

Solomon G, Nigist M, Kassa B (2001). Seasonal variation of ticks on calves at Sebeta in Western Shoa zone. Ethio. Vet. J. 7:17-30.

Tessema T, Gashaw A (2010). Prevalence of ticks on local and crossbred cattle in and around Asella town, southeast Ethiopia. Ethio. Vet. J. 14(2): 79-89.

Turkson PK, Botchey M (1999). Acaricide resistance in the cattle tick, *Amblyomma variegatum*, in the coastal savanna zone of Ghana. Ghana J. Agric. Sci. 32: 199-204.

19

Study of *Giardia* spp., *Cryptosporidium* spp. and *Eimeria* spp. infections in dairy cattle in Algeria

N. Ouchene[1], N. A. Ouchene-Khelifi[1], F. Zeroual[1], A. Benakhla[1] and K. Adjou[2]

[1]Département des Sciences Vétérinaires, Université d'El Tarf, BP73, 36000, El Tarf, Algérie.
[2]UMR Bipar, Unité de pathologie du bétail. École Nationale Vétérinaire d'Alfort7, Avenue du Général-de-Gaulle. 94704 Maisons-Alfort, France.

The prevalence of *Giardia* spp., *Cryptosporidium* spp and *Eimeria* spp. in dairy cattle was studied on sixteen farms in three regions at north-eastern Algeria. 690 fecal samples were collected (330 cows and 360 calves aged 1 day to 12 months), were processed by a concentration technique (to examine them for *Giardia* spp. cysts and *Eimeria* spp. oocysts), followed by a modified Ziehl-Neelsen staining to examine them for *Cryptosporidium* spp. oocysts. *Giardia* spp., *Cryptosporidium* spp. and *Eimeria* spp. were reported in 50.75 and 62.5% of all farms, respectively. The prevalence of *Giardia* spp. in calves and adults was 14.72 and 11.81%, respectively with a high level in calves aged 4 to 12 months compared to other age groups ($p < 0.05$). *Cryptosporidium* spp. was observed principally in calves 4 to 30 days old compared to other classes age of calves ($p < 0.01$) and the overall prevalence was 26.11% in calves and 18.78% in adults. *Eimeria* spp. oocysts were excreted by 12.77% of the calves and 4.54% of adults. *Giardia* spp. and *Cryptosporidium* spp. were detected more frequently in diarrheic calves than in non-diarrheic calves ($p < 0.01$). The results demonstrated the presence of these parasites in various age groups of dairy cattle in Algeria which could have important economic and welfare implications.

Key words: *Giardia* spp., *Cryptosporidium* spp., *Eimeria* spp., diarrhea, calves, cows, Algeria.

INTRODUCTION

Giardia spp, *Cryptosporidium* spp and *Eimeria* spp are intestinal protozoan parasites affecting cattle and many vertebrates. Their transmission is essentially by fecal-oral route (Cacciò et al., 2006; Mage and Reynal, 2003). Giardiasis is a zoonosis parasite caused by *Giardia duodenalis* in cattle. It manifests clinically by diarrhea, stunting and wasting (Le Drean-Quenech'du, 2003). The excretion of cysts in feces may last several weeks unlike *Cryptosporidium parvum*, *Giardia duodenalis* affects older calves (O'Handley et al., 1999). Cryptosporidiosis is mainly due to *C. parvum*. This species is a zoonosis and a dominant digestive tract pathogen in young ruminants. The principal characteristic of this parasitic infection is the large infectivity and the great difficulty of treatment or prevention in the absence of truly effective medicaments. During the first two weeks of life, the calf can shed thousands of oocysts of *C. parvum*, while adult individuals remain mostly asymptomatic carriers (Chartier and Paraud, 2010).

Eimeria bovis and *Eimeria zuernii* are considered to be

the most pathogenic species in cattle. After infection, the appearance of the diarrhea usually begins after 2 weeks. The disease usually occurs in older calves from 3 weeks to 18 months (Mage and Reynal, 2003). Several studies in the world determined the prevalence of infection of these three parasites in cattle (O'Handley et al., 1999; Fayer et al., 2000; Gül et al., 2008; Huetink et al., 2001). However, in Algeria the frequency of these three protozoa in cattle is poorly understood, especially for *Giardia* spp. and *Eimeria* spp. This study is a contribution to the study of the epidemiology of these three protozoa in dairy cattle in three regions (Bejaia, Setif and Souk Ahras) in Algeria.

MATERIALS AND METHODS

Study area

This study was realized in three areas in North-eastern Algeria: Bejaia, Setif and Souk Ahras.

Animals and samples

During a period of January, 2010 to December, 2010, a study was conducted in 16 dairy farms selected in a random manner. Cattle were the only species bred in these farms. A single fecal sample was collected from each animal. A total of 690 fecal samples were collected including 360 calves (aged one day to 12 months) and 330 cows (adults > 24 months). Samples of 47 calves were diarrheal and no adult was found with diarrhea (liquid faeces). The number of samples in the region of Bejaia was 223 (99 calves and 124 cows) realized in 5 farms. In the Setif region, the number was 407 samples (201 calves and 206 cows) belong to 8 farms. At the region of Souk Ahras, this number was 60 calves, belonging to 3 farms. These samples were collected in sterile vials directly from the rectum of animals (or digital rectal in younger animals) and sent to the laboratory in a cooler (at a temperature of 4°C approximately) for a period of one to six hours for analyses of suspected parasites.

Laboratory analysis

The samples were analyzed directly upon their arrival at the laboratory. Each sample was analyzed by the technique of Ritchie simplified by Allen and Ridley (1970). A drop of the sediment was placed between slide and cover slide to search *Giardia* spp. and *Eimeria* spp. The examination was conducted under an optical microscope (200 × 400). To search for *Cryptosporidium* spp. a fecal smear was prepared from the sediment obtained by the method previously described and then the smear was stained by the technique of Ziehl-Neelsen modified by Henriksen and Pohlenz (1981). The examination was conducted under an optical microscope (400 × 800).

Statistical analysis

The χ^2 test (95% confidence interval) was used to study the influence of the age of the calves and the diarrhea on the prevalence excretion of *Giardia* spp. *Cryptosporidium* spp. and *Eimeria* spp. The same test was also used to investigate the influence of different types of infections (singles-infections or associations) of these three parasites on the prevalence excretion (significant results when $p < 0.05$).

RESULTS

In all sixteen dairy farms concerned by this study *Giardia* spp., *Cryptosporidium* spp. and *Eimeria* spp. were found in 8 (50%), 12 (75%) and 10 farms (62.50%), respectively. With calves aged between 1 day to 12 months, the search for *Giardia* spp revealed that 14.72% (53/360; 95% confidence interval) subjects were excreting. This prevalence varied in different farms from 12.12 to 28.88%. The *Cryptosporidium* spp. prevalence in the different farms varied from 15.78 to 42.22% with the overall of 26.11% (94/360; 95% confidence interval). By contrast, overall *Eimeria* spp. prevalence in calves was the lowest at 12.77% (46/360; 95% confidence interval) (with a range of prevalence on different farms from 5.55 to 27.77%). In adult cows, 11.81% (39/330; 95% confidence interval), 18.78% (62/330; 95% confidence interval) and 4.54% (15/330; 95% confidence interval) were excreting *Giardia* spp, *Cryptosporidium* spp. and *Eimeria* spp., respectively (Table 1). In different farms, the prevalence varied from 2.08 to 25.68% for *Giardia* spp, 7.14 to 50% for *Cryptosporidium* spp and 8.33 to 20.83% for *Eimeria* spp.

Giardia spp., *Cryptosporidium* spp. and *Eimeria* spp. were found in calves after 8, 4 and 12 days old, respectively and in adults (Table 1). The study of the relationship between the animal age and the prevalence of *Giardia* spp. showed that calves aged between 4 to 12 months were the most affected (25.86%; 30/116; 95% confidence interval) compared to all others age groups ($p < 0.05$). Calves aged 4 to 30 days (36.70%; 51/139; 95% confidence interval) have a higher prevalence of *Cryptosporidium* spp. compared to other age groups ($p < 0.01$) and calves aged 1 to12 months (15.71%; 33/210; 95% confidence interval) have been the most excretory of *Eimeria* spp. cysts ($p < 0.05$).

Giardia spp., *Cryptosporidium* spp. and *Eimeria* spp. were demonstrated in 36.17% (17/47; 95% confidence interval), 68.08% (32/47; 95% confidence interval) and 19.14% (9/47; 95% confidence interval) of the diarrhea calves, respectively (Table 2). The presence of *Giardia* spp. and *Cryptosporidium* spp. in calve samples was significantly associated with diarrhea ($p < 0.01$). By contrast, no significant relationship was observed between the excretion of *Eimeria* spp. and diarrhea. Also, diarrheic calves aged 4 to 30 days were the most excretory of *Cryptosporidium* spp. oocysts with a prevalence of 88% ($p < 0.01$) against a prevalence of 45.45% among those aged from 1 to 12 months. Of the 47 diarrheic calves, 46.80% were revealed excreting oocysts of *Cryptosporidium* spp. without the presence of *Giardia* spp. and *Eimeria* spp. This prevalence was significantly high compared to other types of associations ($p < 0.01$).

Table 1. Prevalence of *Giardia* spp, *Cryptosporidium* spp and *Eimeria* spp infections in dairy cattle according to age in three areas in North-eastern Algeria: Bejaia, Setif and Souk Ahras.

Age	No. of sample	Number of animals infected (%; 95% confidence interval)		
		Giardia spp.	*Cryptosporidium* spp.	*Eimeria* spp.
1-3 days	11	0 (-)	0 (-)	0 (-)
4-7 days	23	0 (-)	8 (34.78)	0 (-)
8-14 days	35	3 (8.57)	11 (31.42)	3 (8.57)
15-21 days	39	3 (7.69)	15 (38.46)	5 (12.82)
22-30 days	42	4 (9.52)	17 (40.47)	5 (11.9)
1-3 months	94	13 (13.82)	22 (23.40)	16 (17.02)
4-12 months	116	30 (25.86)	21 (18.10)	17 (14.65)
Adults	330	39 (11.81)	62 (18.78)	15 (4.54)

Table 2. Prevalence of *Giardia* spp., *Cryptosporidium* spp. and *Eimeria* spp. infections in diarrheic and non-diarrheic calves (aged 01 day to 12 months).

Aspect of the feces	No. of calves infected (%)		
	Giardia spp.	*Cryptosporidium* spp.	*Eimeria* spp.
Diarrheal feces (n=47)	17 (36.17)	32 (68.08)	9 (19.14)
Non diarrheal feces (n=313)	36 (11.5)	62 (19.8)	37 (11.82)
Total (n=360)	53 (14.72)	94 (26.11)	46 (12.77)

Table 3. Prevalence of these three parasites infections in diarrheic calves (aged 1 day to 12 months) according to the types of infections.

Infection type (single-infection or association)	No. of diarrheal samples	%
Cryptosporidium spp. only	22	46.80
Giardia spp. only	9	19.14
Eimeria spp. only	0	0
Cryptosporidium spp. + *Giardia* spp.	4	08.51
Cryptosporidium spp. + *Eimeria* spp.	5	10.63
Giardia spp. + *Eimeria* spp.	3	6.38
Cryptosporidium spp. + *Giardia* spp. + *Eimeria* spp.	1	02.12

Thus, only three calves (6.38%) excreted cysts of *Giardia* spp. and *Eimeria* spp. alone. In addition, a single calf (2.12%) was reported excretory of these three parasites at the same time (Table 3).

DISCUSSION

Several studies in the world determine the prevalence of these three parasites in cattle farms (O'Handley et al., 1999; Fayer et al., 2000; Gül et al., 2008; Huetink et al., 2001). The results reached by these authors show widely varying rates, some of which are consistent with our observations. For *Giardia* spp., Ruest et al. (1998) reported that 45.7% of all dairy farms were positive. Quílez et al. (1996), on their part, found that 53.3% of farms are infected with *Giardia* spp. These results are close to ours (50%) concerning *Cryptosporidium* spp., Olson et al. (1997) reported that 80% of farms were infected and according to Quílez et al. (1996), this prevalence was 63.3%. In our study, *Cryptosporidium* spp. was observed in 75% of all farms. Concerning *Eimeria* spp., we reported that 62.5% of farms were infected and the results obtained by different studies report that the infection prevalence had been ranging from 43.2 to 100% of the farms tested (Bourguignon et al., 2010; Lassen et al., 2009; Richard, 2006).

In cattle, the prevalence of *Giardia* spp. can range from

9 to 73% (Olson et al., 1997; Mcallister et al., 2005). In our study this prevalence is 14.72% in calves (aged 1 day to 12 months) and 11.81% in adults. These proportions are close to the results presented by Gow and Waldner (2006) on the calves (22.6%) and adults (17%). Again, Quílez et al. (1996) reported that the prevalence in calves was 18.23 and 2.2% in adults.

The excretion of *Giardia* spp. in calves can begin on the fourth day after birth (Wade et al., 2000). In this study, *Giardia* spp. was observed at the age of 8 days, which is similar to the results highlighted by Bjorkman et al. (2003) which reported that excretion begins at the age of nine days. We noted in this study also that the highest *Giardia* spp. prevalence was reported in calves aged 4 to 12 months ($p < 0.05$). A study showed that the highest prevalence was observed in calves aged 4 to 5 months (Huetink et al., 2001).

On the other hand, we found that the prevalence of *Cryptosporidium* spp. in calves and adults was, respectively 26.11 and 18.78%. In Algeria, the prevalence of cryptosporidiosis in cattle is poorly understood. The studies focused in the area of Mitidja, revealing the excretion rate of *Cryptosporidium* spp. in calves ranging from 21.21 to 28.44% (Akam et al., 2004; Khelef et al., 2007; Khelef et al., 2002). Quilez et al. (1996) in a study conducted in Spain, reported that the prevalence in calves from 3 days to 24 months of age is 20.97% and in adults is 17.8%. These results are similar to ours.

The relatively high prevalence (18.78%) recorded during our study in adults is due to the fact that these animals are cows from calving, and in a very contaminated environment with *Cryptosporidium* spp. The same observation was established by other authors (Huetink et al., 2001; Quilez et al., 1996; Olson et al., 1997). After the detection of *Cryptosporidium* spp. in the feces of calves of 4 days of age, we found that the calves were infected immediately after birth. This is consistent with the studies of several authors (Quilez et al., 1996; Castro-Hermida et al., 2002).

In addition, we noted that the highest rate of the excretion was observed in calves aged 4 to 30 days ($p < 0.01$). Such remark confirms the observation noted by other authors (Huetink et al., 2001; Olson et al., 1997). The prevalence of bovine coccidiosis varies from 3.1 to 31.3% depending on the species involved (Kennedy and Kralka, 1987). In this study, *Eimeria* spp. was isolated from 12.77% of all calves. Thus, calves aged 1 to 12 months have been the most infected ($p < 0.05$). This result was supported by Hansen and Perry (1994) in calves of this age group. Oocysts of *Eimeria* spp. were observed in our study from the twelfth day of age. This result corroborates the observation from Chibunda et al. (1997).

On the other hand, we noted that the prevalence of *Eimeria* spp. is 4.54% for adults a result quite similar to that reported by Fayer et al. (2000) which is 5.26%. Of the 360 calves concerned by this study, 47 (13.05%) of them had diarrhea and 36.17% of these diarrheal calves excreted *Giardia* spp. and this excretion was significantly associated with diarrhea ($p < 0.01$). This result is identical to that obtained by McAllister et al. (2005) who showed that 36% of cases of giardiasis in diarrheic calves and excretion of *Giardia* spp. was significantly associated with diarrhea. But some authors have observed that the excretion of *Giardia* spp. in calves has no relationship with diarrhea (Huetink et al., 2001; Quilez et al., 1996; Adjou, 2011).

Cryptosporidiosis has a major role in neonatal diarrhea calves (Chartier and Paraud, 2010). The immune system of calves during the first days of life is still immature, hence the high sensitivity to cryptosporidiosis. Several authors have noted a higher prevalence of the parasite in the feces of diarrheic calves (Huetink et al., 2001; Quilez et al., 1996; Adjou, 2011). In our study, *Cryptosporidium* spp. was more observed in diarrheic calves than in those who were not and especially for those aged less than 30 days ($p < 0.01$). The prevalence reported in diarrheal and non-diarrheal calves are 68.08 and 19.08%, respectively. These values are similar to those observed by Sevinc et al. (2003), who found 63.92 and 9.85% in diarrheal and non-diarrheal calves, respectively.

For *Eimeria* spp. we found that 19.14% of diarrheic calves were excreting, but no significant relationship was observed with the presence or absence of diarrhea. The same observation was reported by Lassen et al. (2009) and Quílez et al. (1996). However, Alemayehu et al. (2013) in Ethiopia and Chibunda et al. (1997) in Tanzania noted that the excretion of *Eimeria* spp. was very significantly associated with diarrhea. The study of the association between these three parasites in calves shows that diarrheal calves tend to excrete *Cryptosporidium* spp. alone than in combination with the other two parasites ($p < 0.01$). The same observation was represented by Adjou (2011) and Quilez et al. (1996).

Conclusion

From this study, we conclude that infection of cattle in Algeria by these three protozoa seems very important. These parasites were observed in young calves and adults. Adult cows asymptomatic porters are important sources of infections. *Cryptosporidium* spp. mainly affects young calves under 30 days of age. However, *Giardia* spp. and *Eimeria* spp. hit older calves, 4 to 12 months and 1 to 12 months, respectively. The excretion of *Cryptosporidium* spp. and *Giardia* spp. was significantly related to the presence of diarrhea.

Conflict of Interest

The authors declare that they have no conflict of interests.

REFERENCES

Adjou K (2011). Moins de cryptosporidies lors de vaccination contre les virus et bactéries. Le Point Vétérinaire, NO. 315.

Akam A, Khelef D, Kaidi R, Othmani A, Lafri M, Tali-Maamar H, Rahal K, Tahrat N, Chirila F, Cozma V, Abdul-Hussain MS (2004). Fréquences d'isolement de *Cryptosporidium parvum*, *d'Escherichia coli* K99 et de *Salmonella* spp. chez les veaux diarrhéïques et non diarrhéïques dans six fermes laitières de la Mitidja d'Algérie (Résultats préléminaires). Sci. Parasitol. 5(1-2):13-21.

Alemayehu A, Nuru M, Belina T (2013). Prevalence of bovine coccidia in Kombolcha district of South Wollo, Ethiopia. J. Vet. Med. Animal Health. 5(2):41-45.

Allen VH, Ridley DS (1970). Further observations on the formol ether concentration technique for faecal parasites. J. Clin. Pathol. 23:545-546.

Bjorkman C, Svensson C, Christensson B, De Verdier K (2003). *Cryptosporidium parvum* and *Giardia intestinalis* in calf diarrhoea in Sweden. Acta. Vet. Scand. 44:145-152.

Bourguignon JM, Mathieux S, Medina B (2010). Enquête sur la prévalence de la coccidiose dans les exploitations bovines en Wallonie. Renc. Rech. Ruminants. 3:17.

Cacciò SM, Andrew Thompson RC, Mclauchlin J, Smith HW (2006). Unravelling *Cryptosporidium* and *Giardia* epidemiology. Trends Parasitol. 21:43-437.

Castro-Hermida JA, González-Losada YA, Mezo-Menéndez M, Ares-Mazás E (2002). A study of cryptosporidiosis in a cohort of neonatal calves. Vet. Parasitol. 106:11-17.

Chartier Ch, Paraud C (2010). La cryptosporidiose des ruminants. Bull. GTV 52:109-118.

Chibunda RT, Muhairwa AP, Kambarage DM, Mtambo MM, Kusiluka LJ, Kazwala RR (1997). Eimeriosis in dairy cattle farms in Morogoro municipality of Tanzania. Prev. Vet. Med. 31(3-4):191-197.

Fayer R, Trout JM, Graczyk TK, Lewis EJ (2000). Prevalence of *Cryptosporidium*, *Giardia* and *Eimeria* infections in post-weaned and adult cattle on three Maryland farms. Vet. Parasitol. 93:103-112.

Gow S, Waldner C (2006). An examination of the prevalence of and risk factors for shedding of *Cryptosporidium* spp. and *Giardia* spp. in cows and calves from western Canadian cow–calf herds. Vet. Parasitol. 137(1-2):50-61.

Gül A, Çiçek M, Kilinç Ö (2008). Prevalence of *Eimeria* spp., *Cryptosporidium* spp. and *Giardia* spp. in Calves in the Van Province. Türk. Parazitol. Derg. 32(3):202-204.

Hansen J, Perry B (1994). The epidemiology, diagnosis and control of helminthes parasites of ruminants. International Laboratory for Research on Animal Diseases (ILRAD), Nairobi, Kenya.

Henriksen SA, Pohlenz JFL (1981). Staining of *Cryptosporidia* by a modified Ziehl-Neelson technique. Acta Vet. Scand. 22:594-596.

Huetink REC, Vander Giessen JWB, Noordhuizen JPTM, Ploger HW (2001). Epidemiology of *Cryptosporidium* and *Giardia duodenalis* on a dairy farm. Vet. Parasitol. 102:53-57.

Kennedy MJ, Kralka RA (1987). A survey of *Eimeria* spp. in cattle in central Alberta. Can. Vet. J. 28:124-125.

Khelef D, Akam A, Kaidi R, Abdulhussein MS, Suteu E, Cozma V (2002). Evaluation comparative des méthodes de détection de l'oocyste de *Cryptosporidium parvum* dans les selles des veaux. Sci. Parasitol. 1:22-27.

Khelef D, Saib MZ, Akam A, Kaidi R, Chirila V, Cozma V, Adjou KT (2007). Epidémiologie de la cryptosporidiose chez les bovins en Algérie. Rev. Med. Vet. 158:260-264.

Lassen B, Viltrop A, Raaperi K, Jarvis T (2009). *Eimeria* and *Cryptosporidium* in Estonian dairy farms in regard to age, species, and diarrhea. Vet. Parasitol. 166:212-219.

Le Drean-Quenech'du S (2003). Parasitisme. L'importance de la giardiose. Acta Vét.1654:11-12.

Mage C, Reynal PH (2003). Les coccidioses bovines : La clinique et l'excrétion parasitaire, Journées Nationales des GTV, Nantes.

Mcallister TA, Olson ME, Fletch A, Wetzstein M, Entz T (2005). Prevalence of *Giardia* and *Cryptosporidium* in beef cows in southern Ontario and in beef calves in southern British Columbia. Can. Vet. J. 46:47-55.

O'Handley RM, Cockwill C, Mcallister TA, Jelinski M, Morck DW, Olson ME (1999). Duration of naturally acquired giardiasis and cryptosporidiosis in dairy calves and their association with diarrhea. J. Am. Vet. Med. Assoc. 214:391-396.

Olson ME, Guselle NJ, O'Handley RM, Swift ML, Mcallister TA, Jelinski MD, Morck DW (1997). *Giardia* and *Cryptosporidium* in dairy calves in British Columbia. Can. Vet. J. 38:703-706.

Quilez J, Sanchez-Acedo C, Del Cacho E, Clavel A, Causape AC (1996). Prevalence of *Cryptosporidium* and *Giardia* infections in cattle in Aragon (Northeastern Spain). Vet. Parasitol. 66:139-146.

Richard A, Rizet C, Reynaud A, Valognes A (2006). Prévalence des coccidies sur veaux Charolais à l'étable dans l'Allier. Renc. Rech. Ruminants 4:13.

Ruest N, Faubert GM, Couture Y (1998). Prevalence and geographical distribution of *Giardia* spp. and *Cryptosporidium* spp. in dairy farms in Quebec. Can. Vet. J. 39:697-700.

Sevinc F, Irmak K, Sevinc M (2003). The prevalence of *Cryptosporidium parvum* infection in the diarrhoiec and non- diarrhoeic calves. Rev. Méd. Vét. 154(5):357-361.

Wade SE, Mohammed HO, Schaaf SL (2000). Prevalence of Giardia sp., Cryptosporidium parvum and *Cryptosporidium muris* (*C. andersoni*) in 109 dairy herds in five counties of southeastern New York. Vet. Parasitol. 93:1-11.

20

The prevalence of intestinal helminthic infections and associated risk factors among school Children in Lumame town, Northwest, Ethiopia

Mengistu Wale[1], Melaku Wale[2] and Tesfu Fekensa[1]

[1]Ethiopian Biodiversity Institute, P. O. Box 30726 Addis Ababa, Ethiopia.
[2]College of Science, Bahar Dar University P. O. Box 26 Bahir Dar, Ethiopia.

This study was conducted on prevalence of intestinal helminthes infection and their associated risk factors among school children from a rural and a semi urban setting in Lumame town, Northwest, Ethiopia. A cross-sectional parasitological study was conducted to determine the prevalence and associated risk factors of intestinal helminthes infection. A total of 402 students' stool samples were taken and processed with direct wet mount and formalin ether concentration techniques from December to January 2011/2012. A structured questionnaire was prepared to assess the association of intestinal helminthes infection with socio-demographic and socioeconomic variables. The data collected was analyzed using χ^2 test and logistic regression ($p < 0.05$ was considered as statistically significant). The overall prevalence rate for at least one intestinal helminthes infection was 54.5%. Of which *Ascaris lumbricoides* (32.6%) was the dominant followed by hookworm (12.2%); the others were minor cases. High rate infection (*A. lumbricoides*) was recorded among students who had dirty finger nails, large family, habit of eating undercooked vegetable, walking barefoot, and had no latrine than their respective counterpart. Such relatively high prevalence rate of helminthes infection in the study area could be used as a baseline for the concerned bodies to launch de-worming intervention.

Key words: Intestinal helminthes, prevalence, school children, Lumame town.

INTRODUCTION

Intestinal helminthes infections are among the most common and neglected public health problems in many developing countries including Ethiopia. Worldwide, more than 3.5 billion people are infected with intestinal worms (Luong, 2003), and an estimated 4.5 billion individuals are at risk of soil-transmitted helminths infection (Uneke, 2010). Majority of infected people live in resource-poor settings, 80% of these in sub-Saharan Africa (Jamison et al., 2006). Epidemiological surveys have revealed that poor sanitary conditions such as open field defecation and faecal contamination of water bodies are the most important factors leading to intestinal worm infestation, while the spread is due to lack of personal hygiene (Okyay et al., 2004; Hung et al., 2005; Abebe et al., 2001).

In Ethiopia, the prevalence and distribution of intestinal helminth infection among school children varies from place to place (Mengistu and Birhanu, 2004; Girum, 2005). Likewise, Yared et al. (2001) reported that the low economic standard, poor sanitation, ignorance of simple health promotion factors favor the wide distribution of intestinal helminth infections. Of all types of disease in the country, helminthiasis is the second most common cause of outpatient morbidity next to malaria (Yared et al., 2001).

Although several studies have been conducted on the distribution and prevalence of intestinal parasite in Ethiopia, there are still several localities for which epidemiological information is not available. One of such localities is Lumamae town. Thus, the present investigation was undertaken to determine the prevalence and associated risk factors of intestinal helminth infection, to provide recent and valuable information and to set baseline data for those who are working in the prevention and control of intestinal helminth infection among school children in the area.

MATERIALS AND METHODS

Study area

Lumame is located 260 km away from the capital city, Addis Ababa with an altitude of 1850 m above sea level and an average temperature of 16.25°C. The area is predominantly rural about 89% and most residents live in villages as agriculturalists (AFEO, 2011).

Study design and period

Institution based cross-sectional study was conducted among Lumame primary school children (grade 1 to 8) from December to January 2011/2012.

Study population

The study population was students attending Lumame elementary school in 2011/2012 academic year. The total population enrolled by that academic year was 2865.

Sample size determination and sampling techniques

The sample size was determined using single population proportion formula by considering the prevalence of intestinal parasites in primary school children 95% confidence interval (CI), and design effect that give a final sample size of 422. Proportional allocation for each grade was determined and the desired sample was obtained by systematic sampling technique using the class roster formula (Daniel et al., 1999):

$$n = z^2_{\alpha/2}\, p\, (1-P)/d^2$$

where n = total number of sample size required, Z = Z statistics for a level of confidence ($Z_{\alpha/2}$ = 1.96), d = marginal error for 95% confidence level, p = expected prevalence of intestinal helminthes in the study area. Since, there was no previous study in the area; P value of 0.5 was taken.

Data collection and laboratory procedures

Data about the socio demographic characteristics and other associated factors were collected using a semi structured based questionnaire. Onsite training was given for interviewers. The interviewers also inspected whether they wore shoe or not. Approximately 2 g of stool specimen was collected using clean, tightly corked, leak proof containers. Small amount of the sample was analyzed using wet mount technique and the remaining portion was concentrated using formal-ether concentration technique and examined microscopically (Adefioye et al., 2011).

Statistical analysis

All statistical analysis was performed using Statistical Package for Social Sciences (SPSS) for window version 16 statistical package. Descriptive statistics was used to present prevalence of intestinal helminth infection as percentage and proportion. The association between student's socio-demographic or socioeconomic variables and prevalence of parasitism was analyzed using Chi-square test. Logistic regression analysis was used to determine the independent effect of the variables by calculating the strength of the association between infection and risk factors using odds ratio (OR) and 95% confidence interval (CI). P-value < 0.05 was considered as statistically significant.

RESULTS

Socio-demographic and other characteristics of study subjects

A total of 422 students were sampled from grade 1 to 8 in this study. Of these, 402 (95.3%) of them positively responded for intestinal helminthes examination by providing fresh stool. The age of study subjects range from 6 to 19 with mean age of 11.67 years.

Students were grouped into three age groups (6 to 10, 11 to14, and 15 to 19 years old). Age group 6 to 10 year accounts for 39.5%, while the remaining group was 40.8 and 19.7%, respectively. Students were further grouped into two grade levels (grades 1 to 4 and 5 to 8) and grade 5 to 8 group accounts for 51.2% of the study subjects. From a total of 402 study participants, 121 (30%) were from nearby rural area and 281 (70%) were urban settings. The proportion of male and female participants was nearly equal 206 (51.2%) and 196 (48.8%), respectively.

Half of student's mothers were illiterate and concerning the occupation, 42.3% students mother were housewives; the remaining, 31.3, 23.2 and 3.2% earn their living as government employees, merchants and daily laborers, respectively.

More than half of the study participants (52.7%) used pipe water, while 39.6 and 7.7% used rivers/streams and well water as source of drinking, respectively. Regarding latrine availability and defecation habit, 61.4% of the study participant had latrine, and 60.7% defecate in latrine only while the rest 39.3% in open field.

Regarding shoe wearing habit, 43.5% wore shoe always, 40.1% sometimes and 16.4% not at all. On assessment about family size, 42.3% of the study subjects

Table 1. Single, double and triple intestinal helimnth infections.

Types of intestinal helminth infections	Frequency (n = 402)	Percentage
A. lumbricoides	107	26.6
Hookworm species	31	7.7
Trichuristrichuria	7	1.7
Hymnolepis nana	18	4.5
S. stercoralis	9	2.2
Taenia species	8	2
E. vermicularis	12	3
A. lumbricoides plus Hookworm species	14	3.5
A. lumbricoides plus *T. trichuria*	4	1
T. trichuria plus *S. stercoralis*	3	0.75
A. lumbricoides plus *E. vermicularis*	2	0.5
A. lumbricoides plus Hookworm plus *T. trichuria*	4	1
Single	192	47.8
Double	23	5.7
Triple	4	1
Total	219	54.5

subjects had family size range from 1 to 3, 26.6% had 4 to 6 member and 31.1% above 6.

Concerning food habit, 37.1% of the study subjects eat raw or under cooked meat and 38.1% study subjects eat unwashed or under cooked vegetables while the remaining 62.9 and 61.9% did not such a habit, respectively. About 96% of the study subjects washed their hands before eating a meal. However, observation of the students hand cleanliness at the time of interview revealed that 32.8% of the total had dirty material in their finger nails. Furthermore, 74.1% of the study subjects did not have a habit of swimming.

Prevalence of intestinal helminth infections

Based on stool sample diagnosis, the prevalence intestinal helminth infections in the study area were identified. Generally, seven types of intestinal helminth species were encountered: *Ascaris lumbricoides*, Hookworm (*Necator americanus* or *Ankylostoma duodenale*), *Trichuris trichuria, Sterongloides stercoralis, Enterobius vermicularis*, *Taenia* species and *Hymenolepis nana* (Figure 1). Out of 402 stool specimen examined, 219 (54.5%) were positive for one or more intestinal helminth infections (Table 1).

Variation was observed in prevalence rate of each intestinal helminth species. The prevalence of *Taenia* spp. was the lowest as compared with other intestinal helminthes encountered. *A. lumbricoides* (32.6%) was the most predominant helminth infection followed by Hookworm (12.2%), *T. trichuria* (4.5%), *H. nana* (4.5%), *E. vermicularis* (3.3%), and *S. stercoralis* (3%).

Nearly half (47.8%) of the study subjects had a single helminth infections, while 5.7 and 1% had double and triple helminth infections, respectively (Table 1). The most frequent combination of helminth infection identified was co-infection of *A. lumbricoides* and Hook worm (3.5%), followed by *A. lumbricoides* and *T. trichuria* (1.0%) and the least being *E. vermicularis* and *A. lumbricoides* (0.5%).

There was a statistical significant association between age group, mother's occupation, family size, educational status of mothers, and availability of latrine, shoe wearing habit, defecation habit, presence or absence of dirty matter on finger nails, eating raw/unwashed vegetable with overall intestinal helminth infection ($p < 0.05$) (Table 2).

Helminth infection varied significantly with grade level of study subjects (χ^2=9.305, df=1, p=0.002). Grade level 1 to 4 students harbor more helminth infection, 55.7% of the total infected students than grade level 5 to 8. The overall prevalence of intestinal helminth infection with respect to age group was statistically significant (χ^2=14.827, df=2, p=0.001). Children between the age group 6 to 10 years old were more affected by helminth infection than the other age groups. Prevalence decreased in age group 11 to 14 years old and the least being 15 to 19 age groups (Table 2).

Risk factors associated with intestinal helminth infections

Risk factor analysis was performed for all variables that were significantly associated with any of the intestinal helminth infection from Chi-square test analysis. Multivariate logistic regression analysis on intestinal helminth infection showed that there was no significant difference in infection rate between grade levels, age groups, mother's education, mother's occupation, defecation

Table 2. Association of socio-demographic and other characteristics with intestinal helminth infection.

Variable	Intestinal helminth infection		χ^2	P- value
	Positive n (%)	Negative n (%)		
Age group				
6 – 10	105 (66)	54 (34)		
11 – 14	74 (45.1)	90 (45.9)	14.827	0.001
15 -19	40 (50.6)	39 (49.4)		
Grade level				
1 – 4	122 (62.2)	74 (37.8)	9.305	0.002
5 – 8	97 (47.1)	109 (52.9)		
Gender				
Male	115 (52.8)	91 (44.2)	0.309	0.578
Female	104 (53.1)	92 (46.9)		
Residence				
Urban	145 (51.6)	136 (48.4)	3.114	0.082
Rural	74 (61.2)	47 (38.8)		
Source of drinking water				
Stream/rivers	110 (69.2)	49 (30.8)		
Pipe water	88 (41.5)	124 (58.5)	30.43	0
Well water	21 (67.7)	10 (32.3)		
Latrine availability				
Present	97 (39.3)	150 (60.7)	59.73	0
Absent	122 (78.7)	33 (21.3)		
Defection habit				
Latrine	98 (40.2)	146 (59.8)	51.288	0
Open field	121 (76.6)	33 (23.4		
Shoe wearing habit				
Always	67 (38.3)	108 (61.7)		
Sometimes	98 (60.9)	63 (39.1)	41.047	0
Not at all	54 (81.8)	12 (18.2)		
Hand washing				
Yes	208 (53.9)	178 (46.1)	1.369	0.242
No	11 (68.8)	5 (31.2)		
Eating raw meat				
Yes	83 (57.7)	66 (44.3)	0.144	0.705
No	136 (53.8)	117 (46.2)		
Eating unwashed vegetables				
Yes	117 (76.5)	36 (23.5)	48.177	0
No	102 (41.0)	147 (59.0)		
Swimming				
Yes	59 (56.7)	45 (43.3)	0.287	0.648
No	160 (53.7)	138 (46.3)		

Table 2. Cont'd.

Dirty finger nails				
Present	114 (86.4)	18 (13.6)	80.574	0
Absent	105 (38.9)	165 (61.1)		
Mothers education				
Illiterate	138 (68.7)	63 (31.3)	32.59	0
Literate	81 (40.3)	120 (59.7)		
Family size				
1 – 3	42 (24.7)	128 (75.3)		
4 – 6	74(69.2)	33 (30.8)	109.357	0
> 6	103 (82.4)	22 (17.6)		
Mothers occupation				
House wife	109 (64.5)	60 (35.5)		
Merchant	57 (60.0)	38 (40)	36.969	0
Daily laborer	10 (76.9)	3 (23.1)		
Government employed	43(34.4)	82(65.6)		

habit and source of drinking water ($p > 0.05$). None of these were risk factors for helminth infection in this study (Table 3). On the other hand, presence of dirty material in the finger nails was the potential risk factor (OR = 7.925 for dirty versus clean finger nails; 95% CI 3.983 – 15.766). Study subjects with dirty material in finger nails were 7.925 times more likely to be infected. Similarly, the model showed that having large family was a risk factor for helminth infection (OR = 6.949 for family member >6 versus 1 to 3, $P < 0.000$ and OR = 4.784 for family member 4 to 6 versus 1 to 3, $p < 0.000$).

Regarding feeding habit, study subjects who eat unwashed or undercooked vegetables and fruit were more likely exposed to helminth infection than those who did not eat. The difference was significant (OR = 4.095 for eating versus not eating; 95% CI: 2.176 – 7.704).

The model also showed that shoe wearing habit was a risk factor for helminth infection. Students who did not wear shoe were at higher risk (OR = 0.36 for those who did not wear shoe versus those who wear always; 95% CI: 0.147 - 0.884).

Furthermore, the model identified the lack of latrine as a risk factor for helminth infection. The difference was significant (OR = 0.307 for none latrine versus latrine; 95% CI: 0.097 – 0.973). Students who did not have latrine were 0.307 times more likely to be infected than those that had latrine.

Relationship between intestinal helminth species and some selected variables

Although, age groups, grade levels, source of drinking water and defecation habit and mother's occupation were not identified as risk factor for helminth infection. They were significantly associated with intestinal helminth infection. Except *E. vermicularis* and *T. trichuria* each helminth species identified in this study revealed significant difference with respect to age group (χ^2). Prevalence of hookworm and *Taenia* spp. were significantly increased with age, whereas *A. lumbricoides*, *H. nana* and *S. stercoralis* significantly decreased (Table 4).

The two predominant helminth infections in this study, *A. lumbricoides* and Hookworm, were significantly associated with source of drinking water while the rest of the helminthes did not. Source of drinking water significantly associated with *A. lumbricoides* and Hookworm infection among students (χ^2= 15.928, df = 2, p = 0.000) and (χ^2= 7.81, df = 2, p = 0.02), respectively.

Study subjects who defecated in open field harbor more helminth infections than those that defecated in latrine (76.6% versus 40.2% from the total helminth infection). Defecation habit was significantly associated with *A. lumbricoides* and Hookworm infections among study subjects (χ^2= 16.267, df = 1, p = 0.000) and (χ^2= 9.245, df = 1, p = 0.002), respectively (Figure 2).

Logistic regression analysis revealed shoe wearing as risk factors for hookworm infection. Study subjects who did not wear shoe were 3.28 times at risk of infection thanthose who wore shoe always and the difference were statistically significant. Although, the overall helminth prevalence did not differ significantly with respect to residence (Rural versus Urban), hookworm infection was significantly different between urban and rural students and being rural was a risk factor. Rural student were 3.070 times more exposed to hookworm infection than urban students (Table 5).

Table 3. Multivariate logistic regression analysis showing the impact of selected risk factors on intestinal helminth infections.

Risk factor	Helminth infection		Adjusted OR	95% CI	P-value
	Positive	Negative			
Age group					
6 -10	105	54	2.238	0.687 - 7.292	0.181
11 - 14	74	90	0.814	0.366 – 1.808	0.613
15 - 19	40	39	1	-	-
Grade level					
1 – 4	122	74	0.68	0.26 – 1.777	0.680
5 – 8	97	109	1		
Latrine availability					
Present	97	150	1	0.097 – 0.973	0.045
Absent	122	33	0.307		
Defecation habit					
Open field	121	33	1.131	0.377 – 3.395	0.826
Latrine	98	146	1		
Shoe wearing habit					
Always	67	108	1	-	-
Sometimes	98	63	0.236	0.088 – 0.628	0.004
Not at all	54	12	0.36	0.147 – 0.884	0.026
Source of drinking water					
Stream/rivers	110	49	1.259	0.607 – 2.612	0.536
Well water	21	10	1.288	0.416 – 3.984	0.66
Pipe water	88	124	1	-	-
Eating undercooked vegetable					
Yes	117	36	4.095	2.176 – 7.704	0.000
No	102	147	1		
Dirty material in finger nails					
Present	114	18	7.925	3.983 – 15.766	0.000
Absent	105	165	1		
Mothers education					
Illiterate	138	63	0.88	0.446 – 1.736	0.713
Literate	81	120	1		
Family size					
1 – 3	92	128	1	-	-
4 – 6	74	33	4.784	2.373 – 9.646	0.000
>6	103	22	6.949	3.272 – 14.756	0.000
Mothers occupation					
House wives	109	60	0.383	0.067 – 2.184	0.258
Merchants	57	38	0.358	0.06 – 2.127	0.280
Daily laborers	10	3	0.247	0.04 – 1.513	0.130
Government employees	43	82	1	-	-

OR stands for Odds Ratio, 95% CI for the 95 percent confidence interval.

Table 4. The relationship between intestinal helminth infections and age groups.

Age group (years)	Al [n (%)]	Hk [n (%)]	Tt [n (%)]	Tsp [n (%)]	St [n (%)]	Hn [n (%)]	Ev [n (%)]
6-10	67 (42.1)	14 (8.8)	4 (2.5)	0 (0)	10 (6.3)	13 (8.2)	7 (4.4)
11-14	49 (29.9)	16 (9.8)	12 (7.3)	3 (1.8)	1 (0.8)	3 (1.8)	7 (4.3)
15-19	15 (19)	19 (24.1)	2 (2.5)	5 (6.3)	1 (1.3)	2 (2.5)	0 (0)
Total	131 (32.6)	49 (12.2)	18 (4.5)	8 (2)	12 (3)	18 (4.5)	14 (2.6)
Chi-square	12.896	12.993	5.222	10.876	9.996	8.474	3.55
P- value	0.002	0.002	0.073	0.004	0.007	0.014	0.169

Table 5. The risk factors of Hookworm infection with respect to shoe wearing habit and residence.

Risk factor		Hookworm		OR	95% CI	P- value
		Positive [n (%)]	Negative [n (%)]			
Shoe wearing habit	Always	4 (8.2)	171 (48.4)	1	-	-
	Sometimes	15 (30.6)	146 (41.4)	1.12	0.527- 2.377	0.769
	Not at all	30 (61.2)	36 (10.2)	3.28	1.545 – 6.964	0.002
Residence	Urban	23 (46.9)	258 (73.1)	1	1.671 – 5.641	0.000
	Rural	26 (53.1)	95 (26.9)	3.070		

DISCUSSION

Studying the infection prevalence of intestinal helminthes and associated risk factors in different localities is a primary objective to identify high risk communities and select appropriate intervention mechanisms. In line with this view, the present study attempted to assess the prevalence of intestinal helminth infections and risk factors among Lumame elementary school children.

This study identified seven types of intestinal helminth species. The overall prevalence of any helminth infection was 54.5% among students. This result was in line with prevalence rates reported by Teklemariam and Teklemariam (1983) and Olusegum et al. (2011). On the other hand, the present finding was higher than that reported by Girum (2005) and Anosike et al. (2006). The differences in findings among the studies might be due to variations in socio-economic conditions, individual behavioral habits of selected children, the methods employed for stool examination, the sample size taken and the time of study (Mengistu and Berhanu , 2004; Yared et al., 2001).

A. lumbricoides was the most dominant helminth infection followed by hookworm. The result was higher than school children reported by Ihesiulor et al. (2006) and Abebe et al. (2011). However, it was comparable to the reports by Lapiso et al. (2002) and Olusegun et al. (2011). Hookworm species was lower in the study conducted by Mengistu and Birhanu (2004) and Fekadu et al. (2008). But it was in line with studies Mani et al. (2002) and Anosike et al. (2006). The difference in prevalence rate of hookworm could be climate and behavior of study subjects. The relatively low prevalence of *T. trichuria*, *H. nana*, *S. stercoralis*, *Taenia* spp. and *E. vermicularis,* in this study was in agreement with other studies (Yared et al., 2001; Fekadu et al., 2008; Asrat et al., 2011).

In this study, the presence of dirty material in finger nails, having large family and availability of latrine were the main risk factor of *A. lumbricoides* infection. The result is in line with other studies (Dongre et al., 2008; Asrat et al., 2011; Olusegun et al., 2011). This is probably due to insufficient water supplies, poor hygienic practice, poor socio-economic status of the study subjects, and contamination of vegetables with fecal materials in the farm.

A. lumbricoides significantly affected the age group and grade level in the study. Similar findings were reported by Awasthi et al. (2003), Fleming et al. (2006), Inabo and John (2010) and Abebe et al. (2011). This high infection rate of *A. lumbricoides* could be linked to the route of infection being faeco-oral, since children within this age are easily susceptible due to poor level of hygiene and indiscriminately play on fecal contaminated grounds and have a common habit of placing soiled fingers in the mouth. On the other hand, low prevalence rate among older age group and grade level 5 to 8 in this study, was possibly due to the change in attitude, habits, and more awareness regarding to personal hygiene among the older school children.

In addition, children who defecated in open field and used rivers/streams as source of drinking water harbored more *A. lumbricoides* infection than their respective counterparts. This result was consistent with other studies Yared et al. (2001) and Asrat et al. (2011). This is

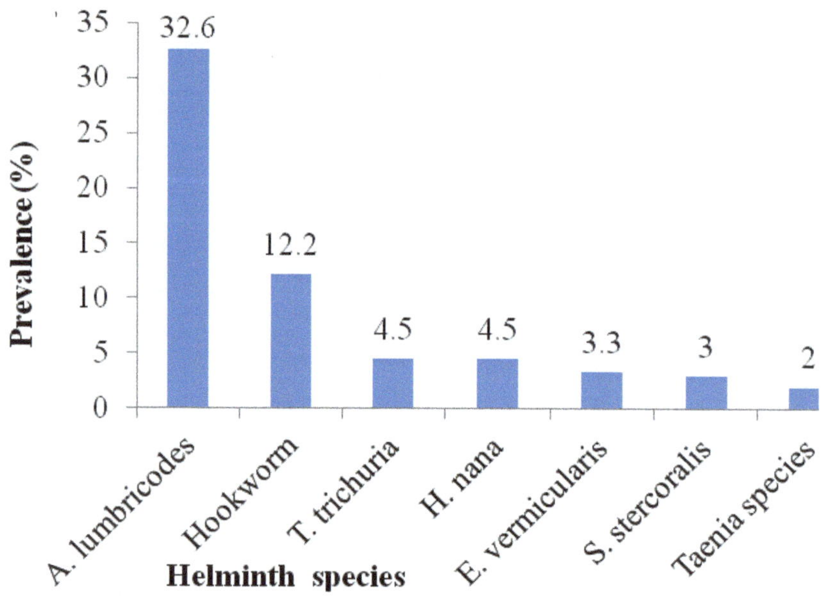

Figure 1. Prevalence of intestinal helminth infection in the study subjects.

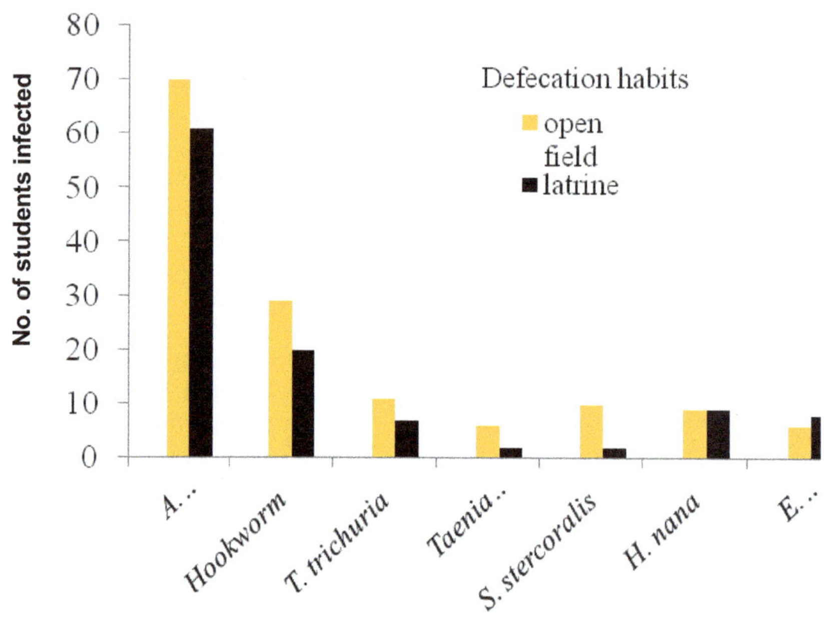

Figure 2. Prevalence of intestinal helminth infections with respect to defecation habits.

the environment and hence favor the transmission.

Similarly, literacy rate of mother and mothers' occupation are found to be important helminth transmission factors. The result was in agreement with findings reported by Okyay et al. (2004) and Hung et al. (2005). This is more likely that parents of children at high level of education provide better sanitation condition for their children than low educational level parents and low level of hygiene and high exposure potentials of daily laborer mothers, respectively.

Hookworm infection, in this study significantly varied with age. This is in agreement with other studies (Anosike et al., 2006; Fleming et al., 2006). High rate of hookworm infection in older aged group probably linked to having high outdoor activities such as helping in farming, playing at the back yard in moist soil, etc., resulting in higher exposure to infective egg and filariform larvae in the soil.

In this study, students who did not wear shoe were more likely to be infected by hookworm infection than who wore shoe always. Similarly, being rural was the risk factor for hookworm infection than urban. This finding was in agreement with Lapiso et al. (2002).

CONCLUSION AND RECOMMENDATION

Intestinal helminth infections were prevalent in varying magnitude among Lumame elementary school children and were important health problem with relatively high overall rate of prevalence (54.5%). The predominant helminth was *A. lumbricoides* followed by Hookworm and the rest were minor cases. The relatively high prevalence rate of intestinal helminth infection in the study area might be the reflection of poor sanitation of the environment, poor personal hygiene, relatively unhygienic water supply, and others.

The presence of dirty material in finger nails was the main risk factor for intestinal helminth infection followed by having large family size, consumption of unwashed or raw vegetable, walking bare foot and lack of toilet in the study area. Therefore, personal sanitary education, having small family size, proper cooking or washing of vegetables before eating, using toilet for defecation, and always wearing shoe were important measures that greatly reduced the predominant intestinal helminth infections prevalent in the study area.

In order to reduce the impact of helminth infection among children, adequate intervention strategies should be designed and implemented. Therefore, the authors recommended that local health sector should collaborate with school health program for delivering health education to enhance the knowledge and attitude of the children against transmission of intestinal helminthes. The local government should improve pipe water supply, latrine facilities and health education to the society since *A. lumbricoides* infection can be significantly reduced by improving these services. Furthermore, the current study did not address the intensity of helminth infection. Hence, further research focusing on intensity infection should be conducted for further status of helminth infection in the study area.

Conflict Interests

The authors declare that there is no conflict of interests.

REFERENCES

Abebe A, Asmamaw A, Zelalem A, Biniam M, Birhan Simon G, Belay G (2011). Soil transmitted helminth and schistosoma *mansoni* infection among school children in Zarima town. BMC Infect. Dis. 11(189):1-7.

Adefioye O, Efunshile A, Ojurongbe O, Akindele A, Adewuyi K (2011). Intestinal Helminthiasis among School Children in Ilie, Osun State, Southwest, Nigeria. Sierra Leone J. Biomed. 3:36-42.

AFEO (2011). Awabel Woreda Finance and Economy development Office statistical report. Unpublished.

Anosike J, Zaccheasu V, Adiyongo C, Abeobi O, Dada E, Oku E, Keke I, Uwaezuoke J, Ama Juoyo O, Obiukwu C, Nwosu D, Ogbusu F (2006). Studies on intestinal worm infestation in central Nigerian Rural Community. J. Appl. Sci. Environ. 10(2):61-66.

Asrat A, Tewodros D, Alemayehu W (2011). Prevalence and risk factors of intestinal parasite among Delgi school children North Gondar, Ethiopia. J. Parasitol. Vect. Biol. 3(5):75-81.

Awasthi S, Bundy D, Savioli L (2003). Helminthes infection. Biol. J. Med. 227:431-433.

Daniel WT (1999). Biostatistics: A foundation for analysis in the health science, 7th ed., New York: John Wiley and Sons Inc., USA.

Dongre A, Deshmukh P, Boratne A, Thaware P (2008). An approach to hygiene education among rural Indian school going children. J. Health Allied Sci. 4:2.

Fekadu D, Beyene P, Amha K (2008). Hookworm species distribution among school children in Asendabo town, Jimma Zone, South West Ethiopia. Ethiop. J. Health Sci. 18(2):53-57.

Fleming F, Brooker S, Geiger S, Caldas I, Hotez P, Bethony J (2006). Synergistic associations between hook worm and other helminth species in rural community in Brazil. Trop. Med. Int. Health 11(1):56-64.

Girum T (2005). The prevalence of intestinal helminthic infections and associated factors among school children in Babile town, Eastern Ethiopia. Ethiop. J. Health Dev. 19(2):140-147.

Hung L, Vries P, Giao P, Binh T, Nam N, Kager P (2005). Anemia, Malaria and Hookworm Infections in a Vietnamese Ethnic Minority. Southeast Asian J. Trop. Med. Publ. Health 36(4):816-821.

Ihesiulor G, Emokpae M, Adeeke S, Samaila A (2006). Soil transmitted helminthiasis among apparently healthy children in Kano Municipality. Afr. J. Clin. Exp. Microbiol. 7(3):77-82.

Inabo H, John H (2010). Intestinal helminthes among children in orhanages in some parts of Kaduna state of Nigeria. Contin. J. Microbiol. 4:31-36.

Jamison D, Berman J, Measham A (2006). Disease control priorities in developing countries. 2nd ed. Wasington (DC), World Bank. pp. 467-479.

Lapiso E, Yared M, Ayele A, Mohammed A, Ashenafi B, Sultan M, Shiferaw T, Solomon S (2002). Prevalence of Hookworm infection and hemoglobin status among rural elementary School children in Southern Ethiopia. Ethiop. J. Health Dev. 16(1):113-115.

Luong TV (2003). De-worming school children and hygiene intervention. Int. J. Environ. Health Res. 13(1):153-159.

Mani T, Rajendran R, Munirathinam A, Sunis I, Abdullah P, Augustin D, Satyanarayana K (2002). Efficacy of co-administration of albendazole and diethylcarbamazine against Geohelminthiases: A study from South India. Trop. Med. Int. Health. 7(6):541-548.

Mengistu L, Birhanu E (2004). Prevalence of intestinal parasites among school children in a rural area close to the Southeast of Lake Langano, Ethiopia. Ethiop. J. Health Dev. 18(2):116-120.

Okyay P, Ertug S, Gultekin B, Onen O, Beser E (2004). Intestinal parasite prevalence and related factors in school children, a Western city sample Turkey. BMC Pub. Health 22:4-64.

Olusegun A, Efunsile A, Ojurongbe O, Akindele A, Adewuyi I, Bolaji O, Adedokun S, Adeyeba A (2011). Intestinal helminthiasis among school children in Ilie, Osum state, Southwest, Nigeria. Sierra Leone J. Biomed. Res. 3(1):36-42.

Teklemariam T, Teklemariam A (1983). Intestinal helminth infection in Lake Ziway Island, Central Ethiopia. Ethiop. Med. 21:14-153.

Uneke CJ (2010). Soil transmitted helminth infections and schistosomiasis in school age children in sub-Saharan Africa: Efficacy of chemotherapeutic intervention since World Health Assembly Resolution. Tanz. J. Health 12(1):1-9.

Yared M, Madouh H, Girma M, Shiferaw T (2001). Intestinal helminthic infection among school children around Lake Awassa area, South Ethiopia. Ethiop. J. Health Dev. 15(1):31-37.

APPENDIX

I. A questionnaire prepared to assess prevalence and risk of intestinal helminth infection on the study subjects (students) among Lumame elementary school.

Part I: Student's school record and Socio-demographic data
1. Name of student____________
2. Student code_______________
3. Grade level ________________
4. Age______________________
5. Sex: 1. Male 2. Female
6. Residence: 1. Urban 2. Rural

Part II: socioeconomic data
1. From where do you get water for drinking? From:
 1. Streams or rivers 2. Pipe water 3. Well water
2. Do your parents have latrine? 1. Yes 2.No
3. Where do you defecate your feces? 1. Open field 2. Latrine only
4. How often do you wear shoes? 1. Always 2. Sometimes 3. Not at all (no shoe)
5. Do you wash your hands before eating meal? 1. Yes 2. No
6. Do you eat raw meat? 1. Yes 2. No
7. Do you eat unwashed or raw vegetables or fruits? 1. Yes 2. No
8. Do you swim in the nearby river? 1. Yes 2. No
9. Is there dirty materials in the finger nails (if not trimmed off)? (Inspected by Interviewer) 1. Yes 2. No
10. What is your mother education level?
a) Unable to read/write (Illiterate) b) Able to read /write (Literate)
11. What is the number of your family? a) 1 - 3 b) 4 –6 c) >6
12. What is four mother's occupation? A. house wife b. merchant c. daily laborer d. government employees

2. Parasitological investigation of the study subjects at the study site health center Laboratory data

Code of study subject________________________
Date of examination__________________________
1. Microscopic Examinations
1.1 Direct wet mount microscopic observation
A. Intestinal helminthes egg or larva observed_______, _______, ________
B. other parasites (protozoa)__________________,___________________
1.2. Formalin-ether concentration method
A) Ova of helminthes parasite observed________________, _____________
B) Other parasites
____________________, ___________________
Name of laboratory investigator:______________________________
Signature: ___________________
Date:______________________

21

Centre for Disease Control and Prevention (CDC) bottle bioassay: A real complementary method to World Health Organization (WHO) susceptibility test for the determination of insecticide susceptibility in malaria vectors

Nazaire Aïzoun[1,2], Rock Aïkpon [1,2], Roseric Azondekon [1,3], Virgile Gnanguenon [1,2], Razaki Osse[1,2], Gil Germain Padonou[1,2] and Martin Akogbéto[1,2]

[1]Centre de Recherche Entomologique de Cotonou (CREC), 06 BP 2604, Cotonou, Bénin.
[2]Faculté des Sciences et Techniques, Université d'Abomey Calavi, Calavi, Bénin.
[3]University of Massachusetts Amherst, Amherst, Massachusetts, USA.

In order to investigate the complementarities between both World Health Organization (WHO) and Centre for Disease Control and Prevention (CDC) methods and the specificities of each method useful for the determination of the insecticide susceptibility in malaria vectors, larvae and pupae of *Anopheles gambiae s.l.* mosquitoes were collected from the breeding sites in Littoral department. WHO susceptibility tests were conducted on unfed female mosquitoes aged 2 to 5 days old with impregnated-papers with deltamethrin (0.05%) and bendiocarb (0.1%), whereas CDC susceptibility tests were conducted with stock solutions of deltamethrin and bendiocarb (12.5 µg per bottle). WHO method and CDC method gave comparable results. There were complementarities between both methods. Both WHO and CDC methods are two complementary tools for the determination of insecticide susceptibility in malaria vectors. However, each method has also its own specificity.

Key words: Complementary tools, insecticide, resistance, malaria vectors.

INTRODUCTION

Monitoring of insecticide resistance is a necessary element of any medium-scale or large-scale deployment of an insecticidal intervention. In 2010, 78 countries reported that they were carrying out insecticide resistance monitoring (World Health Organization (WHO), 2011). Current methods of malaria control are highly dependent on a single class of insecticides, the pyrethroids, which is the only insecticide class used for insecticide treated nets (ITNs) and accounts for approximately 77% of indoor residual spraying (IRS) in terms of spray area covered. The widespread use of a single class of insecticide increases the risk that mosquitoes will develop resistance to it. This risk is of particular concern in sub-Saharan Africa, where insecticidal vector control is being deployed with unprecedented levels of coverage. Resistance to pyrethroids has been reported in 27 countries in sub-Saharan Africa; the point at which this reduces the effectiveness of vector control is still uncertain and may depend on the locally identified resistance mechanism (WHO, 2011).

Routine monitoring of insecticide resistance in the natural populations of vectors helps us to detect early resistance and improve effectiveness of operational control strategies (Aïzoun et al., 2013a). Insecticide resistance in a vector population is initially detected and characterized by using some sort of bioassay to determine whether a particular insecticide is able to control a vector at a given time. Ideally, this fundamental question should be answered before a particular insecticide is chosen and procured for vector control (Brogdon and Chan, 2010). So, insecticide susceptibility in malaria vectors is the first important step in insecticide resistance surveillance. A recent study was carried out to investigate the advantages and drawbacks of both Centre for Disease Control and Prevention (CDC) bottle bioassay and WHO susceptibility test (Aïzoun et al., 2013a).

The aim of this study was to investigate the complementarities between both WHO and CDC methods and the specificities of each method used for the determination of the insecticide susceptibility in malaria vectors.

METHODOLOGY

Study area

The study area is located in Republic of Benin (West Africa) and includes the department of Littoral in the Southern Benin. The study was carried out in Cotonou district more precisely in Ladji location. Ladji is a large village on the outskirts or on the periphery of Cotonou, the capital of Benin. This village floods annually during the rainy season, creating breeding sites for *Anopheles gambiae*. This location is crossed by the Nokoue Lake's streams. The choice of the study sites took into account the economic activities of populations, their usual protection practices against mosquito bites, and peasant practices to control farming pests. These factors have a direct impact on the development of insecticide resistance in the local mosquito vectors. Cotonou is characterized by a tropical coastal guinean climate with two rainy seasons (April to July and September to November). The mean annual rainfall is over 1,500 mm.

Mosquito sampling

A. gambiae s.l. mosquitoes were collected during the rainy seasons (April to July and September to November, 2011) across Cotonou district selected in southern Benin. Larvae and pupae were collected from breeding sites and kept in separated labeled bottles. The samples were reared to adults in the Centre de Recherche Entomologique de Cotonou, Benin (CREC) insectary. *A. gambiae* Kisumu, a reference susceptible strain, was used as a control for the bioassay tests. Susceptibility tests were done simultaneously following WHO and CDC protocols on unfed female mosquitoes aged 2 to 5 days old, reared from the larval and pupal collections. *A. gambiae s.l.* sample was separated into two batches: batch 1 was used for susceptibility tests following the WHO protocol and batch 2 for CDC susceptibility tests. All susceptibility tests were conducted in the CREC laboratory at 25 ± 2°C and 70 to 80% relative humidity.

Testing insecticide susceptibility

WHO protocol

The principle of the WHO bioassay is to expose insects to a given dose of insecticide for a given time to assess susceptibility or resistance. The standard WHO discriminating dosages are twice the experimentally derived 100% lethal concentration (LC100 value) of a reference susceptible strain (WHO, 1998). In this study, two insecticides were tested: deltamethrin (0.05%) and bendiocarb (0.1%). The choice of bendiocarb was justified by its use for IRS campaign under the financial support of the President's Malaria Initiative (PMI) to control *A. gambiae s.l.* populations from Ouemé department in southern Benin (2008 to 2010). We used deltamethrin because it is the insecticide used on PermaNets that are distributed free by the National Malaria Control Programme (NMCP) in the swampy areas of Ouemé (2008 to 2010). Ouemé department has a boundary with Littoral department. An aspirator was used to introduce 20 to 25 unfed female mosquitoes aged 2 to 5 days old from batch 1 into five WHO holding tubes (four tests and one control) that contained untreated papers. WHO susceptibility test was done once without replicate. They were then gently blown into the exposure tubes containing the insecticide impregnated papers. After one-hour exposure, mosquitoes were transferred back into holding tubes and provided with cotton wool moistened with a 10% honey solution. The number of mosquitoes "knocked down" at 60 min and mortalities at 24 h were recorded following the WHO protocol (WHO, 1998).

CDC protocol

The principle of CDC bottle bioassay is to determine the time it takes an insecticide to penetrate an arthropod, traverse its intervening tissues, get to the target site, and act on that site relative to a susceptible control. Anything that prevents or delays the compound from achieving its objective of killing the arthropods contributes to resistance. Diagnostic doses that were applied in the current study were the doses recommended by CDC (Brogdon and Chan, 2010). These doses were checked on the *A. gambiae* Kisumu susceptible reference strain before being applied to field populations. For *A. gambiae s.l.*, the diagnostic dose of 12.5 µg per bottle for both deltamethrin and bendiocarb was used for a diagnostic exposure time of 30 min. The solutions were prepared and the bottles coated according to the CDC protocol (Brogdon and Chan, 2010). Fifteen to 20 unfed female mosquitoes aged 2 to 5

Table 1. Susceptibility of *Anopheles gambiae s.l.* populations to pyrethroids.

Populations	Insecticides	Number tested		% Mortality		Resistance status		P-values
		WHO	CDC	WHO	CDC	WHO	CDC	
Kisumu (Control)	Deltamethrin	92	98	100	100	S	S	P=1
Ladji	Deltamethrin	72	64	100	100	S	S	P=1

days old from batch 2 were introduced into four 250 ml Wheaton bottles coated with insecticide and one control bottle coated with acetone only. CDC susceptibility test was also done once without replicate. The number of dead or alive mosquitoes was monitored at different time intervals (15, 30, 35, 40, 45, 60, 75, 90, 105 and 120 min). This allowed us to determine the total percent mortality (Y axis) against time (X axis) for all replicates using a linear scale.

Data analysis

The resistance status of mosquito samples from batch 1 was determined according to the latest WHO criteria (WHO, 2013) as follows:

1. Mortality rates between 98 to 100% indicate full susceptibility.
2. Mortality rates between 90 to 97% require further investigation.
3. Mortality rates < 90%, the population is considered resistant to the tested insecticides.

The resistance status of mosquito samples from batch 2 was determined according to the CDC criteria (Brogdon and McAllister, 1998; Brogdon and Chan, 2010). The susceptibility thresholds at the diagnostic time of 30 min for pyrethroids and carbamates are:

1. Mortality rate = 100%: the population is fully susceptible.
2. Mortality rate < 100%: the population is considered resistant to the tested insecticides.

Abbott's modified formula was not used in this study for the correction of mortality rates in either the test-tubes or test-bottles because the mortality rates in all controls was always less than 5% (Abbott, 1987). Analysis using Fisher's exact test and test of proportion was performed on the data sets gathered from the locality of Ladji surveyed and from Kisumu to compare each of two tested insecticides and assess the resistance status of each tested *A. gambiae* population using both WHO and CDC methods. The software R-2.15.2. (R Development Core Team, 2011) was used for the statistical analysis. The significance level was set at 5%.

RESULTS

Susceptibility of *A. gambiae s.l.* populations to pyrethroids

The results of 24 h mortality recording after mosquito exposure to WHO impregnated papers with deltamethrin (0.05%) were compared to those recorded with CDC bottles bioassays at the susceptibility threshold (30 min). CDC bottles bioassays were performed with stock solutions of deltamethrin (1.25%) (Table 1). Kisumu strain (control) confirmed its susceptibility status as a reference strain according to both WHO and CDC methods. The percentage of dead mosquitoes recorded with deltamethrin was 100% (92/92) with WHO method and 100% (98/98) with CDC method. *A. gambiae s.l.* populations from Ladji were also susceptible to deltamethrin according to both methods. The percentage of dead mosquitoes recorded was 100% (72/72) according to WHO method whereas the mortality rate recorded with CDC method was 100% (64/64).

Susceptibility of *A. gambiae s.l.* populations to carbamates

The results of 24 h mortality recording after mosquito exposure to WHO impregnated papers with bendiocarb (0.1%) were compared to those recorded with CDC bottles bioassays at the susceptibility threshold (30 min). CDC bottles bioassays were performed with stock solutions of bendiocarb (1.25%) (Table 2). Kisumu strain (control) confirmed its susceptibility status as a reference strain according to both WHO and CDC methods. The percentage of dead mosquitoes recorded with bendiocarb was 100% (93/93) with WHO method and 100% (100/100) with CDC method. *A. gambiae s.l.* populations from Ladji were also susceptible to bendiocarb according to both methods. The percentage of dead mosquitoes recorded was 100% (56/56) according to WHO method whereas the mortality rate recorded with CDC method was 100% (56/56).

Complementarities between both WHO and CDC methods and specificities of each method

The complementarities and specificities of both methods are mentioned in Table 3.

DISCUSSION

A. gambiae s.l. Kisumu populations were susceptible to deltamethrin according to both WHO and CDC methods.

Table 2. Susceptibility of *Anopheles gambiae s.l.* populations to carbamates.

Populations	Insecticides	Number tested		% Mortality		Resistance status		P-values
		WHO	CDC	WHO	CDC	WHO	CDC	
Kisumu (Control)	Bendiocarb	93	100	100	100	S	S	P=1
Ladji	Bendiocarb	56	56	100	100	S	S	P=1

Table 3. Complementarities between both WHO and CDC methods and specificities of each method.

Parameter		Both WHO and CDC methods allow to determine insecticide susceptibility in malaria vectors
WHO and CDC methods	Complementarities	CDC bottle bioassay can be assessed when the results obtained with WHO method require further investigation in order to clarify the resistance status of malaria vectors to insecticides
		Both WHO and CDC protocols require the assessment of insecticide susceptibility tests in malaria vectors before any identification of mechanisms involved in malaria vector resistance to insecticides
		Mosquitoes from the assessment of both WHO and CDC methods can be used for PCR test
		Both WHO and CDC protocols recommend to use female *Anopheles* mosquitoes in the assessment of insecticide susceptibility test in malara vectors
		The purchasse of both WHO and CDC kits useful in the assessment of susceptibility tests are centralized
	Specificities	WHO bioassays utilize cylinder plastic tubes whereas CDC bottles bioassays use 250 ml Wheaton bottles which are made from glass
		WHO susceptibility test uses impregnated-papers with insecticide whereas CDC bottle need to be coated with insecticide by oneself before each bioassay
		WHO cylinder plastic tube test determines directly the percent mortality or mortality rate of malaria vectors to insecticide
		CDC bottle bioassay determines the time it takes an insecticide to penetrate an arthropod, traverse its intervening tissues, get to the target site, and act on that site relative to a susceptible control

Susceptibilirty tests done with these methods using deltamethrin with *A. gambiae s.l.* Ladji populations showed that these populations were also susceptible to this product. Aïzoun et al. (2013b) have recently shown that Ladji is crossed by the Nokoue Lake's streams which sweep and converge the environmental pollutants and pesticide residues in this locality, and these xenobiotics available in larval breeding sites in Ladji may be one of the possible factors selected for pyrethroid resistance in *A. gambiae* populations in this locality. However, *A. gambiae s.l.* Ladji populations were fully susceptible to deltamethrin in the current study. These results showed that *A. gambiae s.l.* populations resistance level varied from one year to another in a same location. So, the results obtained with both methods are comparable. A similar pattern was also obtained with bendiocarb. In fact, both *A. gambiae s.l.* Kisumu and Ladji populations were fully susceptible to bendiocarb according to both methods. These results showed that *A. gambiae s.l.* populations from Southern Benin were still fully susceptible to bendiocarb. Simmilar results were also recently obtained by Aïzoun et al. (2013c) in Seme district in Ouemé Department.

Insecticide susceptibility in malaria vectors can be detected either with WHO method or with CDC method. In addition, when the obtained results with WHO method require further investigation, CDC method can be assessed to clarify the real vector resistance status regarding the used insecticide. Both WHO and CDC protocols required the use of diagnostic doses or concentrations already pre-established for numerous insecticides used in public health against malaria vectors in the assessment of insecticide susceptibility tests in malaria vectors. But in case where these diagnostic doses did not correspond to those of some local vectors, CDC protocol required its own diagnostic doses establishment for these local vectors. Regarding the new PCR-based molecular tests used to detect the presence of *kdr* mutations in *A. gambiae* from Western parts of Africa, this is not a generally recommended practice and the WHO susceptibility tests (or the CDC bottle bioassays) should always be carried out in addition to the molecular assays (WHO, 2013).

In fact, after the assessment of insecticide susceptibility tests with either WHO method or CDC method, mosquitoes from these tests can be used for polymerase chain reaction (PCR)-based molecular tests. In addition, both WHO and CDC protocols required the use of female *Anopheles* mosquitoes only in the assessment of insecticide susceptibility tests in malaria vectors. The use of males is not recommended for resistance monitoring as they are usually smaller and more fragile than females, and therefore tend to have higher control mortalities (WHO, 2013; Brogdon and Chan, 2010). In addition, male *Anopheles* mosquitoes were not able to transmit malaria. For this reason, susceptibility testing is conducted using only female mosquitoes (WHO, 2013; Brogdon and Chan, 2010). In addition, some resistance mechanisms are sex-linked, and one can be misled by using males in the control (Brogdon and Chan, 2010).

WHO assay requires the purchase of all components (WHO kit) from a centralized source and that allows easy comparison of results from one year to another and from one study site to another (Aïzoun et al., 2013a). Test kits and insecticide-impregnated papers are prepared on behalf of WHO by the Universiti Sains Malaysia, which is based in Penang, Malaysia for this purpose. In a similar way, CDC assay requires the purchase of all components (CDC kit) from a centralized source (CDC Atlanta, USA). Test kits and test insecticides should be procured from the CDC for this purpose (WHO, 2013). According to Perea et al. (2009), the bottle assay is a simple, flexible and robust resistance monitoring tool that was able to discriminate pyrethroid (deltamethrin) resistance in mosquito populations as effectively as the WHO assay. Even if there were complementarities between both methods, there were also some specificities between them.

Regarding the specificities of each method, it is important to mention that CDC bottle bioassay determines the time it takes an insecticide to penetrate an arthropod, traverse its intervening tissues, get to the target site, and act on that site relative to a susceptible control. WHO cylinder plastic tube test determines the percent mortality or mortality rates of malaria vectors to insecticide. WHO bioassays utilize cylinder plastic tubes whereas CDC bottles bioassays use 250 ml Wheaton bottles which are made from glass. WHO papers do not need to be treated by oneself before their utilization because they are always ordered in the impregnated form. Conversely, CDC bottles need to be coated with insecticide by oneself before each bioassay. In fact, the shelf-life and re-use of pre-prepared bottles are still not well documented or studied in laboratory conditions (Aïzoun et al., 2013a).

Conclusion

The current study clearly shows that both WHO and CDC methods are two complementary tools for the determination of insecticide susceptibility in malaria vectors. However, each method has also its own specificity. Insecticide susceptibility in malaria vectors can be assessed either with WHO cylinder plastic tube test or CDC bottle bioassay.

Conflict of Interest

The authors declare that they have no conflict of interests.

ACKNOWLEDGEMENTS

The authors thank the Ministère de l'Enseignement Supérieur et de la Recherche Scientifique (MESRS) of Benin for supporting the doctoral training of Nazaire. They also thank Dr William G. Brogdon from CDC Atlanta, USA who supplied them with the reagents used for CDC bioassays. The authors would like to thank Damien Todjinou for providing technical assistance. This research was funded by the President's Malaria Initiative of the U.S. Government through USAID and the Ministère de l'Enseignement Supérieur et de la Recherche Scientifique(MESRS) of Benin.

REFERENCES

Abbott WS (1987). A method of computing the effectiveness of an insecticide. J. Am. Mosq. Control Assoc. 3(2):302-303.

Aïzoun N, Ossè R, Azondekon R, Alia R, Oussou O, Gnanguenon V, Aikpon R, Padonou GG, Akogbéto M (2013a). Comparison of the standard WHO susceptibility tests and the CDC bottle bioassay for the determination of insecticide susceptibility in malaria vectors and their correlation with biochemical and molecular biology assays in Benin, West Africa. Parasit. Vectors 6:147.

Aïzoun N, Aïkpon R, Gnanguenon V, Azondekon R, Oké-Agbo F, Padonou GG, Akogbéto M (2013b). Dynamics of insecticide resistance and effect of synergists piperonyl butoxide (PBO), S.S.S-tributylphosphorotrithioate (DEF) and ethacrynic acid (ETAA or EA) on permethrin, deltamethrin and dichlorodiphenyltrichloroethane (DDT) resistance in two *Anopheles gambiae s.l.* populations from southern Benin, West Africa. J. Parasitol. Vector Biol. 6(1):1-10.

Aïzoun N, Aïkpon R, Gnanguenon V, Oussou O, Agossa F, Padonou GG, Akogbéto M (2013c). Status of organophosphate and carbamate resistance in *Anopheles gambiae sensu lato* from the south and north Benin, West Africa. Parasit. Vectors 6:274.

Brogdon WG, McAllister JC (1998). Simplification of adult mosquito bioassays through use of time-mortality determinations in glass bottles. J. Am. Mosq. Control Assoc. 14(2):159-164.

Brogdon W, Chan A (2010). Guidelines for Evaluating Insecticide Resistance in Vectors using the CDC Bottle Bioassay/ Methods in anopheles research. Second edition. CDC Atlanta USA: CDC technical report. P 343.

Perea EZ, León RB, Salcedo MP, Brogdon WG, Devine GJ (2009). Adaptation and evaluation of the bottle assay for monitoring insecticide resistance in disease vector mosquitoes in the Peruvian Amazon. Malar. J. 8:208.

World Health Organization (WHO) (1998). Report of the WHO Informal Consultation: Tests procedures for insecticide resistance monitoring in malaria vectors, bio-efficacy and persistence of insecticides on treated surfaces. Geneva: World Health Organization: Parasitic Diseases and Vector Control (PVC)/Communicable Disease Control, Prevention and Eradication (CPE). p. 43.

World Health Organization WHO/CDS/CPC/MAL/98.12.

World Health Organization (WHO) (2011). World Malaria Report. Geneva, World Health Organization.

World Health Organization (WHO) (2013). Test procedures for insecticide resistance monitoring in malaria vector mosquitoes. Geneva, World Health Organization.

Cryptosporidium oocysts in *Anodonta* sp. (bivalve mollusc) as indicators of pollution of Tiga Lake ecosystem in Kano State, Nigeria

Ebere V. Ugwoke[1], Jarleth U. Umoh[1], Emmanuel C. Okolocha[1] and Idris A. Lawal[2]

[1]Department of Veterinary Public Health and Preventive Medicine, Ahmadu Bello University, Zaria, Kaduna State, Nigeria.
[2]Department of Veterinary Parasitology and Entomology, Faculty of Veterinary Medicine, Ahmadu Bello University, Zaria, Kaduna State, Nigeria.

Bivalve molluscs are filter feeders and can bioaccumulate oocysts of Cryptosporidium. Tiga Lake in Kano State Nigeria is used for recreational, domestic and agricultural purposes by humans and also serves as a source of drinking water for animals. Bivalve molluscs from the lake are consumed by people. This study was conducted to assess the occurrence of *Cryptosporidium* spp. in Tiga Lake using edible *Anodonta* sp., (fresh water mussels) a bivalve mollusc as sentinel. The samples were examined using modified acid fast staining technique and micrometery of the oocysts. 169 and 150 samples of the molluscs were collected from two locations namely Tashan Idi and Rurum, respectively. The organs examined from each of the molluscs were the gastrointestinal tracts (GIT), gills and haemolymph. The mean oocysts load was higher in the GIT (192.50 ± 173.03) than in the other organs, although the difference was not statistically significant (P > 0.05). Of the two sampling sites, 60 (35.50%) and 40 (26.67%) of the molluscs from Tashan Idi and Rurum respectively were positive for the oocysts. However, the difference was not statistically significant (P > 0.05).The micrometry of the oocysts showed that most of them fell within the size range of 4.0 to 4.3 µm and 4.4 to 4.7 µm suggesting that the oocysts encountered in this work might be those of *Cryptosporidium parvum* and *C. meleagridis* which infect a wide range of animals and also humans. The result of this study reveals that 27.59% of the bivalve molluscs harboured Cryptosporidium oocysts and this may have public health implications if undercooked molluscs are consumed by humans.

Key words: Cryptosporidium, bivalve molluscs, Tiga Lake, bioaccumulate, public health.

INTRODUCTION

Cryptosporidium is a water-borne zoonotic coccidian protozoan parasite. It invades and then replicates within the microvillus region of epithelial cells, lining the digestive and respiratory organs of vertebrates (Nimri and Hijazi, 1994; Fayer et al., 1999; Xiao et al., 2002). Cryptosporidium is transmitted faeco – orally by ingestion of the sporulated oocysts, and most human cryptosporidiosis outbreaks have been associated with water-borne routes of transmission (Solo-Gabriele and Neumeister, 1996). Cryptosporidium parasites are endemic in many domestic and wild life populations, with young animals often shedding over a million oocysts during initial infection. The infective dose of *C. parvum* in humans can be as low as 10 to 100 oocysts (Fayer et al., 1998). The oocyst stages of *Cryptosporidium* spp. are shed in the faeces of animals and humans which may then enter sewage facilities via agricultural runoffs, wastewater discharges and faecal contamination by wildlife or persist for a long time in the environment. These oocysts can contaminate surface water, water used for drinking, recreational activities and shellfish production (Graczyk et al., 1997). This can pose a serious public health problem especially on aquatic life intended for human consumption. Environmental monitoring for *Cryptosporidium* spp. Can be problematic, partly because of the dilution effect that occurs as oocysts are disseminated from terrestrial to aquatic ecosystems, and also because particulate matter can inhibit or interfere with Cryptosporidium detection methods (Feng et al., 2003). Filter-feeding invertebrates such as bivalve molluscs, which can filter over 2 L of water/h/shellfish, can act as a natural concentration system (McMahon, 1991). These bivalves can then be collected and tested for pathogens, providing an indication of water quality (Freire-Santos et al., 2000; Graczyk, 2003; Miller et al., 2005; Tamburrini and Pozio, 1999).

Bivalve molluscs belong to the Phylum Mollusca, Class Bivalvia, Order Unionoida, Family Unionidae and Genus *Anodonta* (Nedeau et al., 2005). They use their gills to capture particulate food from the water. The water current enters the shell from the posterior ventral surface of the animal and then passes upward through the gills and the filtered particles are transported to the mouth. Food entering the mouth is passed to the stomach via cilliary action. Oocysts can be removed by bivalve molluscs from contaminated water and retained on their gills and haemolymph (Freire-Santos et al., 2000; Giangaspero et al., 2007). An approximation of the parasite load of shellfish contaminated naturally indicated that each shellfish could transport more than 10^3 oocysts (Fayer et al., 1999). Numerous edible bivalve species live burrowed in sand or mud and respire by means of siphons which reach to the surface of water. They are eaten as a delicacy in many parts of the world and are also served in place of red meat (www.wikipedia.org.seafood, 1999; Gomez-Bautista et al., 2000). In developing countries, cryptosporidiosis occurs mostly in children younger than 5 years, with peak occurrence of infections and diarrhoea in children younger than 2 years. Fatalities of this disease occur primarily among patients with Acquired Immune Deficiency Syndrome (AIDS) or other system disorders (Bern et al., 2000; Fox and Lytle, 1996; Levy et al., 1997).

Tiga Lake of Kano State, Nigeria, receives agricultural run-offs and is used by livestock and wildlife. These may bring Cryptosporidium into the lake. The lake also has a lot of molluscs that are harvested for human consumption.

Statement of the problem

In Nigeria, there is paucity of literature on the status and prevalence of Cryptosporidium infection from bivalve molluscs. Globally, there is an increase in sourcing of food especially protein and consumption of less cholesterol meat. Varying consumer vogue like roasting, frying and undercooking of molluscs to retain natural taste; pre-contamination of infected molluscs with *Cryptosporidium* oocysts on hands of the handlers and cooking utensils can pose a risk for food borne transmission of Cryptosporidium from bivalve molluscs intended for human consumption (Doyle, 2003).

The paper assessed the occurrence of Cryptosporidium oocysts in Tiga Lake of Kano State, Nigeria using *Anodonta* sp. as sentinel to determine: the occurrence of Cryptosporidium oocysts in *Anodonta* sp., the load of Cryptosporidium oocysts in *Anodonta* sp., and the sites where Cryptosporidium oocysts are mostly found in the molluscs (gills, haemolymph and gastrointestinal tract).

METHODOLOGY

Tiga lake is situated in Kano State between Latitude 11° 151' to 11° 29' North and Longitude 8° 16' and 8° 38' East. It is a man-made lake in Northern Nigeria with a surface area of 17,806 hectares and was the second largest man-made lake before the creation of the Jebba and Shiroro Reservoirs in 1983 in Nigeria (Ita et al., 1983; Galkowski and Galkowski, 1980). The lake was formed primarily for irrigation purposes with hydroelectric power generation as a secondary consideration and with recreation and conservation as other objectives. Other activities include, cattle grazing and supply of drinking water and human activities including washing, bathing, swimming and fishing (Galkowski and Galkowski, 1980). Sampling sites in the lake was Tashan Idi and Rurum which are tributaries of the lake. Bivalve molluscs are some of the fauna found in the lake.

A total of 319 *Anodonta* sp. were collected by hand picking between the periods of December 2009 through April 2010 by convenience random sampling from the two sampling sites in Tiga Lake (169 molluscs from Tashan Idi and 150 from Rurum). Fifty samples were collected weekly. Each collection was kept on ice and transported to the Parasitic Zoonoses Laboratory of the Department of Veterinary Public Health and Preventive Medicine, Ahmadu Bello University Zaria, Nigeria immediately for examination. The anterior and posterior muscles of each mollusc were cut with a scalpel inserted gently below the anterior and posterior ends of the molluscs (Graczyk et al., 1997; Rowett, 1970). The shell was opened. The haemolymph, approximately 0.5 ml, was collected with a pipette from the adductor muscle into a clean well labeled test tube. The gills and GIT were excised from each bivalve and each of these organs was placed individually in 2 ml of phosphate buffered saline (PBS) pH 7.4 in an appropriately labeled test tube, washed and left for 30 min. The organs were then vortexed using a vorter mixer and removed. A drop of the sediment from each mollusc was carefully smeared on a properly labeled clean glass slide. The slides were air-dried overnight and stained using modified acid fast staining technique. Oocysts from the positive samples were further measured with a micrometer.

The modified acid fast staining technique was conducted as Nielson and Ward (1999). The prepared slides were examined by covering about 8 to 12 fields with a binocular microscope using x40 objective under a bright field.Measurement of the oocysts from the positive samples was estimated by using a calibrated microscope, knowing the size of Cryptosporidium oocysts to be between 4 to 6 µm (Sréter et al., 2000). A calibrated eye piece was inserted into the micrometer; the slide containing the oocysts was placed on the stage of the microscope and focused while viewing from the eye piece. The oocyst was moved towards the calibration with help of the adjustment knob of the microscope until the calibration was in alignment with the oocyst. The zero point was placed at the tip of the oocysts and measurement was taken from one end of the

Table 1. Prevalence of individual parasite in 200 cats in both Ode-Irele and Oyo communities.

Parasite	Oyo community (%) N = 100	Ode-Irele community (%) N = 100	Both communities (%) N = 200
Hookworm spp.	69.0	45.0	57.0
Toxocara cati	64.0	33.0	48.5
Isospora spp.	48.0	13.0	30.5
Physaloptera spp.	26.0	28.0	27.0
Toxascaris leonina	22.0	25.0	23.5
Ollulanus tricuspis	33.0	10.0	21.5
Pseudophyllidia spp.	16.0	8.0	12.0
Trichuris spp.	15.0	2.0	8.5
Taenia spp.	13.0	2.0	7.5
Dipylidium caninum	10.0	0.0	5.0
Spirocerca lupi	4.0	0.0	2.0
Schistosoma spp.	0.0	4.0	2.0
Capillaria spp.	3.0	1.0	2.0

Table 2. Prevalence (%) of gastrointestinal parasites relative to age and sex of cats.

Age (months)	No. examined	No. infected	Prevalence (%)
0-6	92	89	96.7
7-12	36	32	88.9
13-18	35	29	82.9
19-24	18	13	72.2
25-36	14	6	42.9
>36	5	2	40.0
		P = 0.000	
Sex			
Male	120	105	87.5
Female	80	66	82.5
		P > 0.05	

oocyst to the other end. Each calibration represented a micrometer.

This was done in the Helminthology Laboratory of the Department of Veterinary Parasitology and Entomology, Ahmadu Bello University, Zaria, NigeriaOne drop (approximately 0.02 ml) of the prepared sediment was placed on a clean slide. An approximate estimate of the oocysts was calculated as follows: Total volume of solution divided by 0.02 ml multiplied by number of oocysts present in a drop of the sample expressed in mills; as modified by Mc master oocysts counting technique (Soulsby, 1968).

The oocysts from the samples of the gills, GIT and haemolymph were enumerated separately to determine the concentration and where the oocysts are trapped most, in the different organs.One-way analysis of variance was used to compare the mean number of oocysts detected in the different organs. The t test was used to determine the difference in mean oocysts load according to the site of collection and values of $P < 0.05$ were considered significant.

RESULTS

The overall infection rate of 319 *Anodonta* sp. obtained from Tiga Lake of Kano State was 88 (27.59%). Table 1 shows that 100 of the 957 organs of the molluscs examined were infected with Cryptosporidium oocysts. Out of 319 gastrointestinal tracts (GIT) of molluscs examined, 40 (12.54%) were positive. Thirty one (9.72%) of 319 gill samples were positive while 29 (9.02%) of the 319 samples of haemolymph examined were positive for *Cryptosporidium* oocysts.The contamination rate in the GIT (12.54%) was higher than in the other organs (Table 1). However, the difference was not statistically significant ($P > 0.05$). The mean oocyst load was higher in GIT (192.50) than in any of the organs, though the difference was not statistically significant ($P > 0.05$) (Table 1). Table 2 shows that the single organ contamination with Cryptosporidium oocysts was morecommon than the multiple organ contamination.Out of 169 samples of the *Anodonta* sp. from Tashan Idi examined, 60 (35.50%) were contaminated with Cryptosporidium oocysts, and

Table 3. The prevalence (%) and intensity (I) of gastrointestinal parasites in relation to age of cats at Oyo and Ode-Irele Communities.

Parasite	Age (months)											
	0-6		7-12		13-18		19-24		25-36		>36	
	Prevalence[b] (%)	I (±SEM)	Prevalence[b] (%)	I (±SEM)	Prevalence[b] (%)	I (±SEM)	Prevalence[b] (%)	I (±SEM)	Prevalence[b] (%)	I (±SEM)	Prevalence[b] (%)	I (±SEM)
Hookworm spp.	68.5	199.6 ± 36.7	52.8	233.9 ± 93.3	60.0	226.9 ± 75.3	44.4	203.3 ± 76.3	21.4	30.0 ± 19.8	0.0	0.0 ± 0.0
Toxocara cati	62.0	641.1 ± 203.2	52.8	372.8 ± 187.3	25.7	109.7± 57.9	44.4	71.1 ± 29.1	28.6	54.3 ± 30.5	0.0	0.0 ± 0.0
Isospora spp.	43.5	-	19.4	-	28.6	-	5.6	-	14.3	-	20.0	-
Physaloptera spp.	28.3	103.9 ± 42.8	33.3	40.0 ± 15.0	28.6	14.3 ± 6.6	16.7	4.4 ± 2.6	21.4	21.4 ± 11.9	0.0	0.0 ± 0.0
Toxascaris leonina	20.7	34.6 ± 12.7	36.1	66.1 ± 30.7	25.7	14.3 ± 7.2	11.1	15.6 ± 14.4	21.4	35.7 ± 29.9	20.0	16.0 ± 16.0
Ollulanus tricuspis	29.3	42.8 ± 15.8	19.4	88.3 ± 56.1	14.3	16.6 ± 8.6	22.2	40.0 ± 19.1	0.0	0.0 ± 0.0	0.0	0.0 ± 0.0
Pseudophyllidia spp.	9.8	65.2 ± 44.7	8.3	5.0 ± 2.8	2.9	18.3 ± 10.9	11.1	1.1 ± 1.1	0.0	0.0 ± 0.0	0.0	0.0 ± 0.0
Trichuris spp.	15.2	13.7 ± 6.5	11.1	28.9 ± 22.5	14.3	1.1 ± 1.1	5.6	3.3 ± 2.4	0.0	0.0 ± 0.0	0.0	0.0 ± 0.0
Taenia spp.	12.0	184.1 ± 123.8	8.3	95.0 ± 88.3	2.9	0.6 ± 0.6	11.1	12.2 ± 11.1	0.0	0.0 ± 0.0	0.0	0.0 ± 0.0
Dipylidium caninum	6.5	6.9 ± 3.8	8.3	33.9 ± 28.5	0.0	0.0 ± 0.0	0.0	0.0 ± 0.0	7.1	4.3 ± 4.3	0.0	0.0 ± 0.0
Spirocerca lupi	2.2	0.4 ± 0.4	5.6	4.4 ± 3.9	0.0	0.0 ± 0.0	0.0	0.0 ± 0.0	0.0	0.0 ± 0.0	0.0	0.0 ± 0.0
Schistosoma spp.	3.3	0.8 ± 0.5	0.0	0.0 ± 0.0	2.9	2.9 ± 2.9	0.0	0.0 ± 0.0	0.0	0.0 ± 0.0	0.0	0.0 ± 0.0
Capillaria spp.	3.3	19.6 ± 15.8	0.0	0.0 ± 0.0	0.0	0.0 ± 0.0	0.0	0.0 ± 0.0	7.1	2.9 ± 2.9	0.0	0.0 ± 0.0
% Infected Cats	96.7		88.9		82.9		72.2		42.9		40.0	
		92[a]		36[a]		35[a]		18[a]		14[a]		5[a]

[a]Total number of cats examined in each age group.
[b]Prevalence was calculated by dividing the number of cats infected with individual parasite by the total number of cats in each age group.

and of the 150 samples of molluscs collected from Rurum, 40 (26.67%) was contaminated with the oocysts. The difference in contami-nation rate between the two locations (Tashan Idi and Rurum) was not statistically significant ($P > 0.05$) (Table3). Oocysts with the size range of 4.0 to 4.3 µm were more in number 35 (45.45%), followed by 17 (22.08%) oocyst that fell in the range of 4.4 to 4.7 µm (Table 4).

DISCUSSION

The presence of Cryptosporidium oocysts in the bivalve molluscs examined (27.59%) in Tiga Lake of Kano State showed that the molluscs are indicators of faecal contamination as seen with the findings of Gomez-Couso (2006) in which 184 mussel sample (29.30%) were positive for Cryptosporidium oocysts, though this was statistically significant, isolation and detection sometimes play a significant role in the recovery and enumeration of the oocysts (Downey and Graczyk, 2007).The gastrointestinal tracts (GIT) of the molluscs examined were more contaminated with the oocysts than other organs. This is probably because most of the filtered particles from the surface water end up in the GIT for digestion before being circulated to other parts of the body (Sreter et al., 2000). However, Downey and Graczyk (2007) found that haemolymph examined harboured more oocysts. They attributed this to the fact that shortly after exposure, oocysts may be caught in the gills and other tissues of the digestive system but with time, the haemocytes scavenge these particles in an attempt to clear them from the gills and digestive tracts. Though the recovery of the oocysts from the GITs was higher than the haemolymph and gills, this was not statistically significant ($P > 0.05$) indicating that these organs have an equal chance of being contaminated. Multiple-organ contamination observed in this study agrees with the findings of Fayer et al. (1997) in which oysters removed *Cryptosporidium parvum* oocysts from artificially contaminated water and retains them in haemocytes, gills and within the body for at least one month. Multiple organ contamination indicates a higher chance of contracting cryptosporidial oocysts since these oocysts are trapped all over the body of the molluscs. Also filter feeders such

Table 4. Prevalence (%), intensity (I) and range of parasites eggs in the feaces of cats of different sexes.

Parasite	Male (N = 120)			Female (N = 80)			Both sexes (N = 200)		
	%	I (±SEM)	R	%	I (±SEM)	R	%	I (±SEM)	R
Hookworm spp.	57.5	214.5 ± 36.2	0 - 2380	56.2	163.3 ± 44.9	0 - 2240	57.0	194.0 ± 28.1	0 - 2380
Toxocara cati	45.8	349.7 ± 140.9	0 - 12300	52.5	454.0 ± 139.7	0 - 7080	48.5	391.4 ± 101.2	0 - 12300
Physaloptera spp.	29.2	87.0 ± 33.0	0 - 2620	23.8	18.0 ± 6.1	0 - 400	27.0	59.4 ± 20.1	0 - 2620
Toxascaris leonina	21.7	33.0 ± 10.5	0 - 820	26.2	37.0 ± 14.5	0 - 1020	23.5	34.6 ± 8.5	0 - 1020
Ollulanus tricuspis	24.2	52.8 ± 19.6	0 - 1860	17.5	26.0 ± 11.7	0 - 840	21.5	42.1 ± 12.7	0 - 1860
Pseudophyllidia spp.	14.2	48.3 ± 34.2	0 - 4040	8.8	13.0 ± 6.41	0 - 340	12.0	34.2 ± 20.7	0 - 4040
Trichuris spp.	11.7	19.3 ± 8.4	0 - 800	3.8	1.0 ± 0.6	0 - 40	8.5	12.0 ± 5.1	0 - 800
Taenia spp.	7.5	33.7 ± 26.7	0 - 3180	7.5	207.0 ± 142.2	0 - 8200	7.5	103.0 ± 59.2	0 - 8200
Dipylidium caninum	5.0	12.5 ± 8.8	0 - 1020	5.0	5.3 ± 3.3	0 - 240	5.0	9.6 ± 5.4	0 - 1020
Schistosoma spp.	0.0	0.0 ± 0.0	-	5.0	2.3 ± 1.4	0 - 100	2.0	0.90 ± 0.6	0 - 100
Capillaria spp.	0.0	0.0 ± 0.0	-	5.0	23.0 ± 18.1	0 - 1400	2.0	9.2 ± 7.3	0 - 1400
Spirocerca lupi	2.5	1.7 ± 1.2	0 - 140	0.0	0.0 ± 0.0	-	1.5	1.0 ± 0.7	0 - 140
Isospora spp.	-	-	-	-	-	-	-	-	-

as bivalve molluscs may be of value as biological monitors for the presence of Cryptosporidium oocysts and other water borne pathogens in river and marine water and in particular contaminating bivalves used for consumption which has obvious public health implications (Chalmers et al., 1997).

In Tiga Lake where large scale fishing activities takes place, bivalve molluscs which are in abundance are used as baits to catch fish. Contaminated molluscs when used for such purposes may serve as a route for cryptosporidial oocysts infection in fishes. This is possible as Graczyk et al. (1996) reported an infection in fishes, amphibians and reptiles. Thus aquatic animals other than bivalve molluscs might be potential sources of waterborne Cryptosporidium oocysts.

The two sampling sites investigated, showed that molluscs harvested at Tashan Idi had a higher number of molluscs positive for the oocysts than Rurum. This may be attributed to the activities of the surrounding villages such as swimming, bathing, farming and fishing. Also cattle owners around bring their herds for grazing and drinking of water. These activities may serve as routes in which Cryptosporidium oocysts are discharged into the water body. These agree with findings of Fayer et al. (1997) and CDC (2000) in which oocysts can contaminate surface water directly from human faeces, urine, recreational activities and indirectly from runoffs.

In micrometery of the oocysts most of them fell within the range of 4.0 to 4.8 µm, suggesting that the oocysts might be *C. parvum* or *Cryptosporidium meleagridis*. These species are of zoonotic importance and their sizes fall in line with that of Sreter et al. (2000) and Xiao et al. (2004).

To date there are no published reports of human cryptosporidiosis cases resulting from the consumption of bivalve molluscs, despite a growing body of scientific literature on the recovery of Cryptosporidium oocysts (Gomez-Bautista et al., 2000; Schets et al., 2007). The reason for this discrepancy is unclear. One explanation could be that infection caused by consumption of contaminated molluscs are not being diagnosed or reported. The long incubation period, that is, 7 to 10 days makes it difficult to associate the infection with a particular exposure (Graczyk and Schwab, 2000).

Conclusion

This study has established the presence of Cryptosporidium oocysts in *Anodonta* sp. in Tiga Lake of Kano State. It is established that these molluscs can effectively remove and retain oocysts of Cryptosporidium from contaminated water bodies. Due to the zoonotic potentials of the infection, threats to human life should not be underestimated especially when considering susceptible groups such as newborn infants, the elderly, patients on immunosuppressive drugs, people infected with HIV who are at a higher risk of infection. There is a need for a more in depth molecular investigation to highlight and identify the Cryptosporidium species recovered from *Anodonta* sp.

ACKNOWLEDGEMENTS

We greatly appreciate the contributions of the staff of the laboratory of the Department of Veterinary Public health and Preventive Medicine, Ahmadu Bello University, Zaria, Nigeria, Mr. B.O. Odoba and Mrs. J.C. Atawodi and the staff of the Helminthology Unit of the Department of Veterinary Parasitology and Entomology, Ahmadu Bello University, Zaria, Nigeria. Dr. D. Adebote, Mr. I. k. Gazama, Prof. J. k. Balogun, Prof. T. Akenova and Prof. P. A. Audu of Department of Biological Sciences, Ahmadu Bello University, Zaria, Nigeria, for their assistance in various aspect of the research work.

REFERENCES

Bern C, Hernandez B, Lopez MB, Arrowood MJ, De Merida AM, Klein RE (2000). The contrasting epidemiology of *Cyclospora* and *Cryptosporidium* among outpatients in Guatemala. Am. J. Trop. Med. Hyg. 63:231-235.

Centers for Disease Control and Prevention, (CDC). (2000). Guidelines for the preventing opportunistic infections among hematopoietic stem cell transplant recipients. Morb. Mortal. Wkly. Rep. 49(10):1–128.

Chalmers RM, Sturdee AP, Mellor P, Nicholson V, Lawalor F, Kenny F, Timpson P (1997). *Cryptosporidium parvum* in environmental samples in the Sligo area, Republic of Ireland: a preliminary report. Lett. Appl. Microbiol. 25:380-384.

Downey AS, Graczyk TK (2007). Maximizing Recovery and Detection of *Cryptosporidium parvum* oocysts from spiked Eastern oysters *(Crassostrea virginica)* samples. Appl. Environ. Microbiol. 73(21):6910–6915.

Doyle E (2003). FRI Briefings. Foodborne Parasites. A Review of the Scientific Literature. Food Research Institute, University of Wisconsin UW Madison WI 53706. pp. 9-10.

Fayer R, Farley CA, Lewis EJ, Trout JM, Graczyk TK (1997). Potential role of the Eastern oyster, *Crassostrea virginica*, in the epidemiology of *Cryptosporidium parvum*. Appl. Environ. Microbiol. 63:2086-2088.

Fayer R. Trout JM Jenkins MC (1998). Infectivity of *Cryptosporidium parvum* oocysts stored in water at environmental Temperatures. J. Parasitol. 84:1165-1169.

Fayer R, Graczyk TK, Lewis EJ, James MT, Farley AC (1999). Survival of Infectious *Cryptosporidium parvum* oocysts in Seawater and Eastern Oysters (*Crassostrea virginica*) in the Chesapeake Bay. Appl. Environ. Microbiol. 64:1070 -1074.

Feng YY, Ong SL, Hu JY, Song LF, Tan XL, Ng WJ (2003). Effect of particles on the recovery of *Cryptosporidium* oocysts from source water samples of various turbidities. Appl. Environ. Microbiol. 69:1898-1903.

Fox KR, Lytle DA (1996). Milwaukees cryptosporidiosis outbreak-investigation and recommendations. J. Am. Water Works Assoc. 88:87-94.

Freire-santos F, Citeiza-loiprez AM, Viergrazza- Castibullamcon CA, Aves-Mlazsaise E, Allwamasz S, Garcia-martin O (2000). Dctoction of *Cryptosporidium* oocysts in Bivalve Molluscs Destined for Human Consumption. J. Parasitol. 56:850-856.

Galkowski E, Galkowski K (1980). Main projects Data of Dams and Reservoirs. Water Resources and Engineering Construction Agency, Kano State.

Giangaspero A, Berrilli F, Brandorusio O (2007). *Giardia* and *Cryptosporidium* and public health: the epidemiological scenario from the Italian perspective. Parasitol. Res. 101:1169-1182.

Gomez-Bautista M, Ortega- Lopez IM, Tabarez E, Lopez-Rodas V, Costas E (2000). Detection of infectious *Cryptosporidium parvum* oocysts in mussels, (*Mytilus galloprovincialis*) and cockles *(Cerastoderma edule)*. Appl. Environ. Microbiol. 66:1866-1870.

Gomez-couso H, Mendez-Hermida F, Gastro-Hermidia JA, Ares-Mazas E (2006). Giardia in shellfish farming areas: detection in mussels, river water and waste water. Vet. Parasitol. 133:13-18.

Graczyk TK, Schwab KJ (2000). Food-borne infections vectored by molluscan shellfish. Curr. Gastroenterol. Rep. 2:305-309.

Graczyk TK, Fayer R, Cranfield MR (1996). *Cryptosporidium parvum* is not transmissible to Fish, Amphibians or Reptiles. J. Parasitol. 82:748-751.

Graczyk TK (2003). Human waterborne parasites in molluscan shellfish. J. Parasitol. 89:557–561.

Graczyk TK, Fayer R, Cranfield MR (1997). Zoonotic transmission of *Cryptosporidium parvum*: implications for water borne cryptosporidiosis. Parasitol. Today 13:348-351.

Ita EO, Mohammed A, Omonukoba WS, Bankole NO, Awojebi S (1983). A preliminary report on the immediate Post-impoundment fishery survey of Shiroro reservoir, Niger State Nigeria. In: Annual Report, Kainji, Lake Research Institute. pp. 25-29.

Levy J, Espanol-Boren T, Thomas C, Fischer A, Tovo P, Bordigoni P, Resnick I, Fasth A, Baer M, Gomez L, Sanders EA, Tabone MD, Plantaz D, Etzioni A, Monafo V, Abinun M, Hammarstrom L, Abrabamsen T, Jones A, Finn A, Klemola T, DeVries E, Sanal O, Peitsch, MC, Notarangelo LD (1997). Clinical spectrum of X-linked hyper-IgM syndrome. J. Pediatr. 131:47–54.

McMahon RF (1991). Mollusca: Bivalvia. In: Thorp JH, Covich AP (eds.), Ecology and classification of North American invertebrates. Academic Press, Inc.

Miller WA, Atwill ER, Gardner IA, Miller MA, Fritz HM, Hedrick RP, Melli AC, Barnes NM, Conrad PA (2005). Clams (*Corbicula fluminea*) as bioindicators of fecal contamination with *Cryptosporidium* and *Giardia* spp. in freshwater ecosystems in California. Int. J. Parasitol. 35(6):673-684.

Nedeau E, Smith AK, Stone J (2005). Freshwater Mussels of the Pacific Northwest. Pacific Northwest Native Freashwater Mussel Workgroup, Vancouver; Washington. P 45.

Nielson EK, Ward LA (1999). Enhanced Detection of *Cryptosporidium parvum* in Acid Fast Stain Protocol. J. Vet. Diagn. Investig. 11(6):567-569.

Nimri LF, Hijazi SS (1994). *Cryptosporidium*. A cause of gastroenteritis in preschool children in Jordan. J. Clin. Gastroenterol. 19:288-291.

Rowett HGQ (1970). Dissection Guides for invertebrates, Seventh edition. John Murray, London. pp. 45-50.

Schets F, Berg HV, Engels G, Lodder W, de RodaHusman AM (2007). *Cryptosporidium* and *Giardia* in commercial and non-commercial oysters (*Crassostreagigas*) and water from the Oosterschelde, The Netherlands. Int. J. Food Microbiol. 113:189-194.

Solo-Gabriele H, Neumeister M (1996). U.S. outbreaks of Cryptosporidiosis. J. Am. Water Works Assoc. 88:76-86.

Soulsby EJL (1968). Nematoda. In: Soulsby EJL (ed.), Helminths, arthropods and protozoa of domesticated animals, Sixth edition of Monnig's Veterinary Helminthology and Entomology). Bailliere Tindall and Cassell, London. P 789.

Sréter T, Kovács G, da Silva AJ, Pieniazek NJ, Széll Z, Dobos-Kovács M, Márialigeti K, Varga I (2000). Morphologic, Host Specificity, and Molecular Characterization of a Hungarian *Cryptosporidium meleagridis* Isolate. Appl. Environ. Microbiol. 65:1455-1458.

Tamburrini A, Pozio E (1999). Longterm Survival of *Cryptosporidium parvum* Oocysts in Seawater and in Experimentally Infected Mussels. (*Mytilus galloprovincialis*). Int. J. Parasitol. 29:711-715.

Wikipedia (1999). www.wikipedia.org.seafood Last Modified May 1999.

Xiao L, Sulaiman IM, Morgan CM, Zhou L, Atwill ER, Tischler MI, Zhang X, Fayer R, Lal AA (2002). Host adaptation and host parasite co-evolution in *Cyptosporidium*. Int. J. Parasitol. 52:1773- 1785.

Xiao L, Fayer R, Ryan U, Upton SJ (2004). *Cryptosporidium* Taxonomy: Recent Advances and Implications for Public Health. Clin. Microbiol. Rev. 17:72-97.

Parasitological evaluation of potential candidate vaccines in *Schistosoma mansoni*-infected mice

Samia E. Etewa[1], Naglaa F. Abd El-Aal[1], Sara A. Abdel-Rahman[1] and Mahmoud A. El-Shafey[2]

[1]Department of Medical Parasitology, Faculty of Medicine, Zagazig University, Zagazig, Egypt.
[2]Department of Clinical Pathology, Faculty of Medicine, Zagazig University, Zagazig, Egypt.

This study aims to assess the prophylaxis of different single and combined crude antigens used as candidate vaccines in an experimental model of *Schistosoma mansoni* infection. Cercarial antigen preparation (CAP), soluble worm antigen preparation (SWAP), soluble egg antigen (SEA) and combined antigens (CAP + SWAP + SEA) against schistosomiasis combined with either Freund's adjuvant or with Bacillus Calmette-Guérin (BCG) were tested for vaccine efficacy in Albino mice by detecting their effects on worm burden and tissue egg count (liver and intestine). The data obtained showed that, the combined antigens was the most protective with significant reduction in the worm burden (90.28%), and tissue egg load (liver and intestine: 93.67 and 93.98%), respectively. Thus, combining these different antigens (CAP, SWAP and SEA) provides augmentation of the protective immunity compared to each component administered individually. The combination therefore stands out as a potential candidate worth considering in the development of a feasible vaccine against schistosomiasis.

Key words: *Schistosoma mansoni*, vaccine, worm burden, egg count.

INTRODUCTION

Schistosomiasis is the most important of the human helminthiases associated with subtle but persistent morbidities and mortality (Lambertucci et al., 2007). Schistosomiasis infects approximately 700 million people in 74 endemic countries. Those infected are 207 million, with 85% of the cases occurring in Africa (WHO, 2010).

Current schistosomiasis control strategies are mainly based on Praziquantel (PZQ) , but in spite of decades of mass treatment, the number of infected people remains constant (Hotez et al., 2010). Furthermore, efficient drug delivery can require a substantial infrastructure to regularly cover all parts of an area of endemicity, this can make chemotherapy an expensive and often impractical approach, although there is not yet clear-cut evidence for the existence of PZQ resistant schistosome strains, decreased susceptibility to the drug has been observed (Doenhoff and Pica-Mattoccia, 2006; Gryseels et al., 2006; James et al., 2009). The best long-term strategy to control schistosomiasis is through immunization with an anti-schistosomiasis vaccine combined with drug treatment (Martin et al., 2012).

Vaccination can be targeted either towards the prevention of schistosome infection or to the reduction of parasite fecundity. A reduction in worm numbers is the gold standard for antischistosome vaccine development, but because schistosome eggs are responsible for both pathology and transmission, a vaccine targeted at parasite fecundity and egg viability is also relevant (McManus and Loukas, 2008).

The selection of a suitable adjuvant to aid in the stimulation of the appropriate immune response is a critical step in the path to the development and employment of successful antischistosome vaccines, and the approaches are needed to be assessed carefully (Khalifa et al., 2011).

The Bacillus Calmette-Guérin (BCG), a live attenuated

Mycobacterium bovis strain that has been used for the prevention of human tuberculosis for decades, is considered a promising candidate for the development of live vector systems for the delivery of foreign antigens to the immune system (Ohara and Yamada, 2001). Several advantages are associated with the use of BCG as an antigen-presenting system, including its known adjuvant properties, its ability to elicit humoral or cellular immune responses toward heterologous antigens, its thermo-stability which eliminates the need for a cold chain, and most importantly the possibility of obtaining an efficient immune response by using a single dose (Himmelrich et al., 2000).

Schistosoma mansoni vaccine must be safe, effective, stable, easily administered, and affordable by the target population. The goal in developing the schistosome vaccine under discussion is to produce a vaccine that protects people against infection and the consequent morbidity and mortality associated with *S. mansoni* (Todd and Colley, 2002).

The aim of this study was to design an experimental and a comprehensive model system consisting of individual and combined antigens that can provide prophylaxis to the infection by *S. mansoni*, through parasitological evaluation of these different vaccines by detecting their effects on worm burden and tissue egg deposition (liver and intestine). This system may have the advantage of resisting *S. mansoni* infection.

MATERIALS AND METHODS

The procedures were performed at the laboratories of Parasitology, Faculty of Medicinc, Zagazig University and Theodor Bilharz Research Institute, Giza, Egypt. This is a case control experimental study.

Infective cercariae

Laboratory bred *Biomphalaria alexandrina* snails were purchased from the Shistosome Biological Supply Unit, Theodore Bilharz Research Institute (Giza, Egypt). After exposure to light for at least 4 h, *S. mansoni* cercariae shed from the snails were used to infect the experimental mice through the subcutaneous route.

Experimental animals

One hundred and ten laboratory bred Albino male mice (aged 6 to 8 weeks and weighing 18 to 20 g each), were used in this study. Mice were fed on standard diet with free accessibility to water at the Schistosome Biological Supply Program (SBSP), Theodor Bilharz Research Institute, Giza, Egypt.

Ethical aspects

All procedures related to animal experimentation in the present study met the International Guiding Principles for Biomedical Research Involving Animals as issued by the International Organizations of Medical Sciences and approved by ethics committee of the Faculty of Medicine, Zagazig University.

Schistosomal antigens preparation

Cercarial antigen preparation (CAP), soluble worm antigen preparation (SWAP) and soluble egg antigen (SEA) were prepared at the SBSP at Theodor Bilharz Research Institute (TBRI). CAP was prepared according to the method of Carter and Collely (1979), SWAP was prepared according to the method of Salih et al. (1978) and SEA was prepared according to Boros and Warren (1970).

Adjuvants preparation

Freund's adjuvant (Adj) {Freund's complete adjuvant (FCA) and Freund's incomplete adjuvant (FIA)} was obtained from Sigma Chemical Co., St Louis, Mo, USA and emulsified in phosphate-buffered saline (PBS) at a ratio of 2:1 (v/v).

BCG was obtained from Veterinary Serum and Vaccine Institute, Bacteria Diagnostic Research Department, Abasia, Cairo, and emulsified in PBS at a ratio of 2:1 (v/v).

Vaccination procedure

The total protein content of each antigen preparation was determined by the method of Bradford (1976). The vaccination schedule was performed according to the method of Nabih and Soliman (1986). Each mouse was sensitized with an initial subcutaneous injection of 200 µl of the extracted antigen with total antigen concentration contained 30 µg protein. After two weeks, a second subcutaneous injection of 200 µl of the same antigen was given containing 20 µg proteins; hence, each mouse received a total antigen dose of 50 µg protein.

The antigen was combined with either Freund's adjuvant (complete for the first injection, incomplete for the boost) subcutaneously at a 1:1 ratio (v/v), that is boosted twice at 2-weeks interval (Smithers et al., 1989), or with BCG 100 µl intradermally with the booster dose for each mouse (Boulanger et al., 1999).

Infection procedure

Infection was done by subcutaneous injection into each mouse (vaccinated and unvaccinated) with ±80 *S. mansoni* cercariae suspended in 0.2 ml solution after 3 weeks from the first vaccination (Peters and Warren, 1969).

Experimental design

Mice were divided into eleven groups (10 animals each). Three control groups: (i) infected only, (ii) infected and supported by Freund's adjuvant and (iii) infected and supported by BCG. Eight groups were vaccinated by different antigens: CAP, SWAP, SEA and combined (cocktail) antigens (CAP + SWAP + SEA); these antigens were supported by Freund's adjuvant or with BCG as shown in Table 1.

Assessment of the vaccine preparations effects

Animals were sacrificed by cervical dislocation 10 weeks post infection and parasitological evaluation were done.

Parasitological study

Worm burden recovery: Perfusion of adult worms from the liver and porto-mesenteric system was performed 10 weeks after infection according to Duvall and Dewitt (1967). The degree of protection (% reduction of worm) by the various vaccination preparations was calculated as follows:

P (%R) = C – V/C × 100

where P is the percentage protection, C is the mean number of parasites recovered from infected mice and V is the mean of parasite recovered from vaccinated mice (Tendler et al., 1986).

Tissue egg load: One gram of liver and of intestine from each mouse was weighed and put into a test tube containing 2 ml of 5% KOH, then left for 18 h at room temperature for complete digestion without egg destruction. The second day, all test tubes were put in the incubator at 37°C for 6 h. Each tube was shaken well for 5 min, and then 0.1 ml of the digest was examined microscopically for counting *S. mansoni* eggs. The process of counting was repeated five times, to calculate the mean number of ova per 0.1 ml. Mean number of ova per gram of tissue (liver and intestine) = mean number of ova in 0.1 ml × 100 according to the procedure by Cheever (1968).

Oogram pattern: Following perfusion for the recovery of the schistosomes, fragments of the intestine (from the middle loop of intestine till the terminal ileum) from each animal were separated and transferred to the Petri dishes containing saline (0.9%). The intestines were opened lengthwise and the excess mucus was removed. One-centimeter fragments were weighed and placed between a glass slide and a glass cover. The preparation was inverted and pressed on a rubber surface padded with filter paper. The percentages of immature, mature, and dead ova in the small intestines were computed from a total of 100 eggs per intestinal segment and classified according to the categories previously defined by Pellegrino et al. (1962).

Statistical analysis

Data were entered, checked and analyzed using statistical computer program Statistical package for Social Sciences (SPSS version 16 windows). Data are expressed as the mean±standard deviation (SD). Comparison between the mean values of different parameters in the studied groups was performed using one way analysis of variance (ANOVA) test, with paired (t) test for comparison between means of two groups.

RESULTS

One hundred and ten laboratory bred albino male mice, each mouse was sensitized with an initial subcutaneous injection of the extracted antigen. After two weeks, a second subcutaneous injection of the same antigen. The antigen was combined with either Freund's adjuvant (complete for the first injection, incomplete for the boost) subcutaneously, that is, boosted twice at 2-weeks interval, or with BCG intradermally with the booster dose for each mouse. Infection was done by subcutaneous injection into each mouse (vaccinated and unvaccinated) with ±80 *S. mansoni* cercariae after 3 weeks from the first vaccination.

The effect of vaccination with different schistosomal antigens depend on the degree of the protection in mice infected with *S. mansoni* cercariae; protection was investigated by comparing the results of the parasitological evaluation (worm burden, tissue egg count, and oogram studies), in the different groups vaccinated with CAP, SWAP, SEA or combined compared to that in the control groups [Mice group (10 mice) of combined + BCG antigens died early before the time of sacrificing (two mice died 2 weeks post infection, 5 mice died 4 weeks post infection and the rest died 5 weeks post infection). The reason was not known, but we suggested that, this may be due to the higher level of activation of immune system, it need more investigations].

Worm burden

The result showed that the most effective antigen with significant reduction in the mean worm burden at p<0.001 was combined + Freund's percent reduction of 90.28%, followed by SEA with BCG with percent reduction of 69.91%, SWAP with BCG with percent reduction of 66.67%, then SEA + Freund's with percent reduction of 64.35% and SWAP + Freund's with percent reduction 62.5%. While, the least effective vaccines with significant reduction at p<0.01 are noticed in CAP with BCG with percent reduction of 56.02% followed by CAP + Freund's with percent reduction of 51.85%. On the other hand, insignificant reduction at p>0.05 detected in C2 and C3 with percent reduction of 13.43 and 5.09%, respectively compared to C1 (the control infected group) (Figure 1) [percent reduction (%R) = % protection = comparing mean worm burden in vaccinated (V) and control (C) where % = (C-V)/C×100.

Tissue egg load

The result detected that the most effective antigen with significant reduction in tissue egg load (the mean egg count/gram tissue – liver and intestine) in combined + Freund's with percent reduction of 93.67 and 93.98%, respectively, followed by SEA + BCG with percent reduction of 72.22 and 73.93%, SWAP + BCG with percent reduction of 69.52 and 69.95%, then SEA + Freund's with percent reduction of 64.78 and 66.78% and SWAP + Freund's with percent reduction of 59.66 and 60.68% (Table 2 and Figure 2).

While, the least effective antigens are CAP + BCG with percent reduction of 56 and 55.79% followed by CAP + Freund's with percent reduction of 46.37 and 48.4%. On the other hand, insignificant reduction detected in C2 and C3 with percent reduction of 1.56 and 9.02 and 0.515 and

Table 1. Animal groups.

Group	Explanation
Control group	C1 (control infected group): Mice were unvaccinated and infected by ±80 *S. mansoni* cercariae.
	C2 (control infected + BCG): Mice were injected by BCG and infected by ±80 *S. mansoni* cercariae.
	C3 (control infected + Freund's): Mice were injected by Freund's adjuvant and infected by ±80 *S. mansoni* cercariae acting as.
Vaccinated group	CAP+BCG: Mice were vaccinated by CAP+BCG and infected by ±80 *S. mansoni* cercariae.
	CAP+Freund's: Mice were vaccinated by CAP+Freund's and infected by ±80 *S. mansoni* cercariae.
	SWAP+BCG: Mice were vaccinated by SWAP+BCG and infected by ±80 *S. mansoni* cercariae.
	SWAP+Freund's: Mice were vaccinated by SWAP+Freund's and infected by ±80 *S. mansoni* cercariae.
	SEA+BCG: Mice were vaccinated by SEA+BCG and infected by ±80 *S. mansoni* cercariae.
	SEA+Freund's: Mice were vaccinated by SEA+Freund's and infected by ±80 *S. mansoni* cercariae.
	Combined+BCG: Mice were vaccinated by cocktail antigen preparation (CAP+SWAP+SEA)+BCG and infected by ±80 *S. mansoni* cercariae.
	Combined+Freund's: Mice were vaccinated by cocktail antigen preparation (CAP+SWAP+SEA) + Freund's and infected by ±80 *S. mansoni* cercariae.

Table 2. The effect of vaccination with different antigens on the tissue egg load (mean egg count/gram tissue – liver and intestine) and oogram of the studied mice groups.

Group	Tissue egg load (Mean egg count/gram tissue)						Oogram		
	Liver eggs			Intestinal eggs			Immature (%)	Mature (%)	Dead (%)
	Mean±SD	R (%)	t test (p)	Mean±SD	R (%)	t test (p)	Mean±SD	Mean±SD	Mean±SD
C1	1029.8±74.28	-	-	1173.90±115.53	-	-	27.3±2.4	67.5±3.2	5.2±0.9
C2	1024.5±20.88	0.515	3.14	1106.0±225.26	5.78	0.729	27.2±2.1	67.1±2.9	5.9±1.2
C3	1013.7±24.35	1.56	0.688	1068.0±317.51	9.02	0.879	26.9±2.2	66.4±2.8	6.7±1.4
CAP+BCG	453.1±121.39***	56	10.29	518.9±24.00***	55.79	15.59	18.1±1.8*	45.2±2.4*	36.7±2.1**
CAP+Freund's	552.3±46.37***	46.37	14.53	605.7±34.05***	48.4	13.87	22.4±1.9	47.7±2.6*	29.9±2.3*
SWAP+BCG	313.9±43.97***	69.52	65.31	352.8±24.45***	69.95	21.852	13.4±1.7**	40.8±2.2	45.8±2.5**
SWAP+Freund's	415.4±27.25***	59.66	22.93	461.6±33.59***	60.68	17.33	17.3±1.8**	42.9±2.6*	39.8±2.3**
SEA+BCG	286.6±42.22***	72.22	34.19	306.00±130.06***	73.93	18.621	12.7±1.6**	39.7±2.1*	47.6±2.7**
SEA+Freund's	362.7±25.84***	64.78	24.34	390.00±140.24***	66.78	16.132	16.2±1.5**	41.2±2.1*	42.6±2.4**
Combined+Freund's	65.2±21.22***	93.67	44.81	70.70±15.97***	93.98	29.28	5.6±1.2***	2.4.2±1.9***	70.2±2.3***

*Significant difference from infected control at $p<0.05$. **High significant difference from infected control at $p<0.01$. ***Very high significant difference from infected control at $p<0.001$.

5.78%, respectively compared to C1 (control infected group).

The most effective antigen with remarkable changes inthe oogram with significant reduction in immature ova and mature ova and significant increase in dead ova in combined + Freund's, followed by SEA + BCG, SWAP + BCG, then SEA + Freund's and SWAP + Freund's, respectively. The least effective antigens are CAP + BCG with

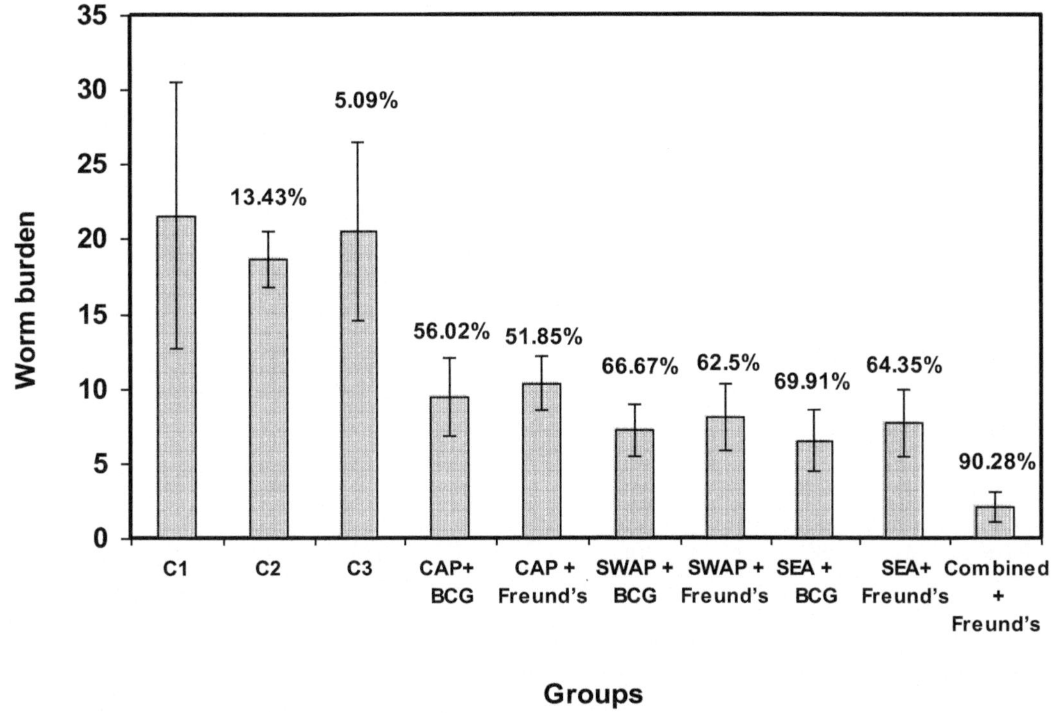

Figure 1. The effect of vaccination with different antigens on the mean worm burden of the studied mice groups [% = %R (% protection)].

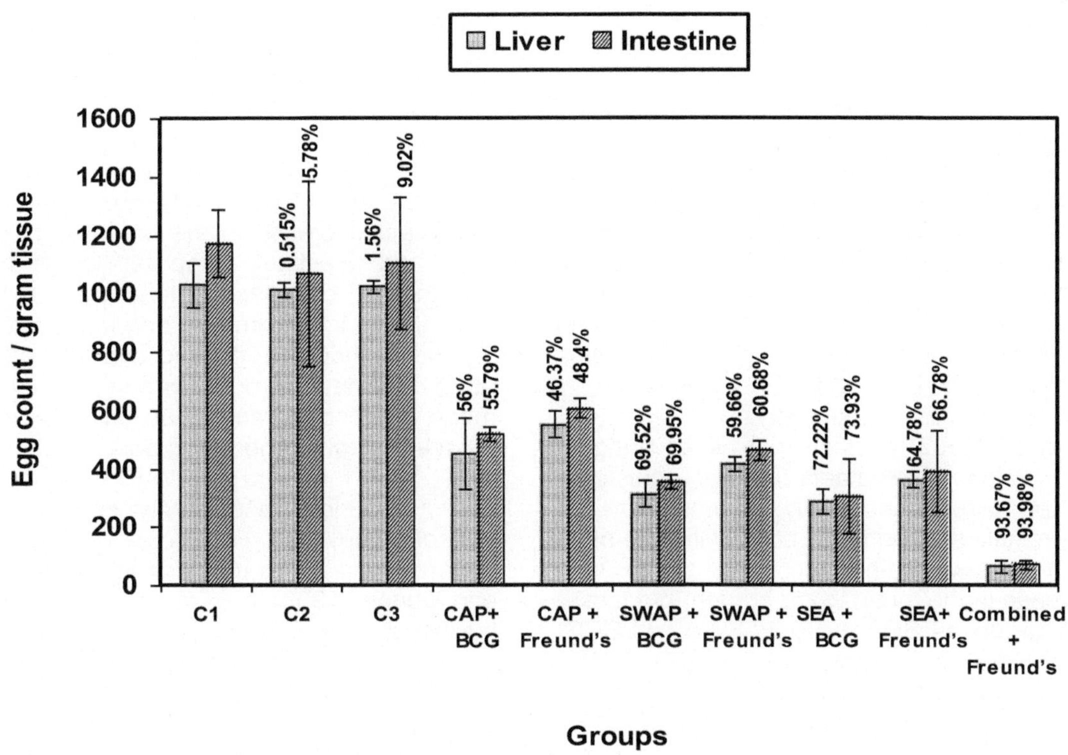

Figure 2. The effect of vaccination with different antigens on the tissue egg load (liver& intestine) of the studied mice groups. %= %R (% protection).

significant reduction in immature ova and significant reduction in mature ova and high significant increase in dead ova and CAP + Freund's with significant reduction in mature ova and significant increase in dead ova. On the other hand, insignificant oogram changes detected in C2 and C3 compared to C1 (Table 2 and Figure 2).

DISCUSSION

Development of a vaccine for schistosomiasis is far fetched; it has been targeted as a priority by researchers and the World Health Organization (WHO, 2000).

Waine and McManus (1999) decided that the attenuated vaccines for schistosomiasis are considered neither safe nor practicable for human use, and therefore, other approaches must be considered. In the current experimental study, the crude antigens were used in mice, because it was suggested that these antigens were good representative to all *Schistosoma* stages present inside the human body. In addition, Ashour et al. (2004) reported that schistosomes are multicellular parasites with differentiated tissues, and therefore with antigen complex.

A vaccine that induces even a partial reduction in worm burdens could considerably reduce pathology and limit parasite transmission (Chitsulo et al., 2004). Since the levels of protection elicited by a single antigen were low, development of novel cocktail vaccine formulations is necessary to enhance protection levels to at least induce protection ranges from 70 to 80% (Khalifa et al., 2011).

In this context, the results of the current study showed that combined antigens, displayed the best protection with the highest level of decrease in the mean worm burden compared to unvaccinated infected control group (C1). The result was closely related to Ismail (2005) who reported that the reduction in worm burden was 88% in mice which were vaccinated with combined antigens (Adj + CAP + SWAP + SEA) compared to unvaccinated infected control group. In addition, Ashour et al. (2004) detected that worm burden protection percentage was 60.91, 78.16 and 64.77% for (SEA + CAP), (SEA + SWAP) and (CAP + SWAP), respectively.

Nascimento et al. (2002) decided that as an antigen, candidate for vaccine development against schistosemiasis should be given consistently at least 40% of protection. The results support those of Nascimento et al. (2002) in which there were other vaccinated groups with significant decrease in the mean worm burden, but lesser than combined group as SEA + BCG, SWAP + BCG, SEA + Freund's, SWAP + Freund's, CAP + BCG, CAP + Freund's, respectively, compared to C1 (infected control group).

In addition, Ismail (2005) concluded that percent reduction in worm burden was 87.3, 71.5 and 76.4% for CAP, SEA, and SWAP, respectively. Although Soliman et al. (2008) used different animal model, our results were closely related to them, they tested *Schistosoma mansoni* SWAP and cercarial antigen (CAP) alone to detect their efficacy against the infection in hamsters, results showed that the worm reduction percentages were 53.8 and 56.4% for CAP and SWAP, respectively.

Our results are higher than those of Ashour et al. (2004) who found that worm burden protection percentage was 42.5, 58.33 and 53.33% for CAP, SEA, and SWAP, respectively; this may be attributed to using of Freund's adjuvant and BCG in our study. Also, the protective level obtained by SEA in our result was higher than that of El-Ahwany et al. (2012) who recorded 49% decrease in worm load on administration of SEA prior to infection.

Human studies of vaccine development are instructive for not only identifying the few antigens directly and exclusively associated with resistance, but also for indicating which of these components can be formulated with adjuvants to generate protective responses in animal models (Bergquist et al., 2008). In this context, we used two different adjuvants BCG and Freund's.

Traditional approaches had declared that Freund's adjuvant is the best to be used when antigens are first being assessed as potential candidate vaccines in the mouse model (McManus and Loukas, 2008). Our results showed that although C2 induced higher levels of protection than C3, but both were statistically insignificant compared to C1 (control infected group), confirming that these substances alone have no effect against *S. mansoni* when used alone, become more effective when used with different antigens. This finding agreed with El-Marhoumy et al. (2009) who reported that BCG acting as a potent adjuvant in association with larval antigens, but was not protective when administrated alone.

In this work, combined (cocktail) antigens, showed significant decrease in the mean egg count/gram tissue (liver and intestine) compared to C1 (control infected) with marked changes in the oogram including significant reduction in immature ova and mature ova and very high significant increase in dead ova. In addition, there were other vaccinated groups with significant decrease in the mean egg count/gram tissue and with oogram changes, but lesser than combined group as SEA + BCG, SWAP + BCG, SEA + Freund's, SWAP + Freund's, CAP + BCG, CAP + Freund's, respectively, compared to C1 (infected control group).

Our results are in confidence with the results of Fallon and Dunne (1999) who reported that reduction in worm burden should result in a corresponding decrease in the number of deposited eggs. Our results also, agreed with Ismail (2005) who reported marked oogram changes with reduction in total ova count in (liver and intestine), 98.02, 96.31, 90.27, 79.11 and 78.93% in mice vaccinated with cocktail antigens, CAP, SWAP, SEA and Freund's adjuvant, respectively. Besides, Teixeira de Melo et al. (2010)

emphasized that immunization of mice with adult worm tegument (Smteg) together with Freund's adjuvant induced a Th1 type of immune response associated with a significant reduction in eggs trapped in the liver by 65%.

All mice of combined+BCG group died before scarifying, the reason was not known, further investigations will be needed, and this may be attributed to the higher level of activation of immune system as a direct consequence of vaccination.

In this study, administration of single or combined antigens prior to infection resulted in decreased worm load, hepatic and intestinal ova together with change in oogram pattern. This could be due to enhancement of immune response or would be acting as assort of primary infection that somewhat hinders the challenge one.

The dream of a vaccine against schistosomiasis remains a powerful driving force in research (El-Marhoumy et al., 2009). Obviously, further research is required on the development of novel adjuvant vehicles as well as cocktail vaccine formulations to enhance protection levels with the eventual aim of 100% worm reduction (Siddiqui et al., 2011).

Conclusively, the present study has shown that combining different antigens (cocktail antigens) {CAP, SWAP and SEA} provide augmentation of the protective immunity when compared with each component administered individually, and considered a convenient mean to overcome many of the problems associated with the successful implementation of vaccine schedule of schistosomiasis. Adding adjuvants like BCG and Freund's provide more power to the candidate vaccine. This further suggest that immune interference resulting from immunizations with antigenically distinct vaccine targets that may be an important consideration in the development of multicomponent vaccine preparations, as the need for *Schistosoma* vaccines is now more pressing than ever.

ACKNOWLEDGEMENT

This research was financially supported by Zagazig University project. The authors thank the director Prof. Dr. Samia E. Etewa for her cooperation and kind help to accomplish this work. Thanks to Prof. Dr. Soheir S. Mahmoud Professor of parasitology, Theodor Bilharz Research Institute (TBRI), Cairo, Egypt for her kind help and support.

REFERENCES

Ashour AA, Ahmed SA, Maghraby AS, Zahran HG (2004). Immunoprophylactic effect of single and mixed schistosomal antigens on *Schistosoma mansoni* infected mice. Egypt J. Hosp. Med. 14:86-103.

Bergquist R, Utzinger J, McManus DP (2008). Trick or treat: the role of vaccines in integrated schistosomiasis control. PLoS Negl. Trop. Dis. 2(6):e244.

Boros DL, Warren KS (1970). Delayed hypersensitivity type III: granuloma formation and dermal reaction induced and elicited by a soluble factor isolated from *Schistosoma mansoni* eggs. J. Exp. Med. 132(3):488-507.

Boulanger D, Warter A, Sellin B, Lindner V, Pierce RJ, Chippaux JP, Capron A (1999). Vaccine potential of a recombinant glutathione S-transferase cloned from *Schistosoma haematobium* in primates experimentally infected with a homologous challenge. Vaccine 17(4):319-326.

Bradford MM (1976). A rapid and sensitive method for the quantitation of microgram qualities of protein utilizing the principle of protein dye binding. Anal. Biochem. 72:248-254.

Carter CE, Collely DG (1979). Partial purification of Schistosoms *mansoni* soluble egg antigen with Con-Sepharase chromatography. J. Immunol. 122:2204-2209.

Cheever AW (1968). Aquantitative post-mortem study of schistosomiasis *mansoni* in man. Am. J. Trop. Med. Hyg. 7(1):38-64.

Chitsulo L, Loverde P, Engels D (2004). Schistosomiasis. Nat. Rev. Microbiol. 2(1):12–13.

Doenhoff MJ, Pica-Mattoccia L (2006). Praziquantel for the treatment of schistosomiasis: its use for control in areas with endemic disease and prospects for drug resistance. Expert Rev. Anti Infect. Ther. 4(2):199-210.

Duvall RH, Dewitt WB (1967). An improved perfusion technique for recovering adult *Schistosoma* from laboratory animals. Am. J. Trop. Med. Hyg. 16(4):483–484.

El-Ahwany E, Bauiomy IR, Nagy F, Zalat R, Mahmoud O, Zada S (2012). T Regulatory cell responses to immunization with a soluble egg antigen in *Schistosoma mansoni*-infected mice. Korean J. Parasitol. 50(1):29–35.

El-Marhoumy SM, El-Nouby KA, Emara MA, Abou-Rayia DM (2009). An experimental study for evaluating the efficacy of cercarial vaccine against *Schistosoma mansoni*. J. Egypt Soc. Parasitol. 39(3):917-932.

Fallon PG, Dunne DW (1999). Tolerization of mice to *Schistosoma mansoni* egg antigens causes elevated type 1 and diminished type 2 cytokine responses and increased mortality in acute infection. J. Immunol. 162(7):4122-4132.

Gryseels BK, Clerinx PJ, Kestens L (2006). Human schistosomiasis. Lancet 368(9541):1106-1118.

Himmelrich H, Lo-Man R, Winter N, Guermonprez P, Sedlik C, Rojas M, Monnaie D, Gheorghiu M, Lagranderie M, Hofnung M, Gicquel B, Clement JM, Leclerc C (2000). Immune responses induced by recombinant BCG strains according to level of production of a foreign antigen: malE. Vaccine 18(24):2636-2647.

Hotez PJ, Engels D, Fenwick A, Savioli L (2010). Africa is desperate for praziquantel. Lancet 376(9740):496-498.

Ismail OA (2005). Study of the efficacy of adult worm, cercarial and egg antigens in protection against experimental intestinal schistosomiasis. MD. Thesis. Faculty of Medicine, Suez Canal University.

James CE, Hudson AL, Davey MW (2009). An update on P-glycoprotein and drug resistance in *Schistosoma mansoni*. Trends Parasitol. 25(12):538-547.

Khalifa RMA, Elnadi NA, Omran EK, Abdel-Tawab RA (2011). Immunological response and the probability of production of vaccine for schistosome parasites. Egypt J. Med. Sci. 32(2):547-570.

Lambertucci JR, Silva LC, Amaral RS (2007). Guidelines for the diagnosis and treatment of schistosomal myeloradiculopathy. Rev. Soc. Bras. Med. Trop. 40(5):574-581.

Martin VP, Pinheiro CS, Figueiredo BCP, Assis NRG, Morais SB, Caliari MV, Azevedo V, Castro-Borges W, Wilson RA, Sergio C, Oliveira SC (2012). Vaccination with enzymatically cleaved GPI-Anchored proteins from *Schistosoma mansoni* induces protection against challenge infection. Clin. Dev. Immunol. 2012:962538.

McManus DP, Loukas A (2008). Current Status of Vaccines for Schistosomiasis. Clin. Microbiol. Rev. 21(1):225-242.

Nabih I, Soliman AM (1986). Studies on fresh water snails, specific intermediate host for schistosomiasis. II. Isolationof total protein from native and irradiated snails. Cell Mol. Biol. 32:315- 317.

Nascimento E, Leão IC, Pereira VR, Gomes YM, Chikhlikar P, August T, Marques E, Lucena-Silva N (2002). Protective immunity of single and multi-antigen DNA vaccines against schistosomiasis. Mem. Inst. Oswaldo Cruz 97 Suppl. 1:105-109.

Ohara N, Yamada T (2001). Recombinant BCG vaccines. Vaccine 19:4089-4098.

Pellegrino J, Oliveira CA, Faria J, Cunha AS (1962). New approach to the screening of drugs in experimental schistosomiasis *mansoni* in mice. Am. J. Trop. Med. Hyg. 11:201-215.

Peters AP, Warren KS (1969). A rapid method of infecting mice and other laboratory animals with *Schistosoma mansoni* subcutaneous injection. J. Parsitol. 55:558- 563.

Salih SY, Bartlett A, Voller A (1978). Detection of antibodies by enzyme immunoassay in human *schistosoma mansoni* infections: a clinical and chemotherapeutic study. Tropenmed. Parasitol. 29(4):409-412.

Siddiqui AA, Bilal A, Siddiqui,BA, Ganley-Leal L (2011). Schistosomiasis vaccines. Hum. Vaccin. 7(11):1192–1197.

Smithers SR, Hackett F, Ali OP, Simpson AJG (1989). Protective immunization of mice against *Schistosoma mansoni* with purified adult worm surface membrane. Parasit. Immunol. 11:301-318.

Soliman MF, El Shenawy NS, El Arabi SE (2008). *Schistosoma mansoni:* melatonin enhances efficacy of cercarial and soluble worm antigens in the induction of protective immunity against infection in the hamster. Exp. Parasitol. 119(2):291-295.

Teixeira de Melo T, Michel de Araujo J, Do Valle Durães F, Caliari MV, Oliveira SC, Coelho PM, Fonseca CT (2010). Immunization with newly transformed *Schistosoma mansoni* schistosomula tegument elicits tegument damage, reduction in egg and parasite burden. Parasit. Immunol. 32(11-12):749-759.

Tendler M, Pinto RM, Oliveira LA, Gebara G, Katz N (1986). *Schistosoma mansoni*: vaccination with adult worm antigens. Int. J. Parasitol. 16(4):347-352.

Todd CW, Colley DG (2002). Practical and ethical issues in the development of a vaccine against schistosomiasis *mansoni*. Am. J. Trop. Med. Hyg. 66(4):348–358.

Waine GJ, McManus DP (1999). Schistosomiasis vaccine development--the current picture. Bioassays 19(5):435-438.

WHO (2000). Prospects for immunologic intervention in human schistosomiasis. 1-17 Technical Report Series, Geneva, Swizerland.

WHO (2010). Schistosomiasis Fact Sheet N°115; www.who.int/mediacentre/factsheets/fs115/en/index.html; Updated February 2010.

Infection prevalence of intestinal helminths and associated risk factors among schoolchildren in selected kebeles of Enderta district, Tigray, Northern Ethiopia

Ashenafi Teklemariam[1], Tadesse Dejenie[2] and Zewdneh Tomass[3]

[1]Department of Biology, College of Natural and Computational Science, Welkitie University, Ethiopia.
[2]Department of Biology, College of Natural and Computational Science, Mekelle University, Ethiopia.
[3]Department of Biology, College of Natural and Computational Science, Wolaita Sodo University, Ethiopia.

Lack of sanitary facilities and associated poor personal and environmental hygiene predispose people to the risk of intestinal parasite infections Thus, the objective of this survey was to assess the prevalence of intestinal helminth infections and associated risk factors among primary school children in selected kebeles of Enderta district, Tigray, northern Ethiopia. A cross sectional parasitological study consisting of 480 school children was conducted between February and March, 2011. Interview and microscopic examination of stool specimens of the children were made. Data was entered and analyzed by statistical packages for social sciences (SPSS) version 16.0 soft ware. The overall prevalence for at least one intestinal helminthic infection was 41.46% (199/480). The most dominant helminths were *Schistosoma mansoni* (23.13%), *Hymenolepis nana* (11.46%), *Ascaris lumbricoides* (5.83%) and *Enterobius vermicularis* (5.42%). Male children were at high risk for intestinal helminth infection than females ($P < 0.014$). Hand washing habit before meal ($P = 0.0001$), cleanness of fingernails ($P = 0.0001$) and swimming habit of the children ($P = 0.0001$) were significantly associated with intestinal helminth infections. Provision of health education, improving personal hygiene, potable drinking water supply as well as school based deworming should be the major focus area of the regional as well as Federal Health institutions.

Key words: Intestinal helminths, prevalence, risk factors, school children, Tigray.

INTRODUCTION

In developing countries, epidemiological determinants including lack of sanitary facilities and associated poor personal and environmental hygiene predispose people to the risk of intestinal parasite infections (Brooker et al.,

2008). Soil-transmitted helminths (STHs) and schistosomes are among the most prevalent helminth infections of humans worldwide. In Ethiopia, both STHs and *Schistosoma mansoni* are widely distributed among school children (Mengistu and Berhanu, 2004; Girum, 2005; Tadesse and Beyene, 2009; Tadesse and Tsehaye, 2010). Both *S. mansoni* and *Schistosoma haematobium* species are endemic in Ethiopia. However, infection caused by *S. mansoni* has a wide geographical distribution in the country (Berhanu et al., 2002).

Although, various studies have tried to address the problems encountered by intestinal helminthes in Tigray, northern Ethiopia (Tadesse and Beyene, 2009; Tadesse and Tsehaye, 2008, 2010; Tadesse et al., 2009, 2010), none of them tried to correlate the prevalence of the parasitic infections with associated risk factors.

Neglected tropical diseases are still a major public health concern for developing countries such as Ethiopia. Soil-transmitted helminthiasis and schistosomiasis are among the five most common neglected diseases (Hotez et al., 2006). However, with the development of more effective treatments in the past 20 years, several implementation strategies have attempted to control, eliminate and even eradicate these diseases with varying degrees of success (Hotez et al., 2006). Key determinants of success have been how effectively these treatments and strategies have been integrated into health systems of endemic countries, and the existence, reach, and quality of health systems. Most neglected tropical disease interventions use community-directed mass drug administration as a strategy where formal health systems are weak or absent. The evidence from the success story of community-directed treatment in the control of neglected tropical diseases (NTD), such as lymphatic filariasis and schistosomiasis, were effective (Gyapong et al., 2001; Massa et al., 2008). Thus, such survey could contribute in collecting baseline data to be used for mass drug administration strategy in Tigray Region.

In addition, to our knowledge, no parasitological study, was conducted in Enderta. The objective of this study was, therefore, to assess the prevalence of major intestinal helminth parasites of human (*S. mansoni* and soil transmitted helminths) and associated risk factors among primary school children in selected Kebeles of Enderta district, northern Ethiopia.

MATERIALS AND METHODS

Study area description

The study was conducted in selected Kebeles of Enderta district which is located in the south-eastern zone of Tigray. According to central statistical agency of Ethiopia (2005), the population of the district is 123,421 (62,039 males and 61,382 females). The district has about 25,751 households in 17 Tabia and 68 Kebeles. This cross sectional survey intestinal helmint parasite infections was carried out among primary school children found in three primary schools (Merb-Miti, Maykeyah and Romanat) between February to March, 2011. The study population consisted of all children attending their class in these primary schools. In 2011, the total student population in the selected schools was 3,202 (1,613 males and 1,589 females). The primary schoolchildren population is expected to represent the community because schoolchildren were used as an index for assessing community prevalence (Guyatt et al., 1999).

Sample size and sampling technique

Sample size was estimated using the statistical formula of sample size calculation $n = p\ (1 - p)\ z^2 / d^2$, where, n = required sample size, z = confidence level at 95% which is standard value of 1.96, p = estimated prevalence of intestinal parasite and d = marginal error at 5%, standard value of 0.05 (Daniel, 1995). Since the overall prevalence of intestinal helminths was not known for this study area, prevalence (p) was taken to be 50% and this gave the minimum sample size of 384. In order to minimize the likelihood of non compliance or possible drop out, 25% of the sample size was added to the normal sample size. The selected total sample size was only 15% of the total student population in the selected schools. Thus, 480 schoolchildren were selected for this study. To select the study subjects, the students were first stratified according to their educational level (grade 1 to 8). The sample size was then proportionally allocated and the required numbers of students from each class were then drawn using systematic random sampling technique.

Ethical considerations

The objectives of the study were explained to the school teachers and students at the time of sample collection. They were also briefed on impact of helminths, mode of transmission and related issues at the time. Prior to sample collection, informal consent was sought from the school directors, and ethical approval were obtained from Institutional Research Ethics Review Committee of the college of Health Science, Mekelle University (HRERC: Ref.No: CHS/1679/Dn/16; Date: 17/01/2011). Positive individuals were treated with appropriate doze of anthelminthic drugs.

Data collection

A pre-tested questionnaire, which was developed and modified based on known potential risk factors, was used to gather demographic, socio-economic and behavioral data. During the time of interview, children were also checked for condition of their finger nails. Interview questionnaire were administered to the children in their mother tongue, Tigrinya. Finally, accuracy and completeness of all the questionnaires were checked at the end of each data collection day. Prior to stool sample collection, children were guided on how to bring their stool samples not to mix with soil and urine. Then immediately after interview, each child was provided with a dry, clean and leak proof stool cup, labeled with the identification number of each child and applicator stick. Samples were preserved in 8 ml of 10% formalin solution, and transported to the Microbiology and Parasitological Laboratory of the department of Biology, Mekelle University, for processing and parasitic examination.

Table 1. Gender related prevalence of intestinal helminths among schoolchildren in selected Kebeles of Enderta district, Tigray, northern Ethiopia, 2011.

Helminths infection	Gender		Total (n=480) [No. (%)]
	Males (n=252) [No. (%)]	Females (n=228) [No. (%)]	
A. lumbricoides	9 (3.57)	19 (8.33)	28 (5.83)
Hookworm	2 (0.78)	-	2 (0.42)
T. trichiura	1 (0.40)	-	1 (0.21)
S. mansoni	70 (27.77)	41 (17.98)	111 (23.13)
E. vermicularis	10 (3.99)	16 (7.02)	26 (5.42)
S. stercoralis	1 (0.40)	-	1 (0.21)
H. nana	28 (11.11)	27 (11.84)	55 (11.46)
Taenia Spp	-	3 (1.32)	3 (0.63)

A. lumbricoides = Ascaris lumbricoides, T.trichiura = Trichuris trichiura, S.mansoni = Schistosoma mansoni, E.vermicularis = Enterobius vermicularis, S.stercoralis = Strongyloides stercoralis, H.nana = Hymenolepis nana

Formalin-ether concentration technique was used for laboratory examination of the collected samples (World Health Organization (WHO), 1991). The results were properly recorded in the table form prepared for the purpose. Stool examination was conducted by the help of experienced medical laboratory technicians.

Data analysis

The data collected through interview based questionnaires and parasitological examinations were initially fed into Microsoft Excel 2003 software and then copied to SPSS software, version 16 for analysis. Chi-square test was performed to assess the presence of any significant differences between age, gender and location. Logistic regression was performed to determine the independent effect of the variables by calculating the strength of the association between infection and risk factors using odd ratio (OR) at 95% confidence interval (CI). Crude OR was estimated by univariate analysis and adjusted OR was then estimated by multivariate logistic regression analysis. P-value less than 0.05 (5%) was considered to be statistically significant.

RESULTS

Prevalence of intestinal helminth infections

From a total of 480 stool samples examined, 41.46% (199) were positive for one or more of intestinal helminths. Eight species of intestinal helminths were identified with variable prevalence: *S. mansoni* (23.13%), *Hymenolepis nana* (11.46%), *Ascaris lumbricoides* (5.83%), *Enterobius vermicularis* (5.42%), *Taenia* species (0.63%), hookworm (0.42%), *Trichuris trichiura* (0.21%) and *Strongyloide stercoralis* (0.21%). The prevalence of soil transmitted helminths (STHs) was 6.67%. Among 199 positive individuals, 109 (22.71%) were males and 90 (18.75%) were females. The majority (172; 35.83%) of infected individuals had single infection and 27 (5.63%) were infected with more than one intestinal helminth parasites. The most frequent combinations of helminths infection identified were co-infection of *S. mansoni* and *H. nana* (1.46%; 7/480), followed by *S. mansoni* and *E. vermicularis* 1.25% (6 out of 480). Gender related difference in prevalence rate of intestinal helminthic infections was observed in *S.mansoni* (χ^2 = 6.46, P = 0.011), where male children were found to be more infected (27.77%) compared to females (17.98%) (Table 1).

Age related intestinal helminthic infections were only observed in *S.mansoni* infection (χ^2 = 23.48, P = 0.0001). The age group 15 to 19 years was the most affected (32.97%) followed by the age group 10 to 14 years old (24.68%) (Table 2). The prevalence of intestinal helminthic infections showed significant difference among the selected schools (χ^2=41.25, P=0.0001). The least intestinal helminth parasitic infection was recorded in Mykeiya (26.19%) and the highest was in Romanat primary school (58.60%). The pattern of *S. mansoni* was also similar to intestinal helminthic infection and significantly varied among the three schools (χ^2 = 80.63, P = 0.0001) (Figure 1).

Risk factors associated with intestinal helminths infection

The distributions of socio-demographic and behavioral risk factors of the subjects are shown in Tables 3 and 4. In univariate analysis, significant association were seen between intestinal helminth infection and gender (COR = 0.652; 95%CI: 0.500, 0.851; P = 0.002), family occupation (COR = 0.776; 95%CI: 0.682, 0.883 agriculture versus non agriculture; P = 0.0001), availability of latrine (COR = 0.473; 95%CI: 0.374 to 0.597; P = 0.0001), family size (COR = 0.834; 95%CI: 0.725 to 0.961 for

Table 2. Age related prevalence of intestinal helminths among schoolchildren in selected kebeles of Enderta district, 2011.

Helminths infection	Age groups				Total (n=480) [No. (%)]
	5-9 (n=80) [No. (%)]	10-14 (n=308) [No. (%)]	15-19 (n=91) [No. (%)]	20-24 (n=1) [No. (%)]	
A. lumbricoides	7 (8.75)	15 (4.87)			28 (5.83)
Hookworm	-	2 (0.65)			2 (0.42)
T. trichiura	1 (1.25)	-	-	-	1 (0.21)
S. mansoni	4 (5.00)	76 (24.68)	30 (32.97)	1 (100)	111 (23.13)
E. vermicularis	6 (7.50)	15 (4.87)	5(5.50)	-	26 (5.42)
S. stercoralis	1 (1.25)	-	-	-	1 (0.21)
H. nana	15 (18.75)	33 (10.71)	7 (7.69)	-	55 (11.46)
Taenia Spp	-	3 (0.97)	-	-	3 (0.63)

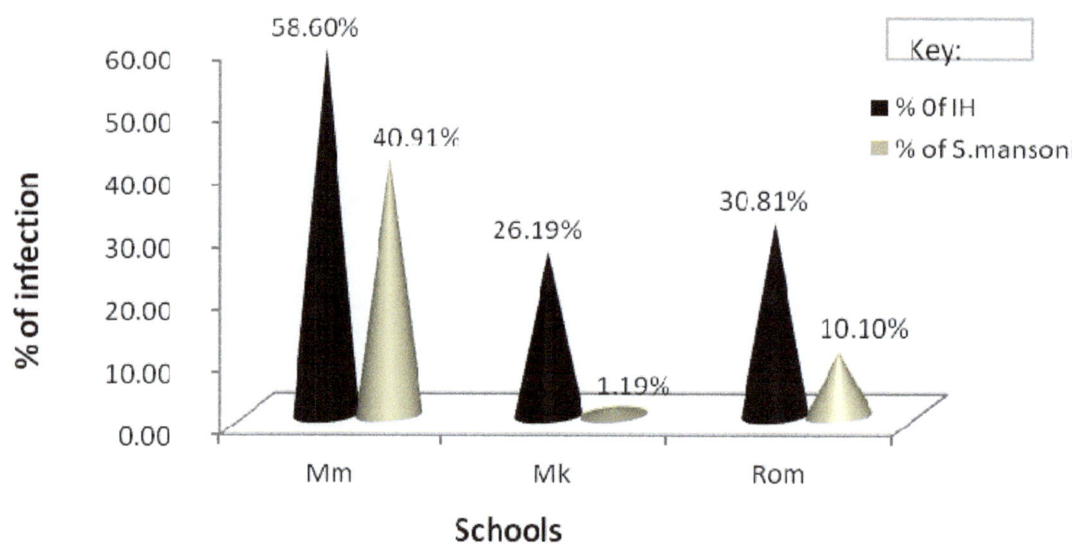

Figure 1. Prevalence rate of intestinal helminths among primary schools in selected Kebeles of Enderta district, 2011. Mm= Merb-Miti, Mk= May Keyah, Rom= Romanat.

family size > 5; P = 0.012) but no significant association with the source of water (COR = 0.879; 95%CI: 0.651 to 1.187 for pipe user versus river/stream; P = 0.401). In addition, hand washing habit (COR = 0.503; 95%CI: 0.394, 0.641; P = 0.0001), latrine usage habit (0.515; 95%CI: 0.406, 0.652; P = 0.0001), habit of swimming (COR = 1.855; 95%CI: 1.336 to 2.574; P= 0.0001), and cleanness of finger nails (COR= 1.153; 95%CI: 0.917 to 1.450; P = 0.0001) were significantly associated with intestinal helminthic infections.

Risk factors independently associated with intestinal helminth infections

The final model (multivariate logistic regression) analysis indicated that male children were at high risk of infection (AOR = 1.835; 95%CI: 1.133 to 2.971; P = 0.014). Children who had not habit of washing their hand always were more likely to be infected with intestinal helminth infections (AOR = 0.434; 95%CI: 0.277 to 0.681; P = 0.0001) than those who washed their hand always before meal. Children who had habit of swimming (AOR = 6.113; 95%CI: 3.604 to 10.368; P = 0.0001) and dirty material in their hand fingernails (AOR = 3.133; 95%CI: 1.923 to 5.104; P = 0.0001) were more likely to be infected by intestinal helminth infection. The final model analysis also indicated that there were no statistically significant association between intestinal helminth infection and family occupation (P = 0.087), latrine distribution (P = 0.427), family size (P = 0.088), latrine usage habit (P = 0.198).

Table 3. Univariate analysis for socio-economic risk factors associated with intestinal helminths infections among schoolchildren in selected Kebeles of Enderta district, 2011.

Risk factor	Intestinal helminths		COR	95%CI	P-value
	Positive [No. (%)]	Negative [No. (%)]			
Sex					
Male	109 (43.3)	143 (56.7)	0.652	0.500-0.851	0.002
Female	90 (39.5)	138 (60.5)			
Family occupation					
Agriculture	137 (43.2)	180 (56.8)	0.776	0.682-0.883	0.0001
Non agriculture	62 (38.0)	101 (61.9)			
Availability of latrine					
Present	104 (32.1)	220 (67.9)	0.473	0.374-0.597	0.0001
Absent	95 (60.9)	61 (39.1)			
Source of water					
Pipe	119 (38.5)	190 (61.5)	0.879	0.651-1.187	0.401
River/stream	80 (46.8)	91 (53.2)			
Family size					
≤ 5	142 (37.7)	235 (62.3)	0.834	0.725-0.961	0.012
>5	57 (55.3)	46 (44.7)			

COR=Crude Odd Ratio, 95%CI= ninety five percent confidence interval

Table 4. Univariate analysis for behavioral risk factors associated with intestinal helminths among schoolchildren in selected Kebeles of Enderta district, 2011.

Risk factor	Intestinal helminths		COR	95%CI	P-value
	Positive [No. (%)]	Negative [No. (%)]			
Hand washing habit					
Always	98 (33.4)	195 (66.6)	0.503	0.394-0.641	0.0001
Sometimes/not at all	101 (54.0)	86 (46.0)			
Latrine usage habit					
Present	104 (34.0)	202 (66.0)	0.515	0.406-0.652	0.0001
Absent	95 (54.6)	79 (45.4)			
Habit of swimming					
Present	102 (65.0)	55 (35.0)	1.855	1.336-2.574	0.0001
Absent	97 (30.0)	226 (70.0)			
Dirty materials in the hand fingernails					
Present	158 (53.6)	137 (46.4)	1.153	0.917-1.450	0.0001
Absent	41 (22.2)	144 (77.8)			

DISCUSSION

The present study revealed the occurrence of eight species of intestinal helminths of public health importance among school children in the study area. The overall prevalence roles of helminths observed in the present study was found to be higher than earlier reports from different parts of the region: southern and central zones of Tigray (Tadesse and Beyene, 2009), Hintalo Wejerat (Tadesse and Tsehaye, 2008) and from eastern Ethiopia (Girum, 2005). Further, the present finding was in agreement with that reported from south Wello (Assefa et al., 1998). However, the present rates were far lower than those reports from rural areas close to south east lake Langano (Mengistu and Berhanu, 2004), near Lake Tana (Melakeberhan et al., 1993), Lake Awassa (Yared et al., 2001) and Chilga district, northwest Ethiopia (Leykun, 2001). The differences in observed prevalence among the studies might be due to varied factors, including socio-economic conditions, individual behavioral habits of selected children, the methods employed for stool examination, the sample size taken as well as the time of study conducted. Further, Mengistu and Berhanu (2004) stated that the distribution and prevalence of various species of intestinal parasites also vary from region to region because of several environmental, social and geographical factors. In addition, in the present study area there was no school based deworming activity in the area. Thus, this may attribute to the observed high prevalence of helminth infections in the present study area.

In the present study, *S. mansoni* was the dominant helminth infection. The result was higher than reported from southern and central zone of Tigray (12.4%) (Tadesse and Beyene, 2009). However, the prevalence of *S. mansoni* (21.25%) was similar with the results obtained from south east of Lake Langano (Mengistu and Berhanu, 2004). High prevalence of *S. mansoni* in the present study may be attributed to the presence of irrigation schemes near to the study areas and high water contact of the children to the nearby water bodies which enhanced the transmission. Increases in Schistosomiasis transmission associated with water development projects is well documented (Tadesse and Beyene, 2009; Leykun, 1999). Similarly, Bethony et al. (2006) stated that frequent contact with contaminated water create favorable condition for the transmission and wide spread of intestinal helminths infection.

The higher predominance of males to *S. mansoni* in the present study is consistent with reports from elsewhere in same region and eastern Ethiopia (Girum, 2005; Tadesse and Tsehaye, 2008; Tadesse and Beyene, 2009). The difference in *S. mansoni* infection among sexes can be explained in behavioral differences among sexes in which males have high frequency of water contact habit. In the present study, assessment of sex related water contact habit (swimming habit) revealed that males do have significantly high swimming habit compared to females (X^2 = 50.78, P = 0.000). Thus, the water contact through swimming could be one reason for the observed *S. mansoni* infection difference among sexes. The age dependent prevalence of *S. mansoni* infection obtained in the present study is consistent with reports from the same region (Tadesse and Bayene, 2009). Accordingly, significantly higher infection was reported in the age group 15 to 19 and 10 to 14 compared to the 5 to 9 age group which was the least affected group. The low infection rate of *S. mansoni* among 5 to 9 age groups might be as a result of low water contact behavior of the children.

The prevalence of soil transmitted helminth infection reported in this study was comparable with reports from Hintalo Wejerat, northern Ethiopia (7.5%) (Tadesse and Beyene, 2009), but lower than result obtained from Southern and Central zone of Tigray (13.9%) (Tadesse and Beyene, 2009) and Babile town, eastern Ethiopia (14.2%) (Girum, 2005). Further, the result of this study was lower than reported from different part of Nigeria; South west Nigeria (83%) (Ibidapo and Okwa, 2008) and Ebenebe town, Nigeria (87%) (Chukwuma et al., 2009). The low prevalence of soil transmitted infection in the present study area might be as a result of the local climatic condition which is an important determinant factor for transmission of various intestinal helminth infections. In fact climate with adequate moisture and warmth temperature is very crucial for larval development of STHs in the soil (Brooker et al., 2006; Brooker and Michael, 2000). This idea was also supported by Tadesse and Beyene (2009) in that geohelminth infection was high in southern Tigray than central Tigray since the moist condition and hot temperature in the southern part was favorable for the preservation of infective helminth. In addition, the low prevalence of STHs in the present study might be because of the time of sample collection. It is well recognized that wet or damp soil favors the eggs of helminths and rainy season enables them to thrive more than dry season (WHO, 2002).

The present study also tried to assess the possible association of intestinal helminth infection with socio-demographic and behavioral risk factors among school children. The analysis in this study indicated that male children were at high risk for intestinal helminth infections compared to female children. Similar observations are found in different parts of Ethiopia and other tropical countries (Ikon and Useh, 1999; Adhikari et al., 2004; Ukpai and Ugwu, 2003; Girum, 2005, Tadesse and Bayene, 2009). This gender difference of helminthic infection might be because of the difference in gender behavior. The males are more likely to involve in outdoor activities than females because of social and religious restriction and the outdoor environment including agricultural areas are common place for defection by males during working time, therefore contamination of soil in the areas would facilitate the transmission of intestinal helminth. But contrasting results were reported in different areas in which females had a relatively high prevalent rate than males (Adefemi and Musa, 2006; Rajini and Gurdip, 2010). Further, Adefemi and Musa (2006) stated that the influence of gender on the prevalence of helminthic infection is inconclusive since it may or may not play a role depending on the regional and environmental factors. Thus, more systematic research works are required to elucidate this.

In the present study, high degree of association between hand washing habit and cleanness of the finger nails with intestinal helminth infections is well recognized. This was in agreement with earlier reports in different parts of Ethiopia (Ali et al., 1999; Girum, 2005; Yeshambel et al., 2010). This is probably due to insufficient water supplies, poor hygienic practice, and socio-economic status. Further, Filippo et al. (1998) stated that promoting hand-washing and improvement of general personal hygiene could bring reduction in the level of intestinal helminthic infection. Thus, lack of personal hygiene might increase the probability of exposure to intestinal helminths infection. There is also evidence from Gurez valley of Jammu and Kashmir state, India that children with better personal hygiene had lower prevalence of infecting by helminthic infection than those children living in poor hygienic condition (Showkat et al., 2010).

Although the distributions of better toilet facilities in each individual house hold as well as in each school could have reduced effect on the prevalence of intestinal helminth infections, the logistic regression analysis in the present study indicated that the availability of toilate (latrine) and the type of latrine was not significantly associated with the prevalence of intestinal helminthic infection. Similar results were reported in different areas of Ethiopia (Girum, 2005; Tadesse and Beyene, 2009).

Tashikuka et al. (1995) described that the use of pit latrine was associated with a lower risk of acquisition of intestinal helminth compared to open bush, but where the provision of pit latrine is not accompanied with adequate supply of water, the chance of faecal contamination becomes higher. Furthermore, Ali et al. (1999) stated that improper usage and poor quality hygiene of the toilet have been shown to influence acquisition of intestinal helminths. Thus, the mere availability of a toilet, unless properly used, does not guarantee protection against intestinal helminth infection.

Conclusion

This study added to the knowledge of the epidemiology of intestinal parasitic infections in Ethiopia in general and revealed that hand washing habit before meal, swimming habit of the children and availability of dirty materials in the fingernails of the schoolchildren were found to be significantly associated with intestinal helminths infection in the study area. The information may be used in design and application of control strategies.

ACKNOWLEDGEMENTS

The authors would like to appreciate the cooperation of Merb-Miti, Maykeyah and Romanat primary school communities. We thank mekelle University for funding the research and laboratory facility supply.

Conflict of interest

The authors declare that there is no conflict of interest.

REFERENCES

Adefemi S, Musa O (2006). Intestinal helminthes infestation among pupils in rural and urban communities of Kwara state, Nigeria. Afr. J. Clin. Exp. Microbiol. 7:217-228.

Adhikari M, Bomjan R, Khatri B, Joshi R, Dhakal P, Lekhak B (2004). Intestinal helminthic infections among school children in Kathmandu valley. J. Nepal Health Res. Council 5:17-21.

Ali I, Mekete G, Wdajo N (1999). Intestinal parasitism and related risk factors among students of Asendabo elementary and junior secondary school, South Western Ethiopia. Ethiop. J. Health Dev. 13:157-161.

Assefa T, Woldemicael T, Dejenie A (1998). Intestinal parasitism among students in three localities in south Wello, Ethiopia. Ethiop. J. Health Dev. 12:231-235.

Berhanu B, Medhin G, Berhe N, Abebe F, Gebre T, Michael S, Gundersen G (2002). Epidemiologic studies on intestinal Schistosomiasis in Wondo Genet, Sothern Ethiopia. Ethiop. Med. J. 41:333-44.

Bethony J, Brooker S, Albonico M (2006). Soil-transmitted helminth infections: Ascariasis, Trichuriasis, and hookworm. Lancet 367:1521-1532.

Brooker S, Michael E (2000). The potential of geographical information systems and remote sensing in the epidemiology and control of human helminth infections. Adv. Parasitol. 47:245-288.

Brooker S, Archie C, Clements A, Bundy D (2006). Global epidemiology, ecology and control of soil-transmitted helminth infections. Adv. Parasitol. 62:221-261.

Brooker S, Hotez P, Bundy D (2008). Hookworm-Related Anaemia among Pregnant Women: A Systematic Review. Plos Neglected Trop. Dis. 2:274-291.

Chukwuma C, Ekejindu M, Agbakoba R, Ezeagwuna D, Anaghaw C, Nwosu D (2009). The prevalence and risk factors of geohelminths infections among primary school children in Ebenebe Town, Anambra state, Nigeria. Middle East J. Sci. Res. 4:211-215.

Daniel W (1995). Biostatistics a foundation for analysis in the health science. In: Statistical analysis (6 th ed.), New York: John Willey and sons Inc, USA. p. 155.

Filippo C, Patrizio P, Abdel L, Hamdan A, Pietro I, Youshri S, Marzio B (1998). Knowledge, perceptions and behavior of mothers toward intestinal helminths in Upper Egypt: implications for control. Health Policy Plan. 13:423-432.

Girum T (2005). The prevalence of intestinal helminthic infections and association risk factors among schoolchildren in Babile town, eastern Ethiopia. Ethiop. J. Health Dev. 19:140-147.

Guyatt H, Brooker S, Donnelly C (1999). Can prevalence of infection in school aged children be used as an index for assessing community prevalence? Parasitology 118:257-268.

Gyapong M, Gyapong JO, Owusu-Banahene G (2001). Community-directed treatment: the way forward to eliminating lymphatic filariasis as a public health problem in Ghana. Ann. Trop. Med. Parasitol. 95:77-86.

Hotez PJ, Ottesen E, Fenwick A, Molyneux D (2006). The neglected tropical diseases: the ancient afflictions of stigma and poverty and the prospects for their control and elimination. Adv. Exp. Biol. Med. 582:22-33.

Ibidapo C, Okwa O (2008). The prevalence and intensity of soil transmitted helminths in a rural community, Lagos Suburb, South west Nigeria. Internet J. Agric. Biol. 10:89-92.

Ikon G, Useh C (1999). Epidemiology of gastrointestinal helminthosis among pupils in urban and sub-urban communities of, Nigeria. J. Med. Lab. Sci. 8:1-6.

Leykun J (1999). Intestinal helminth infections in schoolchildren in Gonder town and surrounding areas, northwest Ethiopia. Ethiop. J. Sci. 22:209-220.

Leykun J (2001). Soil-transmitted helminth infections and *Schistosomiasis mansoni* in schoolchildren from Chilga District, Northwest Ethiopia. Ethiop. J. Health Sci. 11:79-87.

Massa K, Magnussen P, Sheshe A, Ntakamulenga R, Ndawi B, Olsen A (2008). Community perceptions on the community-directed treatment and school-based approaches for the control of schistosomiasis and soil-transmitted helminthiasis among school-age children in Lushoto District, Tanzania. J. Biosoc. Sci. 41:89-105.

Melake-Berhan D, Wondwossen H, Tsefaye W, Elis G, Sissay Y, Tariku A, Tibabu D (1993). Intensity of intestinal parasite infection in a small farming village near Lake Tana, Ethiopia. Ethiop. J. Health Dev. 7:27-31.

Mengistu L, Berhanu E (2004). Prevalence of intestinal parasites among schoolchildren in a rural area close to the southeast of Lake Langano, Ethiopia. Ethiop. J. Health Dev. 18:116-120.

Rajini K, Gurdip S (2010). Epidemiology and control of Schistosomiasis and other intestinal parasitic infections among school children in three rural villages of south Saint Lucia. J. Vector Borne Dis. 47:228-234.

Showkat AW, Fayaz A, Showkat AZ, Ayesha A, Zubair A, Pervaiz A (2010). Intestinal Helminthiasis in Children of Gurez Valley of Jammu and Kashmir State, India. J. Global Infect. Dis. 2:91-94.

Tadesse D, Beyene P (2009). Irrigation practices and Intestinal helminth infections in Southern and Central Tigray. Ethiop. J. Health Dev. 23:48-56.

Tashikuka J, Scott M, Gary-Donald K (1995). *Ascaris lumbricoides* infection and environmental risk factors in an urban African setting. Ann. Trop. Med. Parasitol. 89:505-514.

Ukpai O, Ugwu C (2003). The prevalence of gastro-intestinal tract parasites in primary school children in Ikwuano Local Government Area of Abia State, Nigeria. Nig. J. Parasitol. 240:129-36.

World Health Organization (WHO) (2002). Prevention and Control of Schistosomiasis and Soil-Transmitted Helminthiasis. World Health Organization, Technical Series Report 912. Geneva.

World Health Organization (WHO) (1991). Basic laboratory methods in medical parasitology. Geneva, Switzerland.

Yared M, Mamdouh H, Girma M, Shiferaw T (2001). Intestinal helminthic infection among children at Lake Awassa Area, South Ethiopia. Ethiop. J. Health Dev. 15:31-38.

Yeshambel B, Girmay M, Alemayehu A, Berhanu E, Charloette H, Atalay A, Andrea V, John B, Gail D (2010). Prevalence and risk factors for soli transmitted helminth infection in mothers and their infants in Butajira, Ethiopia: A population based study. BioMed Central Public Health 2:10-12.

25

A preliminary characterization of an invariant 52kDa surface antigen in *Trypanosoma evansi* that serve as a diagnostic marker

E. M. El Hassan

Department of Microbiology and Parasitology, King Faisal University, P.O. Box 400, Al-Ahsa 31982, Kingdom of Saudi Arabia.

The study was carried out to examine the ways that *Trypanosoma evansi* components interact with the host, by investigating the components of the parasite which acted as antigens during infection, and to study the dynamic of some of these antigens during infection. Such approach could help in identifying new diagnostic reagents, a better understanding of existing diagnostic methods and target antigen for production of vaccines. The antigenic components of intact and trypsin-treated *T. evansi* were identified using sodium dodecyl sulfate polyacrylamide gel electrophoresis (SDS-PAGE) and immunoblotting against sera from infected rabbits and rabbits immunized with a range of soluble parasite materials. A trypsin sensitive component of ~52 kDa which was cleaved from the parasite by the process of trypsinisation and produced a potent reaction to infection and immunization sera was selected for further study. This antigenic component was purified by electro-elution from acrylamide gels and monospecific antiserum was produced and used in an antigen-capture enzyme linked immunosorbent assay (ELISA). This antigen was found to be preserved between different variants and populations of *T. evansi* and reached a detectable level in the circulation of infected animals before the remission of the first parasitemia and was cleared from the circulation following successful chemotherapy. The antigen represents a good candidate for the diagnosis of infection with *T. evansi* and for developing a serologically-based test for monitoring the effectiveness of chemotherapy.

Key words: *Trypanosoma evansi*, invariant antigens, immunoblotting, dynamic, diagnosis.

INTRODUCTION

Most of the parasites in the sub-genus *Trypanozoon* are of veterinary and medical importance, causing wide-spread disease throughout many tropical and sub-tropical countries. *Trypanosoma evansi*, the causative agent of the disease most commonly known as surra, was the first pathogenic trypanosome described. Infection with this parasite has an extremely wide geographical range in countries with hot and warm-temperate climates (Hoare, 1972). It infects a range of economically important live-stock including equids, camels, cattle, buffaloes, goats, sheep and pigs (Luckins, 1988), but the principal affected hosts are equines and dromedaries (Hoare, 1972).

Scientific endeavour over the past 100 years has yielded large amounts of information on many aspects of the parasitic trypanosomes including morphology, infectivity, behavior of intact parasites and host/parasite interaction, and has been the subject of extensive reviews (Hoare, 1972; Antoine-Moussiaux et al., 2009; Morrison et al., 2010). Our knowledge on many aspects of host/parasite interaction and immunology in particular, is mainly gleaned from studies based only on whole parasite extracts. Such extracts consist of a complex mixture of molecules each potentially capable of interacting with the host in a particular way. A full understanding of the way in which trypanosomes interact with the host must take into account the nature and behavior of individual parasite components and the way in which they interact with the host. Such information could help in identifying new diagnostic reagents, provide better insight into host/parasite interaction and provide indicators of infection, progress of the disease, effectiveness of chemotherapy and target antigens for production of vaccines.

Although the variant antigens of Salivarian trypanosomes constitute the major part of the surface, invariant surface antigens have also been reported. Infected mammalian hosts produce antibodies against minor and invariant surface antigens (Burgess and Jerrells, 1985; Radwanska et al., 2000). These invariant molecules are suggested to be immunogenic and may be present as circulating invariant antigens during infection (Burgess and Jerrells, 1985). These invariant surface antigens, reported mainly in *Trypanosoma brucei* are glycoproteins and referred to as invariant surface glycoproteins (ISGs); their abundance is said to be > 100-fold lower than the variant surface glycoprotein (VSG) (Overath et al., 1994). These antigens are considered to be arranged in between the VSG molecules on the surface of the parasite. These invariant surface proteins are necessary for nutrient uptake signaling and structural integrity, and are protected from the adaptive immune system of the mammalian host by the VSG in an unknown way (Schwede and Carrington, 2010).

The 145 and 140 kDa molecules are invariant surface glycoproteins which act as receptors for low-density lipoprotein (LDL) which is required for trypanosome growth (Hide et al., 1989; Coppens et al., 1988; Rolin et al., 1990). Other ISGs function as glucose transporters (Parsons and Nielson, 1990; Bringuad and Baltz, 1992). ISG100 is a polytopic protein partly resident within the endosomal system (Nolan et al., 1997; Pal et al., 2002). ISGs with as yet unknown function have been identified and include a 77 kDa protein isolated from coated endocytotic vesicles of *T. brucei* (Webster and Shapiro, 1990). Two invariant surface glycoproteins of 65 and 75 kDa have been identified by Ziegelbauer and Overath (1992) in the bloodstream stage of *T. brucei*. These particular ISGs having large extracellular domains similar in size to the VSGs and are distributed over the entire surface of the trypanosome (Ziegelbauer and Overath, 1993), were initially thought to be the most abundant surface proteins with the exception of VSG (Overath et al., 1994). However, with the exception of a weak reaction with anti-ISG 75 antibodies, in live trypanosomes, these ISGs were not considered to be accessible to antibodies (Ziegelbauer et al., 1992) and are not involved in protection against reinfection. However, Lanca et al. (2011) were able to partially protect Balb/C mice against a lethal dose of *T. brucei* by intramuscular immunization with a single dose of a DNA plasmid encoding ISG 65. Although ISG65 is expressed on the trypanosome cell surface, an intracellular pool of this protein was also detected (Chung et al., 2004).

Definitive diagnosis of *T. evansi* infection is still based on microscopic demonstration of the parasite, an insensitive method which depends mainly on the level of parasitemia. A sensitive parasite detection method, mini-anion-exchange centrifugation method, is available but is rarely used for routine diagnosis (Lanham and Godfrey, 1970; Gutierrez et al., 2004). Currently, tests for indirect diagnosis of *T. evansi* infection through detection of specific antibodies in mammalian hosts are widely used. All these antibody detection tests can not differentiate between current and previous infections. Many of these tests are based on the native VSG of the predominant variable antigen type (VAT) RoTat 1.2 of *T. evansi*. These include a card agglutination test for trypanosomiasis for *T. evansi* (CATT/*T. evansi*) (Bajana and Hamers, 1988), an immune trypanolysis (TL) assay (Van Meirvenne et al., 1995), a latex agglutination test for *T. evansi* (Verloo et al., 1998) and an enzyme-linked immunosorbent assay for *T. evansi* (ELISA/*T. evansi*) (Verloo et al., 2000). Antibody-ELISA based on crude *T. evansi* extract was also used to detect anti-*T. evansi* equine antibodies (Reyna-Bello et al., 1998). Recently an antibody detection ELISA independent of VSG and based on recombinant invariant surface glycoprotein 75 (rISG75) derived from *T. b. gambiense* LiTat 1.3 has been developed for diagnosis of *T. evansi* infection (Tran et al., 2009). This test showed a sensitivity of 94.6% and a 100% specificity, and demonstrated an almost perfect agreement with the previous tests.

Soluble trypanosome antigens circulating in body fluids of infected animals have been detected using enzyme-linked immunosorbent assay antigen-ELISA. The test is based on a polyclonal antiserum raised against crude trypanosomal antigens or monoclonal antibodies to membrane antigens of procyclic trypanosomes of *T. brucei* or a cell membrane antigen of *T. evansi* (Nantulya et al., 1989; Olaho-Mukani et al., 1993; Rayulu et al., 2007). These antigen capture assays are capable of detecting trypanosome antigens released into bloodstream by dying parasites during infection and are very sensitive compared to parasitological methods.

However, it has also been reported that the Ag-ELISA may produce false negative results, possibly due to insufficient destruction of trypanosomes during multiplication phase to produce detectable levels of antigens in the blood antigens in the blood of infected animals (Waitumbi and Nantulya, 1993). A test that can detect an invariant surface antigen accessible to antibodies during infection might overcome this phenomenon. Molecular methods for specific detection of *T. evansi* infection have also been used. Claes et al. (2004) used polymerase chain reaction based on sequenced gene coding for *T. evansi* RoTat 1.2 VSG which was present in all *T. evansi* strains (Urakawa et al., 2001).

The dynamic aspects and the fate of individual trypanosome antigens during infection are not well understood. Such information might provide better insight into host/parasite interaction and could help in identifying new diagnostic reagents. Nantulya and Lindqvist (1989) in developing antigen ELISA for diagnosis of bovine trypanosomiosis using monoclonal antibodies to membrane antigens of procyclic trypanosomes detected differences in the dynamics of circulating trypanosome antigens. *Trypanosoma vivax* and *Trypanosoma congolense* antigens were detectable between 10 to 12 days following tsetse challenge, while *T. brucei* antigens were detectable between 8 to 14 days. Following treatment of the infected cattle with Berenil, *T. vivax* and *T. congolense* antigens were cleared from the circulation within two weeks but the rate of clearance of *T. brucei* antigens was slower compared to the other two species. The host species has also been found to influence the dynamics of trypanosome antigens. El Amin et al. (1993) during an evaluation of antigen-ELISA for the diagnosis of *T. evansi* infection using the anti-brucei group-specific antigen detected the antigen in camels 2 to 4 days after infection, while in goats the parasite antigens were detected after 7 days of infection.

Thammasart et al. (2001) using antigen-detection ELISA were able to detected *T. evansi* antigens 10 to 14 days in most of the cattle experimentally infected with this parasite. The antigens disappeared from the circulation 3 days after treatment. While in some cattle, Ag-ELISA was not able to detect the parasite antigens despite positive hematocrit (HCT) results and they attributed these findings to variation in the pathogenicity of the parasite strains and in the balance of parasitemia and antibody levels in the circulation. The antigenic components of *T. evansi* were investigated in this study and the dynamic of some of these antigens was studied in search for an invariant surface antigen accessible to antibodies during infection to overcome the false negative results encountered in Ag-ELISA.

MATERIALS AND METHODS

Trypanosome

Two stocks of *T. evansi* were originally isolated from naturally infected camels in Al-Ahsa area, Saudi Arabia, propagated in mice and cryopreserved until used in this study. They include one population designated (TR 2257) and two variants of another population (TR 2165 & TR 2222).

Experimental animals

Mice, rabbits and rats were used as experimental models to expand the trypanosomes, to produce infection serum and polyclonal antibodies against *T. evansi* materials and to investigate the dynamics of the 52 kDa *T. evansi* antigen during infection, respectively.

Separation of trypanosome from mice blood

Trypanosomes expanded in mice were separated from infected blood by ion-exchange chromatography on Diethylaminoethyl (DEAE) cellulose (DE52, Whatman Biochemical, UK) as described by Lanham and Godfrey (1970). The separated trypanosomes were washed three times in phosphate saline glucose (PSG, pH 8.0) by centrifugation at 2650 g for 20 min at 4°C.

Preparation of trypanosome lysate

Crude soluble extract was prepared by re-suspending the pellet of 2.2×10^9 DEAE column separated trypanosomes in an equal volume of PSG and then subjected to three cycles of freezing to -80°C for 10 min, followed by thawing to room temperature for 10 min. The lysate was then centrifuged at 10,000 g for 45 min at 4°C, supernatant removed, its protein concentration determined and the lysate aliquoted into 100 µl volumes and stored at -20°C. Whole trypanosome extract and trypsinised trypanosomes extract were prepared through detergent lysis (Anderson and Blobel, 1983) using sodium dodecyle sulphate (SDS) sample buffer. 2.2×10^9 DEAE column separated trypanosomes were diluted with an equal volume of SDS sample buffer and heated to 100°C for 5 min. After cooling to room temperature and centrifugation at 10,000 g for 5 min, they were aliquoted into 200 µl volumes and stored at -20°C. Trypsinisation of trypanosomes was carried out by incubating the trypanosomes with trypsin (20 µl/ml at a concentration of 5 mg/ml) for 75 min at 37°C and the trypsin neutralized by incubation with trypsin inhibitor (20 µl/ml at a concentration of 5 mg/ml in distilled water) for further 5 min at 37°C. The trypanosomes were then washed by centrifugation at 5,000 g for 5 min in PSG.

Electron microscopy of intact and trypsinised trypanosomes

Electron microscopy was carried out according to the method of Garnett et al. (1978), with minor modification. After post-fixation in 1% osmium tetroxide and washing, the trypanosomes were dehydrated in ascending series of acetone, then infiltrated in araldite mix: analar acetone and infiltrated in araldite mix and accelerator and embedded in araldite mix: accelerator. Sixty nm thick sections of embedded trypanosomes were cut and mounted on 200 mesh copper grids and stained with uranyl acetate and lead citrate before being viewed and photographed using a Philip 400 transmission electron microscope at 100 kV.

Raising antisera

Infection serum to intact and trypsinised trypanosomes, polyclonal anti-*T. evansi* crude soluble extract antibodies and polyclonal anti-52 kDa monospecific antibodies were raised in rabbits. Rabbit inoculated with trypsinised trypanosomes received 3 intravenous injections of 1×10^7 freshly prepared trypanosomes in 1 ml PSG on days 1, 28 and 57. On each occasion, two mice were also inoculated with 1×10^7 trypsinised trypanosomes in 0.5 ml PSG to serve as infection controls. Parasitemia in both rabbits and mice was monitored daily. The immunization regime for production of polyclonal antibodies was adapted from Harlow and Lane (1988). Anti-52 kDa IgG fraction was prepared using the commercially available ImmunoPure^{R} IgG kit (Pierce Chemical Company, USA) utilizing immobilized protein A columns (Protein A affinatyPak^{TM} columns).

Electrophoresis and Western blotting

Fifty microliter of freeze-thawed soluble extract containing 45 µg protein, 50 µl of each of SDS-solubilized whole trypanosome extract and trypsinised trypanosomes each equivalent to 1.1×10^8 trypanosomes were subjected separately to SDS-PAGE (Laemmli, 1970). 7 to 20% gradient poly-acrylamide gels were used and protein banding pattern visualized by Coomassie Blue stain. Western blots (Towbin et al., 1979) were prepared by electro-blotting separated antigens in unstained gels onto nitrocellulose paper. Antigens were detected by Western blotting against infection serum to intact and trypsinised trypanosomes and antisera raised against soluble extract.

Monitoring the specificity and reactivity of anti-52 kDa antibodies

Specificity and reactivity by antibody-ELISA

Serum collected from rabbits immunized with 52 kDa antigen, purified by electro-elution from stained gels, was tested by antibody-ELISA against freeze-thawed soluble extracts of homologous (TR 2165) and heterologous (TR 2222 and 2257) *T. evansi* populations coated at a dilution of 1/80. Each trypanosome extract was coated into 30 wells of the ELISA plate. Pre-immunization serum and PBS/Tween were also included in the plate as negative controls. All sera were tested at a dilution of 1/1000. The test serum, pre-immunization serum and PBS were each added to 10 wells of the three trypanosome extracts. Mean optical density (OD) values of test samples in each trypanosome extract was compared with the mean OD values of pre-immunization serum and PBS. Sample with an OD value of more than two standard deviations (2SD) above the mean was considered positive.

Specificity and reactivity by Western blotting

The antiserum raised to the 52 kDa antigen was tested by immunoblotting against whole trypanosome extract from homologous (TR 2165) and heterologous (TR 2222 and 2257) *T. evansi* populations. The serum samples were diluted to 1/50 in blocking buffer, while the conjugate was used at a dilution of 1/500.

Labeling of anti-52kDa IgG with horseradish peroxidase (HRP)

IgG fraction was conjugated to HRP using the modified periodate method as described by Wilson and Nakane (1978). The conjugate reactivity was tested against the freeze-thawed antigens of the homologous *T. evansi* population at a dilution of 1/80 using direct ELISA. The conjugated antibody was tested at 2-fold dilutions starting from 1/50 to 1/3200 and the reaction was visualized using TMB substrate.

ELISA development for *T. evansi* antigens

Sensitivity of conjugated anti-52kDa IgG for antigen detection

The sensitivity of the labeled IgG preparation was tested against titrated *T. evansi* soluble extract (1.8 mg/ml) along with titrated normal rabbit serum (NRS) as control over a 2-fold dilution range from 1/50 to 1/6400 in PBS/Tween. Each dilution of trypanosome extract and NRS was run 6× on the plate. The labeled IgG preparation (5 mg/ml) was used at a dilution of 1/200 in PBS/Tween containing 1% (v/v) NRS. Results were expressed as an antigen titer equivalent to the last antigen dilution that showed an absorbance value greater than 2SD above the mean of the NRS at the same dilution.

Standardization of the ELISA test

Un-labeled IgG preparation to the 52 kDa antigen diluted from 1/50 to 1/6400 was tested against freeze-thawed *T. evansi* extract or normal rabbit serum (NRS). The trypanosome extract and NRS were tested in duplicates at a dilution of 1/50. The remaining area of the plate was incubated with PBS/Tween. The anti-52 kDa IgG preparation labeled with HRP diluted 1/200 in PBS/Tween containing 1% (v/v) NRS was used as a conjugate. Mean OD values of trypanosome extract greater than 2SD above the means of corresponding NRS and PBS were considered positive.

Evaluation of the developed assay

After the establishment of the ELISA assay under *in vitro* conditions the assay was subsequently evaluated for its ability to detect the antigen *in vivo* in a rabbit inoculated intravenously with 10 mg of freeze-thawed *T. evansi* soluble extract. Serum samples as a source of antigen were collected from the rabbit one day before inoculation, at 0.5, 1, 2, 4, and 24 h post-inoculation, and then daily until the end of the experiment 6 days later. Antigen-ELISA was performed by coating ELISA microtiter plate with un-labeled anti-52 kDa antibodies at a dilution of 1/200 and then the antibody was reacted with a duplicate of undiluted serum of each of the serum samples. HRP-labeled anti-52 kDa antibody was used as a conjugate at a dilution of 1/200. The cut-off point was calculated according to the formula: Cut-off point = (mean + 3s) + 10% (mean = mean of the mean values of all low color wells; S = standard deviation of these mean values).

The dynamics of the 52 kDa *T. evansi* antigen during infection

Ten male rats were used in this experiment, 4 of which served as un-infected controls. The remaining 6 rats were infected intraperitoneally with 1×10^4 *T. evansi* (TR 2165) trypanosomes per rat. Parasitemia was monitored daily until the end of the experiment on day 17 post-infection using the microhaematocrit centrifugation technique. Tail blood was collected daily from the infected and control rats as a source of serum. All infected rats were injected intraperitoneally with a single therapeutic dose of 3.5 mg/kg Berenil (Al-Mohammed, 2008) on day 8 post-infection to release the antigen from killed trypanosomes for monitoring its dynamic within the host. Effectiveness of chemotherapy was measured using both parasitemia and antigenemia. The developed antigen-ELISA was used to monitor the level of the 52 kDa antigen in the blood of the infected rats. Pooled serum collected daily from the each infected or un-infected rats was diluted to 1/3 before use in the assay. All serum samples were run in duplicates. The cut-off point was calculated according to the formula:

Cut-off point = (mean + 3s) + 10% (mean = mean of the mean values of all negative control sera;

S = standard deviation of these mean values). Serum sample showing an absorbance value above the cut-off point was considered positive for the antigen.

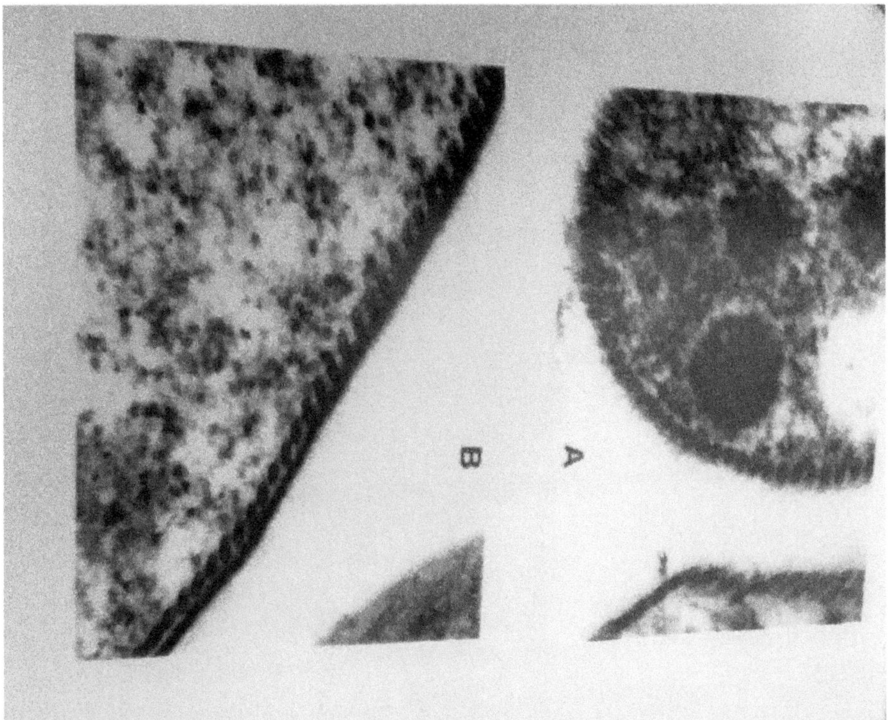

Figure 1. Transmission electron micrograph of section through surface of body of *T. evansi* (TR 2165). 1 mm ≡ 10 nm.
(A) 75 min trypsin-treated trypanosomes lacking a surface coat and (B) Intact non-treated trypanosomes possessing a surface coat.

RESULTS

Trypsinisation of *T. evansi*

Electron microscope sections prepared from untreated trypanosomes were seen to possess a thick (20 nm), compact and dense surface coat overlying the plasma membrane (Figure 1B). In the sections prepared from trypsinised trypanosomes, the surface coat was removed, leaving behind the plasma membrane which was sometimes difficult to resolve but the microtubules appeared normal (Figure 1A).

Protein profile of *T. evansi* materials

Slight difference was seen in the number of the protein bands resolved by Coomassie blue stain of the whole trypanosome extract and the soluble extract. The highest number of individual proteins were seen in the whole trypanosome extract, with a total of 39 individual protein bands with molecular weight ranging from 158.6 kDa to about 9.1 kDa (Figure 2, lane 2). Despite the equal amount of trypanosome proteins in the three extracts, two of these protein bands with molecular weight of 146 and 142 kDa were absent from the soluble extract which contained 37 proteins (Figure 2, lane 3). Otherwise both extracts had similar banding pattern with high molecular weight components in the range of 158.6 to 125.1 kDa, and showed the lowest staining intensities. Of the 39 bands in the whole extract, only 32 polypeptide bands were visible by Coomassie blue staining of the trypsinised trypanosomes extract in the size range of 96 to 9.1 kDa (Figure 2, lane 4). Seven of the protein bands present in the whole trypanosome extract were absent from the trypsinsed trypanosome extract. Five of these bands were in the high molecular weight part of the gel (158.6 to 125 kDa), with the other two at 58.6 and 52 kDa. Furthermore only few proteins dominated by a 42 kDa component showed a similar staining intensity to that of the other two extracts. The staining intensity of the other proteins in the trypsinised trypanosomes extract appeared to be lower than that of the other two extracts despite equivalent amounts of trypanosomes.

Immunological characterization of *T. evansi* proteins

Pooled infection serum collected from the rabbit infected with whole trypanosomes extract for up to 21 days after infection recognized a total of 17 antigenic components ranging from 172 to 21.2 kDa (Figure 3, lane 1) in the three trypanosomal preparations used in this study. Of

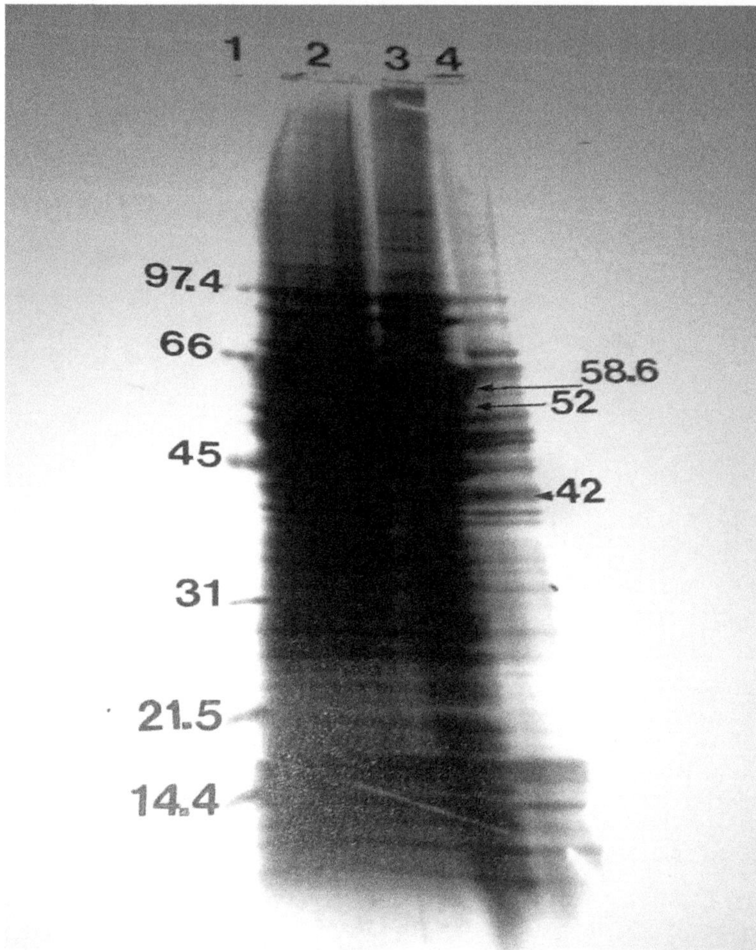

Figure 2. Total protein profile of *T. evansi* (TR 2165) fractionated on SDS-PAGE and stained with Coomassie Blue.
(1) Molecular weight markers; (2) whole cells extract; (3) soluble extract; (4) trypsinised trypanosomes extract.

the 39 protein components present in the whole trypanosome extract, 16 reacted with the infection serum. Twelve of these antigens were also recognized by this serum in the soluble extract. Five antigens of 65.5, 58.6, 31.5, 30 and 21.2 kDa present in the whole extract were absent from the soluble extract, with one 44 kDa component present only in the soluble extract. The infection serum recognized 7 components in the trypsinised extract, with corresponding antigens in both whole and soluble extracts while a 65.5 kDa component was common to the trypsinised and whole extracts. Hyperimmune serum raised to soluble extract recognized a total of 20 components in the three trypanosomal preparations with a molecular weight ranging from 166.1 to 26.6 kDa (Figure 3, lane 2). Ten antigens recognized during 21 days of infection were also recognized by the hyperimmune serum. The majority of the antigens recognized by the hyperimmune serum were present in the soluble extract with 16 components recognized by the serum. Seven of these 16 antigens were also present in the whole cell extract and one 58.6 kDa component was present only in the whole extract. The hyperimmune serum recognized 4 of the 16 components in the trypsinsed trypanosome extract which also contained two additional components of 54.7 and 26.6 kDa. Serum raised to trypsinised trypanosomes did not recognize any protein component on any of the three trypanosomal extracts (Figure 3, lane 3).

Specificity and reactivity of anti-52kDa antibodies

Antibodies raised to the 52 kDa antigen when tested by ELISA reacted with both the homologous (TR 2165) and heterologous (TR 2222 and 2257) populations of *T. evansi* (Table 1). As shown in the table each of the soluble extracts showed absorbance levels greater than 2SD above the mean of the negative control serum and

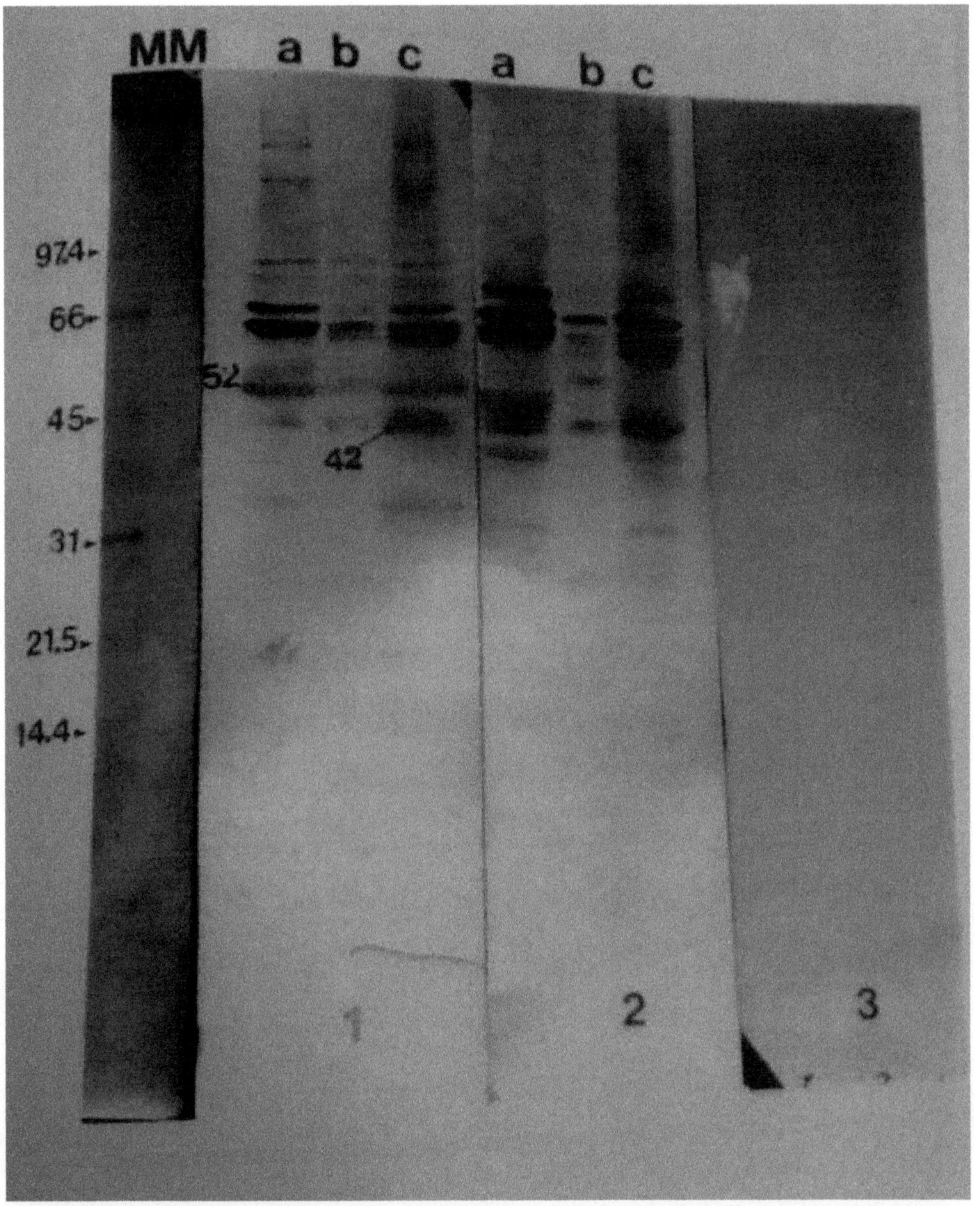

Figure 3. *T. evansi* (TR 2165) proteins identified by infection serum from rabbits infected with intact living trypanosomes and trypsin-treated trypanosomes and polyclonal serum to soluble extract. MM = molecular weight marker.
(1) Infection serum to intact trypanosomes; (2) serum raised against soluble extract; (3) infection serum against trypsin-treated trypanosomes; (a) soluble extract; (b) trypsin-treated trypanosomes and (c) whole cell extract.

PBS. The *T. evansi* heterologous population TR 2222 showed the highest absorbance (0.717) followed by TR 2257 (0.668) while the homologous population of *T. evansi* (TR 2165) showed an absorbance value of 0.630. In immunoblots, the antibodies recognized a single protein band of molecular weight ~52 kDa in the homologous population of *T. evansi* (TR 2165) (Figure 4, lane 2). While in the heterologous population of *T. evansi* (TR 2222 and 2257), two components were recognized: a major component of ~52 kDa and a minor 33 kDa component (Figure 5, lanes A and B).

HRP conjugate reactivity

HRP conjugate reacted with the homologous *T. evansi* antigens when tested by a direct ELISA which titrated out with increasing conjugate dilution (Figure 6). The absorbance values ranged from 1.713 at a dilution of 1/50 to 0.105 at a dilution of 1/3200. The conjugate showed an

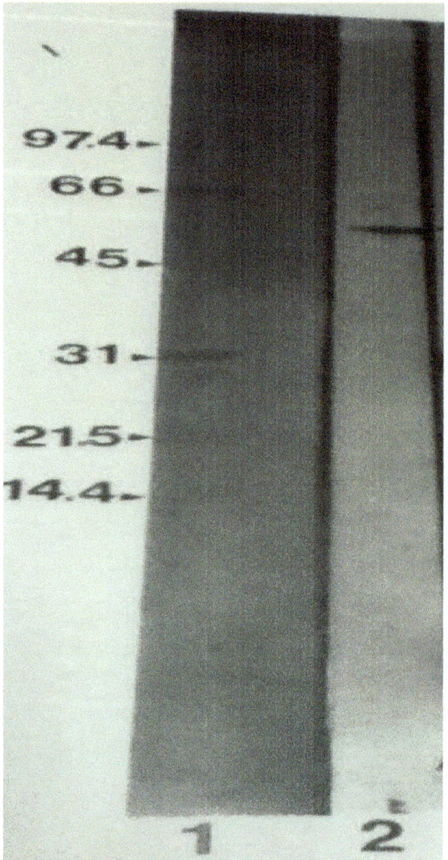

Figure 4. Western immunoblotting of homologous *T. evansi* population (TR 2165) against serum to the 52kDa antigen. (1) Molecular weight markers and (2) antiserum to 52 kDa antigen.

estimated titer of 1/200 at an absorbance value of 1.0.

Sensitivity of conjugated anti-52 kDa IgG for antigen detection

The peroxidase labeled anti-52k Da antibodies produced an antigen end-point titer of 1/400 along with an antigen binding ratio of 2.2 (Table 2).

Standardization of the ELISA test

The assay utilizing 1% NRS in PBS/Tween as a blocking agent in the buffer used to dilute HRP-conjugate was able to differentiate between the antigen and the NRS. Results for the antigen and NRS at 1/50 dilution is shown in Table 3.

Evaluation of the developed assay

The anti-52 kDa antibodies detected antigens in the

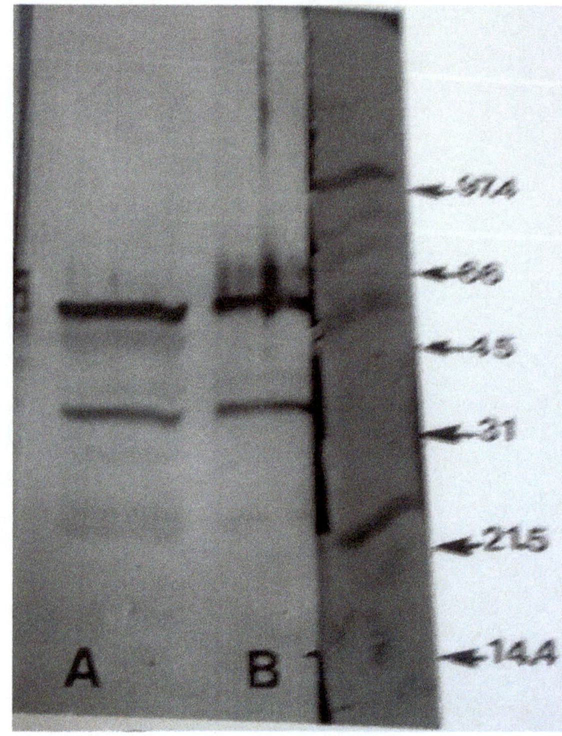

Figure 5. Western immunoblotting of heterologous *T. evansi* populations (TR 2222 and 2257) against serum to the 52 kDa antigen. (A) *T. evansi* TREU 2222 and (B) *T. evansi* TREU 2257.

blood of the inoculated rabbit by 2 h after inoculation. Thereafter, there was a gradual decline in the antigen titer which reached pre-inoculation level at 72 h post-inoculation (Figure 7).

The dynamics of the 52 kDa *T. evansi* antigen during infection

Parasitemia

The parasites were first detected in the tail blood of the rats at 4 days after infection. Thereafter, their number increased reaching a peak on day 8 post-infection. Following treatment with Berenil on day 8, the parasites disappeared from the circulation the next day and were not detected up to the time that the experiment was terminated at 17 days post-infection (Table 4).

Antigenemia

The antibodies to the 52 kDa antigen detected the antigen in the blood of rats by day 6, reaching peak activity on day 8 post-infection. After treatment on day 8 the activity declined, reaching negative control level on day 17 post-infection at the end of the experiment (Table 4).

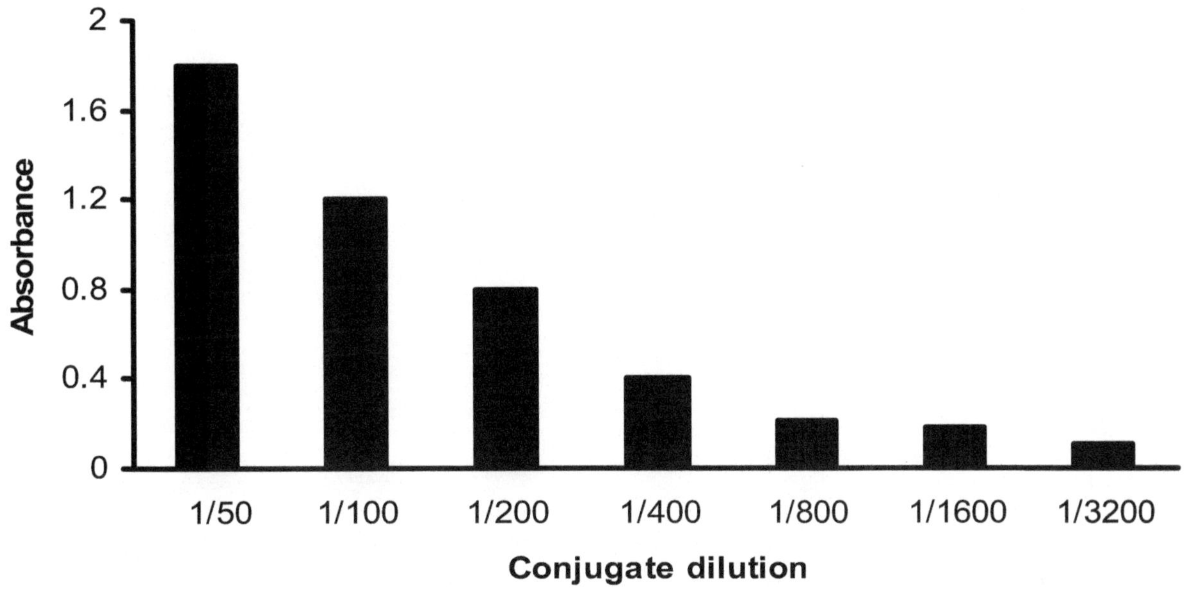

Figure 6. Reaction of labeled anti-52 kDa antibodies to homologous *T. evansi* population by direct ELISA.

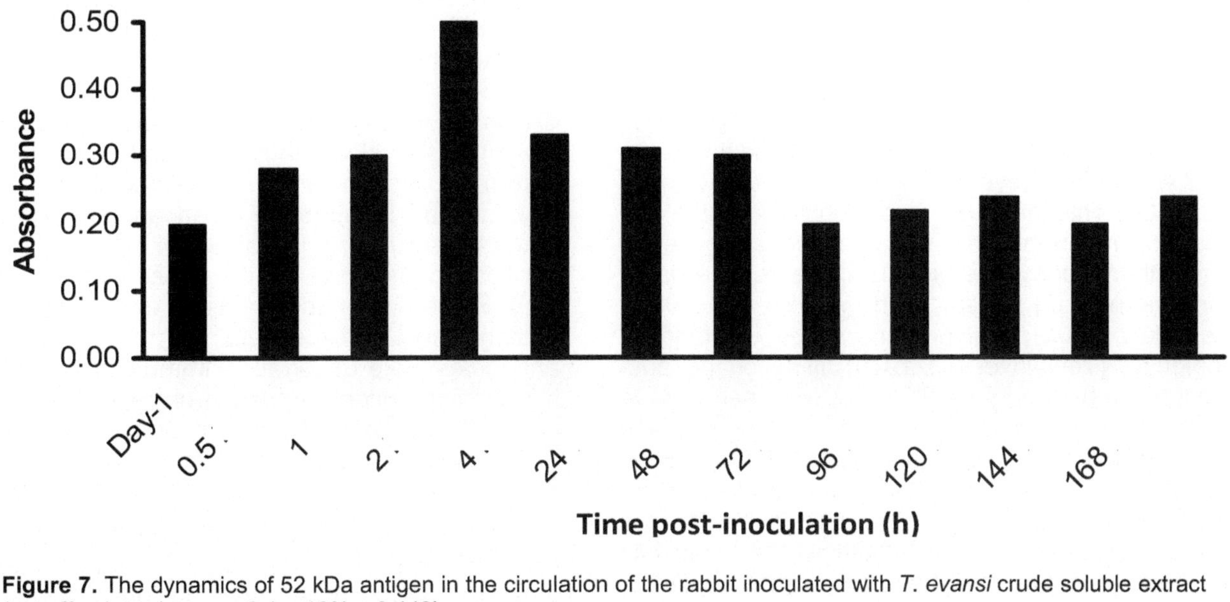

Figure 7. The dynamics of 52 kDa antigen in the circulation of the rabbit inoculated with *T. evansi* crude soluble extract (cut-off point = (mean + 3s) + 10% = 0.443).

DISCUSSION

The whole trypanosomes extract contained the largest number of the parasite proteins compared to the other two trypanosomal preparations used in this study. Two of these proteins with molecular weight of 146 and 142 kDa were not released by aqueous solubilisation as they were absent from the soluble extract, suggesting a possible membrane association as detergent lysis is known to solubilize the membrane proteins by replacement of the lipid bilayer, with a micelle of detergent (Garner et al., 2008). Two molecules with similar molecular weights (145 and 140 kDa) were reported in *T. brucei* (Coppens et al., 1992). The trypsinised trypanosomes extract contained the lowest number of the parasite proteins, with 7 bands absent from this extract compared to the whole extract. Although trypsin digestion is reputed to cleave mainly the VSG from the parasite surface (Mehlert et al., 2002), other proteins were reported to be cleaved during trypsinisation (Frommel et al., 1988). Possibly, in the present

Table 1. Specificity and reactivity of anti-52kDa antibodies tested by enzyme linked immunosorbent assay (ELISA).

Sample	TR 2165	TR 2222	TR 2257
Test serum	0.630±0.004	0.717±0.004	0.668±0.009
NRS	0.179±0.005	0.178±0.006	0.178±0.005
PBS	0.166±0.005	0.166±0.006	0.166±0.006
2SD + mean	0.189	0.187	0.188

Table 2. Sensitivity of conjugated anti-52kDa IgG for antigen detection.

Antigen end point titer	Absorbance at end-point titer		
	Antigen	NRS	Antigen binding ratio
1/400	0.537 ± 0.017	0.246 ± 0.018	2.2

study, these molecules possessed trypsin sensitive sites, arginine and lysine (Brown et al., 2002), and were released from the parasite together with the surface coat. Of these molecules, a 58.6 kDa protein fall within the 53 to 64 kDa range reported for VSG molecules of *T. evansi* (Uzcanga et al., 2004).

Twenty seven components of *T. evansi* were found to be antigenic in the present study depending on the extraction procedure. The molecular weight of these antigens ranged from 172 to 21.2 kDa although the upper molecular weight limit revealed by Coomassie blue stain was 158.6 kDa. This difference is possibly due to the fact that Western immunoblotting is a more sensitive detection system than Coomassie blue. Coomassie blue has a detection limit of 1 µg protein (Harlow and Lane, 1988), while as little as 100 pg of protein can be detected by immunoblotting (Towbin et al., 1979). Nine of the antigens recognized by the infection serum were sensitive to trypsin as they were absent from the trypsinised trypanosomes extract. This finding suggests that these antigens possess trypsin sensitive sites and are present in the surface of the parasite as the trypanosomes were viable and motile after the trypsinisation. Surface antigens other than the VSG have been reported in the trypanosomes and were used in protection experiments and as diagnostic reagents (Lanca et al., 2011; Tran et al., 2009).

Serum raised against trypsinised trypanosomes in the present study did not react with any of the three trypanosomal preparations. The trypsinised trypanosomes, unlike the intact trypanosomes, did not replicate within the host body as indicated by their absence in the circulation of the rabbit inoculated with these trypsinised trypanosomes. This would limit the amount of antigens released from them compared to the continuous stimulation of the immune system in animals infected with the intact trypanosomes (Shapiro and Murray, 1982). This lack of antibody response to trypsinised trypanosomes suggests that the surface coat antigen might be important in the initiation of the humoral immune response against the non-surface antigens. This might take place through direct activation of B lymphocytes without the involvement of antigen-presenting cells such as macrophages. The role of macrophages in the presentation of trypanosomes antigens has been reported negligible (Paulnock et al., 1988; Namangala et al., 2000).

One of the trypsin sensitive antigens with molecular weight of 52 kDa reacted potently with both infection serum and serum raised to the soluble extract and was also a good immunogen when administered as soluble extract, making it suitable for use in a range of purification methods. In the present study, this antigen was purified from the parasite extract through electro-elution from Coomassie stained polyacrylamide gels and antibodies were raised to it. Immunization of rabbits with this purified antigen elicited strong antibody responses as revealed by immnuoblotting and ELISA. This result indicates that this antigen retained its immunogenicity and confirms the usefulness of the SDS-PAGE for the purification of antigens for immunization as reported by previous workers (Harlow and Lane, 1988).

The 52 kDa antigen was found to be conserved between different populations and stocks of *T. evansi*. Immunoblotting with the serum recognized the target antigen as a single component in the homologous population but as two components of 52 and 33 kDa in the heterologous *T. evansi* populations. This 33 kDa component might be a breakdown product or another component only present in the heterologous population carrying similar epitopes to the 52 kDa. These findings indicate that the 52 kDa represents an invariant antigen. Invariant antigens had been reported from different species and populations of trypanosomes, some of them have been shown to be located on the surface of the parasite (Overath et al., 1994), while others are non-surface associated antigens (Muller et al., 1992). Evidence

Table 3. Labeled anti-52 kDa antibodies at a dilution of 1/200 with PBS/Tween containing 1% NRS.

Coating antibody	1/50	1/100	1/200	1/400	1/800	1/1600	1/3200	1/6400
Antigen	0.994±0.003	0.900±0.003	0.763±0.004	0.652±0.003	0.656±0.003	0.640±0.004	0.459±0.004	0.460±0.001
NRS	0.294±0.003	0.247±0.004	0.178±0.003	0.159±0.001	0.142±0.001	0.134±0.001	0.106±0.003	0.103±0.001
PBS	0.290±0.005	0.288±0.005	0.179±0.005	0.151±0.003	0.126±0.003	0.138±0.004	0.105±0.003	0.117±0.003
2SD + mean OD	0.298	0.326	0.179	0.167	0.156	0.142	0.106	0.130

NRS: normal rabbit serum. SD: standard deviation, OD: optical density.

Table 4. Parasitemia and antigenemia in the experimental rats.

Days after infection	Tryps/20 fields	52 kDa OD	Control serum OD*	Days after infection	Tryps/20 fields	52 kDa OD*	Control serum OD*
1	0	0.257	0.189	10	0	1.159*	0.234
2	0	0.286	0.249	11	0	0.407*	0.239
3	0	0.249	0.220	12	0	0.401*	0.234
4	10	0.286	0.215	13	0	0.432*	0.238
5	22	0.283	0.231	14	0	0.417*	0.212
6	41	0.562*	0.255	15	0	0.421*	0.221
7	85	0.894*	0.259	16	0	0.397*	0.207
8	200	1.346*	0.265	17	0	0.292	0.209
9	0	1.078*	0.212	-	-	-	-

*Positive samples. Mean optical density (OD) of control sera = 0.229 ± 0.02. Cut-off point = 0.229 + 0.06 + 0.0289 = 0.318.

from the present study suggests that the 52 kDa antigen is surface associated as it was cleaved from the parasite by the process of trypsinisation.

The antibodies produced to this antigen were used in the present study to develop an immunoassay for the detection of the antigen in serum of infected animals. The antigen-capture ELISA developed in this study was able to detect *T. evansi* antigens in serum. When soluble extract of the parasite was inoculated in an aqueous form intravenously into a rabbit, the 52 kDa antigen was detected within two hours post-inoculation due to its circulation with blood throughout the animal body. Un-aggregated protein antigens are known to distribute evenly throughout the animal's blood before being cleared by the phagocytic cells (Tizard, 1992). Thereafter the level of the antigen declined, reaching pre-inoculation level within 3 days either as a result of dilution or phagocytic clearance.

The 52 kDa antigen was detected early in the circulation of *T. evansi* infected rats in this study. Such early appearance might be allied with the antigen's surface-association, making it one of the first antigens that would be accessible to antibodies early during infection. The detection of this antigen during low parasitemia levels in the blood of the infected rats would possibly suggests its detection during chronic phase of the disease when parasitemia is low. Although the level of this antigen remained high for two days following treatment, a sharp decline was seen on the second day. The rapid clearance of the antigen from the circulation suggests that it could have been circulating as antigen-antibody complexes. Such complexes are known to be rapidly cleared by phagocytic cells (Tizard, 1992) and elimination of antigens from the circulation by these mononuclear phagocytic cells is inefficient when antibodies have low affinity for the antigen. The peak level of antigenemia generally coincided with the highest levels of parasitemia. It is likely that the

ELISA result reflects the presence of free target antigen in the circulation of the infected rats rather than antigen appearing in the blood sample due to breakdown of parasites after blood sampling, since the antigen persisted in the circulation for eight days after complete disappearance of trypanosomes from rat's blood. Previous studies had shown that trypanosomes at concentrations of approximately 1×10^9 trypanosomes/ml would be needed in the blood sample to reach the absorbance levels obtained with serum from the rats. These levels were never reached during the course of infection in the present study.

The clearance of the 52 kDa antigen from the circulation after successful chemotherapy suggests a possible role in serologically-based assay for measuring the effectiveness of chemotherapy. This antigen, being an invariant antigen accessible to antibodies during infection and can reach a detectable level in the circulation before the remission of the first parasitemic wave, is a possible candidate for the diagnosis of infection at least in areas where *T. evansi* is the only trypanosome species present.

Further inquiry in its presence in other trypanosome species is recommended for investigation of cross reactivity. The surface association and accessibility of this antigen to antibodies during early stages of infection warrants further investigation in its viability in vaccine production. Some of these invariant surface antigens were involved in partial protection against infection (Lanca et al., 2011).

ACKNOWLEDGEMENT

The author is greatly indebted to Deanship of Scientific Research of King Faisal University whose financial support (DSR 140204) made this study possible.

Conflict of Interest

The authors declare that they have no conflict of interests.

REFERENCES

Al-Mohammed HI (2008). Comparative In Vivo Activities of Diminazene, Suramine, Quinapyramine and Homidium Bromide on *Trypanosoma evansi* infection in Mice. Sci. J. King Faisal Univ. (Basic Appl. Sci.) 9(1):139-147.

Anderson DJ, Blobel G (1983). Immunoprecipitation of proteins from cell-free translations. Methods Enzymol. 96:111-120.

Antoine-Moussiaux N, Büscher P, Desmecht D (2009). Host-Parasite Interactions in Trypanosomiasis: on the way to an antidisease strategy. Infect. Immun. 77(4):1276-1284.

Bajyana SE, Hamers R (1988). A card agglutination test (CATT) for veterinary use based on an early VAT RoTat 1/2 of *Trypanosoma evansi*. Ann. Soc. Belg. Med. Trop. 68:233-240.

Bringuad F, Baltz T (1992). A potential hexose transporter gene expressed predominantly in the bloodstream form of *Trypanosoma brucei*. Mol. Biochem. Parasit. 52(1):111-122.

Brown AJ, Sun L, Feramisco JD, Brown MS, Goldstein JL (2002). Cholesterol addition to ER membranes alters conformation of SCAP, the SREBP escort protein that regulates cholesterol metabolism. Mol. Cell 10:237-245.

Burgess DE, Jerrells T (1985). Molecular Identity and Location of Invariant Antigens on *Trypanosoma brucei rhodesiense* Defined with Monoclonal Antibodies Reactive with Sera from Trypanosomiasis Patients. Infect. Immun. 50(3):893-899.

Chung W, Carrington M, Field MC (2004). Cytoplasmic targeting signals in transmembrane invariant surface glycoproteins of trypanosomes. J. Biol. Chem. 279(52):54887-54895.

Claes F, Radwanska M, Urakawa T, Majiwa PAO, Goddeeris B, Büscher P (2004). Variable Surface Glycoprotein RoTat 1.2 PCR as a specific diagnostic tool for the detection of *Trypanosoma evansi* infections. Kinetoplastid Biol. Dis. 3:3

Coppens I, Bastin PH, Opperdoes FR, Baudhuin P, Courtoy PJ (1992). *Trypanosoma brucei brucei*: antigenic stability of its LDL receptor and immunological cross-reactivity with LDL receptor of the mammalian host. Exp. Parasitol. 74(1):77-86.

El Amin GA, Hassan EM, El Bashir MOA, Luckins AG (1993). Prevalence of trypanosomiasis in pastoral camels in the Butana plains, Mid-Eastern Sudan. In: Improving the diagnosis and control of trypanosomiasis and other vector-borne diseases of African livestock using immunoassay methods. IAEA-TECDOC--707;ISSN 1011-4289.

Frommel TO, Seyfang A, Balber AE (1988). *Trypanosoma brucei* sspp.: Cleavage of variant specific and common glycoproteins during exposure of live cells to trypsin. Exp. Parasitol. 66:(2)213-224.

Garner AE, Smith DA, Hooper NM (2008). Visualization of detergent solubilization of membranes: Implications for the isolation of the rafts. Biophys. J. 94(4):1326-1340.

Henley GL, Lee CM, Takeuchi A. (1978). Electron microscopy observations on Trypanosoma brucei: Freeze-cleaving and thin-sectioning study of the apical part of the flagellar pocket. Z. Parasitenkd 55:(3)181-187.

Gutierrez C, Corbera JA, Doreste F, Buscher P (2004). Use of miniature anion exchange centrifugation technique to isolate *Trypanosoma evansi* from goats. Ann. NY. Acad. Sci. 1026:149-151.

Harlow E, Lane D (1988). Antibodies a laboratory manual. Cold Spring Harbor Laboratory.

Hoare CA (1972). The trypanosomes of mammals (A zoological monograph). Blackwell, Oxford, Scientific Publications.

Laemmli UK (1970). Cleavage of structural proteins during the assembly of the head of bacteriophage T4. Nature 227(5259):680-685.

Lanca ASC, de Sousa KP, Atouguia J, Prazeres DMF, Monterio GA, Silva MS (2011). *Trypanosoma brucei*: Immunization with plasmid DNA encoding invariant surface glycoprotein gene is able to induce partial protection in experimental African trypanosomiasis. Exp Parasitol 127(1):18-24.

Lanham SM, Godfrey DG (1970). Isolation of salivarian trypanosomes from man and other mammals using DEAE cellulose. Exp. Parasitol. 28:521-534.

Luckins AG (1988). *Trypanosoma evansi* in Asia. Parasitol. Today 4(5):137-142.

Mehlert A, Bond CS, Ferguson MAJ (2002). The glycoforms of a *Trypanosoma brucei* variant surface glycoprotein and molecular modeling of a glycosylated surface coat. Glycobiology 12(10):607-612.

Morrison LJ, Mclellan S, Sweeney L (2010). Role of parasite genetic diversity in differential host responses to *Trypanosoma brucei* infection. Infect. Immun. 78(3):1096-1108.

Muller N, Hemphill A, Imboden M (1992). Identification and characterization of two repetitive non-variable antigens from African trypanosomes which are recognized early during infection. Parasitology 104(1):111-120.

Namangala B, Brys L, Magez S, De Baetselier P, Beschin A (2000). *Trypanosoma brucei brucei* infection impairs MHC class II antigen presentation capacity of macrophages. Parasite Immunol. 22(7):361-370.

Nantulya VM, Bajyana-Songa E, Hamers R (1989). Detection of circulating trypanosomal antigens in Trypanosoma evansi-infected animals using a *T. brucei* group-specific monoclonal antibody. Trop.

Med. Parasitol. 40:263-266.

Nantulya VM, Lindqvist KJ (1989). Antigen-detection enzyme immunoassays for the diagnosis of *Trypanosoma vivax*, *T. congolense* and *T. brucei* infections in cattle. Trop. Med. Parasitol. 40:267-272.

Nolan DP, Jackson DG, Windle HJ, (1997). Characterization of a novel, stage-specific, invariant surface protein in *Trypanosoma brucei* containing an internal, serine-rich, repetitive motif. J. Biol. Chem. 272(46):29212-29221.

Olaho-Mukani W, Munyua WK, Mutugi MW, Njogu AR (1993). Comparison of antibody-and antigen-detection enzyme immunoassays for the diagnosis of *Trypanosoma evansi* infections in camels. Vet. Parasitol. 45(3-4):231-240.

Overath P, Chaudhri M, Steverding D, Ziegelbauer K (1994). Invariant surface proteins in bloodstream forms of *Trypanosoma brucei*. Parasitol. Today 10(2):53-58.

Pal A, Hall BS, Nesbeth DN, Field HI, Field MC (2002). Differential Endocytic Functions of *Trypanosoma brucei* Rab5 Isoforms Reveal a Glycosylphosphatidylinositol-specific Endosomal Pathway. J. Biol. Chem. 277:9529-9539.

Parsons MN, Nielsen B (1990). Active transport of 2-deoxy-D-glucose in *Trypanosoma brucei* procyclic forms. Mol. Biochem. Parasit. 42:197-204.

Paulnock DM, Smith C, Mansfield JM (1988). Antigen presenting cell function in African trypanosomiasis. In: Schook LB, Tew J (eds.), Antigen Presenting Cells: Diversity, Differentiation, and Regulation. Alan R. Liss New York. pp. 135-144.

Radwanska M, Magez, S, Michel A, et al (2000). Comparative analysis of antibody responses against HSP60, invariant surface glycoprotein 70, and variant surface glycoprotein reveals a complex antigen-specific pattern of immunoglobulin isotype switching during infection by *Trypanosoma brucei*. Infect. Immun. 68(2):848-860.

Rayulu VC, Singh A, Chaudhri SS (2007). Monoclonal antibody based immunoassays for the detection of circulating antigens of *Trypanosoma evansi* in buffaloes. Ital. J. Anim. Sci. 6(2):907-910.

Reyna-Bello A, Garcia FA, Rivera M, Sansó B, Aso PM (1998). Enzyme-linked immunosorbent assay (ELISA) for detection of anti-Trypanosoma evansi equine antibodies. Vet. Parasitol. 80(2):149-157.

Schwede A, Carrington M (2010). Bloodstream form trypanosome plasma membrane proteins: antigenic variation and invariant antigens. Parasitology 137(4):2029-2039.

Shapiro SZ, Murray M (1982). African trypanosome antigens recognized during the course of infection in N'dama and Zebu cattle. Infect. Immun. 35(2):410-416.

Thammasart S, Kanitpun R, Saithasao M, Kashiwazaki Y (2001). Preliminary studies by ELISA on the antigen and antibody dynamics in the early stages of experimental infections with *Trypanosoma evansi* in cattle. Trop. Anim. Health Prod. 33:189-199.

Tizard I (1992). Macrophages and antigen processing. In: Tizard I (eds.), Veterinary Immunology. WB Saunders Company, Philadelphia. pp. 27-40.

Towbin H, Staehelin T, Gordon J (1979). Electrophoretic transfer of proteins from polyacrylamide gels to nitrocellulose sheets: procedure and some applications. P. Natl. Acad. Sci. USA 76:4350-4354.

Tran T, Claes F, Verloo D, De Greve H, Büscher P (2009). Towards a new reference test for Surra in camels. Clin. Vaccine Immunol. 16(7):999-1002.

Urakawa T, Verloo D, Moens L, Büscher P, Majiwa PAO (2001). *Trypanosoma evansi*: cloning and expression in Spodoptera fugiperda insect cells of the diagnostic antigen RoTat 1.2. Exp. Parasitol. 99:181-189.

Uzcanga GL, Perrone T, Noda AG, et al (2004). Variant surface glycoprotein from *Trypanosoma evansi* is partially responsible for the cross-reaction between *Trypanosoma evansi* and *Trypanosoma vivax*. Biochemistry 43(3):595-606.

Van Meirvenne N, Magnus E, Büscher P (1995). Evaluation of variant specific trypanolysis tests for serodiagnosis of human infections with *Trypanosoma brucei gambiense*. Acta Trop. 60:189-199.

Verloo D, Tibayrenc R, Magnus E, Büscher P, Van Meirvenne N (1998). Performance of serological tests for Trypanosoma evansi infections in camels from Niger. J. Protozool. Res. 8:190-193.

Verloo D1, Holland W, My LN, Thanh NG, Tam PT, Goddeeris B, Vercruysse J, Büscher P (2000). Comparison of serological tests for *Trypanosoma evansi* natural infections in water buffaloes from North Vietnam. Vet. Parasitol. 92(2):87-96.

Waitumbi NJ, Nantulya VM (1993). A comparison of the antigen detection ELISA and parasite detection for diagnosis of *Trypanosoma evansi* infections in camels. Vet. Parasitol. 49(2-4):159-178.

Webster P, Shapiro SZ (1990). *Trypanosoma brucei*: a membrane-associated protein in coated endocytic vesicles. Exp. Parasitol. 70:154-163.

Wilson MB, Nakane PP (1978). Recent developments in the periodate method of conjugating horse radish peroxidase (HRPO) to antibodies. In: Knapp W, Holubar K, Wick G (eds.), Immunofluorescence and Related Staining Techniques. Elsevier/North Holland Biomedical, Amsterdam. pp. 215–224.

Ziegelbauer K, Multhaup G, Overath P (1992). Molecular characterization of two invariant surface glycoproteins specific for bloodstream stage of *Trypanosoma brucei*. J. Biol. Chem. 267(15):10797-10803.

Ziegelbauer K, Overath P (1992). Identification of invariant surface glycoprotein in the bloodstream stage of *Trypanosoma brucei*. J. Biol. Chem. 267(15):10791-10796.

Ziegelbauer K, Overath P (1993). Organization of two invariant surface glycoproteins in the surface coat of *Trypanosoma brucei*. Infect Immun. 61:4540-4545.

Laboratory feeding of *Bulinus truncatus* and *Bulinus globosus* with *Tridax procumbens* leaves

O. M. Agbolade , O. W. Lawal and K. A. Jonathan

Department of Plant Science and Applied Zoology, Parasitology and Medical Entomology Laboratory, Olabisi Onabanjo University, P.M.B. 2002, Ago-Iwoye, Ogun State, Nigeria.

Suitability of *Tridax procumbens* leaves in laboratory feeding of *Bulinus truncatus* and *Bulinus globosus* was assessed in comparison with *Lactuca sativa* between September and October, 2011. The snails were collected from Eri-Iope stream in Ago-Iwoye, while *T. procumbens* were collected from the Mini Campus of the Olabisi Onabanjo University, Ago-Iwoye, Ijebu North, Southwestern Nigeria. For *B. truncatus*, fresh, sun-dried and oven-dried *T. procumbens* were used, while only fresh *T. procumbens* were used for *B. globosus*. The mean percentage survivals of *B. truncatus* fed with fresh, sun-dried and oven-dried *T. procumbens* compared with those of the corresponding control snails showed no significant difference (χ^2 = 0.51, 1.85, and 2.21, respectively). *B. truncatus* fed with fresh *T. procumbens* had the highest mean live-weight percentage increase (46.4%) as compared to those fed with sun-dried and oven-dried (χ^2 = 45.65). The mean percentage survival of *B. globosus* fed with fresh *T. procumbens* (79.2%) was similar with that of the control (84.6%) (χ^2 = 0.18). The percentage increase in mean live-weight of *B. globosus* fed with fresh *T. procumbens* (24.0%) was lower as compared to the control (χ^2 = 5.34). This study showed that feeding the studied snails using *T. procumbens* may be profitable in Ijebu North, Southwestern Nigeria.

Key words: *Bulinus truncatus*, *Bulinus globosus*, laboratory feeding, *Tridax procumbens*.

INTRODUCTION

All species of *Bulinus*, including *Bulinus truncatus* and *Bulinus globosus*, are freshwater gastropod snails of the family Planorbidae. *Bulinus* species are widespread in Africa, having been reported from regions such as Northwest Africa, Ethiopia, Sudan, Egypt, Tanzania and Nigeria (Brown, 1994; Chitsulo et al., 2000). *B. globosus*, *B. truncatus* and some other *Bulinus* spp. are established vectors of *Schistosoma haematobium* (Ukoli, 1984; WHO, 1993). Therefore, much attention has been consistently devoted towards understanding the biology of the aforementioned vectors and other vector snails towards sustainable control of human schistosomiasis in tropical Africa including Ijebu North, South-western Nigeria (Adewunmi et al., 1991; Agbolade et al., 2004; Madsen and Stauffer, 2011).

Tridax procumbens is a dicotyledonous plant in the Daisy family and is a commonly widespread weed and pest plant in the tropical, sub-tropical and mild temperate regions of the world (http://en.wikipedia.org/wiki/Tridax_procumbens).

Nevertheless, some studies have revealed some usefulness of *T. procumbens*. It has been reported that the plant is rich in nutritionally valuable substances such as protein, fat, carbohydrate, fibre, sodium, potassium, calcium, carotenoids and flavonoids (Jude et al., 2009). Moreover, it is known that the leaves of *T. procumbens* are eaten by *Archachatina marginata* (Odaibo, 1997), and a recent study showed that domestication of *Lanistes libycus* (an edible freshwater apple snail) using the leaves of *T. procumbens* could be a profitable venture (Agbolade et al., 2010).

In view of the public health significance of *B. truncatus* and *B. globosus*, it is often necessary to culture them in the laboratory for research purposes. Unfortunately, *Lactuca sativa* (lettuce) conventionally used to feed freshwater snails seems scarce and expensive in Ijebu North area, Southwestern Nigeria. This study was therefore designed to assess the suitability of the leaves of *T. procumbens* in the laboratory feeding of *B. truncatus* and *B. globosus*.

MATERIALS AND METHODS

Samples' source and collection methods

B. truncatus and *B. globosus* specimens were collected from Eri-lope stream in Ago-Iwoye. Ago-Iwoye (6° 56' 55" N, 3° 54' 40" E) is the location town of the main Campus of the Olabisi Onabanjo University, and it is located in Ijebu North area of Ogun State, Southwestern Nigeria. The snails were collected by picking them from among submerged and emergent plants, fallen leaves and twigs in the water body. The specimens were transported to the laboratory in a well-aerated container. *T. procumbens* plants were hand-collected from the Mini Campus of the Olabisi Onabanjo University, Ago-Iwoye. *L. sativa* (lettuce) was purchased from a market in Lagos, Nigeria. Snails' collection and culturing were carried out between September and October, 2011.

Culturing method

In the laboratory, the specimens were identified and sorted into species with reference to Brown and Kristensen (1993). Specimens of each species were divided into size classes after initial shell length measurement. For *B. truncatus*, the snails were later divided into three categories according to their diet, placed in bowls labelled fresh, sun-dried, and oven-dried. Each category had three replicates (A, B, C), and one control. Each of the bowls was lined with transparent nylon to make easy the collection of eggs. De-chlorinated water was added to each culture bowl. 20 g of fresh and 10 g of dried *T. procumbens* leaves (for trial bowls) and *L. sativa* (for control bowls) were used in feeding *B. truncatus* snails as appropriate. For *B. globosus*, only fresh *T. procumbens* leaves (for trial bowls) and *L. sativa* (for control bowls) were used. There were four trial replicates (A, B, C, D) and one control.

For both species of *Bulinus*, each of the trial and control bowls was covered with polythene mosquito net held in place with rubber band to prevent snails from escaping. Changing of water and feeding were done every other day. Any mortality was recorded and the specimen was removed immediately on detection. Shell length measurement of the snails was done once a fortnight. The mean live-weight of the snails in each bowl was measured using weighing balance on day 0 (initial weight) and subsequently fortnightly. The final shell length and live-weight measurements were taken at the end of the 7th week.

Statistical analysis

Percentage increase in live-weight and shell length were calculated by comparing the mean values on day 0 with the corresponding mean values at the end of the 7th week using the following formulae:

Live-weight increase (%) = (Final live-weight – Initial live-weight / Initial live-weight) × 100

Shell length increase (%) = (Final length – Initial length / Initial length) × 100

Chi-square (χ^2) was used to compare percentages of survival and increase in live-weight and shell length. For mean live-weight and shell length, the replicates with the least values of percentage increase were used as benchmarks for comparison with the control values.

RESULTS

The survival, live-weight and shell length of *B. truncatus* fed with fresh, sun-dried and oven-dried *T. procumbens* leaves are summarised in Table 1. At the end of the 7th week, the percentages of surviving control snails fed with fresh (84.6%), sun-dried (84.6%) and oven-dried (85.7%) *L. sativa* were not significantly different (χ^2 = 0.01, df = 2, P > 0.05). The mean percentage survivals of *B. truncatus* fed with fresh (75.6%), sun-dried (67.8%) and oven-dried (67.3%) *T. procumbens* were not significantly different (χ^2 = 0.62, df = 2, P > 0.05). The mean percentage survivals of snails fed with fresh, sun-dried and oven-dried *T. procumbens* when compared with those of the corresponding control snails (fresh = 84.6%, sun-dried = 84.6%, oven-dried = 85.7%) showed no statistically significant difference (χ^2 = 0.51, 1.85, and 2.21, respectively; df = 1, P > 0.05 for each category).

The percentage increase in the mean live-weight of *B. truncatus* specimens fed with fresh (70.0%) *L. sativa* was significantly the highest when compared with those fed with sun-dried (42.9%) and oven-dried (42.6%) *L. sativa* (χ^2 = 9.55, df = 2, P < 0.01). Similarly, mean live-weight percentage increase of *B. truncatus* fed with fresh *T. procumbens* (46.4%) was significantly the highest when compared with those fed with sun-dried (14.6%) and oven-dried (3.7%) *T. procumbens* (χ^2 = 45.65, df = 2, P < 0.001). The percentage increases in the mean live-weight of *B. truncatus* fed with fresh, sun-dried and oven-dried *T. procumbens* were significantly lower than those of the corresponding control snails (χ^2 = 4.78, P < 0.05; χ^2 = 6.96, P < 0.01; χ^2 = 16.34, P < 0.001, respectively; df = 1 for each category).

The percentage increases in the mean shell length of *B. truncatus* fed with fresh (8.8%), sun-dried (5.4%) and oven-dried (1.2%) *L. sativa* were not significantly different (χ^2 = 5.65, df = 2, P > 0.05). Likewise, the percentage increases in mean shell length of *B. truncatus* fed with fresh (5.9%), sun-dried (1.8%) and oven-dried (0.8%) *T. procumbens* leaves were not significantly different (χ^2 = 5.16, df = 2, P > 0.05). In addition, comparison of feeding with fresh, sun-dried and oven-dried *L. sativa* (control) and *T. procumbens* (trial) showed no significant differences

Table 1. Survival, live-weight and shell length of *B. truncatus* fed with fresh, sun-dried and oven-dried *T. procumbens* leaves.

Week	Replicate	Fresh			Sun-dried			Oven-dried		
		No. (%) surviving	Mean live-weight (g)	Mean shell length (mm)	No. (%) surviving	Mean live-weight (g)	Mean shell length (mm)	No. (%) surviving	Mean live-weight (g)	Mean shell length (mm)
0	A	24 (100)	0.29	3.50	23 (100)	0.41	5.80	20 (100)	0.51	12.00
	B	26 (100)	0.30	3.40	23 (100)	0.42	5.60	23 (100)	0.54	11.90
	C	24 (100)	0.28	3.50	25 (100)	0.41	6.00	27 (100)	0.55	11.90
	Control	26 (100)	0.30	3.40	26 (100)	0.42	5.60	21 (100)	0.54	12.08
2	A	22 (91.7)	0.33	3.55	21 (91.3)	0.43	5.80	18 (90.0)	0.52	12.01
	B	24 (92.3)	0.34	3.45	21 (91.3)	0.44	5.63	21 (91.3)	0.55	11.93
	C	22 (91.7)	0.31	3.58	22 (88.0)	0.43	6.05	24 (88.9)	0.56	11.93
	Control	25 (96.2)	0.35	3.48	25 (96.2)	0.47	5.68	20 (95.2)	0.60	12.08
4	A	20 (83.3)	0.37	3.60	19 (82.6)	0.45	5.80	16 (80.0)	0.53	12.05
	B	22 (91.7)	0.37	3.50	19 (82.6)	0.46	5.65	19 (82.6)	0.55	11.95
	C	20 (83.3)	0.35	3.65	19 (76.0)	0.44	6.10	21 (77.8)	0.57	11.95
	Control	24 (92.3)	0.41	3.55	24 (92.3)	0.51	5.75	19 (90.5)	0.66	12.15
6	A	18 (75.0)	0.40	3.65	17 (73.9)	0.47	5.80	15 (75.0)	0.53	12.08
	B	21 (80.8)	0.41	3.55	18 (78.3)	0.47	5.68	18 (78.3)	0.56	11.98
	C	19 (79.2)	0.38	3.65	16 (64.0)	0.46	6.15	19 (70.4)	0.57	11.98
	Control	23 (88.5)	0.46	3.63	23 (88.5)	0.56	5.83	18 (85.7)	0.71	12.23
7	A	17 (70.8)	0.44	3.70	16 (69.6)	0.49	5.80	14 (70.0)	0.54	12.10
	B	20 (76.9)	0.44	3.60	17 (73.9)	0.49	5.70	15 (65.2)	0.56	12.00
	C	19 (79.2)	0.41	3.80	15 (60.0)	0.47	6.20	18 (66.7)	0.58	12.00
	Control	22 (84.6)	0.51	3.70	22 (84.6)	0.60	5.90	18 (85.7)	0.77	12.23

in mean shell length percentage increases of *B. truncatus* (χ^2 = 0.57, 1.80, 0.08, respectively; df = 1, P > 0.05 for each category). Table 2 summarises the survival, live-weight and shell length of *B. globosus* fed with fresh *T. procumbens* leaves. The mean percentage survival of *B. globosus* fed with fresh *T. procumbens* (79.2%) was not significantly different from that of control (84.6%) (χ^2 = 0.18, df = 1, P > 0.05). The percentage increase in mean live-weight of *B. globosus* fed with fresh *T. procumbens* (24.0%) was statistically lower as compared to the control (42.9%) (χ^2 = 5.34, df = 1, P < 0.05). However, the percentage increase in the mean shell length of *B. globosus* fed with fresh *T. procumbens* (2.35%) was not significantly different from that of the control (4.71%) (χ^2 = 0.79, df = 1, P > 0.05).

DISCUSSION

The results of this study show clearly that *Bulinus* spp. can be cultured in the laboratory using the free available *T. procumbens* instead of the

Table 2. Survival, live-weight and shell length of *B. globosus* fed with fresh *T. procumbens* leaves.

Week	Replicate	No. (%) surviving	Mean live-weight (g)	Mean shell length (mm)
	A	25(100)	0.49	8.70
	B	25(100)	0.47	8.90
0	C	25(100)	0.50	8.80
	D	26(100)	0.49	8.50
	Control	26(100)	0.49	8.50
	A	24(96.0)	0.53	8.80
	B	23(92.0)	0.52	8.96
2	C	23(92.0)	0.55	8.88
	D	24(92.3)	0.54	8.55
	Control	25(96.2)	0.54	8.60
	A	23(92.0)	0.57	8.90
	B	21(84.0)	0.57	9.05
4	C	21(84.0)	0.61	8.95
	D	22(84.6)	0.59	8.60
	Control	24(92.3)	0.60	8.70
	A	22(88.0)	0.61	9.00
	B	20(80.0)	0.62	9.13
6	C	20(80.0)	0.66	9.03
	D	21(80.8)	0.64	8.65
	Control	23(88.5)	0.65	8.80
	A	21(84.0)	0.65	9.10
	B	19(76.0)	0.62	9.20
7	C	19(76.0)	0.66	9.10
	D	21(80.8)	0.64	8.70
	Control	22(84.6)	0.70	8.90

expensive *L. sativa* with similar results in terms of percentage survival. This may give a sigh of relief for freshwater snails' researchers who may thereby expend their often meagre finance with higher frugality.

The observed laboratory survival of *B. truncatus* and *B. globosus* exclusively on *T. procumbens* leaves diet in this study is the first of its kind from Ijebu North area, Southwestern Nigeria. Literature shows, for instance, that the leaves of *T. procumbens* are consumed by *A. marginata* (Odaibo, 1997). Similarly, the leaves of this plant have earlier been found to better support *L. libycus*, in comparison with some other common dicotyledonous plants, in laboratory cultures (Agbolade et al., 2010). Survival of *B. truncatus* and *B. globosus* fed with *T. procumbens* corroborates the established nutritional values of the plant (Jude et al., 2009). In this study, *B. truncatus* seems to survive most on fresh *T. procumbens* leaves as opposed to *L. libycus* which prefers dried *T. procumbens* leaves. Although, in this study, *B. globosus* was fed only on fresh *T. procumbens* leaves; previously unreported observations suggest that fresh leaves of *T. procumbens* are likely to yield the best result on the survival of this species.

The percentage of shell growth shows that fresh *T. procumbens* leaves yielded relatively the highest increase as compared to dried ones, although the difference is not statistically significant. From the results of this study, it seems drying diminishes the dietary composition of *T. procumbens* leaves, particularly when done in the oven. The difference in the nutritional values of fresh *T. procumbens* leaves and dried ones seems magnified by the recorded percentage of increase in live-weight of *B. truncatus* when fed with fresh leaves. This seems to show that oven-dried leaves are poorer nutritionally. A previous study revealed depreciation of live-weight of *L. libycus* fed with dried *T. procumbens* (Agbolade et al., 2010). Like argued in an earlier paper, drying *T. procumbens* leaves might cause the destruction of many essential nutrients (such vitamins and proteins) needed by the water snails.

By and large, the results of this study have shown that laboratory culturing of *B. truncatus* and *B. globosus* using *T. procumbens* leaves may constitute a great forward step to water snails' researchers in Ijebu North, South-western Nigeria, using this free and easily available plant from the environment. However, for optimal performance the snails should be fed with fresh *T. procumbens*. Nevertheless, further studies (particularly parallel studies) are required to assess the reproductive potentials of *Bulinus* snails cultured using *T. procumbens* leaves.

ACKNOWLEDGEMENT

The authors thank the laboratory staff of the Department of Plant Science and Applied Zoology, Olabisi Onabanjo University, Ago-Iwoye, Nigeria for their unflinching support towards the success of this study.

REFERENCES

Adewunmi CO, Furu P, Christensen NO, Olorunmola F (1991). Endemicity, seasonality and focality of transmission of human schistosomiasis in 3 communities in South-western Nigeria. Trop. Med. Parasitol. 42: 332-334.

Agbolade OM, Akinboye DO, Fajebe OT, Abolade OM, Adebambo AA (2004). Human urinary schistosomiasis transmission foci and period in an endemic Town of Ijebu North, Southwest Nigeria. Trop. Biomed. 21(Suppl): 15-22.

Agbolade OM, Soewu DA, Bolanle OO, Momodu LA, Adegboyegun-King OO, Odubanjo AO, Owode JO, Asimi-Aina AB (2010). Preliminary study on culturing of *Lanistes libycus* using the leaves of some dicotyledonous plants. World J. Zool. 5: 252-256.

http://en.wikipedia.org/wiki/Tridax_procumbens. *Tridax procumbens*. Wikipedia, the free encyclopaedia. Accessed on14th January, 2013.

Brown DS (1994). Freshwater snails of Africa and their medical importance. 2nd Edn. Taylor and Francis Ltd., London, 609p.

Brown DS, Kristensen TK (1993). A field guide to African freshwater snails. West African species. Danish Bilharziasis Laboratory, Denmark, 55p.

Chitsulo L, Engels D, Montressor A, Savioli L (2000). The global status of Schistosomiasis and its control. Acta Trop. 77:41-51.

Jude CI, Catherine CI, Ngozi MI (2009). Chemical profile of *Tridax procumbens* Linn. Pak. J. Nutr. 8:548-550.

Madsen H, Stauffer JR (2011). Density of *Trematocranus placodon* (Pisces: Cichlidae): a predictor of density of the schistosome intermediate host, *Bulinus nyassanus* (Gastropoda: Planorbidae), in Lake Malawi. Ecohealth 8:177-189.

Odaibo AB (1997). Snail and snail farming: Nigerian edible land snails. Stirling-Horden Publishers Ltd., Ibadan, Nigeria, pp.10-11.

Ukoli FMA (1984). Introduction to parasitology in tropical Africa. John Wiley & Sons Ltd., Chichester. pp. 57-59

WHO (1993). The control of schistosomiasis: second report of the WHO Expert Committee. WHO Technical Report Series No. 830. World Health Organization, Geneva, Switzerland. pp. 51-54.

In-vitro efficacy evaluation of amitraz 0.025% and diazinon 0.06% against *Rhipicephalus pulchellus* and *Amblyomma gemma* in Borena pastoral area, Southern rangeland of Ethiopia

Eyob Eshetu[1], Getachew Dinede[2], Matios Lakew[2] and Tadele Tolosa[3]

[1]Department of Parasitology, Yabello Regional Veterinary Laboratory, Oromia Regional State, Yabello, Ethiopia.
[2]Oromia Pastoral Areas Technical and Vocational Education and Training College, Oromia Regional State, Yabello, Ethiopia.
[3]Department of Veterinary Public Health and Veterinary Microbiology, Jimma University College of Agriculture and Veterinary Medicine, Jimma, Ethiopia.

This study was conducted in Borena pastoral area, southern rangeland of Ethiopia to determine the *in vitro* efficacy of amitraz 0.025% and diazinon 0.06% against *Rhipicephalus pulchellus* and *Amblyomma gemma* using modified adult immersion. A total of 180 engorged adult female ticks (ninety ticks of each *A. gemma* and *R. pulchellus* species) were immersed in amitraz 0.025% or diazinon 0.06% at field recommended concentration (treated groups), or in distilled water (control groups) for 10 min and then incubated at 27 ± 1°C for 7 days. The oviposition response of each tick species in both groups was followed. The mean mass of eggs laid by each *A. gemma* and *R. pulchellus* in the treated group and those of untreated groups was compared to estimate the efficacy of each tested acaricide. Amitraz 0.025% has significantly higher ($P < 0.05$) in overall mean percent oviposition control (C% = 95.47) of *A. gemma* and *R. pulchellus* than diazinon 0.06% (C% = 80.9). At this recommended concentration, both amitraz 0.025% and diazinon 0.06% were less efficient in *R. pulchellus* oviposition control (C%= 90.94 for amitraz and C% = 71.41 for diazinon) than *A. gemma* oviposition control (C% = 100 for amitraz and C% = 88.85 for diazinon). The results of the study suggested that amitraz at field recommended concentration provides better efficient *R. pulchellus* and *A. gemma* oviposition inhibition than diazinon.

Key words: Amitraz, diazinon, Borena pastoral area, cattle, efficacy, *in vitro* testing, *Amblyomma*, *Rhipicephalus*.

INTRODUCTION

In tropical Africa, tick and tick borne diseases (TBDs) are economically very important diseases next to trypanosomosis (Belew and Mekonnen, 2011). Among 60 tick species found infesting both domestic and wild animal of Ethiopia, 30 species have been widespread and are important parasites of livestock (Solomon et al.,

2004), and causes significant economic losses to the livestock industry. The economic losses incurred from downgrading of hides and skins are enormous; its export yields foreign earnings of the country, second only to coffee (Sileshi et al., 2001).

Rhipicephalus pulchellus (R. pulchellus) and *Amblyomma gemma (A. gemma)* are the predominant species in arid and drier land (Walker et al., 2000). Similarly, their predominance is reported in the arid and drier land of Borena pastoral area (Solomon and Kaaya, 1998; Regassa, 2001; Solomon and Kaaya, 2001).

The currently available tools for tick control consists of acaricides relying on treatment with different application methods and/or formulations, tick resistant animals, tick vaccines, TBD vaccines and management interventions. The successful implementation of rational and sustainable tick control programme in grazing animals is dependent upon a sound knowledge of the ecology or epidemiology of the parasite as it interacts with the host in specific climatic, management and production environments (Alanr, 2011). Nevertheless, not all developing countries and those in transition may have such information available, due to a lack of human, economic and infrastructural resources (Food and Agriculture Organization (FAO), 2004).

In most situations, however, efficient and reliable control of ticks and TBDs are still based primarily on intensive use of acaricides, often without the local understanding of those responsible factors for tick distribution dynamics (Brito et al., 2011). Since acaricide introduction in South Africa around 1890, tick treatment relying on different application methods have been the main method of tick control in Africa, leading to numerous problems; environmental pollution, development of resistant tick strains and escalating costs (Alanr, 2011; Brito et al., 2011). To alleviate these problems, the most frequently used techniques to detect resistance in cattle tick are: the adult immersion test (AIT), larval packet test (LPT), and larval immersion test (LIT) (Castro-Janer et al., 2009). However, for the success of any tick management strategy, it is necessary to use a test that is practical, quick, economical and reliable to detect presence of resistance in target population (FAO, 2004). Enthusiasm for the development of standard protocols for modified AITs has been driven by the ease with which the test can be done, lack of any special equipment, and the fact that the test can be completed within only 7 days and this test mimicked the field condition better than the Shaw Larval Immersion Test (SLIT) (Jonsson et al., 2007).

Likewise, in Ethiopia, over the past decades ticks are mainly controlled by using variety of acaricides; including organochlorines, organophosphates, carbamates, amidines or synthetic pyrethroids (Sileshi, 2001; Yilma et al., 2001). However, with the most widespread, under or over concentration and frequent use of organochlorines and organophosphates compounds, ticks are likely to develop resistance in Ethiopia (Adamu, 1996; Yilma et al., 2001; Sileshi et al., 2004). In Borena pastoral areas, where amitraz 0.025% and diazinon 0.06% were mostly used, in various circumstances, animal health personnels and livestock herders complained of failure of this two acaricides to kill ticks and toxicity associated with diazinon usage (Borena Zone Pastoral Area Development Office (BZPADO), 2009, 2010). Therefore, continuous studies on dynamics of tick population (Alanr, 2011) with the efficacy status of acaricides against the most abundant and important tick in particular area are necessary to carry out efficient tick control and/or tick burden reduction (Solomon, 2001). The objectives of this study were to evaluate the efficacy of amitraz 0.025% and diazinon 0.06% against field population of engorged adult female *R. pulchilus* and *A. gemma* ticks under *in vitro* condition using AIT.

MATERIALS AND METHODS

Characterizations of study area

The study was conducted between March and April, 2012 at Borena pastoral area of Oromia regional state, Southern rangeland of Ethiopia, located at 565 km south of Addis Ababa, the capital city of Ethiopia. The region has predominantly a semi-arid climate, and physiographically, it is dominated by Savannah vegetation. Animal husbandry in the area is characterized by extensive livestock productions system and seasonal mobility. Cattles are the dominant livestock species (Cooperative for Assistance and Relief Everywhere (CARE), 2009). Cattle are heavily infested with *R. pulchellus* and *A. gemma*. Acaricide application is the main tick control methods in the area and their application seems to be regulated primarily by their availability. Cattle are treated for tick infestation whenever the farmers bring their animals to the veterinary clinics for other complaints; otherwise the pastoralists are treating by themselves. The organophosphates (diazinon 0.06% and amitraz 0.025%) are the two acaricides used by the pastoralist communities in the area to control ticks.

Study methodology and procedures

The fully engorged adult female ticks were collected from 17 different cattle herds at Borena pastoral areas, located roughly between 3 to 5°N and 37 to 42°E, covering an area of 120,000 km^2. The herds were selected on the history basis of communities' complaints on acaricides failure. From each herd, 6 or 8 cattle, based on their infestation level with engorged adult female ticks, a total of 134 cattle, were selected. At each collecting site, cattle were restrained, and a maximum of five adult engorged female ticks of any species (as it is difficult to identify the tick species at collection site) were collected from the sampled cattle (Ducoenez, 2005). The ticks were placed in labeled plastic flasks with small holes and free of acaricide.

The choice of acaricides used was based on their commercial availability and patronage by farmers and veterinary clinic in Borena pastoral area. Thus, amitraz 0.025% was manufactured by Laboratorios Microsules Uruguay S.A. and diazinon 0.06% EC was

Table 1. Mean oviposition of adult A. gemma and R. pulchellus after immersion in amitraz 0.025% and diazinon 0.06% EC at recommended concentration and 7 day incubation.

Treated ticks	Treatment	N	M_1	S	M_2	C%
A. gemma	Amitraz 0.025%	10	2.63	0	0.000	100.00
	Water control	10	2.45	8.67	0.333	0.000
R. pulchellus	Amitraz 0.025%	10	2.83	0.67	0.033	90.94
	Water control	10	2.75	9.67	0.373	0.00
A.gemma	Diazinon 0.06% EC	10	2.56	1.0	0.037	88.85
	Water control	10	2.45	8.67	0.333	0.000
R. pulchellus	Diazinon 0.06% EC	10	2.79	1.33	0.12	71.41
	Water control	10	2.75	9.67	0.373	0.00

N = Number of immersed female ticks; M1 = engorgement weight (g); S = number of tick survived after 7 days incubation; M_2 = average egg mass per treatment group (g).

manufactured by Shandong Luxi Animal Medicine Share Co. Ltd (China). The reconstitution of acaricides was done incognizance to the manufactures recommended concentrations to be used on infested animals using distilled water. The indicated concentration for diazinon is 0.06% while that of amitraz is 0.025%. The formula, $V_1C_1 = V_2C_2$ was used to prepare the concentration of acaricides, where V_1 and V_2 are the volume of the acaricide to be drawn from the stock product and the final volume after reconstitution, respectively, C_1 and C_2 are the stock product concentration and the required final concentration after preparation, respectively. For all the preparations, the final volume was 1000 ml.

The bioassay technique used was the modified AIT, a laboratory protocol first described by Drummond et al. (1973) and modified by FAO (2004). At the laboratory, from the total collected engorged adult female ticks, a sample comprising of 180 ticks (ninety ticks of each *R. pulchellus* and *A. gemma*) with no signs of injury was selected. Both species were carefully washed on a sieve using a clean tap water and dried at room temperature on absorbent paper and each species divided into groups according to their size. Three groups of ten ticks of each species were randomly formed (Group I for amitraz, Group II for diazinon and Group III for control/water), and all ticks were pasted onto double sided adhesive strips on glass petridish, with their ventral sides facing up wards. This set up was then covered and incubated in a dessicator maintained at 72% relative humidity (RH) and temperature of 27°C. Three replicates were used for each acaricide and species of tick as described.

The efficacies of the acaricides were evaluated using the egg laying test (ELT). Thus, the overall mean efficacy was then estimated from all the three replica of trials. ELT method involves the comparison of the egg mass of ticks treated with acaricide and the egg mass of untreated ticks, and finally estimates the percentage control value, using the following formula:

$$\text{Percent control} = \frac{\text{MEC - MET}}{\text{MEC}} \times 100$$

Where, MEC and MET are mass of eggs laid by control ticks and treated ticks, respectively.

Data management and analysis

All collected data were entered into Microsoft Excel 2007 computer program. All statistical analyses were performed using statistical package for social science (SPSS)-Version 19 for windows 2007. Percent control (C%) for each acaricide, obtained with ELT, was used to evaluate effectiveness. Independent sample t-test was used to observe the mean C% difference between the two acaricides. A P-value less than 0.05 at 95% confidence intervals was considered for significance.

RESULTS

The oviposition response of *A. gemma* and *R. pulchellus*, after immersion in amitraz 0.025% and diazinon 0.06% in three replicates was determined and presented on (Table 1). In the trial, none of *A. gemma* treated with amitraz laid eggs, while few of the treated *R. pulchellus* laid small batch of eggs with mean weight of 0.033 g. In contrast, both *A. gemma* and *R. pulchellus* immersed in diazinon 0.06% laid eggs with mean weight of 0.037 and 0.12 g, respectively. However, there was no statistical significance variation ($P > 0.05$) between the two acaricides in the overall oviposition control of each tick species (Table 2). Both tick species in the control group laid relatively large batch of eggs with mean weight of 0.33 g by *A. gemma* and 0.373 g by *R. pulchellus*.

The overall mean C% of amitraz 0.025% and diazinon 0.06%, and their respective standard deviations as well as their minimum and maximum mean efficacy during the three replica of the trial is presented in Table 3. Therefore, although, amitraz showed evidence of greatest effect on oviposition on *A. gemma*, the statistical comparison between the overall mean C% of each acaricide revealed no significant differences ($t = 2.438$, df= 4, 95% confidence interval (CI) = -1.5498 to 23.8498, $P > 0.05$). Moreover, the overall mean C% analysis of the two acaricides on the inhibition of *R. pulchellus* oviposition also showed no statistical significant variation ($t = 2.26$, df = 4, 95% CI = -4.48 to 43.71, $P > 0.05$).

Table 2. T-test analysis of mean percent *A. gemma* and *R. pulchellus* oviposition control between amitraz 0.025% and diazinon 0.06% EC at recommended concentration.

Ticks	Acaricide	C%	N	Mean	SD	t-value	df	95% CI	S
A.gemma	Amitraz	100	3	100.0	0.0	2.438	4	-1.55-23.85	NS
	Diazinon	88.85	3	88.85	7.9				
R. pulchellus	Amitraz	90.94	3	90.94	11.3	2.26	4	-4.48-43.71	NS
	Diazinon	71.41	3	71.33	9.9				

C% = Percent control; N = number of trails; SD = standard deviation; NS = not significant; df = difference.

Table 3. Overall mean percent oviposition control of amitraz 0.025% and diazinon 0.06% EC at field recommended concentration against adult female *A. gemma and R. pulchellus*

Acaricide	Min. efficacy (%)	Max. efficacy (%)	Mean efficacy (±SD%)
Amitraz 0.025%	78.38	100	95.47±8.663
Diazinon 0.06% EC	61.11	97.06	80.09±12.537

Min = minimuim, Max = maximum.

DISCUSSION

Several authors have studied the efficacy of amitraz on different tick populations using AIT. The results have shown different susceptibility levels. In most of the studies, amitraz revealed high degree of tickicidal efficacy that was agreed with the present finding. Similarly, a closely comparable finding, C% of 98 to 100, was reported by Sileshi (2001) at Sebeta, Ethiopia. In South Africa, Sileshi et al. (2002) also reported 100% efficacy for amitraz at the same dilution rate. Moreover, Souza et al. (2003) in Southeast Brazil obtained mean amitraz efficacy of above 95%. The minimum mean efficacy of amitraz observed in the present study (78.38%) is in accordance with that of Mendes et al. (2001) finding and he found an average efficacy of 77.44%.

In contrast, Furlong et al. (2007) found mean efficacy of 47.9% for amitraz. Camillo et al. (2009) also observed the presence of resistance with a low efficacy of Amitraz in some tick populations of RioGrande do Sul State, Brazil. In Northeast region of Brazil, a low efficacy of amitraz, with a control of 40.5 and 30.95% was reported by Santana (2000) and Campos and Oliveira (2005), respectively. In our case, we can assume that the tick populations were susceptible to amitraz.

The efficacy variation between the two presently tested acaricides might be associated with the high sterilization effect of amitraz compared to diazinon when applied at field recommended concentration (Sileshi et al., 2003). The oviposition response variation of each tick species is also most probably associated with prior exposure of these tick species to diazinon, as its usage began over the past 10 years in the study area. Sileshi et al. (2003) in South Africa observed relatively higher level of resistance to diazinon than amitraz.

The use of an acaricide at incorrect concentration is also one of the prime factors which affect the efficacy of an acaricide and causes of tick control failure (Natala et al., 2005; Kirby, 2010; Alanr, 2011). It has been observed during the study period that the pastoralists in the area believed that they needed to increase the concentration of acaricides during the peak tick season to control the excessive tick burdens infesting their cattle. This type of increased acaricide concentration can lead to a higher selection pressure for tick resistance (Brito et al., 2011; Pegram et al., 2000). In the present study, the two acaricides have conserved their efficacy on both tested tick species. However, a right tick control management needs to be pursued in order to avoid any resistance.

ACKNOWLEDGEMENTS

The authors are indebted to the Yabello Regional Veterinary Laboratory for material and logistic supports, and cooperation of the staff members to bring this research to completion.

REFERENCES

Adamu G (1996). Bioassay of acaricide resistance on three common cattle tick species (Ixodidae): a case study in Holetta area central highland. DVM thesis, Addis Ababa University, Faculty of Veterinary Medicine, Debrezeit, Ehtiopia.

Alanr W (2011). Eradication and control of livestock ticks: biological, economic and social perspectives. Royal (Dick) School of Veterinary Studies, University of Edinburgh, Summerhall Place, Edinburgh EH9 1QH, UK.

Belew T, Mekonnen A (2011). Distribution of Ixodid Ticks on Cattle in and Around Holeta Town, Ethiopia. Glob. Vet. 7(6):527-531.

Borena Zone Pastoral Area Development Office (BZPADO) (2009/10). Annual Livestock Population and Diseases Report of Borena Zones Pastoral Districts. Borena Zone Livestock Population Report, Cattle population and diseases.

Brito LG, Barbieri FS, Rodrigo B, Rocha MC, Oliveira S, Elisana S (2011). Evaluation of the efficacy of acaricides used to control the cattle tick, *Rhipicephalus microplus*, in dairy herds raised in the Brazilian southwestern amazon. Vet. Med. Int. 201(6):2-13.

Camillo G, Vogel F F, Sangioni L A, Cadore GC, Ferrari R (2009). In vitro evaluations of acaricides efficiency to bovine's ticks of RioGrande do Sul State, Brazil. *Ciˆencia* Rural 39 (2):490–495.

Campos DA, Oliveira PR (2005). *In vitro* valuation of acaricides efficiency to *Boophilus microplus* (Canestrini, 1887) Acari: Ixodidae) from bovines at the region of Ilh´eus, Bahia, Brazil. Cienc Rural 35(6):386–1392.

CARE-Ethiopia (2009). Value Chain Analysis of Milk and Milk products in Borena Pastoralist Area. Regional Resilience Enhancement against Drought Project. Yonas Business Promotion and Consultancy P LC, Addis Ababa, Ethiopia. pp. 12-25.

Castro-Janer E, Rifran L, Piaggio J, Gil A, Miller R J and Schumaker T T S (2009). in vitro test to establish L50 and discriminating concentrations for fipronil against *Rhipicephalus (Boophilus) microplus* (Acari:Ixodidae) and their standardization. Veterin. Parasitol. J. 162 (31):120-128.

Drummond RO, Ernst SE, Trevino JL, Gladney WJ, Graham OH (1973). *Boophilus annulatus* and *B. microplus*: laboratory tests of insecticides. J. Econ. Entomol. 66(1):130–133.

Ducornez S, Barré N, Miller R J, de Garine-Wichatitsky M (2005). Diagnosis of amitraz resistance in *Boophilus microplus* in New Caledonia with the modified Larval Packet Test. Vet Parasitol. 130:285-292.

Food and Agricultural Organization (2004). Acaricide resistance: diagnosis, management and prevention. Animal Production and Health Division, Agriculture Department, Food and Agriculture Organization of the United Nations, Rome, Italy. pp. 25–77.

Furlong J, Martins JR, Prata MCA (2007). The tick of cattle and resistance: we have to celebrate. A Hora Vet. J.27 (7): 27-34.

Jonsson NN, Miller RJ,Robertson JL (2007). Critical evaluation of the modified Adult Immerssion Test with discriminating dose bioassay for *Boophilus microplus* using American and Australian isolates. Vet. Parasitol.146(7):307-315.

Kirby C (2010) Tick Management Hand book. Biological tick Control, 2nd Ed. The Connecticut Agricultural Experimentation Station, New Haven. pp. 70-71.

Mendes MC, Verissimo CJ, Kaneto C N (2001). Bioassays for measuring the acaricides susceptibility of cattle tick *Boophilus microplus* in S˜ao Paulo state. J. Anim. Plant Sci. 68(2):23-27.

Natala A J, Agyei AD, Awumbila B (2005). Susceptibility of *Amblyomma variegatum* ticks to acaricides in Ghana. Exp. Appl. Acarol. 35 (1):259-268.

Pegram RG, Wilson DD, Hansen JW (2000). Past and present national tick control programs: why they succeed or fail. Annals NY Acad. Sci. P 546.

Regassa A (2001). Tick infestation of Boran cattle in the Boran Province of Ethiopia. Onderstepoort J. Vet. Res. 68(8):41- 45.

Santana VLA (2000). Situation of chemical control of *Boophilus microplus* the sub-regions in the area of forest and wasteland in the state of Pernambuco, based on *in vitro* efficacy of acaricides in engorged females. Dissertation in Veterinary Science, Federal Rural University of Pernambuco, Recife, Pernambuco, Brazil.

Sileshi M (2001) *In vivo* evaluation of Amitraz against ticks under field conditions in Ethiopia. JSAVA 72 (3):44-45.

Sileshi M, Kgasi A Mureithi W, Getachew Z, Tilahun T, Solomon G, Yilma J and Bryson N R (2004) In Vivo and In Vitro Evaluation Of The Efficacy Of Cypermethin High-Cis (Ecotomin®) Against Cattle Ticks In Ethiopia. Ethiop. Vet. J. 8(1):29

Sileshi M, Bryson NR, Fourie LJ, Peter RJ, Spickett AM, Taylor RJ, Strydom T, Kemp DH, Horak I G (2003). Comparison of 3 tests to detect acaricide resistance in *Rhipicephalus (Boophilus) decoloratus* on dairy farms in the Eastern Cape Province, South Africa. JSAVA 74(9):41-44.

Sileshi M, Hussein I, Bedane B (2001). The distribution of ixodid ticks (Acari: Ixodidae) in central Ethiopia. OJVr 68(4):243-251.

Sileshi M, Bryson N R, Fourie L J, Peter RJ, Spickett AM, Taylor RJ, StrydomT, Kemp DH, Horak IG (2002). Acaricide Resistance profiles of single and multi - host ticks from communal and commercial farming areas in the Eastern Cape and North West Provinces of South Africa. MVSc thesis, Department of Veterinary Tropical Diseases, Faculty of Veterinary Science, University of Pretoria, Pretoria.

Solomon G (2001). Problem of Acaricide resistance for the control of ticks. Proceeding of the Ethiopian Society of Animal production 9th Conference.

Solomon G, Kaaya GP (2001). Studies on the development and survival of *Rh. pulchellus* and *A. gemma* under field condition in Ethiopia. Ethiop. Vet. J.5 (1):28

Solomon G, Kaaya GP (1998). Comparison of resistance in three breeds of cattle against African Ixodid ticks. Exp. Appl. Acarol. 2(4):223-230.

Solomon G, Sileshi M, Kaaya GP, Tilahun T, and Yilma J (2004). Prevalence of Ixodid ticks and Trypanosomosis in camels in Southern Ethiopia. Ethiop Vet. J. 8 (2):23-24

Souza AP, Sartor AA, Bellato V (2003). Efficacy of acaricides in dairy cattle herds in southern region of Parana state, Brazil. J. Agron. Vet. Sci. 2 (1): 131.

Walker JB, Keirans JE, Horak IG (2000). The genus *Rhipicephalus* (Acari, Ixodidae): a guide to the brown ticks of the world. Cambridge University Press, Cambridge.

Yilma J, Adamu G, Zerbini E (2001). Bioassay of acaricide resistance on three common cattle tick species at Holeta, Central Ethiopia. Rev. Med. Vet. 152(7):385.

Camel hydatidosis: Prevalence and economic significance in pastoral regions of Ethiopia

Bayleyegn Gizachew , Fikadu Kibru and Biruhtesfa Asrade

College of Veterinary Medicine, Haramaya University, P. O. Box 138, Dire Dawa, Ethiopia.

Camel hydatidosis was studied at Addis Ababa abattoir, Ethiopia to determine the prevalence and financial losses associated. From 501 camels slaughtered, 328 (65.47%) were found harboring hydatid cyst. The prevalence between females and males was statistically significant (χ^2=35.74; P=0.000). Additionally, the disease was significantly different among the age groups (χ^2=18.71; P=0.00) revealing higher prevalence in older animals. In respect to origin, the highest prevalence was observed from Borena (65.67%). The lung (47.90%) was the most frequently affected organ. Majority of the cysts identified were non calcified cysts. 57.78 and 39.10% of the cysts were found to be fertile in the lung and liver, respectively. Of these fertile cysts 68.27 and 60% were viable in the lung and liver, respectively. 212 lungs, 209 livers, 21 spleens and 2 hearts were totally condemned and these results in financial loss of 1089758.8 ETB (61222.4 US Dollar) annually. In conclusion, hydatidosis is highly prevalent in camels slaughtered at Addis Ababa Abattoir resulting to high economic loss due to organ condemnation. Thus, an effort should be made to control and prevent echinococcosis in the camel herding areas.

Key words: Akaki abattoir, camel hydatidosis, financial loss, prevalence.

INTRODUCTION

Pastoralism accounts for the livelihood of 50 to 100 million people in developing countries, while 60% of this population lives in more than 21 African countries confined to the most arid regions of the continents (Sheik-Mohamed and Velema, 1999; UNDP, 2007). Pastoralists are migratory people whose livelihood largely depends on livestock rearing. Among the East Africa countries, Ethiopia has the largest pastoralist population accounting for 10 to 12% of the total population (USAID, 2005). Ethiopian pastoralists virtually depend on livestock for their livelihood, moving seasonally from place to place in search of water and pasture for their animals (Nori, 2005). The dromedary camel (*Camelus dromedarius*), which is adaptable and capable of living in desert areas serves as a source of milk, meat, and draft power for the subsistence of the pastoralists. Ethiopia possesses an estimated amount of 2.3 million of *C. dromedaries* (Central Statistical Agency (CSA), 2008) mainly distributed in the Southern, Eastern and Northeast pastoral regions (Workneh, 2002; Ministry of Information, 2005).

In spite of the great ecological and economical value of the dromedary camel, there is scarce research information on camel diseases when compared with that of other domestic animals (Zeleke and Berkeley, 2000). This reveals that camels may be either carrier, susceptible or suffering from several diseases(Demeke, 1998). In connection to these, parasitic diseases like hydatidosis appear to be the major constraints that are hampering the potential performances of the camels (Getahun and Kassa, 2002; Megersa, 2010). In farm animals, it causes considerable economic losses due to condemnation of edible organs, decreased meat and milk production (Polydorous, 1981; Romazanvoc, 2001). In Ethiopia no work has been done on camel hydatidosis except a prevalence report by Muskin et al. (2011). In this regards, as Ethiopian dromedaries are primarily reared by pastoralists, camel hydatidosis could have significant economic and public health consequences in the regions. To this effect, abattoir based epidemiological studies are needed to show the real picture of the disease and to know its impact on economy. Therefore, the objectives of the present study are to investigate epidemiology and assess economic significance of camel hydatidosis in pastoral regions of Ethiopia.

MATERIALS AND METHODS

Study area and subjects

The study was conducted at Akaki (Addis Ababa) abattoir on apparently healthy slaughtered camels. The camels slaughtered at the abattoir were both male and female that originated from pastoral parts of the country mainly from Borena (Southeastern part of Oromia region) and Afar (eastern regions of Ethiopia).

Post mortem examination

Thorough post mortem examination was carried out by visual inspection, palpation and systematic incision on visceral organs; lung, liver, heart, kidney and spleen according to procedures recommended by Food and Agricultural Organization (FAO, 1994). All organs harboring cysts were partially or totally condemned and were judged according to the guidelines on meat inspection for developing countries (FAO, 1994).

Whenever and wherever the hydatid cysts appeared, the number and the size of the cysts per organ and per animal were counted, measured and recorded. Accordingly, the size of cysts were measured systematically and classified into three based on their diameter as small (<4 cm), medium (4 to 8 cm) and large (>8 cm) (Oostburg et al., 2000). The cysts were randomly selected and collected from different organs and were taken to regional laboratory (Shola, Addis Ababa) for fertility and viability tests (Figure 1).

Cyst viability tests

Non-calcified cysts were randomly selected and collected from different organs and were taken to regional laboratory (Shola, Addis Ababa) using ice box for fertility and viability tests. The cyst was put on clean petri-dish and incised by sterile scalpel blade and the fluid part was poured in another clean petri-dish. The supernatant was discarded and resultant sediment was finally examined for the presence of protoscolesces that appear as white dot under the microscopic field at 40X magnification power. Fertile cyst was subjected to viability test. A drop of 0.1% eosin solution was added to equal volume of protoscolisces in hydatid fluid on petri-dish with principle that viable protoscolisces should completely or partially avoids the dye, while the dead once take it up. Furthermore, infertile cysts were classified as sterile or calcified. Sterile hydatid cysts were characterized by their smooth inner lining usually with slightly turbid fluid in its content. Typically calcified cysts produced a gritty sound feeling upon incision (Soulsby, 1982).

Data management and analysis

The data were entered and coded into Microsoft excel program. Pearson χ^2 was used to test the existence of association between sex, body condition, origin and age groups. P-value less than 0.05 ($P<0.05$) was considered as statistically significant.

Estimation of the financial losses was analyzed on the basis of secondary data (annual slaughter rate) from the slaughter house and the mean retail market price of the offal which was obtained from the abattoir workers and butchers, and associated economic loss was calculated on the annual basis according to Sariözkan and Yalcin (2009).

RESULTS AND DISCUSSION

There are few epidemiological studies published on camel hydatidosis in Ethiopia as well as in other countries. Based on post mortem inspection, the prevalence of camel hydatidosis was 328/501 (65.5%). In this study, the prevalence of hydatidosis in camel is relatively higher when compared to the previous reports in camel, cattle and small ruminants in Ethiopia by Ahmed (1998) (18.6%) in Eastern Ethiopia, Weldemeskel (2001) (18.86%), Mulatu (2013) (30.5%) in Dire Dawa and Jijiga Bitsat (2009) (28.5%) and Muskin et al. (2011) (22.6%), and elsewhere in the world: 39.65, 48.5, 35.25, and 32.85%, were reported in Kuwait (Abdul-Salam et al., 1988), Libya (Ibrahim, 1998), Iran (Ahmadi, 2005), and Saudi Arabia (Mohamed, 2010), respectively. The disease was highly prevalent in female animals 300/328 (91.5%) and was statistically significant ($\chi^2=35.74$; $P=0.00$). The result of the association of the different risk factors to the hydatid cyst is presented in Table 1. Among the age groups, the disease was also found to be in higher prevalence in older animals. The age dependent variation could be due to high probability of the exposure of older animals to the infection during their long existence in life for many years.

This is also true for higher prevalent record in female animals, because during inspection, these animals were found old due to the fact that most pastoralists sell the female animals after they finished their productive age. These results agree with the previous reports by Cleaveland et al. (2007) and Inangolet et al. (2008). In this case, it is also good to consider the chronic nature of cyst development in organs. Since camel can live around 40 years (Kinne et al., 2006), they are more affected than other domestic animals. Regarding the origins, camels from Borena area were more affected with disease than

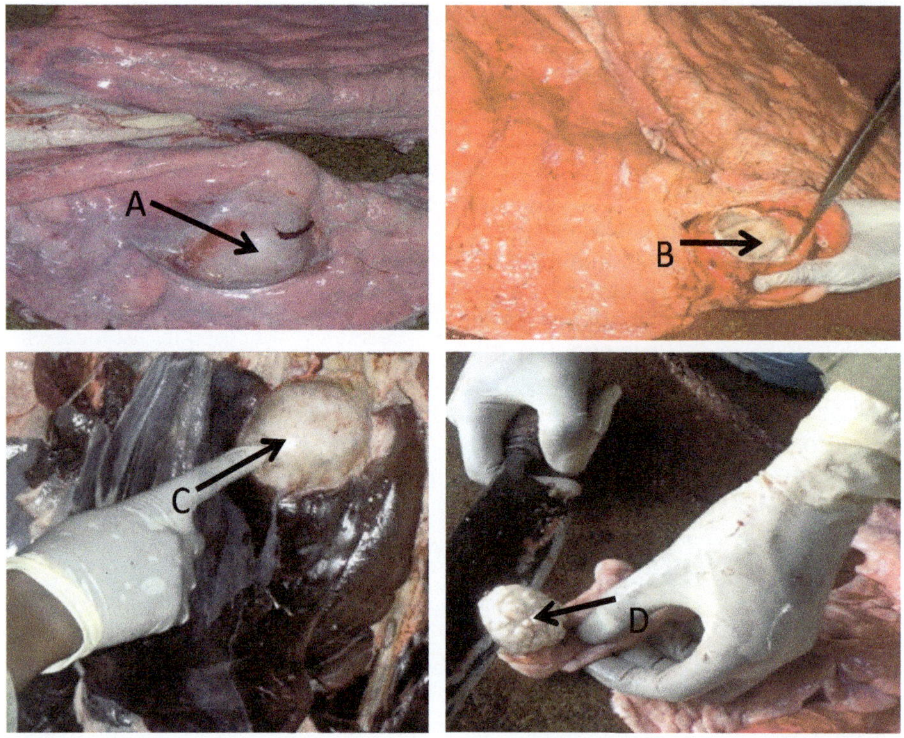

Figure 1. Hydatid cysts on different organs of slaughtered camels. A: None calcified hydatid cyst on the surface of camel lung; B: None calcified hydatid cyst in the lung exposed after incision; C: None calcified hydatid cysts on the surface camel liver; D: calcified hydatid cyst on camel lung.

Table 1. Association host related variables with camel hydatidosis.

Variable	No. of camels slaughtered	No. of positive Animals	% of positive animals	χ^2	P-value
Sex					
Female	423	300	91.5	35.74	0.00*
Male	78	28	8.54		
Age					
< 5	48	21	6.40		
5-10	172	103	31.40	18.2	0.00*
>10	281	204	62.20		
BCS					
Score 1	81	60	18.29		
Score2	38	29	8.84		
Score3	193	129	39.36	9.28	0.054
Score4	124	73	22.26		
Score5	65	37	1.28		
Origin					
Borena	329	233	71.04		
Afar	81	49	14.94	17.24	0.001*
Meiso	66	37	11.28		
Minjar	25	9	2.74		

*Statistically significant.

Table 2. Distribution of hydatid cyst on different organs of slaughtered animals.

Cyst location	No. of infected animal	Percent of infected animal
Lung	94	28.66
Liver	83	25.30
Spleen	2	0.61
Heart	0	0.00
Lung and liver	128	39.02
Lung and spleen	5	1.52
Lung and heart	1	0.31
Liver and spleen	2	0.61
Liver and heart	1	0.31
Lung, liver and spleen	12	3.66

camels from other origins. Theses could be due to the absence of similar environmental and climatic situations in the different areas (Mohamed, 2010).

In this study, it was also possible to characterize the nature of the cyst. Most of the hydatid cysts were recorded in lung and liver of camel. Single and multiple organ pathology with hydatid cysts are depicted in Table 2. The infection rate of cystic echinococcosis in each organ with relation to the cyst size is summarized in Table 3. Majority of the identified cysts were non-calcified. The distribution of calcified and non calcified cysts in relation to age is summarized in Table 4. Regarding the fertility of the hydatid cysts, this study indicated that cysts from the lung showed a higher fertility proportion as compared to cysts collected from livers (Table 5).

The reason for the difference may be attributed to the age of animals. An increased number of fertile cysts tend to occur in older animals (World Health Organization/ Office International des Epizootics (WHO/OIE), 2011); this strongly suggest that fertile cysts, are important factors that can influence the transmission of *Echinococcosis granulosus*, and are most likely found in older animals while the sterile one is found in young animals (Grayesm, 1986).

Furthermore, the extensive practice and roadside slaughter of camels could have contributed to the maintenance of the parasite's cycle. In such a way, dogs have easy access to ingest the offal or organs harboring the cysts and become easily infected. This plays a crucial role in the transmission of the disease and enables the parasite to maintain its life cycle among the hosts. Moreover, pastoralists in east Africa move from place to place with their livestock including camels in search of feed where wild canids are found. This may favor the transmission of the infection between infected canids and susceptible camels and also be considered as risk factors in increment of the prevalence in this study. Moreover, Ethiopia has proclamation for animal disease prevention and control (Federal Negarit Gazeta (FNG), 2002). However, currently there is no documented and implemented rule and regulation for meat inspection in particular for camel meat.

The annual economic loss due to hydatidosis in camel slaughtered at Akaki abattoir was estimated by considering the main cost of organs (lung, liver, heart and spleen) (Sariözkan and Yalcin, 2009). The main price of respective organs at Addis Ababa city was obtained from abattoir workers and butchers during the study period.

These parameters were then fed to the following formula in order to compute the economic loss due to organ condemned, as unfit for human consumption due to cystic echinococcosis thus:

$$X = (AS \times C_{Lu} \times P_{Lu}) + (AS \times C_{Li} \times P_{Li}) + (AS \times C_{Sp} \times P_{Sp}) + (AS \times C_{Kid} \times P_{Kid}) + (AS \times C_{Hr} \times P_{Hr})$$

where AS: estimated mean annual kill; P_{Lu}: percent involvement of the lung; C_{Lu}: local retail price of a lung; P_{Li}: percent involvement of the liver; C_{Li}: local retail price of a liver; P_{Sp}= percent involvement of the spleen; C_{Sp}: local retail price of a spleen; P_{Kid}: percent involvement of the kidney; C_{Kid}: local retail price of a kidney; P_{Hr}: percent involvement of the heart; C_{Hr}: local retail price of a heart; P_{SC}: percent of spleen condemned; C_S: local retail price of the spleen.

$X = (3201 \times 65.47\% \times 10) + (3201 \times 41.72\% \times 800) + (3201 \times 0.39\% \times 35) + (3201 \times 0.00\% \times 80) + (3201 \times 4.19\% \times 0) = 20957 + 1068365.8 + 437 + 0 + 0$

X = 1,089,758.8 ETB (61,222.4 US Dollar)

Conclusively, this study indicated that hydatidosis is a highly prevalent parasitic disease of slaughtered camels at Addis Ababa abattoir. In connection to this, significant

Table 3. Cyst size in different organs of infected camels.

Organ	Small	Medium	Large	Total
	N (%)	N (%)	N (%)	N
Lung	409 (39.10)	850 (57.32)	242 (63.35)	1501
Liver	627 (59.94)	613 (41.34)	136 (35.60)	1376
Spleen	10 (0.96)	19 (1.28)	3 (0.79)	32
Heart	0 (0.00)	1 (0.07)	1 (0.26)	2
Total	1046 (100.00)	1483 (100.00)	382 (100.00)	2911

Table 4. Calcification of cysts.

Age (Years)	No. cyst Examined	Non-calcified cyst	Calcified cyst
		N (%)	N (%)
≤5	19	10 (52.63)	9 (47.37)
5-10	152	115 (75.66)	37 (24.34)
>10	217	183 (84.33)	34 (15.67)
Total	388	308 (79.38)	80 (20.62)

Table 5. Cyst fertility and viability in relation to organ affected.

Cyst location	No. of cyst examined	N(%)			
		Sterile	Fertile	Viable	Dead
Lung	180	76(42.22)	104(57.78)	71(68.27	33(31.73)
Liver	128	78(60.93)	50(39.10)	30(60.00)	20(40.00)
Total	308	154(50.00)	154(50.00)	101(65.58)	53(34.42)

financial losses were recorded due to condemnation of edible organs from the domestic markets. Therefore, an effort should be made to control and prevent camel hydatidosis and further identification and characterization of causative agent would be useful towards the efforts made to control the disease.

ACKNOWLEDGEMENTS

The authors appreciate the cooperation of all workers of Addis Ababa abattoir enterprise and laboratory technicians of regional veterinary laboratory (Shola, Addis Ababa) and Haramaya University for the financial support.

REFERENCES

Abdul-salam JM, Farah MA (1988). Hydatidosis in camels in Kuwait. Parasitol. Res. 74:267-270.

Ahmadi NA (2005). Hydatidosis in camel (*C. dromedaries*) and their potential Role in the Epidemiology of *Echinococcus granulosus*, Iran. J. Helminthol. 79:119-125.

Ahmed I (1998). Preliminary Investigation on Major Disease of Camels of Eastern Ethiopia: Abattoir and Field Survey. DVM Thesis, Addis Ababa University, Faculty of Veterinary Medicine, Debre Ziet, Ethiopia.

Cleaveland S, Shaw DJ, Mfinanga SG, Shirima G, Kazwala RR, Eblate E, Sharp M (2007). *Mycobacterium bovis* in Rural Tanzania: Risk Factors for Infection in Human and Cattle Populations. Tuberculosis 87(1):30–43.

CSA (2008). Federal Democratic Republic of Ethiopia Central Statistical Agency. Agricultural Sample Survey Report on Livestock and Livestock Characteristics, Addis Ababa. Stat. Bull. 2:16.

Federal Negarit Gazeta (FNG, 2002). The Federal Democratic Republic of Ethiopia. Animal Diseases Prevention and Control Proclamation No. 267/2002. 8th Year No 14 Addis Ababa.

Food and Agricultural Organization (1994). Guidelines for Training, Food and Agricultural Organization of the United Nations Rome, Italy. P 34.

Grayesm S (1986). Meat Hygiene. 8th Edition. Bailliere, Tindall, London. P 517.

Inangolet FO, Demelash B, Oloya J, Opuda-Asibo J, Skjerve E (2008). A cross-sectional study of Bovine Tuberculosis in the transhumant and agro-pastoral cattle herds in the border Areas of Katakwi and Moroto Districts, Uganda. Trop. Anim. Health Prod. 40:501–508.

Kinne J, Johnson B, Jahans KL, Smith NH, Ul-Haq A, Wernery U (2006). Camel Tuberculosis-a case report. Trop. Anim. Health Prod. 38:38.

Megersa B (2010). An epidemiological study of major camel diseases in the Borana Lowland, Southern Ethiopia. DCG Report No. 58.

Ministry of Information (MOI, 2005). Export products of Ethiopia. Press release of Ministry of Information, Department of Press and Audiovisual. Addis Ababa, Ethiopia.

Mohamed MI (2010). Study of cystic echinococcosis in slaughtered animals in Al Baha Region, Saudi Arabia: Interaction between some Biotic Factors. Acta. Trop. 113(1):26-33.

Muskin S, Hailu D, Moti Y (2011). Infection Rates, Cyst Fertility and Larval Viability of Hydatid Disease in Camels (*Camelus dromedarius*) from Borena, Kereyu and Harar Areas of Ethiopia. Glob. Vet. 7(6):518-522.

Mulatu M, Biruk M, Tassew H, Ashwani K (2013). Bovine Hydatidosis in Eastern Part of Ethiopia. Momona Ethiop. J. Sci. 5(1)107-114.

Nori M (2005). Sustainable camel milk production and commercialization in Somalia. Future Livestock Systems APS course, Wageningen-4/2005, The Netherlands.

Oostburg BFJ, Vrede MA, Bergen AE (2000). The occurrence of polycystic echinococcosisin Suriname. Ann. Trop. Med. Parsitol. 94(3):247-252.

Polydorous K (1981). Animal health and economic case study: Echinococcosis with reference to Cyprus. Bull. Off. Int. Epizoot. 93(5):981-992.

Romazanvoc F (2001). Cestode zoonosis: Echnococcosis and Cysticercosis an emergent and global problem. IOS Press, Netherlands. pp. 34-57.

Sariözkan S, Yalçin C (2009). Estimating the production losses due to Cystic Echnococcosis in ruminants in Turkey. Vet. Parasitol. 163: 330-334.

Sheik-Mohamed A, Velema JP (1999). Where health care has no access; the nomadic populations of Sub-Sahara Africa. Trop. Med. Int. Med. 4:695-707.

Soulsby EJL (1982). Helminths, Arthropods and protozoa of domesticated animal. 7th edition, London: Baillereire Tindal, pp. 807-809.

UNDP (2007). Between a rock and hard place: Armed violence in Africa pastoral communities. United Nations Development Programme report 2007.

USAID (2005). Multi-sectoral interventions in pastoralist communities: Horn of Africa, 2005.

Weldemeskel M, Issa A, Mersie L, Potgieter LND (2001). Investigation of Parasitic disease of One-humped Camel (*Camelus dromedarius*) in Eastern Ethiopia. J. Camel Pract. Res. 8:77-81.

Workneh N (2002). Socio-economic importance of camel in Ethiopia: An overview. A Paper Presented on the International Workshop on Camel Research: Formulating Research Agenda for the Next Decade, Wed, Medani, Sudan, pp. 9-12.

World Health Organization/Office International des Epizootics (2001). In: Eckert J, Gammel M, Mellin F, Powlosski ZS (Ed.), WHO/OIE Manual of Echinococcosis in Humans and Animals: A Public Health Problem of Global Concern. World Organization of Animal Health. Paris, France. pp. 1-126.

Woldemeskel M, Issa A, Mersie L, Potgieter LND (2001). Investigation of Parasitice Disease of One-humped Camel (*Camelus dromedarius*) in Eastern Ethiopia. J. Camel Pract. Res. 8:77-81.

Zeleke M, Berkeley Y (2000). Camel Herd Health and Productivity in Eastern Ethiopia selected Semi-Nomadic Households. Revue. Elev. Med. Vet. Pays Trop. 53(2): 213-217.

Phototropism of sand flies species (Diptera: Psychodidae) collected in a rural locality in Central Morocco

Khadija Lahouiti[1], Abdelhakim El Ouali Alami[2], Asmae Hmamouch[1] and Khadija Bekhti[1]

[1]Laboratory of Microbial Biotechnology, Department of Biology, Faculty of Sciences and Technology, University Sidi Mohamed Ben Abdellah, Fez, Morocco.
[2]Regional Laboratory of Epidemiological Diagnosis and Environmental Health, Entomology Unit, Hospital Al Ghassani, Fez, Morocco.

An inventory of a year was achieved using two traps techniques (light and sticky trap) in Oulad Aid (center of Morocco), a focus of cutaneous leishmaniasis (CL), to understand better the phototropism of circulating species of the sandflies dominance according to their physiological states (total number of sandflies collected are to be given). The most abundant species were respectively *Phlebotomus papatasi* (50.52%), *Phlebotomus sergenti* (24.7%), *Phlebotomus pernicuosis* (9.69%), *Phlebotomus longicuspis* (5.54%), *Sergentomyia falax* (5.22%), *Sergentomyia minuta* (3.64%), *Sergentomyia antenata* (0.4%) and *Sergentomyia dreyfussi* (0.2 %). With sticky traps, *P. papatasi* constituted 59% of collected samples against 23% for *P. sergenti*, 12% for *P. pernicuosis* and 6% for *P. longicuspis*. However with light traps, phototropism of *P. sergenti* was high and represented 56% of collected samples against 36% for *P. papatasi* and only 4% for *P. pernicuosis* and *P. longicuspis*. The trap technique influenced the dominance of females collected according to their physiological state. Non engorged females were dominant (77.73%) when sticky traps were used but gorged and gravid females were, respectively dominant (32.62 and 21.98%) when light traps were used. Statistical analysis showed that significantly more number of sand flies were obtained using sticky traps, compared with that of light traps. Sticky traps were found to be effective, as the sand flies entering the dwellings either for blood meal or mating got trapped, while light traps yielded relatively lesser number of sand flies species. Perhaps the sand flies which are influenced by the light were attracted towards the light traps. This information may be necessary for designing intervention measure and evaluation of the impact on sand fly prevalence in a CL focus area.

Key words: Sand flies, Psychodidae, phototropism, leishmania, Morocco.

INTRODUCTION

Leishmaniasis are endemic in Morocco and remain a great health problem which represents the second most common vector-borne disease in the Mediterranean region. On a large scale, worldwide, the leishmaniasis infects between 1.5 and 2 million people each year (World Health Organization, 2007). It is estimated that leishmaniasis disease has an undesirable morbid effect corresponding to 2.34 million years of life (World Health Organization, 2010). In Morocco, the vectors involved in the transmission of the disease are: Phlebotomus (larroussius) ariasi Tonnoir 1921 (Rioux et al., 1984a; Rhajaoui, 2011) and Phlebotomus (Larroussius) pernicuosis Newstead 1911 (Guernaoui et al., 2005). These species are found throughout the Rif mountains and lowlands of pre-Rif, where visceral leishmaniasis due to *Leishmania* (*Leishmania*) *infantum* is abundant (Nejjar et al., 1998); *Phlebotomus* (*larroussius*) *longicuspis* Nitzulescu 1930 is considered as the vector of sporadic cutaneous leishmaniasis due to *L.* (*Leishmania*) *infantum* in the semi-arid belt (Dereure et al., 1991) and even in the arid belt (Dereure et al., 1986).

Phlebotomus (*Paraphlebotomus*) *sergenti* Parrot 1917 was found infested by *Leishmania* (*Leishmania*) *tropica* (Ajaoud et al., 2013), responsible for the dry skin which is very recurrent in the semi-arid belt in Western slope of the Atlas from the province of Taza to Essaouira on the Atlantic coast (Guilvard et al., 1991); *Phlebotomus* (*Phlebotomus*) *papatasi* Scopoli 1786 is dominant in the zoonotic cutaneous leishmaniasis focus due to *Leishmania* (*Leishmania*) *major* localized in the arid south of the Atlas mountains that extend from the province of Jrada east to Tata province west (Rioux et al., 1986a; Boussaa, 2008; Guernaoui, 2011).

Morocco is also jointly involved in leishmaniasis control programme which is based on the detection and monitoring of vectors' density. Moulay Yacoub, located about 10 km from the city of Fez, is monitored for skin focus (Fellah et al., 2007). For implementing effective vector control strategy to suppress the leishmania infection, the investigations were carried out on the seasonal abundance and behavior of sandflies, mainly the females, depending on the nature of the technique used in the capture in the province of Moulay Yacoub, which is a known focus of cutaneous leishmaniasis (Lahouiti et al., 2013).

MATERIALS AND METHODS

Studied area

Douar Ouled Aid is located (34° 05' N, 4° 45' W, 345 m MSL) about 35 km to the N-N-east of the city of Fez. It is part of the province of Moulay Yacoub known as a new "microfocus" of human cutaneous leishmaniasis due to *L. tropica* (Fellah et al., 2007). The area is rural and is located on one of South Rif hills and has about 709 inhabitants; the majority of houses are built with clay and straw.
Regions depend on rainfed agriculture and breeding livestock which are considered the only activity for the population of the region. Vegetation is dominated by *Olea européa, Zizyphus lotus, Eucalyptus sp, Cylindropuntia imbricata* and *Opuntia sp.*

Sandfly collection

Light traps "C. D. C. Miniature Light Trap" (Mulhern, 1942; Sudia and Chamberlain, 1988) was used for sandfly collection both indoor and outdoor (Figure 1). This device is composed of a motor assuring the functioning of a small fan to maintain a continuous low intensity suction, an ampoule of 0.3 A, with a transparent plastic cylinder inside which the fan-motor-lamp is located, a cage covered with a very fine mesh cloth piece and a very flat metal cover that covers the whole device and protects the camera from rain and projections. Insects attracted by the lights will be drawn and collected in the collection bottle. Light traps are placed from 6 pm to 8 am. They are placed inside the house and stables at an average height of 2 m. Sticky traps made of sheets of white A4 paper (21 × 29.7 cm) smeared with castor oil (Rioux et al., 1967) were placed upright in the walls of human dwellings (Figure 2) and stables entries (Figure 3). The oiled sheets are held by a thread (Figure 2) or maintained by a rigid support (Figure 3).

The sticky traps placed from 6 pm to 8 am

The traps are placed in many habitats, intra-domestic (Figures 1 and 2) and peri-domestic (Figure 3) in the cow's, donkey's, horse's and sheep's stables. Period of capture lasted all night, traps were placed at the site in the previous evening (6 pm) and collected in the subsequent morning (8 am). Sand flies obtained in the collection, using light and sticky traps, were transferred in a vial containing 70% alcohol, using fine brush and again preserved in 95% absolute alcohol .The samples were then cleaned with a Marc-André solution before being mounted in Canada Baum (Floch and Abonnenc, 1957). Taxonomic identification is based on different morphological criteria described by Abonnec (1972) and Niang et al. (2000), such as cibarium and pharynx in both male and female, spermathecae in female and the armature of genitalia in male.

Statistical analyses

Results were statistically performed by the software R using the Pearson's chi-squared test comparing percentages. For the entire test, the significance level was 0.05.

RESULTS

The results shown in this paper included data from the entomological survey undertaken in Douar Ouled Aid during a year between April, 2011 and March, 2012 to show the particular phototropism of sand fly species and of females according to their physiological states. Data were subjected for statistical analysis, using a software

R and the Pearson's chi-square test for comparing percentage. For the entire test, the significance level is 0.05.

A total of 3,064 sand flies were obtained during the study, of which 87.8% (N = 2690) were obtained from stick traps and the remaining (12.2%; N = 374) from light traps (Table 1). The number of sand flies captured using sticky traps was greater than the number collected using light traps. The difference was statistically significant between the two traps (χ^2 = 1750, 606, p -value < $2.2e^{-16}$). Of the total sand flies caught, the sand flies of the genus *Phlebotomus* were predominant in the sticky trap collections (86.9%), compared to that of light trap collections (13.1%). Similarly, sand flies of the *Sergentomyia* collected using sticky traps were 95.9% and light traps were 4.1%, indicating a statistically significant difference between the efficacy of the traps in sand fly collection (κ^2 = 18.9182, p -value = 1.364×10^{-05}).

In both types of trap collection, male sand flies were predominant (Table 2) that is, 82% male in sticky trap collections and 61.7% male in light trap collections. However, *Sergentomyia fallax* species were the only females captured using both types of traps. On the contrary, *Sergentomyia antennata* males were abundant in the collections. The difference was statistically significant between the two traps (κ^2 = 2.7778, p-value = 0.09558).

Species composition

Among the total collections (N = 3064), *P. papatasi* was represented at 50.5%. The other species of sand flies obtained in order of abundance are as follows; *P. sergenti* (24.7%), *P. pernicuosis* (9.7%) and *P. longicuspisis* (5.5%). The difference was statistically significant between these species (κ^2 = 52, p-value = 2.995×10^{11}). Phlebotomine's collection was carried out twice a month during one year. The traps were placed from 6 pm to 8 am. The morpho-anatomical identification of species was based on the identification of key sand flies (Abonnec, 1972; Niang et al., 2000). The result shows the number of males (M) and females (F) for each species and for each type of trap.

By examining the distribution of species according to the type of trap, we have noticed that with sticky traps, *P. papatasi* was dominant at 59% (κ^2 = 706.2147, p-value < 2.2×10^{-16}) followed by *P. sergenti* (23%) (x^2 = 308443, p-value < 2.2×10^{-16}), *P. pernicuosis* (12%) (x^2 =216.5634, p-value < 2.2×10^{-16}) and *P. longicuspis* (6%) (x^2 = 70.2229, p-value < 2.2×10^{-16}) (Figure 4). However, with light traps (Figure 5), *P. sergenti* was collected at 56% (x^2 = 15.3725, p-value = 8.826×10^{-05}) followed by *P. papatasi* at 36% and *P. pernicuosis* and *P. longicuspis* at 4%.

The physiological states of the females are given in Table 3. The sticky trap and light traps yielded about 70.44% of non engorged females, 19.64% for engorged females and 9.90% for gravid females. The difference is statistically significant between the three physiological states of females (x^2 = 8.2439, p-value = 0.01621). However, with each trap, this percentage changes, with sticky traps at 77.73% of harvested females which are

Table 1. Distribution and abundance of sand flies inventoried for each trap.

Species	Sticky traps	Llight traps
P.papatasi	1416	132
P.sergenti	553	204
P.pernicuosis	284	13
P.longicuspis	157	13
S. falax	157	3
S.minuta	106	6
S.antenata	14	1
S.dreyfussi	3	2
Total	2690	374

Table 2. Distribution and abundance of sand flies inventoried for each trap according to their sex.

Species	Sticky traps	Llight traps
P. papatasi	1208 M 208 F	72 M 60 F
P. sergenti	483 M 70 F	130 M 74 F
P. pernicuosis	266 M 18 F	12 M 1 F
P. longicuspis	131 M 26 F	10 M 3 F
S. falax	0 M 157 F	0 M 3 F
S. minuta	103M 3F	6 M 0 F
S. antenata	14 M 0 F	1 M 0 F
S. dreyfussi	2 M 1 F	0 M 1 F
Total	2690	373

Table 3. Abundance of female *Phlebotomus* species inventoried in each trap according to their physiological state.

Species	Sticky traps	Light traps
	125 FN	34 FN
P. papatasi	71 GS	18 GS
	12 G	8 G
	57 FN	24FN
P. sergenti	2 GS	27 GS
	11 G	20 G
	15 FN	1 FN
P. pernicuosis	0 GS	0 GS
	3 G	1 G
	23 FN	1 FN
P. longicuspis	3 GS	1 GS
	2 G	1 G
	153 FN	3 FN
S. falax	1 GS	0 GS
	3 G	0 G
	3 FN	0 FN
S. minuta	0 GS	0 GS
	0 G	0 G
	0 FN	0 FN
S. antenata	0 GS	0 GS
	0 G	0 G
	1 FN	1 FN
S. dreyfussi	0 GS	0 FGS
	0 G	1 G

Non engorged (FN), engorged females without (GS), gravid (G), normal females (NF).

non engorged, 15.87% of the females which are engorged and 6.39% which are pregnant. With light traps, 45.39% of collected females are non engorged, 32.62% are engorged and 21.98% are pregnant. The difference between the three physiological states females with these types of traps is statistically significant ($x^2 = 90.0431$, p-value $< 2.2 \times 10^{-16}$).

P. sergenti and *P. papatasi* were the two common species in this study; we have investigated the female's distribution by types of traps according to their physiological states. The results (Figure 6) shows that 6% of attracted *P. papatasi* specimens are pregnant females, 34% are females engorged with blood and 60% are neither pregnant females nor engorged. These percentages were almost the same with light traps (Figure 7). However *P. sergenti* females had a clear trap preference depending on their physiological conditions; with sticky traps (Figure 8), 81% of females were normal, 16% of females were engorged and 3% were pregnant but with light traps (Figure 9), 38% of females were gorged with blood, 28% were pregnant and 34% were normal females.

Phlebotomine's collection was carried out twice a month during one year. The traps were placed from 18 pm to 8 am. The morpho-anatomical identification of species was based on the sand flies key identification (Abonnec, 1972; Niang et al., 2000). The result shows the distribution of *P. papatasi* females collected by light trap according to their physiological states: non engorged (FN), engorged females without (GS) and gravid (G).

Phlebotomine's collection was carried out twice a month during one year. The traps were placed from 18 pm to 8 am.The morpho anatomical identification of species was based on the sandflies key identification (Abonnec, 1972 and Niang et al., 2000). The result shows the distribution of *P papatasi* females collected by trap adhesives according to their physiological states: normal females (NF), engorged females without (GS) and gravid (G).

Phlebotomine's collection was carried out twice a month during one year. The traps were placed from 18 pm to 8 am. The morpho-anatomical identification of species is based on the identification of key sand flies (Abonnec, 1972; Niang et al., 2000). The result shows the distribution of females *P. sergenti* harvested by trap adhesives according to their physiological states: normal females (NF), engorged females without (GS) and gravid (G).

Phlebotomine's collection was carried out twice a month during one year. The traps were placed from 18 pm to 8 am. The morpho-anatomical identification of species was based on the sand flies key identification (Abonnec, 1972; Niang et al., 2000). The result shows the distribution of female *P. sergenti* collected by light trap according to their physiological states: non engorged (NF), engorged females without (GS) and gravid (G).

DISCUSSION

The results reported in this paper shows for the first time the distribution of sand fly species and the physiological state of females according to the techniques of traps used for the collection in this semi-arid rural area of Douar Oulad Aid characterized as an anthropic niche of *L. tropica* (Rhajaoui et al., 2004; Fellah et al., 2007). The density of sand flies captured by sticky traps (87.8%) is higher than the density of sand flies collected by the light

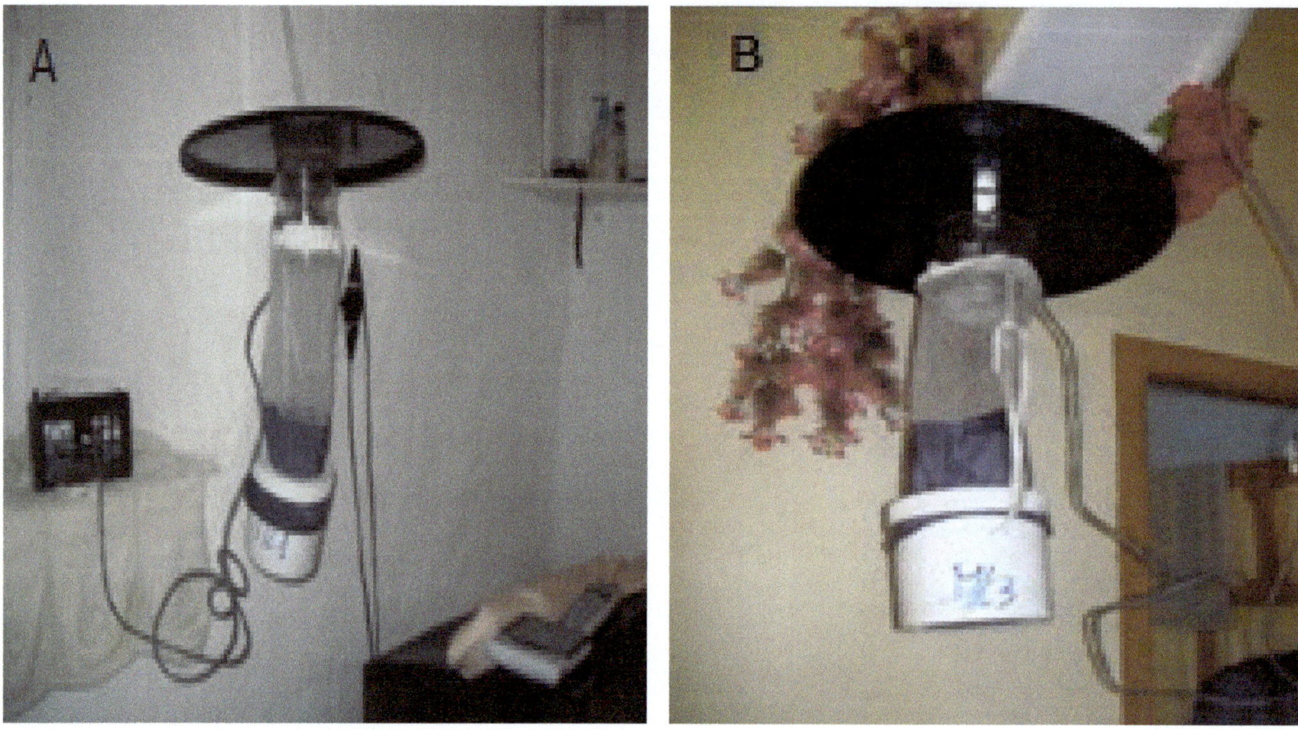

Figure 1. Light traps (A and B) inside houses.

Figure 2. Sticky traps inside the houses.

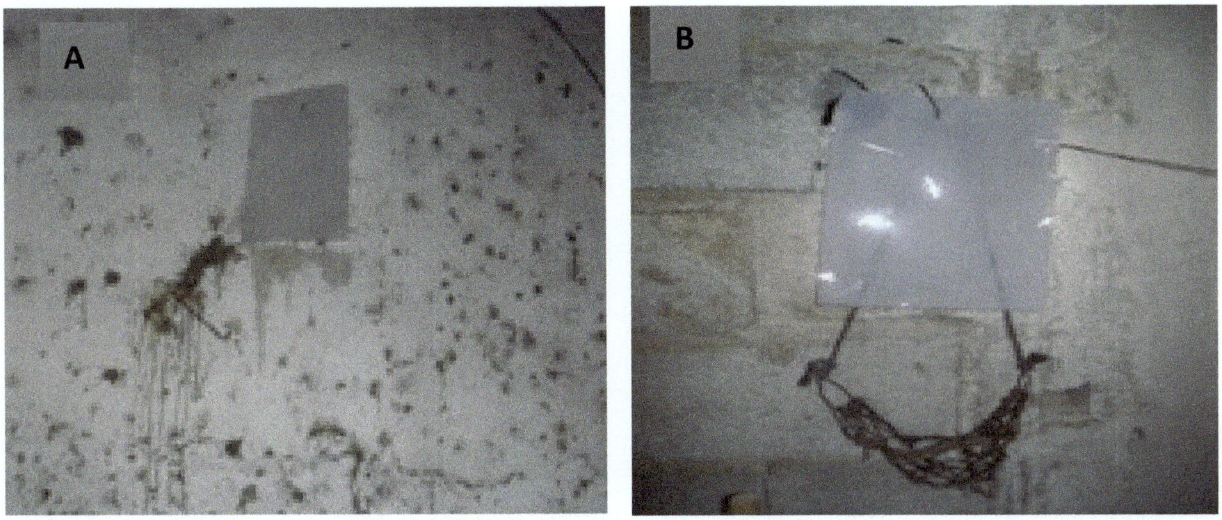

Figure 3. Sticky traps (A and B) inside the stables.

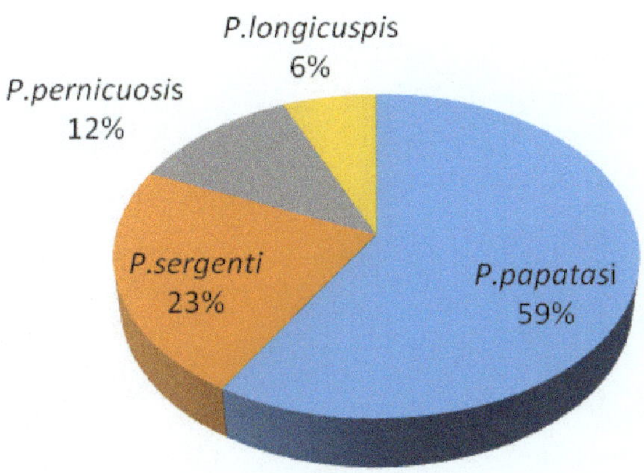

Figure 4. Abundance of species of *Phlebotomus* genus collected by sticky traps (ST).

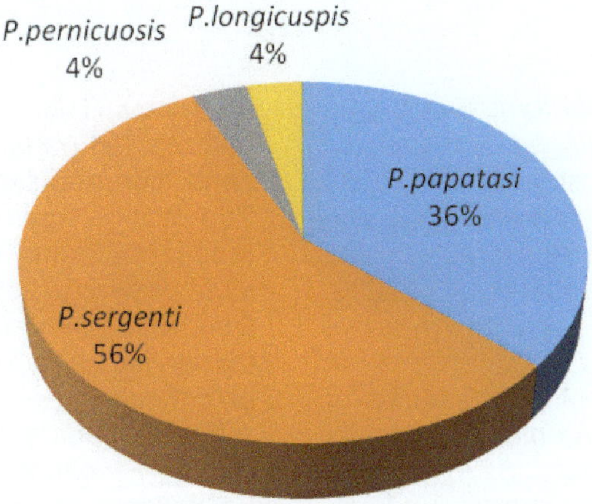

Figure 5. Abundance of species of *Phlebotomus* genus collected by light traps (LT).

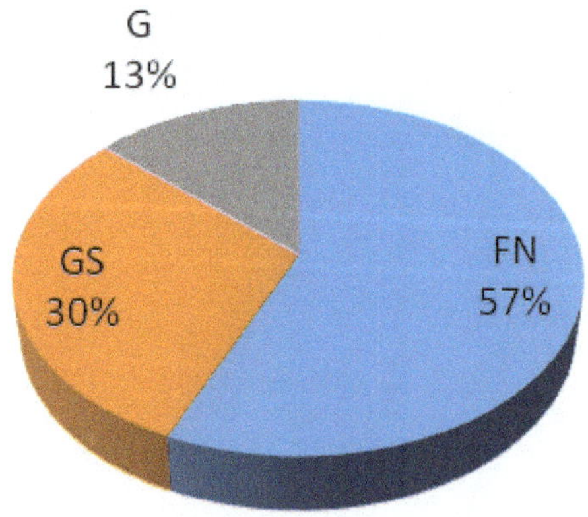

Figure 6. Abundance of female *P. papatasi* collected by light traps (LT) according to their physiological states FN (normal females), GS (engorged females) and G (gravid females).

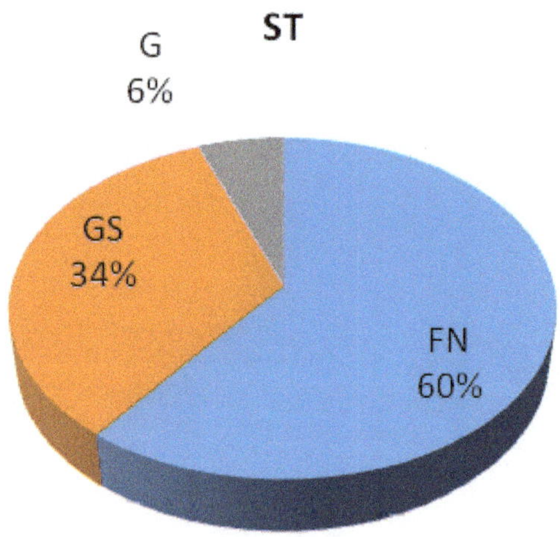

Figure 7. Abundance of female *P. papatasi* by sticky traps (ST) according to their physiological states FN (normal females), GS (females engorged), G (gravid females).

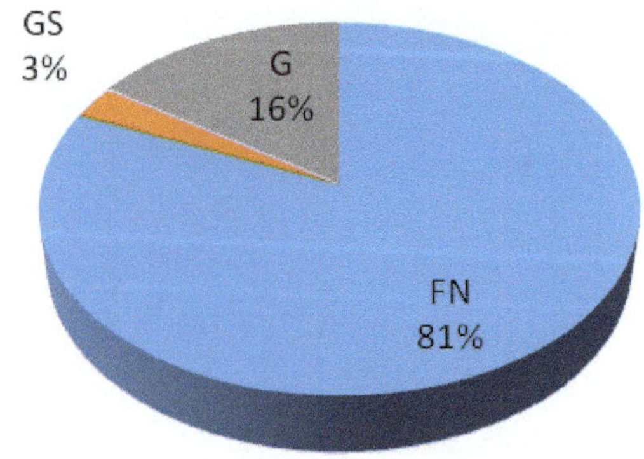

Figure 8. *P.sergenti* females' abundance by sticky traps (ST) according to their physiological states FN (normal females), GS (females engorged) and G (pregnant females).

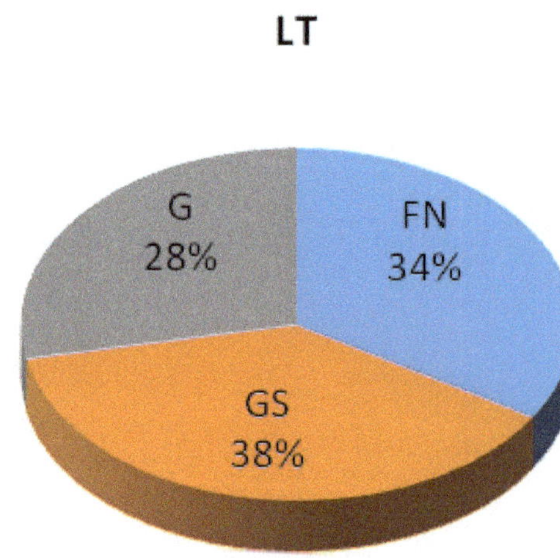

Figure 9. *P. sergenti* females abundance by light traps (LT) according to their physiological states FN (non engorged), GS (engorged females), G (pregnant females).

traps (12.2%). With both traps, *Phlebotomus spp* was collected at 89.46% while *Sergentomyia* was collected at 10.54%. *Phlebotomus spp* is the most abundant, they have been also reported by several authors (Bailly-Choumara et al., 1971; Rioux et al., 1984; Dereure et al., 1991; Raspail et al., 2002; Guernaoui et al., 2000; Boussaa, 2008; Faraj et al., 2013) as the only vectors implicated in the transmission of leishmaniasis in Morocco. Sticky traps captured more of both species of the *Phlebotomus* (86.9%) and *Sergentomyia* genus (94.5%). *Sergentomyia* species (*S. falax, S. minuta, S. antenata, S. dreyfussi*) phototropisms seems undeveloped (5.5%) since its species are not attracted by light light traps. These results confirm the effectiveness of the use of oiled paper as a good sampling and monitoring of sand flies populations technique (Boukraa et al., 2010). The sticky traps have the advantages of not being repellents, being very viscous; they are also soluble in alcohol which facilitates insects' retrieval. However sand flies collected using sticky traps are damaged and the taxonomic study by morpho-anatomical identification can be difficult.

For the entire collection and within the *Phlebotomus* genus, *P. papatasi, P. sergenti, P. pernicuosis* and *P. longicuspis* are the circulating species in the heart of central Morocco, with *P. papatasi* dominating at 56.38%, followed by *P. sergenti* dominating at 27.3%. This result confirms the work of Lahouiti (2013), who reported the dominance of *P. papatasi* in this province and those of Rioux et al. (1984) which reported the widespread that the dominance of a species of sand fly in an outbreak of leishmaniasis depends on used trap technique. It also shows that phototropism of *P. sergenti* species is higher compared to *P. papatasi, P. pernicuosis* and *P. longicuspis* that are captured only at 4%, with light traps contrary to 12% when sticky traps is used. Blood sucking females are the leishmaniasis vectors, the relationship between sand fly-leishmania is close and the study of the distribution by type of trap of females according to their physiological state may be interesting to the extent in which we want to study the vector's infestation or the sand flies trophic power. Our study showed that the sample is composed of 70.44% of normal females (neither gravid nor engorged), with 19.64% of engorged females and 9.90% of pregnant females. However, the abundance of females according to their physiological state in each of the types of traps showed that sticky traps attracts more normal females(77.73%) than engorged females (15.87%) and pregnant females (6.39 %), while light traps attracted 32.62% of engorged females and 21.98% of pregnant females. This original result shows the effectiveness of the use of the technique of light traps in the collection in order to have more females engorged with blood which are necessary to perform more specific studies. The same study is applied to the two most abundant species in this semi-arid outbreak which are *P. papatasi* and *P. sergenti*. The physiological state of *P. papatasi* females is not affected by light, the percentage of distribution of normalfemales, females engorged and gravid females is, respectively 57 to 60, 30 to 34 and 6 to 13% for sticky and light traps which is probably explainable by the lack of infectious activity of this species in this location whose vector is *P. segenti*. The latter species is confirmed as being the only vector of *L. tropica* MON 101 by the works of Rhajaoui (2004) and Fellah et al. (2007) in this area and demonstrated infested by *L. tropica* by Ajouad et al. (2013). We have shown dominance and a high phototropism for the species of *P. segenti*; females also have a net density difference according to the used traps and to their physiological states. Indeed, females engorged and gravid females are attracted to light traps, respectively 38% (GS) and 28% (G), while the sticky traps set at 81% that of normal females and only 3% are females engorged with blood which likely confirms the infectious activity of *P. sergenti* in this focus.

Conclusion

We have shown that the calculated dominance in an outbreak of leishmaniasis depends on the trap chosen. We have also shown that in the locality of Oulad Aid (central Morocco), *P. papatasi* may be dominant if we use sticky traps and *P. sergenti* may be dominant if we use light traps. We have also shown that the sticky traps are effective and therefore, montitoring the population during surveillance sticky traps could be used. While the light traps though yielded a relative lower number of sand flies species, these traps are used in identification, which would also be used to know the species abundance and distribution.

Conflict of Interest

The authors declare that they have no conflict of interests.

REFERENCES

Abonnec E (1972). Les phlébotomes de la région éthiopienne (Diptera, Psychodidae). Paris: ORSTOM, p 289. Multigr. (Mémoires ORSTOM; 55).

Ajaoud M, Es-sette N, Hamdi S, El-Idrissi AL, Riyad M, Lemrani M (2013). Detection and molecular typing of *Leishmania tropica* from *Phlebotomus sergenti* and lesions of cutaneous leishmaniasis in an emerging focus of Morocco. Parasit. Vectors 6:217.

Sudia WD, Chamberlain RW (1988). Battery-operated light trap, an improved model. J. Am. Mosq. Control Assoc. 4(4):536-538.

Bailly-choumara H, Abonnenc E, Pastre J (1971). Contribution à l'étude des phlébotomes du Maroc (Diptera, Psychodidae): données faunistiques et écologiques. Cahiers ORSTOM. Med. Parasitol. Entomol. Series 9(4):431-460.

Boukraa S, Boubidi SC, Zimmer JY, Francis F, Haubruge E, Alibenali-Lounaci Z, Doumandji S (2010). Surveillance des populations de phlébotomes (Diptera: Psychodidae), vecteurs des agents responsables des leishmanioses dans la région du M'Zab-Ghardaïa (Algérie). Entomol. Faunistique 63(3):97-101.

Boussaa S (2008). Epidemiology of leishmanisis in Marrakesh area, Morocco. Effect of urbanization on sandfly distribution and molecular characterisation of there populations. http://www.biomedsearch.com/sci/Epidemiology-leishmanisis-in-Marrakesh-area/0000493621.html

Dereure J, Rioux JA, Gallego M, Périères J, Pratlong F, Mahjour J, Saddiki H (1991). Leishmania tropica in Morocco: infection in dogs. Trans. R. Soc. Trop. Med. Hyg. 85(5):595

Faraj C, Adlaoui E, Ouahabi S, El Kohli M, El Rhazi M, Lakraa L, Ameur B (2013). Distribution and bionomic of sand flies in five ecologically different cutaneous leishmaniasis foci in Morocco. ISRN Epidemiol. http://dx.doi.org/10.5402/2013/145031.

Fellah H, Rhajaoui M, Ouahabi S, Belghiti D, Lyagoubi M (2007). Occurrence of Human Cutaneous Leishmaniasis in Zouagha My Yacoub Province (Morocco). Int. J. Agric. Biol. 9(1):197-198.

Floch H, Abonnenc E (1957). Diptères phlébotomes de la Guyane et des antilles Françaises. Faune de l'Union Française 14:1-207.

Guernaoui S, Pesson B, Boumezzough A, PIchon G (2005). Distribution of.phlebotomine sandflies, of the subgenus *Larroussius*, in Morocco. Med. Vet. Entomol. 19:111-115.

Guilvard E, Rioux JA, Gallego M, Pratlong F, Mahjour J, Martinez-Ortega E, Dereure J, Saddiki A, Martini A (1991). *Leishmaniatropica* au Maroc. III - Rôle vecteur de *Phlebotomus sergenti*. A propos de 89 isolats. Annal. Parasitol. Hum. Comp. 66:96-99.

Lahouiti K, El Oualilalami A, Maniar S, Bekhti K (2013). Seasonal fluctuations of phlebotomines sand fly populations (Diptera: Psychodidae) in the Moulay Yacoub province, centre Morocco: Effect of ecological factors. Afr. J. Environ. Sci. Technol. 7:1028-1036.

Leger N, Depaquit J (2001). Les phlébotomes et leur rôle dans la transmission des leishmanioses. La Revue Française des Laboratoires 338:41-48.

Mulhern, T.D. 1942. New Jersy mechanical trap for mosquito surveys. Circular 421, 1-8 New Jersy Agricultural Experrimental Station, Rutgers University, New Brunswick, New Jersy

Nejjar R, Lemrani M, Malki A, Amarouch H, Benslimane A (1998). Canine leishmaniasis due to *Leishmania infantum* MON-1 in Northern Morocco. Parasite 5:325-30.

Niang AA, Geoffroy B, Anget G, Trouillet L, Killick-kendrick R, Hervy JP, Brunhes J (2000). Les phlébotomes d'Afrique de l'ouest. Logiciel d'identification et d'enseignement l'I.R.D., Montpellier.

Rhajaoui M (2011). Les leishmanioses humaines au Maroc : une diversité nosogéographique; Human leishmaniases in Morocco: A nosogeographical diversity. Pathologie Biologie 59:226-229

Rhajaoui M, Fellah H, Pratlong F, Dedet FJ, Lyagoubi M (2004). Leishmaniasis due to *Leishmania tropica* MON-102 in a new Moroccan focus. Trans. R. Soc. Trop. Med. Hyg. 98(5):299-301.

Rioux JA, Golvan YJ, Croset H, Houin R, Juminer B, Bain O, Tour S (1967). Ecologie des leishmanioses dans le Sud de France. Echantillonnage. Ethologie. Ann. Parasitol. Hum. Comp. 42:561-603.

Rioux JA, Guilvard E, Dereure J, Lanotte G, Denial M, Pratlong F, Serres E, Belmonte A (1986a). . In : Rioux JA (ed.), Leishmania. Infestation naturelle de Phlebotomuspapatasi (Scopoli, 1786) par Leishmania major MON-25. A propos de 28 souches isolées dans un foyer du Sud marocain. In Leishmania. Taxonomie et Phylogenèse. Applications Eco-épidémiologiques. Coll. Int. CNRS/INSERM/OMS. 471-480.

Rioux JA, Jarry DM, Lanotte G, Maazoun R, Killick-Kendrick R (1984a). Ecologie des leishmanioses dans le sud de la France. 18. Identification enzymatique de Leishmania infantum Nicolle, 1908, isolé de Phlebotomus ariasi Tonnoir, 1921 spontanément infesté en Cévennes. Annal. Parasitol. Hum. Comp. 59:331-333.

Rioux J A, Rispail P, Lanotte G, Lepart J (1984). Relations Phlébotomes bioclimats en écologie des leishmanioses. Corollaires épidémiologiques. L'exemple du Maroc. Bull. Soc. Bot. Fr. 131:549-557.

Rispai IP, Dereure J, Jarry D (2002). Risk Zones of Human Leishmaniases in the Western Mediterranean Basin.Correlations between Vector Sand Flies, bioclimatology and Phytosociology. Mem. Inst. Oswaldo Cruz. 97(4):477-83.

Velez ID, Pratlong F, Denial M, Lardi M, Moreno G, Serres E, Lanotte G, Rioux JA (1986b). La leishmaniose viscérale autochtone au Maroc méridional. Présence de Leishmania infantum MON-1 chez le Chien en zone présaharienne. In Leishmania. Taxonomie et Phylogenèse. Applications Eco-épidémiologiques, International colloquium CNRS/INSERM/OMS, July 1984. Institut Méditerranéen d'Etudes Epidémiologiques et Ecologiques, Montpellier, France. 421-425.

30

Altered total cholesterol and triglyceride levels during the course of *Plasmodium falciparum* infection in children

Bouyou-Akotet, M. K., Mawili Mboumba, D.P., Guiyedi, V., Pemba Mihindou, M. and Maryvonne Kombila, M.

Department of Parasitology-Mycology, Faculty of Medicine, Université des Sciences de la Santé, Libreville, Gabon.

Changes of lipid parameter concentrations are observed in patients suffering from malaria. However, there are few data on their evolution during the course of the infection and their relationship with the infection parameters. The levels of triglycerides (TG) and total cholesterol (TC) were measured during and after uncomplicated *Plasmodium falciparum* malaria in 122 children. The relationship with the antimalarial drug treatment, the parasite density and haemoglobin levels was also assessed. Mean TG levels were high before the treatment (1.9 ± 1.2 mmol/L) and at day 3 post-treatment (2.3 ± 1.3 mmol/L), then subsequently decreased to reach normal level from day 7. A negative correlation was found between haemoglobin (Hb) levels and TG (rho = -0.47; *p* < 0.01); previous high parasitaemia at inclusion was associated with a subsequent increase of TG levels after treatment administration. The majority of patients with hypertriglyceridemia were anaemic at day 3 (n = 20/20) and at day 7 (n = 18/19). Only 5 (4.1%) patients had hypertriglyceridemia at day 28. A trend toward a negative correlation between TC levels and parasite density was observed (rho = -0.18; *p* = 0.05). Mean TC concentration was significantly lower at day 0 when all the patients were parasitaemic (3.0 ± 1.0 mmol/L) compared to day 3 (3.7 ± 1.2 mmol/L) (*p* = 0.01). Mean CT was low during the first week and at day 28, 71 (58.2%) children still had a subnormal CT levels. No relationship was found with the type of antimalarial drug. Transient hypertriglyceridemia and hypocholesterolemia are observed during the course of *P. falciparum* infection. High TG levels seem to be related to the malaria related haemolysis.

Key words: Malaria, cholesterol, triglycerides, antimalarial, Gabon.

INTRODUCTION

Lipid parameter changes have been reported during the course of malaria infection. Although the mechanisms involved are not fully understood, some hypotheses have been raised from *in vivo* and *in vitro* studies (Cuisinier et al., 1990; Visser et al., 2013). Some fatty acids are produced and required for the parasites biosynthetic pathways. Extracellular lipids are also implicated in *Plasmodium (P.) in vitro* growth (Vial et al., 2003; Elford et al., 1995). Cholesterol and triglycerides are the most common studied lipid parameters during malaria infection. Concerning cholesterol, it is now admitted that cell membrane cholesterol is implicated in the immune evasion and the pathogenesis of malaria due to *P. falciparum* (Sein and Aikawa, 1998; Bansal et al., 2005). Both parameters have been associated with the intra-erythrocytic parasite development and the maintenance of trophozoïtes stage (Sherman, 2003; Vial et al., 1999). It is also suggested that cholesterol metabolism involved in the parasite growth could be used as target for new treatments and cholesterol and triglyceride serum level modifications during malaria used as diagnostic tools (Kitt et al., 1992; Badiaga et al., 2002; Mitamura et al., 2003).

Controversial data are reported concerning total cholesterol and triglyceride levels during plasmodial infection; hypocholesterolemia and hypercholesterolemia are reported during the acute phase of the infection; hypertriglyceridemia is often described during the first two weeks of the infection (Parola et al., 2004; Ayoola et al., 2012). However, very few studies have examined the *in vivo* relationships between the levels of these lipids and the factors implicated in the pathophysiology of clinical malaria. Liver function disorders and anaemia caused by red blood cell destruction are the main markers of uncomplicated malaria beside fever. They are characterized by a decrease in hemoglobin levels and an increase in liver enzymes. Exogenous lipids used by the parasite could be derived from either the red cell membrane that contains phospholipids or liver where some are synthesized (Haldar et al., 2002). Furthermore, antimalarial treatment has been shown to be responsible for changes in triglycerides and cholesterol levels in patients with malaria (Davis et al., 1995; Bouyou-Akotet et al., 2010). Thus, the present study was designed to analyze the relationship between parasitaemia, total serum cholesterol, triglyceride and hemoglobin levels, in *P. falciparum*-infected patients treated with different antimalarial drugs.

MATERIALS AND METHODS

Study site

This was an analytical study performed at the Department of Parasitology of the Faculty of Medicine of Libreville, the capital city of Gabon. This country is considered hyper- to meso-endemic with a malaria prevalence ranging from 11 to 39.5% (Mawili-Mboumba et al., 2013).

Study population

Data were collected from children aged six months to 10 years suffering from uncomplicated malaria who took part in two clinical trials performed between 2005 and 2006 as described elsewhere (Bouyou-Akotet et al., 2010; Nsimba et al., 2008). These patients were treated either with amodiaquine (AQ), sulfadoxine-pyrimethamine (SP) or artesunate-mefloquine (AM) according to their age and/or weight following World Health Organization (WHO) guidelines. They were included with the following criteria: presence of *P. falciparum* mono-infection, fever or history of fever during the last 24 h, absence of signs of severe malaria, malnutrition or other severe underlying infectious or chronic disease.

Procedures

Eligible children were hospitalized during three days and followed up until day 28 (D28). At inclusion, clinical and socio-demographical data (age, sex and body mass index, BMI) were recorded. Biological data which included malaria diagnosis results, hemoglobin (Hb), triglycerides (TG) and total cholesterol (TC) measurements performed at day 0 (D0) prior to the administration of the antimalarial, day 3 (D3), day 7 (D7) and day 28 (D28) were used for the analysis. According to the design of the clinical trials, Hb was not measured at D3 for AQ and SP treated patients, at D7 for AM patients as well as TC at D7 for AM patients

P. falciparum infection was detected by microscopy using the Lambaréné method (Planche et al., 2001). Briefly, thick smears were prepared as follows: 10 µl of blood were layed on a 10 by 18 mm area of a microscope slide, then dried and stained with 10% giemsa. The parasitaemia was expressed as number of parasites per microliter of blood (p/µl). Smears were read by two experienced technicians using a light microscope (× 100 oil immersion lenses). They were considered negative if no parasite was seen after the examination of at least 100 oil immersion fields in a thick blood smear.

Hb measurements were performed using a Coulter counter (SKTS, Coulter Corporation). Total cholesterol and triglyceride concentrations were determined by using standard enzymatic procedures on an automatic clinical chemistry analyser (Visual, Biomerieux®).

Definitions

Anaemia was defined as an Hb concentration below 11 g/dl, hypocholesterolemia or low total cholesterol when serum cholesterol level was below 4.0 mmol/L and hypertriglyceridemia when serum triglyceride level was higher than 1.88 mmol/L.

Ethical considerations

The study was approved by the public health ministry of Gabon. The clinical trials performed for AQ, SP and AM evaluation were all approved by the National Ethics Committee. None of the children required additional blood collection for this study and data were analyzed for the patients with available blood smears, Hb and biochemistry results. The children's parents or guardians were informed about the study, and their oral consent was required prior to data use.

Data analysis

Demographic and laboratory data of patients were entered into Epi-info version 6.0 (February 9, 2005 CDC Atlanta) database and

Table 1. Distribution of biological parameters according to the day of follow up and the treatment.

Parameter	D0	D3	D7	D28	p
Parasitaemic	122 (100.0)	38 (31.1)	0 (0.0)	0 (0.0)	<0.01
Anaemic [n (%)]	109 (89.3)	34* (100.0)	107(87.1)	69 (56.6)	<0.01
Triglycerides					
Hypertriglycéridemia [n (%)]	48 (39.3)	70 (57.4)	19 (21.6)	5 (4.1)	<0.01
Mean TG concentration, mmol/L (±SD)					
Study population	1.9 (±1.2)	2.3 (±1.3)	1.4 (±0.5)	1.0 (±1)	<0.01
-AQ	1.8 (±0.7)	1.9 (±0.8)	1.3(±0.6)	0.9 (±0.4)	0.03
-AM	1.8 (±0.9)	2.7 (±1.8)	1.3 (±0.5)	0.8 (±0.7)	<0.01
-SP	2.0 (±1.6)	2.4 (±1.1)	1.5 (±0.6)	1.3 (±0.6)	<0.01
Total cholesterol					
Hypocholesterolemia [n (%)]	109 (86.9)	83 (68.0)	(51.0)	71 (58.2)	<0.01
Mean TC concentration, mmol/L (±SD)					
Study population	3.0 (±1.0)	3.7 (±1.2)	4.0 (±1.1)	3.9 (±1.9)	0.03
-AQ	3.7 (±0.6)	3.8 (±1.3)	3.9(±1.1)	3.5 (±0.7)	NS
-AM	2.6 (±0.6)	3.5 (±0.9)	**	4.3 (±1.0)	<0.01
-SP	3.4 (±1.1)	3.9 (±1.2)	4.1 (±1.1)	4.0 (±0.9)	<0.01

*Hb levels available only for AM patients, **Biological tests not performed.

exported into Stata 9.2 (Stata Corporation, College Station, TX USA) for analysis. Median parasitaemia was estimated only in the group of patients with positive blood smears. Variables were summarized as frequencies, means ± standard deviation and median (25th to 75th percentiles) for the parasite density. Differences between groups were assessed using chi-squared or Fisher's exact tests for proportions, Student's t-test and analysis of variance (ANOVA) or Kruskal-Wallis test or Wilcoxon as appropriate for continuous variables. The Spearman rank test was used to identify the correlations. A *p*-value less than 0.05 was considered significant.

RESULTS

Total cholesterol and triglyceride levels were measured in sera from 122 children followed up during 28 days post-treatment. Their mean age and body mass index (BMI) were 35 ± 15 months, 15 ± 2 kg/cm^2 and 52 (43%) were girls. According to the treatment, 40 (33%) received AQ, 48 (39%) SP and 34 (28%) AM.

Evolution of parasite density and Hb levels

At D0, the parasite density of AQ and SP patients was 12500 (3810 to 25040) parasites/µl (p/µl); it was 50225 (17150 to 171500) p/µl in the AM group. At D3, 1 (0.3%) AM, 14 (35%) AQ and 23 (5%) SP treated children were still parasitaemic ($p < 0.01$). From D7, there was no detected microscopic infection in the whole study population. The mean Hb level remained below 12 g/dl during the follow-up. It significantly increased in the whole study population, varying from 8.6 ± 1 g/dl at D0 to 10.8 ± 0.9 g/d at D28 ($p < 0.01$). At D3, the mean Hb was 8.6 ± 1.0 g/dl in the group of patients treated with AM; at D7, it was 9.3 ± 1.4 g/dl and 9.1 ± 0.9 g/dl in the AQ and SP groups, respectively.

Relationship between of lipid levels and the type of antimalarial drug

Both lipid parameters, TC and TG, were modified at inclusion and their profile was subnormal or normal from D7 within each group of treatment (Table 1). Mean TG significantly decreased between D0 and D3 and reached a normal level from D7. TG concentration was the highest at D3 in AM patients (Table 1). Mean CT was the lowest at D0 when all the patients were parasitaemic, it significantly increased until D28 (Table 1). Overall, no change in TC and TG was related to any specific antimalarial.

P. falciparum infection and lipid parameters

Patients with malaria had a higher mean level of TC at D3 compared to D0 ($p < 0.01$) (Table 2). Indeed, a non-

Table 2. Lipid levels (mmol/L) according to the presence of malaria and anaemia.

Parameter	*P. falciparum* infection			Anaemia		
	Yes	No	P	Yes	No	P
DAY 0, n	**122**	**0**		**109**	**13**	
Total cholesterol, mean (±SD)	3.0 (1.0)	NA	-	2.3 (1.0)	3.3 (0.1)	0.15
Hypocholesterolemia,n(%)	106 (86.9)	NA	-	96 (88.1)	10 (76.9)	0.23
Triglycerides, mean (±SD)	1.9 (1.2)	NA	-	2.0 (1.2)	1.3 (0.6)	0.01
Hypertriglyceridemia, n(%)	48 (39.3)	NA	-	45 (41.3)	3 (23.1)	0.23
DAY 3, n	**38**	**84**		**34***	**0**	
Total cholesterol, mean (±SD)	3.8 (1.1)	3.7 (1.3)	0.4	3.5 (0.9)	NA	-
Hypocholesterolemia,n(%)	22 (72.6)	61 (57.9)	0.3	26 (76.5)	NA	-
Triglycerides, mean (±SD)	2.0 (1.2)	1.8 (1.3)	0.3	2.7 (1.8)	NA	-
Hypertriglyceridemia, n(%)	20 (52.4)	50 (59.5)	0.4	20 (58.8)	NA	-
DAY 7, n	**0**	**88****		**69**	**19**	
Total cholesterol, mean (±SD)	NA	4.0 (1.1)	-	4.0 (1.1)	4.3 (1.4)	0.6
Hypocholesterolemia,n(%)	NA	45 (51.1)	-	41 (59.4)	4 (21.0)	0.01
Triglycerides, mean (±SD)	NA	1.4 (0.6)	-	1.5 (1.2)	1.3 (0.8)	0.3
Hypertriglyceridemia, n(%)	NA	19 (21.6)	-	18 (26.1)	1 (0.05)	0.03
DAY 28,n	**0**	**122**		**69**	**53**	
Total cholesterol, mean (±SD)	NA	3.9 (1.9)	-	3.8 (1.0)	4.0 (0.9)	0.3
Hypocholesterolemia,n(%)	NA	71 (58.2)	-	43 (62.3)	28 (32.8)	0.01
Triglycerides, mean (±SD)	NA	1.0 (1.0)	-	1.2 (1.3)	0.9 (0.4)	0.04
Hypertriglyceridemia, n(%)	NA	5 (4.1)	-	45 (41.3)	3 (23.1)	0.04

significant negative correlation was found between the parasite density and TC level at D0 (Rho = -0.18; p = 0.05), whereas it was Rho = -0.02 (p = 0.8) at D3 when the parasitaemia was the lowest, 140 (80 to 400) p/µl versus 16250 (4500 to 37500) p/µl at D0. Moreover, hypocholesterolemia decreased from 86.9% at day 0 to 68.0% at day 3 (p < 0.01) (Table 1). Patients treated with AM who had the highest median parasite density (50225 versus 12500 p/µl) also had the lowest mean CT before treatment administration (Table 1). No relationship was found between mean TG levels and *P. falciparum* parasite density at inclusion: there was no correlation (Rho = 0.1; p = 0.3) and no difference in parasite density between the group of patients with hypertriglyceridemia (19250 [4100 to 67725] p/µl) and those with normal TG (13750 [4000 to 34200] (p = 0.5). However, a trend toward a positive correlation between parasite density and TG concentration existed at D3 (Rho = +0.27; p = 0.07). D0 parasitaemia and TG at D3 in the same patients were strongly correlated (Rho = +0.34; p < 0.01). Likewise, at D3, the mean TG concentration was the highest among the AM patients who had the highest mean parasitaemia at D0 before the treatment administration (Table 1).

Anaemia and lipid parameters

There was a significant negative correlation between HB level and TG (Rho = -0.41; p < 0.01) at D0, it varied from Rho = -0.22 (p = 0.1) at D3, Rho = -0.30 at D7 to Rho = -0.14 (p = 0.1) at D28 when none of the children were parasitaemic. Furthermore, from D0 to D7, a drop of 0.7 to 1.1 g/dl of the mean Hb levels was observed in patients with hypertriglyceridemia (8.2 to 8.6 g/dl), while it was less than 0.6 g/dl in those with a normal TG level (8.9 to 9.5 g/dl) (p < 0.04) (Figure 1). Among the 34 and the 88 patients seen at D3 and D7, respectively, the near total of those with hypertriglyceridemia (n = 20/20 at D3 and n = 18/19 at D7) were anaemic (Table 2). CT levels were not significantly correlated to Hb concentration with a Rho varying from -0.02 to -0.13 (p > 0.1). Likewise, mean Hb remained comparable throughout the follow-up in case of hypocholesterolemia (8.9 to 10.8 g/dl) or normal cholesterol (9.0 to 10.9 g/dl) (Figure 1b), although the frequency of hypocholesterolemia remained important

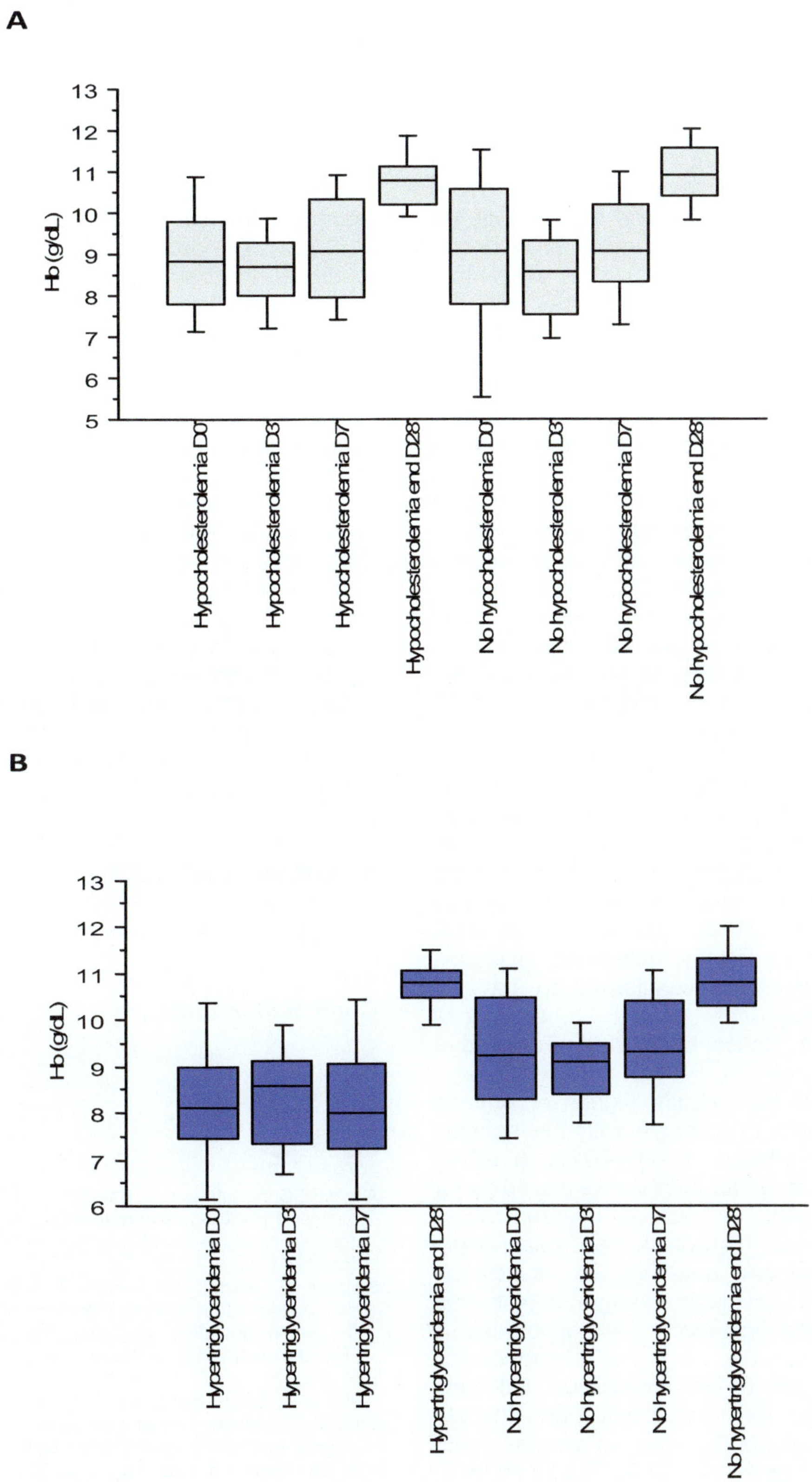

Figure 1. Box plots of hemoglobin levels during the follow-up according to the presence or the absence of hycholesterol emia (A) and hypertriglyceridemia (B).

after the disappearance of parasites (Table 2).

DISCUSSION

The present study confirmed that *P. falciparum* infection is characterized by hypertriglyceridemia and hypocholesterolemia. The decline of TC and the raise of TG observed have been shown to be more pronounced during malaria episode (Visser et al., 2013). In agreement with previous reports, these modifications that are accentuated during the parasitaemic phase of the disease are transient and present in both complicated and uncomplicated malaria compared to other febrile diseases (Batista et al., 1996; Badiaga et al., 2002). Few studies have analyzed the impact of the treatment on both lipid parameter profiles. Their evolution is not specific to any antimalarial drug, although the normalization of mean total cholesterol and TG levels were concomitant to the decrease of the parasite density and prevalence observed from day 7 post-treatment. TC levels seem to slowly normalize as highlighted by the borderline mean level (3.9 mmol/L) determined at day 28. Other authors reported a period of two weeks to six months after antimalarial treatment for TC normalization, whereas TG usually normalize within two weeks (Kim et al., 2008).

The study population was composed of patients who participated to three clinical trials, thus without any significant co-infection. One can assume that the lipids changes observed are specific to malaria, although they exist in other diseases. Nevertheless, other risk factors for hypocholesterolemia and hypertriglyceridemia such as ethnicity, socioeconomic status, type of alimentation, other hematological disorders or infectious diseases should be also recorded and considered to have a complete interpretation of these observations (Hoffmeister et al., 2001; Bansal et al., 2005; Feingold et al., 1993).

TG levels are found significantly higher in malaria patients, their level seem to increase with the severity (Maguire et al., 1991; Mfonkeu et al., 2010). It is not known if the malaria parasite produces essential lipids for its development or if it induces host lipid hyperoxidation. However, the present data highlight a clear relationship between high TG level and anaemia, an association which was stronger after antimalarial drug administration in patients with high parasite density. One hypothesis is that these TG are derived from the phospholipids released by the red blood cell membrane following hemolysis. Although this way has never been investigated, hypertriglyceridemia is described in diseases with haemophagocytosis (Visser et al., 2013). During malaria, oxidative derivate production is one of the host mechanisms of defense against the parasite. This oxidative stress induces peroxidation that is deleterious for the erythrocyte membrane; it is also responsible for an increase of TG which is correlated to lipidic peroxides (Vial et al., 1999).

The occurrence of hypocholesterolemia during plasmodial infection is supposed to be partly host specific due to an acute phase reaction and parasite related factors (Kasturi et al., 2002). HDL cholesterol fraction is the major lipid source for the parasite growth (Faucher et al., 2002). The decrease of cholesterol level seems to be related to the parasite multiplication during its life cycle. Indeed, a mature schizonte is composed of eight to 32 merozoïtes which require cholesterol for their membrane formation, their replication and for the maintenance of their infectivity (Sinnis et al., 1996). Thus the parasite that is able to adapt to low lipid environment uses other lipids sources such as hepatocytes where some lipids are synthesized (Imrie et al., 2004). The fact that the frequency of hypocholesterolemia decreased when the proportion of *P. falciparum* infected decreased is in favor with the use of this lipid by the parasite.

In conclusion, transient cholesterol and triglycerides variations are common during the course of malaria. These modifications last at least two weeks. Hypocholesterolemia is associated with the presence of malaria parasite, whereas hypertriglyceridemia seems to be related to the hemolysis of infected red blood cells. These lipid parameters changes are not related to a specific antimalarial drug.

ACKNOWLEDGEMENT

We thank all the children parents/guardians.

Conflict of interest

None to declare.

REFERENCES

Ayoola OO, Whatmore A, Balogun WO, Jarrett OO, Cruickshank JK, Clayton PE (2012). Maternal malaria status and metabolic profiles in pregnancy and in cord blood: relationships with birth size in Nigerian infants. Malar. J. 11:75.

Badiaga S, Barrau K, Parola P, Brouqui P, Delmont J (2002). Contribution of nonspecific laboratory test to the diagnosis of malaria in febrile travelers returning from endemic areas: value of hypocholesterolemia. J. Travel. Med. 9:117-121.

Bansal D, Bhatti HS, Sehgal R (2005). Role of cholesterol in parasitic infections Lipids. Health Dis. 9(4):10.

Batista JL, Vervoot T, Van Der Stuyft P, Wery M (1996). Changes in plasma lipid levels as a function of Plasmodium falciparum infection in Sao Tome. Parasite 3:335-340.

Bouyou-Akotet MK, Ramharter M, Ngoungou EB, Mabika Mamfoumbi M, Pemba MihindouM, MissinouMA,, Kurth FM, Bélard S, Agnandji ST, Issifou S, Cambon N, Heidecker JL, Kremsner PG, Kombila M (2010). Efficacy and Safety of Paediatric Artesunate-Mefloquine New

Drug Formulations for the Treatment of Uncomplicated Falciparum Malaria in Gabon. Wien. Klin. Wochenschr. 122:173-8.
Cuisinier-Raynal JC, Bire F, Clerc M, Bernard J, Sarrouy J (1990). Le paludisme humain: syndrome de dysglobulinémie-hypocholestérolémie. Med. Trop. 50:91-95.
Davis TM, Supanaranond W, Pukrittayakamee S, Crawley JC, Villaiwanna N, White NJ (1995). The effect of plasma free fatty acids and long-chain triglycerides on glucose metabolism in uncomplicated falciparum malaria. Trans. R. Soc. Trop. Med. Hyg. 89:511-515.
Faucher JF, Ngou Milama E, Anoumou Missinou M, Ngomo R, Kombila M, Kremsner PG (2002). The impact of malaria on common lipid parameters. Parasitol. Res. 88:1040-3.
Feingold KR, Krauss RM, Pang M, Doerrler W, Jensen P, Grunfeld C (1993). The hypertriglyceridemia of acquired immunodeficiency syndrome is associated with an increased prevalence of low density lipoprotein subclass pattern B. J. Clin. Endocrinol. Metab. 76:1423-1427.
Haldar K, Mohandas N, Samuel BU, Harrison T, Hiller NL, Akompong T, Cheresh P (2001). Protein and lipid trafficking induced in erythrocytes infected by malaria parasites. Cell. Microbiol. 4(7):383-395.
Hoffmeister A, Rothenbacher D, Bode G, Persson K, Marz W, Nauck MA, Brenner H, Hombach V, Koenig W (2001). Current infection with Helicobacter pylori, but not seropositivity to Chlamydia pneumoniae or cytomegalovirus, is associated with an atherogenic, modified lipid profile. Arterioscler. Thromb. Vasc. Biol. 21:427-432.
Imrie H, Ferguson DJ, Carter M, Drain J, Schiflett A, Hajduk SL, Day KP (2004). Light and electron microscopical observations of the effects of high-density lipoprotein on growth of *Plasmodium falciparum* in vitro. Parasitology 128:577-584.
Kasturi H, Mohandas N, Benjamin US, Travis H, Hiller NL, Akompong T, Cheresh P (2002). Protein and lipid trafficking induced in erythrocytes infected by malaria parasites. Cell. Microbiol. 4:383-395.
Kim JS, Oh JS, Chang EA, Bae SY, Nam DH, Lee CH, Kim HC, Klein TA, Lim CS, Kim YK (2008). Alteration of platelet counts and lipid profiles after treatment of acute Plasmodium vivax. Acta. Trop. 106:39-43.
Kitt EM, Diridl G, Lenhart V, Neuwald C, Tomasits J, Pichler H, Bauer K (1992). HDL cholesterol as a sensitive diagnostic parameter in malaria. Wien. Klin. Wochenschr. 104:21-24.
Maguire PA, Prudhomme J, Sherman IW (1991). Alterations in erythrocyte membrane phospholipid organization due to the intracellular growth of the human malaria parasite, Plasmodium falciparum. Parasitology 102:179-186.
Mawili-Mboumba DP, Bouyou Akotet MK, Kendjo E, Nzamba J, Medang MO, Mbina JR, Kombila M (2013). Increase in malaria prevalence and age of at risk population in different areas of Gabon. Malar. J. 2(12):3.
Mfonkeu JB, Gouado I, Kuate HF, Zambou O, Combes V, Grau GE, Zollo PH (2010). Biochemical markers of nutritional status and childhood malaria severity in Cameroon. Br. J. Nutr. 104:886-892.
Mitamura T, Palacpac NM (2003). Lipid metabolism in Plasmodium falciparum-infected erythrocytes: possible new targets for malaria chemotherapy. Microbes Infect. 5(6):545-552.
Nsimba B, Guiyedi V, Mabika-Mamfoumbi M, Mourou JR, Ngoungou E, Bouyou-Akotet MK, Loembet R, Durand R, le Bras J, Kombila M (2008). Sulphadoxine/pyrimethamine versus amodiaquine for treating uncomplicated childhood malaria in Gabon: a randomized trial to guide national policy. Malar. J. 7:31-8.
Parola P, Gazin P, Patella F, Badiaga S, Delmont J, Brouqui P (2004). Hypertriglyceridemia as an indicator of the severity of falciparum malaria in returned travelers: a clinical retrospective study. Parasitol. Res. 92:464-466.
Planche T, Krishna S, Kombila M, Engel K, Faucher JF, Ngou-Milama E, Kremsner PG (2001). Comparison of methods for the rapid laboratory assessment of children with malaria. Am. J. Trop. Med. Hyg. 65(5):599-602.
Sinnis P (1996). The malaria sporozoite's journey into the liver. Infect. Agents Dis. 5(3):182-189.
Vial HJ, Eldin P, Martin D, Gannoum L, Calas M, Ancelin ML (1999). Transport of phospholipid synthesis precursor and lipid trafficking into malaria-infected. Novartis Found Symp. 226:74-88.
Visser BJ, Wieten RW, Nagel IM, Grobusch MP (2013). Serum lipids and lipoproteins in malaria - a systematic review and meta-analysis. Malar. J. 12:442.

Determination of insecticidal effect (LCD_{50} and LCD_{90}) of organic fatty acids mixture (C8910+silicone) against malaria vectors

James C. Dunford[1] , Robert A. Wirtz[2], William G. Reifenrath[3], Aneika Falconer[2], Laura N. Leite[2] and William G. Brogdon[2]

[1]Navy and Marine Corps Public Health Center Detachment, Centers for Disease Control and Prevention, Atlanta, GA, 30341, USA.
[2]Centers for Disease Control and Prevention, Atlanta, GA, 30341, USA.
[3]Stratacor, Inc., Richmond, CA, 94804, USA.

Malaria vectors have acquired widespread resistance to several insecticides; thus, there is a critical need for the development of alternative insecticides for use in vector control programs. The mosquito toxicity of a novel insecticide/repellent consisting of medium-chain carbon fatty acids (C8910) was examined. Determination of LCD_{50} and LCD_{90} was made against six colony-reared *Anopheles* species using probit analysis on mortality data generated by Centers for Disease Control and Prevention bottle bioassays. Eight different concentrations of C8910+silicone oil provided an LCD_{50} ranging from 55.4 (44.2 to 65.9) in *Anopheles minimus* to 132.6 (92.8 to 301.3) in *Anopheles dirus*. Similarly, LCD_{90} varied from 138.5 (107.9 to 207.9) to 1228.8 (449.8 to 21400), respectively. Further development of C8910 and similar compounds could provide vector control specialists novel, environmentally-safe insecticides for controlling insect disease vectors.

Key words: LCD_{50}, LCD_{90}, bottle bioassay, C8910, silicone, malaria, *Anopheles*.

INTRODUCTION

There is an increasing need for alternative insecticides and repellents to control disease-transmitting arthropods. The lack of available, novel insecticide classes for vector control has left experts with a limited set of chemicals to manage insecticide resistance. The emergence of vector resistance to insecticides currently deployed in malaria control programs, for example, can severely impact the effectiveness of principle interventions such as indoor residual spraying (IRS) and long-lasting insecticide-treated bed nets (Ochomo et al., 2013). Since the introduction of synthetic pyrethroids more than 30 years ago, no new mosquito adulticide classes have been approved for vector control by the World Health Organization (WHO) (Nauen, 2007). Because of this, compounds unrelated to the four classes (that is, organochlorines, organophosphates, carbamates, and pyrethroids) need to be investigated.

Research using plant essential oils and animal by-products as alternative compounds for arthropod control and repellency is growing (Isman, 2006; Chansang and Mulla, 2008; Hieu et al., 2010; Mng'ong'o, 2011). Plant oils and fatty acids are common in nature and have been shown to provide insecticidal or repellency effects against a variety of arthropods (Dolan et al., 2007; Chansang, 2008; Mullens et al., 2009; Hieu et al., 2010; Cantrell et al., 2011; Ali et al., 2012). C8910, a mixture of medium-chain octanoic (CH3(CH2)6COOH), nonanoic (CH3(CH2)7COOH) and decanoic (CH3(CH2)8COOH) acids in equal parts, discovered and developed by Stratacor, Inc., (Richmond, CA), has shown promise as both an insecticide and repellent (at higher doses) when formulated appropriately with carriers such as silicone oil or kaolin clay (Reifenrath, 2001). The average molecular masses of the three fatty acids (C8 = 144.2; C9 = 158.2; C10 = 172.3 g/mol) are less than half of dichloro-diphenyltrichloroethane (DDT) (=354 g/mol); thus, provide more than twice the number of fatty acid molecules compared to the same weight of DDT for example. Octanoic and decanoic acids are derived from palm kernel oil or coconut oil and nonanoic acid is derived from tallow; all of these products are inexpensive commodity chemicals. These fatty acids have been approved by the United States Food and Drug Administration (US FDA) as food additives in the U.S. since 1965 and are categorized as "Generally Recognized as Safe" (Reifenrath, 2001). The combination of these acids has shown to have selective repellent and/or toxic effect against mosquitoes and a variety of flies (Mullens et al., 2009; Reifenrath, 2006). Whereas, spiders, bumble bees, honey bees, and wasps are relatively insensitive (Reifenrath, 2001).

C8910 has shown repellency against house flies, biting flies, ticks, ants, sand flies, and mosquitoes [*Aedes aegypti* (L.)] (Reifenrath, 2006); however, no formal studies have been conducted to determine effective concentrations and insecticidal effects of C8910 against malaria vectors. In order to begin to determine an effective concentration of C8910 comparable to WHO Pesticide Evaluation Scheme (WHOPES) approved insecticides, the Centers for Disease Control and Prevention (CDC) bottle bioassay procedure was used to determine the LCD_{50} and LCD_{90} of this novel insecticide/repellent against six colony-reared, susceptible (to all insecticide classes) *Anopheles* species (*Anopheles dirus* Peyton and Harrison, *Anopheles farauti* Laveran, *Anopheles freeborni* Aitken, *Anopheles gambiae* Giles, *Anopheles minimus* Theobald, and *Anopheles stephensi*) representing species known from different geographical regions (Southeast Asia, South Pacific, North America, Africa, Southeast Asia/India, and India/Middle East, respectively). The primary objectives of this study were to determine the insecticidal effects, if any, against male and female *Anopheles* species and to establish baseline effective concentration data that can be used to further develop C8910 for controlling malaria vectors.

MATERIALS AND METHODS

Bottle bioassays

Insecticide susceptibility tests were performed at CDC Atlanta, GA using the CDC bottle bioassay under normal room conditions (23° C; RH ~50%) following protocols established by Brogdon and Chan (2010). Bottle bioassays were used to evaluate a range of concentrations to determine an effective concentration of C8910. Dose-fixing concentrations ranging from 15 to 120 µg C8910/bottle (250 ml glass Wheaton bottle; inside surface area = 370 cm^2) were used to compile mortality data. An initial stock solution of 10 mg (= 10.7 µl) of 15% C8910 + 85% silicone oil (silicone oil is a benign carrier and used with acetone allowed the C8910 to adequately treat and dry on glass bottles) in 100 ml acetone (carrier and drying agent) was made and provided a concentration (= dose) of 15 µg C8910 + silicone/bottle. Additional concentrations of 30, 45, 60, 75, 90, 105, and 120 µg C8910 + silicone/bottle were sequentially made after bioassays using lower concentrations of C8910 indicated mortality was not achieved. Brogdon and Chan (2010) recommend that it is best to align the dose so that 100% of mosquitoes are dead by 1 h or less. For the purposes of this study, we used a relatively low concentration of C8910 (15% C8910/85% silicone oil) tested previously (Reifenrath, 2001, 2006; Mullen et al., 2009) and conducted mortality counts up to 2 h. In order to provide comparative results to other WHOPES-approved insecticides, mortality data for the 30 min time interval (Brogdon and Chan, 2010) were used for LCD_{50}/LCD_{90} determination. Dose-fixing followed the procedures recommended by Robertson et al. (2007). Specifically, it is recommended that for estimation of LCD_{50} and LCD_{90}, a minimum of 4 to 5 doses using at least 200 test insects per dose are required. All C8910 + silicone/acetone stock solutions were stored at room temperature in a dark cabinet.

1 ml of stock solution for each concentration was deposited in each bottle to coat the inside surface and was allowed to air dry for approximately 24 h. One ml of acetone was used to treat the inside of the control bottle. Four treated bottles and one control bottle were used for each assay and each assay was considered one replicate. A total of five replicates (= bioassay rounds) for each concentration were performed. Replicates were not all performed on the same day. When coating/treating the inside surface of the bottles, bottles were slowly turned and rolled with the cap on to evenly treat the inside surface of the bottle and cap, and were checked a few minutes later and rolled again to make sure no 'puddling' occurred on the bottom of the bottle (doing this allowed for more thorough drying). Bottles were placed on their sides with caps off and left to dry overnight. Drying the bottles overnight ensured that the C8910/acetone mixture was dry before introducing mosquitoes and may have resulted in loss of volatiles, thus providing a conservative protocol.

Two to five day old, colony-reared *Anopheles* species (suscepti-ble to all insecticide classes) were introduced to bottles and ob-served for up to 2 h with mortality counts made at 15 min intervals.

Table 1. LCD_{50} and LCD_{90} (C8910+silicone oil /bottle) of *Anopheles* sp. tested after 30 min exposure.

Species	LCD_{50}	Upper / Lower	LCD_{90}	Upper / Lower	Slope±SEM
An. dirus	132.592	301.263 92.889	1228.8	214.00 449.80	1.325±0.088
An. farauti	118.170	192.867 100.074	243.855	1249.959 164.114	4.073±0.273
An. freeborni	119.867	162.566 98.495	628.577	1599.870 372.223	1.781±0.101
An. gambiae	91.762	110.020 79.852	242.266	424.510 178.755	3.040±0.130
An. minimus	55.444	65.916 45.167	138.475	216.910 107.891	3.224±0.110
An. stephensi	112.576	212.715 86.986	329.757	2472.405 186.821	2.746±0.131

The upper and lower 95% confidence limits are also shown (Statistical analysis by PoloPlus 2.0). *An* = *Anopheles.*

The six species tested herein have been in colony at their current location (CDC, Roybal Campus, Atlanta, GA) since 2010; overall age of colonies and susceptibility testing history was not determined. Mosquitoes were randomly aspirated out of rearing containers and gently aspirated into bottles, and no attempt was made to separate (or count) males from females. Because mosquitoes were two to five days old, a reasonable assumption can be made that approximately 50% of each sex was represented in each test bottle and on average for all bioassay rounds. Approximately 15 to 25 mosquitoes were introduced to each bottle and the timer was started once all five bottles had a mosquito population. For counting purposes during assays, a mosquito was considered dead if it was unable to stand (Brogdon and Chan, 2010). During some of these assays, a few mosquitoes at lower concentrations would recover to a certain degree only to eventually die. These individuals were counted as alive at said time interval if they appeared to revive. To ensure a mosquito counted as dead at earlier time intervals did not recover fully, assays were conducted up to two hours. There was no category for moribund individuals. The bottle was tapped a few times before a final determination of dead or alive was made.

Determination of LCD_{50} and LCD_{90}

Percent mortalities, obtained for each concentration at 30 min, were plotted in a log-probit graphic using Polo-plus 2.0 (LeOra Software Company®, Petaluma, California; 2005). Percent mortalities and LCD_{50}/LCD_{90} analyses were calculated by combining the total individuals and number of individuals (assuming each sex was represented equally) that responded (= died) during each assay (that is, total individuals and total that responded in each bottle were analyzed using the aggregate of the four treated bottles per assay). Control totals were entered into Polo-plus as the aggregate of the five replicates for all eight concentrations. Parameters for data files analyzed by Polo-plus were as follows: probit model, natural response, and concentrations converted to logarithms. The LCD_{50} and LCD_{90} were then obtained together with 95% confidence upper and lower limits.

RESULTS

LCD_{50} and LCD_{90} and percent mortality

The C8910 + silicone formulation evaluated herein did not provide significant mortality at the concentrations tested; however, percent mortalities do increase over time and at higher concentrations. Eight different concentrations of C8910 + silicone oil provided an LCD_{50} ranging from 55.4 (44.2 to 65.9) in *An. minimus* to 132.6 (92.8 to 301.3) in *An. dirus*. Similarly, LCD_{90} varied from 138.5 (107.9 to 207.9) to 1228.8 (449.8 to 21400), respectively (Table 1). C8910 worked best against *An. minimus* and *An. gambiae*; both showed lower values of $LCD_{50/90}$ accompanied by narrow 95% CI when compared to $LCD_{50/90}$ estimates for the other four vector species tested. *Anopheles dirus*, *An. farauti*, *An. stephensi* and *An. freeborni* showed higher values of $LCD_{50/90}$. Mortality percentages at 30 min after being exposed to the higher C8910 doses was close to 50% in *An. dirus*, *An. freeborni*, *An. gambiae*, and *An. stephensi.* PoloPlus

Table 2. Average percent mortality of *An. dirus* (up to 120 min) in bottles treated with C8910+silicone oil.

Concentration (µg/bottle)	15	30	45	60	75	90	105	120
15 min	3.8	5.8	10.4	7.6	22.0	26.0	22.0	34.8
30 min	17.0	13.6	24.2	31.6	43.2	44.6	38.6	54.0
45 min	31.0	22.2	33.0	48.0	57.2	60.4	51.6	70.3
60 min	37.2	31.6	44.8	56.8	67.8	74.0	65.0	79.3
75 min	46.6	42.8	58.6	64.0	76.8	80.2	73.4	85.9
90 min	49.6	48.0	66.8	70.8	85.0	86.2	81.4	88.5
105 min	60.0	56.4	73.4	74.4	86.8	90.0	88.6	91.2
120 min	63.4	63.8	76.0	79.6	89.8	92.8	90.4	92.8
Control (at 120 min)	1.0	1.7	1.9	0	0	0	1.0	0.0
N	5	5	5	5	5	5	5	5

N =Total number of assays (1 assay=4 treated bottles and 1 control bottle).

Table 3. Average percent mortality of *An. farauti* (up to 120 min) in bottles treated with C8910+silicone oil.

Concentration (µg/bottle)	15	30	45	60	75	90	105	120
15 min	1.0	0.8	3.2	3.4	11.6	16.6	19.8	45.2
30 min	1.8	3.0	8.4	11.8	22.2	27.4	32.6	68.2
45 min	4.6	8.2	13.0	17.8	20.4	44.4	44.6	81.1
60 min	7.0	12.8	18.6	23.0	36.0	52.2	57.0	86.1
75 min	9.8	15.6	22.4	31.4	44.4	64.0	70.4	88.5
90 min	11.6	20.0	25.6	36.4	51.4	72.4	80.0	92.0
105 min	16.0	24.4	31.0	43.8	60.6	80.0	88.4	93.4
120 min	19.6	31.0	35.4	51.8	67.2	86.4	92.4	96.9
Control (at 120 min)	1	2.7	2.8	2.2	0.9	0	3.4	4.2
N	5	5	5	5	5	5	5	5

N=Total number of assays (1 assay=4 treated bottles and 1 control bottle).

regression lines showing percent response at 30 min for each species are included in Figures 1 to 6. Tables 2 to 7 show percent mortalities for each concentration at each time interval up to 120 min.

DISCUSSION

We evaluated a relatively low concentration C8910 formulation here to mirror previous studies. Mortality percentages (Tables 2 to 7) show that the relatively low concentration of C8910 used for these bioassays took longer than 30 min to provide greater than 90% mortality (for most species this was not achieved until 105 min or longer); however, PoloPlus software was able to predict LCD_{90} for test subjects (Table 1), showing that higher concentrations of C8910 + silicone is required to achieve LCD_{90}. This study provides and initial attempt to determine the insecticidal effects of one C8910 formulation; higher concentrations should be evaluated to demonstrate stronger C8910 efficacy. The senior author of this work has noted that pure C8910 (33% of each carbon) will incapacitate colony-reared *An. gambiae* within 5 min and kill all mosquitoes in test bottles by 30 min at concentrations of 500 µg C8910/bottle (unpublished data). Further tests against wild *Anopheles* populations are required, as colony strains may have reduced vigor. Concentrations of C8910 required for sufficient control of wild populations may be higher than those observed or predicted by PoloPlus herein. While we tested here both male and female mosquitoes to determine if C8910 + silicone demonstrated any sex-linked susceptibility differences, it is recommended that future studies use females only or sexes are tested separately; however, we did not note any differences in overall mortality between male and females during testing. Aïzoun et al. (2014) conducted CDC bottle bioassays using both sexes and noted that even though male *An. gambiae* were smaller

Table 4. Average percent mortality of *An. freeborni* (up to 120 min) in bottles treated with C8910+silicone oil.

Concentration (µg/bottle)	15	30	45	60	75	90	105	120
15 min	0.9	7.4	9.6	16.4	21.1	23.8	40.6	34.8
30 min	5.6	15.2	23.5	29.5	35.1	33.1	56.4	52.9
45 min	9.2	23.0	32.2	39.2	41.2	40.5	68.6	74.5
60 min	13.7	29.6	37.7	43.5	49.9	51.5	74.9	75.8
75 min	16.2	38.4	42.2	47.3	57.1	57.4	80.3	81.3
90 min	19.9	42.3	48.0	53.8	63.1	64.7	81.7	84.8
105 min	25.0	47.0	52.3	57.4	68.7	68.6	83.5	85.7
120 min	27.5	49.5	53.9	60.2	71.5	71.6	85.4	89.0
Control (at 120 min)	0.87	5.8	6.1	0	2.8	1	1	4
N	5	5	5	5	5	5	5	5

N=Total number of assays (1 assay=4 treated bottles and 1 control bottle)

Table 5. Average percent mortality of *An. gambiae* (up to 120 min) in bottles treated with C8910+silicone oil.

Concentration (µg/bottle)	15	30	45	60	75	90	105	120
15 min	0.0	0.6	7.9	23.7	19.3	33.7	41.5	42.0
30 min	0.0	3.1	23.8	35.3	38.2	54.5	55.8	58.0
45 min	1.1	6.3	32.6	45.0	46.7	67.7	67.9	65.0
60 min	3.6	11.8	40.3	52.2	53.2	73.7	71.8	72.8
75 min	5.2	14.9	44.0	56.0	55.9	78.1	76.0	80.6
90 min	7.5	18.6	46.2	59.8	58.7	80.6	77.9	83.5
105 min	8.8	21.0	47.9	60.8	60.1	81.6	82.4	89.3
120 min	45.6	23.9	51.9	62.5	63.1	84.4	85.3	92.9
Control (at 120 min)	3.6	8.6	2.8	1	2.1	5.0	2.8	12.1
N	5	5	5	5	5	5	5	5

N=Total number of assays (1 assay=4 treated bottles and 1 control bottle).

Table 6. Average percent mortality of *An. minimus* (up to 120 min) in bottles treated with C8910+silicone oil.

Concentration (µg/bottle)	15	30	45	60	75	90	105	120
15 min	0.0	2.2	19.7	38.6	50.0	49.0	58.1	69.4
30 min	1.2	13.5	47.6	63.6	72.1	68.4	72.7	89.1
45 min	5.9	25.5	58.5	72.5	83.4	75.5	78.8	94.1
60 min	8.3	32.9	62.7	80.2	86.7	82.0	84.9	96.2
75 min	9.6	40.5	64.4	82.3	92.4	86.5	89.7	98.2
90 min	10.0	46.7	69.6	83.7	94.6	91.5	91.2	98.4
105 min	13.6	50.3	74.9	83.9	96.3	93.4	92.6	98.6
120 min	15.3	53.8	79.8	86.0	97.8	95.6	92.8	99.3
Control (at 120 min)	0.0	0.0	3.0	0	2.0	0.0	0.9	0.0
N	5	5	5	5	5	5	5	5

N=Total number of assays (1 assay=4 treated bottles and 1 control bottle).

smaller and more fragile in appearance, there was no difference in susceptibility to permethrin. C8910's effect is fairly rapid on both sexes (within minutes of initial exposure, especially at higher concentrations), and the first signs of effect include rapid wing beats with individuals making only brief flights or hovering close to the bottom of the bottle. Time to death (that is, individuals without any movement) may take several hours, especially when

Table 7. Average percent mortality of *An. stephensi* (up to 120 min) in bottles treated with C8910+silicone oil.

Concentration (μg/bottle)	15	30	45	60	75	90	105	120
15 min	0.4	0.7	0.9	13.8	18.1	16.8	31.7	29.2
30 min	0.9	3.6	7.9	29.3	49.7	37.0	56.8	53.6
45 min	1.5	7.5	18.6	52.8	59.8	52.1	74.8	66.8
60 min	3.4	13.4	25.2	61.7	74.2	65.9	84.4	78.2
75 min	10.3	21.4	34.3	68.4	82.4	74.0	87.5	84.5
90 min	14.3	26.9	44.7	75.1	86.7	79.8	90.8	88.7
105 min	16.7	32.0	48.8	80.1	89.3	83.3	94.4	91.4
120 min	17.9	40.1	52.4	83.0	90.7	87.3	94.6	94.2
Control (at 120 min)	2.0	0.8	2.7	0.8	0	0	2.8	0.8
N	5	5	5	5	5	5	5	5

N=Total number of assays (1 assay=4 treated bottles and 1 control bottle).

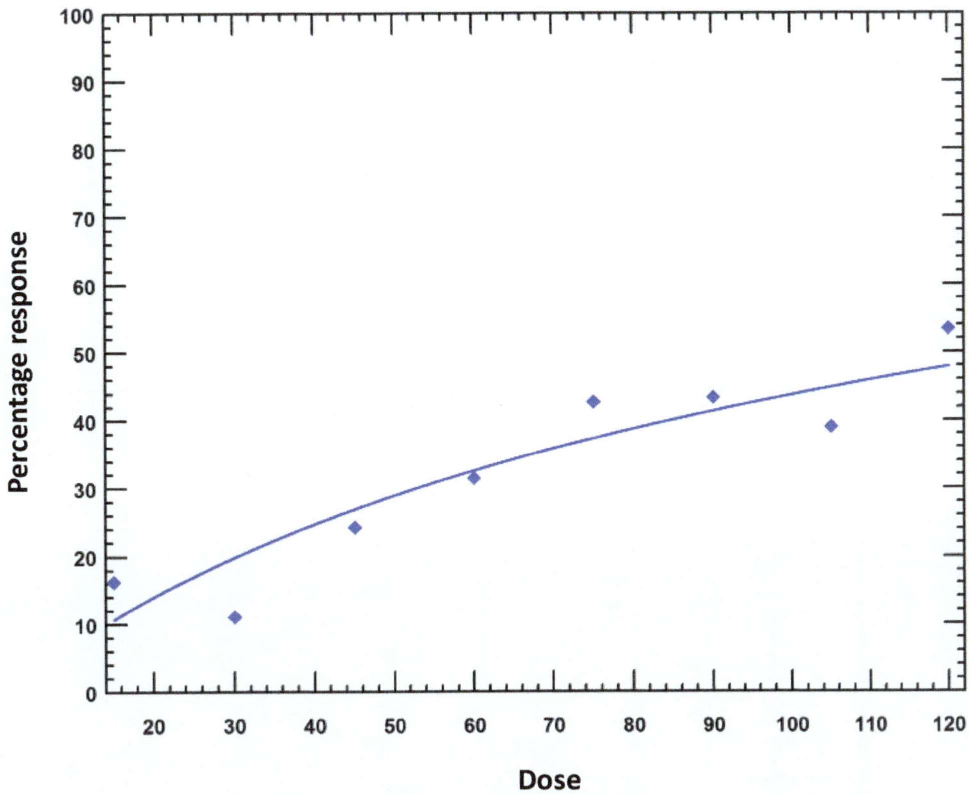

Figure 1. PoloPlus regression line showing percent response (= mortality) of *An. dirus* to C8910+silicone oil at 30 min.

exposed to the lower concentrations tested here. Several individuals were routinely observed attempting to fly after appearing 'dead' when the bottle was tapped/disturbed. Mosquitoes tested against the concentrations used during this study appeared to have an extended moribund period between knockdown and death but do eventually die as no individuals in treated bottles were alive 24 h post exposure. At higher concentrations of C8910, individuals were incapacitated at within 5 to 10 min following exposure.

C8910 in its basic form is an oily substance with a slight coconut odor, similar to sun tan lotion products.

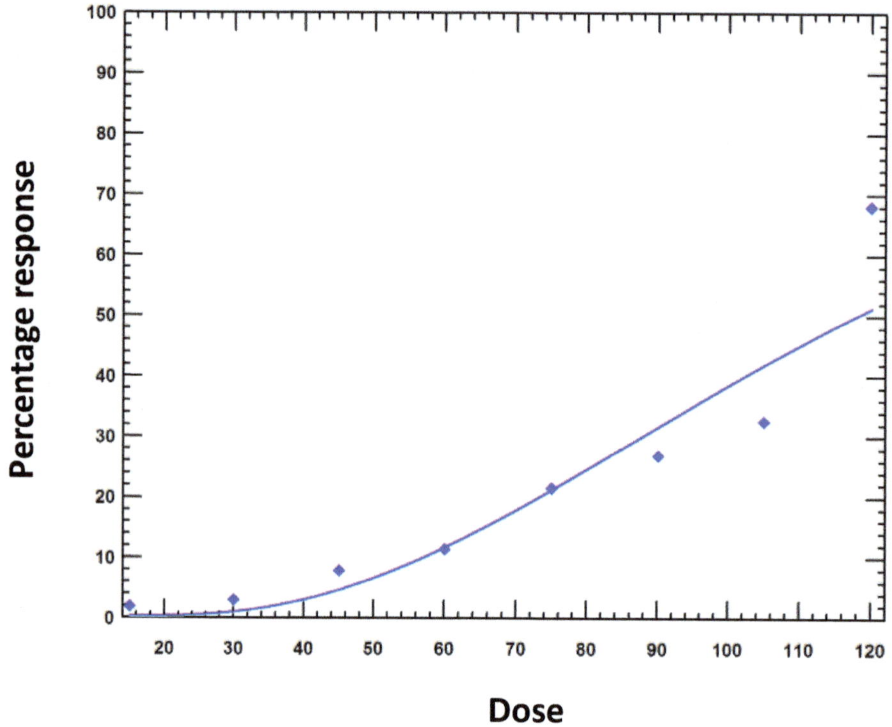

Figure 2. PoloPlus regression line showing percent response (=mortality) of *An. farauti* to C8910+silicone oil at 30 min

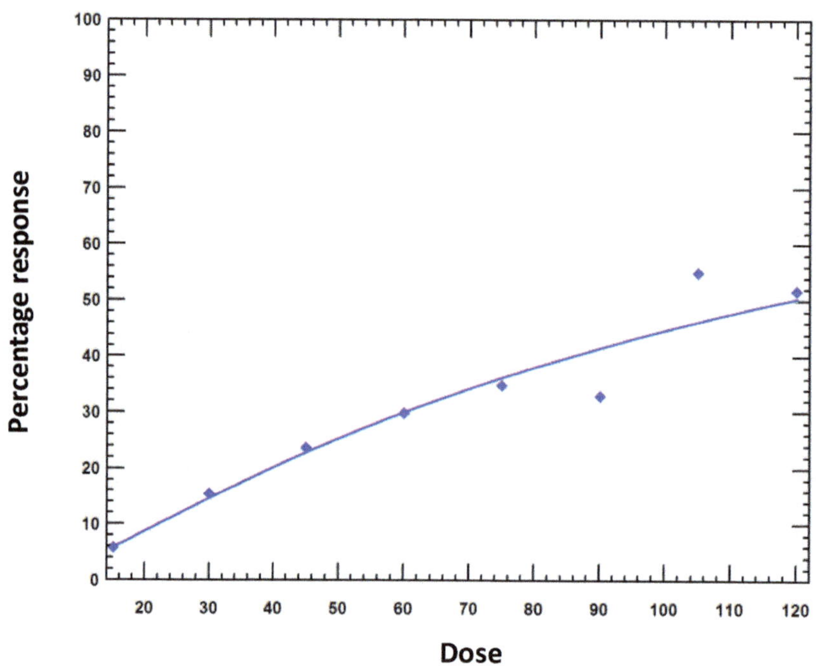

Figure 3. PoloPlus regression line showing percent response (=mortality) of *An. freeborni* to C8910+silicone oil at 30 min.

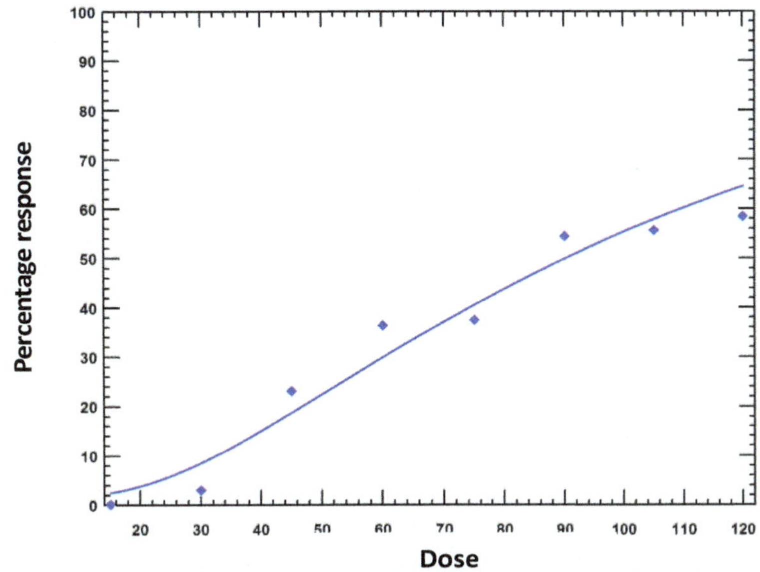

Figure 4. PoloPlus regression line showing percent response (=mortality) of *An. gambiae* to C8910+silicone oil at 30 min.

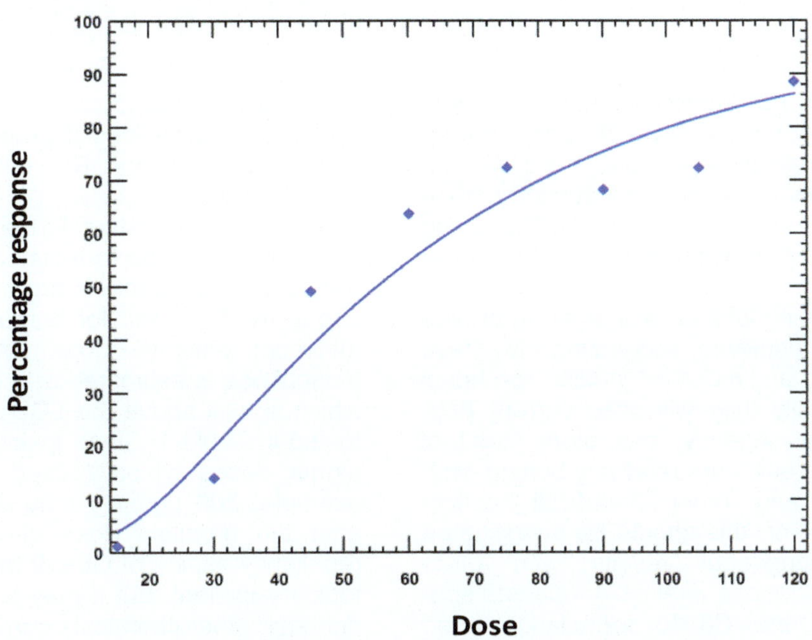

Figure 5. PoloPlus regression line showing percent response (=mortality) of *An. minimus* to C8910 at 30 min.

There is no apparent degradation of efficacy when C8910 is stored at room temperature for long periods of time (Dunford and Reifenrath, personal observations). However, further laboratory studies as well as field studies will be required to evaluate the residual efficacy of C8910. Because the fatty acids are semi-volatile organic compounds, it is thought that controlled release of the compounds would prolong their residual activity. Preliminary bioavailability evaluations using microencapsulated C8910 against *Ae. aegypti*, *Musca domestica* L.,

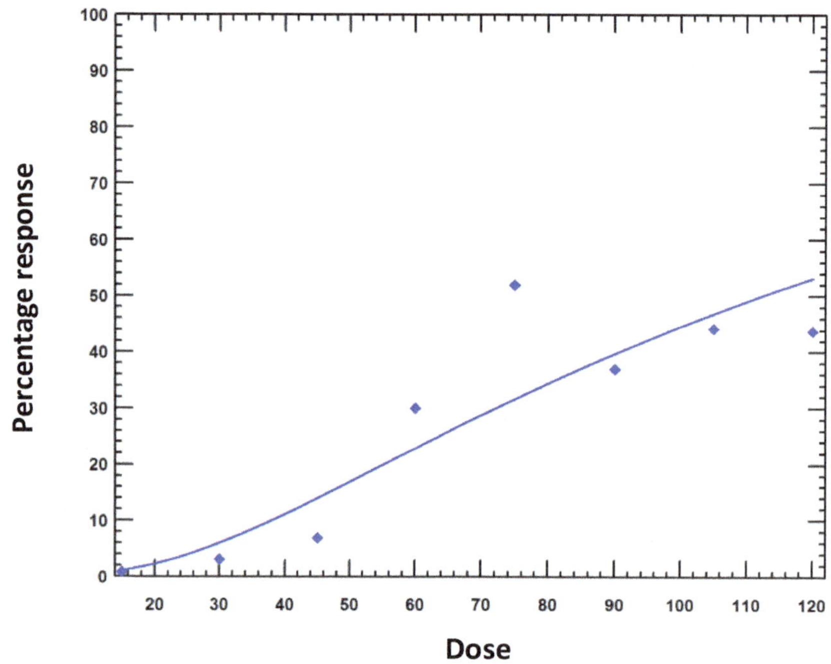

Figure 6. PoloPlus regression line showing percent response (=mortality) of *An. stephensi* to C8910+silicone oil at 30 min.

and *Blatella germanica* (L.) on treated filter paper showed significant mortality against *Ae. aegypti* at doses of 43.1 μg/cm^2 at 24 h, while house flies (431 μg/cm^2) and cockroaches were more resistant to the encapsulated C8910 (Reifenrath, unpublished data). Residual studies using controlled release C8910 formulations need to be further addressed.

The fatty acids comprising C8910 have been approved by the US FDA as food additives, suggesting they have low mammalian toxicity, an important quality for novel insecticides and probability they will pass current EPA (Environmental Protection Agency) regulatory statutes. Environmentally, the fatty acids are relatively benign, with low toxicity to bees and other beneficial insects (Reifenrath, 2001); however, this should be investigated further. When comparing our results with other WHOPES-approved insecticides against *Anopheles* species, concentrations of the C8910 formulation used herein required roughly 11 to 98 times (various pyrethroids) to 1.5 to 12 times (DDT) the diagnostic doses listed in Brogdon and Chan (2010) to achieve comparable mortality percentages; however, higher concentrations should be tested and may prove comparable to other insecticides. An alternative testing method, such as the WHO tube assay, should also be used to further determine if mosquitoes die after initial exposure and knockdown. Further development of C8910 formulations and associated production costs should be compared to traditional insecticides if products containing C8910 are used in the field for IRS (indoor residual spray) and/or treating bed nets.

C8910 is also being further developed for repellency attributes. It can be formulated for direct use on the skin, and topical formulations are being advanced for livestock use in the U.S. and for human use in South Africa. The minimum effective dose (MED) for C8910 to repel mosquitoes is estimated at a surface dose of 25 μg/cm^2, which is well above the LCD_{90} for all *Anopheles* species tested in Table 1. Since toxicity is dependent on contact, higher doses typically used for topical repellency (for example, 300 μg/cm^2) may exert mosquito toxicity only after the repellent dose decays below the MED. The repellency aspect of C8910 may not only provide a novel topical repellent, but it may also be a viable synergist to enhance other chemicals currently being used on treated bed nets. Synergistic interactions between insecticides and repellents have been shown to increase the residual life of impregnated materials and improve the control of pyrethroid-resistant mosquitoes (Pennetier et al., 2007). Pennetier et al. (2007) demonstrated a significant increase in residual efficacy by combining the repellents deet and KBR 3023 with pirimiphos methyl against *An. gambiae*. An earlier study also suggested a significantly higher efficacy was attributed to the combination of deet with propoxur against pyrethroid-resistant *Ae. Aegypti* (Pennetier, 2005).

At present, the mode of action for C8910 has not been investigated, although preliminary studies suggest it is a respiratory inhibitor. The saturated fatty acids are seemingly simple organic compounds but have complex biological effects (Rioux and Legrand, 2007). Fatty acids have been shown to decrease respiratory activity and to uncouple oxidation from phosphorylation in isolated tissue preparations (Scholefield, 1963). Respiratory inhibition increased 2.5 fold for each carbon atom increase in chain length up to a maximum of 12 carbons (lauric acid). A further increase in carbon chain length results in a corresponding decrease in inhibitory activity. This inhibitory effect could be the result of surface tension effects on cell membranes as well as direct effects on enzyme systems. Insects that are sensitive to the toxic effect of low doses of C8910 tend to be repelled at higher doses, while other insects are insensitive to both repellent and toxic effects (Reifenrath, unpublished data). However, further research will be necessary to determine if the reactions are mechanistically related. C8910 shows promising characteristics for use in public health vector control. Future research directions involving C8910 should include the following: 1) description of the mode of action, 2) efficacy testing using stronger concentrations, 3) testing against resistant and other non-resistant mosquito species, to include wild populations, 4) determination of residual duration of controlled release C8910 and mortality results on building surfaces and fabrics, 5) potential use as a repellent against malaria vectors, 6) efficacy testing against other medically important arthropods such as filth flies, biting flies, fleas, kissing bugs, bed bugs, lice, and ticks, and 7) negative effects on non-target insects, birds, and mammals. Further development of C8910 and similar compounds may provide vector control specialists effective, environmentally-safe insecticides for controlling arthropod disease vectors, as well as a sustainable alternative to a growing insecticide resistance crisis.

ACKNOWLEDGEMENTS

Laboratory assistance was kindly provided by Dr. David Hoel (CDR, US Navy, and Marine Corps Public Health Detachment, CDC), Cristiane Gasparetto and Rachel Cruz (Broward College, CDC Worksite Experience Interns), and Jake A. Marika (University of Florida, CDC Worksite Experience Intern). *Anopheles* species used in testing were obtained from the CDC Malaria R & D Laboratory Unit Insectary, from colonies expertly maintained by Ira Goldman, Doug Nace, and Tyrone Williams. Earlier drafts of this manuscript were greatly improved by two anonymous reviewers. Funding to hire laboratory assistants to conduct research at CDC was generously provided by Pulcra Chemicals (Rock Hill, SC) and Emery Oleochemicals (Selangor, Malaysia).

Conflict Interests

The authors declare that there is no conflict of interests

REFERENCES

Aïzoun N, Aïkpon R, Azondekon R, Asidi A, Akogbéto M (2014). Comparative susceptibility to permethrin of two *Anopheles gambiae s.l.* populations from Southern Benin, regarding mosquito sex, physiological status, and mosquito age. Asian Pac. J. Trop. Biomed. 4:312-317.

Ali A, Cantrell C, Bernier U, Duke S, Schneider J, Agramonte N, Khan I (2012). *Aedes aegypti* (Diptera: Culicidae) biting deterrence: structure-activity relationship of saturated and unsaturated fatty acids. J. Med. Entomol. 49:1370-1378.

Brogdon W, Chan A (2010). Guideline for Evaluating Insecticide Resistance in Vectors Using the CDC Bottle Bioassay, 2nd edition. Center for Global Health, Division of Parasitic Diseases and Malaria. Atlanta, GA. P 30.

Cantrell C, Duke A, Khan I (2011). Identification of mosquito biting deterrent constituents from the Indian folk remedy plant *Jatropha curcas*. J. Med. Entomol. 48:836-845.

Chansang U, Mulla M (2008). Field evaluation of repellents and insecticidal aerosol compositions for repelling and control of *Siphunculina funicola* (Diptera: Chloropidae) on aggregation sites in Thailand. J. Am. Mosq. Contr. 24:299-307.

Dolan MC, Dietrich G, Panella NA, Montenieri JA, Karchesy JJ (2007). Biocidal activity of three wood essential oils against *Ixodes scapularis* (Acari: Ixodidae), *Xenopsylla cheopis* (Siphonaptera: Pulicidae), and *Aedes aegypti* (Diptera: Culicidae). J. Econ. Entomol. 100(2):622-5.

Hieu T, Kim S, Lee S, Ahn Y (2010). Repellency to *Stomoxys calcitrans* (Diptera: Muscidae) of plant essential oils alone or in combination with *Calophyllum inophyllum* nut oil. J. Med. Entomol. 47:575-580.

Isman M (2006). Botanical insecticides, deterrents, and repellents in modern agriculture and an increasingly regulated world. Ann. Rev. Entomol. 51:45-66.

LeOra Software (2005). *Polo-Plus*, POLO for Windows, LeOra Software, 1007 B St., Petaluma, CA.

Mng'ong'o F, Sambali J, Eustachkius S, Rubanga J, Magoma J, Ntamatungiro A, Turner E, Nyogea D, Ensink J, Moore S (2011). Repellent plants provide affordable natural screening to prevent mosquito house entry in tropical rural settings-results from a pilot efficacy study. PLOS ONE 6:1-11.

Mullens B, Reifenrath W, Butler S (2009). Laboratory trials of fatty acids as repellents or antifeedents against houseflies, horn flies, and stable flies (Diptera: Muscidae). Pest Manag. Sci. 65:1360-1366.

Nauen R (2007). Insecticide resistance in disease vectors of public health importance. Pest Manag. Sci. 63:628-633.

Ochomo E, Brogdon W, Gimnig J, Ouma C, Vulule J, Walker E (2013). Pyrethroid resistance in *Anopheles gambiae* s.s. and *Anopheles arabiensis* in Western Kenya: phenotypic, metabolic and target site characterizations of three population. Med. Vet. Entomol. 27:156-164.

Pennetier C, Corbel V, Boko P, Odjo A, N'Guessan R, Lapied B, Hougard J (2007). Synergy between repellents and non-pyrethroid insecticides strongly extends the efficacy of treated nets against *Anopheles gambiae*. Malar. J. 6:38.

Pennetier C, Corbel V, Hougard J (2005). Combination of a non-pyrethroid insecticide and a repellent: a new approach for controlling knockdown-resistant mosquitoes. Am. J. Trop. Hyg. 72:739-744.

Reifenrath WG (2001). Natural insect and arthropod repellent. US

Patent 6306, 415B1.
Reifenrath WG (2006). New Repellent Combination against Flies and Mosquitoes. Final Report, USAMRMC Award No. W81XWH-04-1-0787.
Rioux V, Legrand P (2007). Saturated fatty acids: simple molecular structures with complex cellular functions. Curr. Opin. Clin. Nutr. Metab. Care 10:752-758.
Robertson J, Russell R, Preisler H, Savin N (2007). Bioassays with Arthropods. CRC Press, Boca Raton, FL. pp. 55-69.
Scholefield PG (1963). Fatty acids and their analogues. In: Hochster RM, Quastel JH (eds.), Metabolic Inhibitors, Vol. 1. Academic Press, New York. pp. 153-172.

Urinary schistosomiasis among pre-school and school aged children in two peri-urban communities in Southwest Nigeria

Babatunde, T. A., Asaolu, S. O. and Sowemimo, O. A .

Department of Zoology, Obafemi Awolowo University, Ile-Ife, Nigeria.

A cross-sectional study was conducted between March and April, 2010 among pre-school and school aged children in two peri-urban communities in Osun State, Southwest Nigeria. Urine samples were collected from the pre-school and school aged children, tested for microhaematuria using reagent strips, processed and examined for *Schistosoma haematobium* ova. Out of 274 pupils examined, 132 (48.2%) had infection, with no statistically significant difference ($P > 0.05$) in infection between male (48.6%) and female pupils (47.6%). The prevalence of infection increase significantly with age ($P < 0.05$), with the peak (93.3%) of infection recorded in pupils aged 15 to 16 years and the lowest infection (10.0%) in pupils aged 3 to 4 years. There was no statistically significant association ($P > 0.05$) between intensity in male pupils (156.0 ± 34.5/10 ml) and female pupils (141.7 ± 29.5/10 ml). The prevalence of pupils with microhaematuria was 65.0% and it increased significantly with age ($P < 0.001$). The conclusion drawn from the study is that to reduce the transmission of *S. haematobium* in endemic communities, health education and provision of potable water are advocated.

Key words: *Schistosoma haematobium*, microhaeamturia, prevalence, urinary schistosomiasis, school-aged children, Nigeria.

INTRODUCTION

Schistosomiasis remains an important public health problem globally, with approximately 779 million estimated to be at risk (Hotez, 2009). Within sub-Saharan Africa, Nigeria is the country with the highest prevalence of human schistosomiasis; about 29 million in 2008 (Hotez and Kamath, 2009). Urinary schistosomiasis is a human disease condition which is caused by infection of the trematode *Schistosoma haematobium*. The parasite is found in the venous plexus draining the urinary bladder of humans (World Health Organization (WHO), 2002). During infection, the parasites deposit terminal spined eggs which clog the venous plexus, impeding blood flow. The eggs of *S. haematobium* provoke granulomatous inflammation, ulceration and pseudo-polyposis of the vesical and ureteral wall. Haematuria is a very common sign of infection but other signs include dysuria, pollakisuria and proteinura. Kidney failure deaths due to urinary tract scarring, deformity of ureters and the bladder caused by *S. haematobium* infection have become less common due to the use of effective drug praziquantel (Gryseels et al., 2006). In Nigeria, urinary schistosomiasis is widespread in both rural and urban communities;

with prevalences ranging between 2 and 90% and the vast majority of cases occurring among the poor and marginalized (Ugbomoiko, 2000; Mafiana et al., 2003; Oladejo and Ofoezie, 2006; Opara et al., 2007). Several studies on schistosomiasis in Nigeria have focused on school-aged children and adults, with little or no information on pre-school children. Recently, urinary schistosomiasis infections have been reported in pre-school children in settlements in Ogun State and Cross River State, all in Nigeria (Opara et al., 2007; Ekpo et al., 2010). However, information on the endemicity of this disease is still lacking in some parts of the country and data collected is still grossly inadequate for planning a credible control programme.

The study was planned with the aim of providing epidemiological data on the status of schistosomiasis infection among pre-school and school aged children of Akinlalu and Ogbagba communities of Osun State Southwest Nigeria. It is hoped that the results to be generated will complement the existing baseline information on the epidemiology of this infection in the country.

MATERIALS AND METHODS

Study area

This cross-sectional study was carried out in the peri-urban/rural communities Akinlalu and Ogbagba both situated in Osun State in Nigeria. Akinlalu (07° 28′N, 04° 15′E) is situated in Ife North Local government Area and Ogbagba (07° 42′N, 04° 15′E) is situated in Ife East Local Government Area both in Osun State. The two communities are in the rainforest belt of Nigeria and the annual rainfall in the region ranges from 1,000 to 4,000 mm. The average maximum and minimum daily temperatures are 32 and 20°C, respectively (Asaolu et al., 1991). The climate is tropical with distinct dry (November to March) and rainy (April to October) seasons (Asaolu et al., 1991). The ecological characteristics of, as well as the socio-cultural and daily economic activities in both communities are similar. Farming and petty trading are the main occupations, and most residents are of Yoruba ethnicity. Neither of the communities has reliable pipe-borne water supply. However, both communities have shared community wells and rivers which are frequently visited for domestic use as well as leisure and religious beliefs.

Study population and design

A school-based cross-sectional study was conducted between March and April, 2010. The study samples were taken by random sampling from pre-school and school age children population attending Surajudeen Primary School, Akinlalu and Sacred Heart Catholic Mission School, Ogbagba Osun State, Nigeria. For schistosomiasis control, WHO recommends that 200 to 250 school aged children should be sampled in each ecological zone (Montresor et al., 1998). A total of 51 pupils from Ogbagba community and 223 pupils from Akinlalu, making a total of 274 children took part in the study. The permission of the community leaders, local government and school authority was obtained before the commencement of the study. In addition, the consent of the parents of the children was also obtained. The study protocol was approved by the Ethical Committee of the Obafemi Awolowo University Teaching Hospital Complex (OAUTHC).

Data collection

Each child was provided with a pre-labelled, wide-mouthed, screw capped container for the collection of a mid-day urine sample. Each urine sample was divided into two portions: a portion of the urine sample was tested for microhaematuria using commercial reagent strips (Medi-test Combur-9; Analyticon Biotechnologies, Lichtenfels, Germany) in accordance with the manufacturers' instructions. The other portion of the urine samples were transported to the parasitology laboratory at the Obafemi Awolowo University, Ile-Ife's Department of Zoology and preserved by adding two drops of formaldehyde. A pretested structured questionnaire was used to collect demographic and socio-economic data.

Laboratory analysis

A 10 ml of the preserved urine samples were passed through a 8 µm pore membrane filter, so that any *S. haematobium* eggs could be trapped on the filter and counted under a light microscope and recorded as eggs/10 ml of urine (WHO, 1983).

Statistical analysis

Version 16.0 of the statistical package for social sciences (SPSS) for windows software package (SPSS Inc, Chicago, IL) was used for all the data analysis, comparisons of prevalence by subject age and gender was made using χ^2 tests. Differences in mean egg counts between dichotomous variables and variables with more than two levels were explored using Student's t-tests and one-way analysis of variance (ANOVA), respectively.

RESULTS

A total of 274 pre-school and school aged children were enrolled for this study and they all submitted urine samples. The age of the children ranged from 3 to 16 years, with a mean (standard deviation, SD) of 9.0 (3.4) years and a median of 9 years. 148 (54.0%) of them were boys while 126 (46.0%) were girls.

Prevalence and intensity of *S. haematobium* infection by age and sex

The overall prevalence of *S. haematobium* infection was 48.2%. The prevalence pattern showed an increasing trend with increasing age, and peaked around 15 to 16 years. The lowest prevalence of 10.0% was recorded among the pre-school children aged 3 to 4 years. There was a significant difference in prevalence of infection between ages (χ^2 = 59.085, df = 6; P = 0.000). However, there was no statistically significant difference (χ^2 = 0.029, df = 1; P = 0.865) in prevalence between the boys 72

Table 1. Prevalence and intensity of *S. haematobium* among the children in the two communities.

Age (years)	No. Examined	No. Infected	Percentage (%)	Mean ± (SEM)
3-4	30	3	10.0	4.13±2.46
5-6	42	13	31.0	183.64±74.36
7-8	57	17	29.8	126.70±54.46
9-10	55	33	60.0	158.93±54.59
11-12	40	27	67.5	215.65±66.79
13-14	35	25	71.4	191.66±55.00
15-16	15	14	93.3	120.20±31.69
Total	274	132	48.2	149.41±23.00
P-value		<0.05		<0.05
Sex				
Male	148	72	48.6	156.00±34.50
Female	126	60	47.6	141.73±29.51
P-value			>0.05	>0.05

SEM : Standard error of mean.

(48.6%) and girls 60 (47.6%) (Table 1). The overall mean intensity of infection was 149.41 ± 23.00 eggs/10 ml. Pre-school children aged 3 to 4 years had the lowest intensity of 4.28 eggs/10 ml while the highest mean intensity of 221.8 eggs/10 ml was recorded in pupils aged 11 to 12 years. There was a significant difference between age and intensity of infection (P < 0.001). The male pupils have a higher intensity (156.00 ± 34.50 eggs/10 ml) than the female pupils (141.73 ± 29.51 eggs/10 ml), however there was no statistically significant difference between age and intensity of infection (P > 0.05) (Table 1).

Micro-haematuria and visible haematuria

Micro-haematuria was detected in 178 (65.0%) of the 274 urine samples, out of which 47(92.2%) were from Ogbagba community while 131 (58.7%) were from Akinlalu. Out of the 178 with micro-haematuria, 86 (31.4%) had gross haematuria, 48 (17.5%) moderate haematuria and 44 (16.4%) mild haematuria. The prevalence of microhaematuria by age showed that the lowest prevalence of 13.3% was recorded in pupils' age 3 to 4 years while the highest of 100% was recorded in the pupils' age 15 to 16 years (Table 2). The prevalence of female pupils 82 (65.1%) with microhaematuria was higher than that of male pupils 96 (64.9%) but there was no statistically significant difference in the prevalence of micro heamaturia between the sexes (P > 0.05) (Table 2). Visible haematuria was observed in 44 (16.1%) of the 274 urine samples, out of which 9 (17.6%) pupils from Ogbagba community showed visible haematuria while 35 (15.7%) pupils were from Akinlalu. Male pupils had a higher prevalence (21.6%) of visible haematuria than female pupils (9.5%). There was a significant difference between the prevalence of visible haematuria between the two sexes (χ^2 = 7.30, df = 1, P = 0.007). The lowest prevalence (6.7%) of visible haematuria was recorded in pupils aged 3 to 4 years while the highest prevalence (46.7%) was recorded in pupils aged 15 to 16 years. There was a significant differences between the prevalence of visible haematuria among the various age groups (χ^2 = 14.804, df = 6, P = 0.022) (Table 3).

DISCUSSION

This study showed that urinary schistosomiasis is endemic in some communities in Osun State and it corroborates with findings of some earlier studies in this geographical region (Oladejo and Ofoezie, 2006; Ugbomoiko et al., 2010). The overall prevalence of urinary schistosomiasis recorded (48.2%) in this study is about four times higher than the National Nigerian mean of about 13% (Ofoezie, 2002). This result is comparable to 47% reported in settlements near Erinle River Dam, Osun State, Nigeria (Oladejo and Ofoezie, 2006) and lower than 62% reported from two communities in Southwest Nigeria (Ugbomoiko et al., 2010), 71.8% from settlements near a dam reservoir, Ogun State, Nigeria (Mafiana et al., 2003) and 58.1% from Ilewo-Orile community in Ogun State (Ekpo et al., 2010). From the results obtained in this study and previous studies as highlighted above, it appears that urinary schistosomiasis is particularly common in the southwest region of Nigeria. This study showed that there was an increasing trend of

Table 2. Prevalence (%) of micro-haematuria in relation to age and sex of children in Akinlalu and Ogbagba communities.

Age (years)	No. Examined	No. and (%) with no haematuria	No. and (%) with mild haematuria	No. and (%) with moderate haematuria	No. and (%) with gross haematuria	No. and (%) with positive haematuria
3 – 4	30	26 (86.7)	1 (3.3)	1 (3.3)	2 (6.7)	4 (13.3)
5 – 6	42	24 (57.1)	6 (14.3)	4 (9.5)	8 (19.0)	18 (42.9)
7 – 8	57	23 (40.4)	12 (21.1)	10 (17.5)	11 (19.3)	33 (57.9)
9 – 10	55	13 (23.6)	11 (20.0)	10 (18.2)	21 (38.2)	42 (76.4)
11 – 12	40	3 (7.5)	5 (12.5)	12 (30.0)	20 (50.0)	37 (92.5)
13 – 14	35	6 (17.1)	5 (14.3)	8 (22.9)	16 (45.7)	29 (82.9)
15 – 16	15	0 (0.0)	4 (26.7)	3 (20.0)	8 (53.3)	15 (100)
P-value		<0.05	<0.05	<0.05	<0.05	<0.05
Sex						
Male	148	52 (35.1)	17 (11.5)	31 (20.9)	48 (32.4)	96 (64.9)
Female	126	43 (34.1)	27 (21.4)	17 (13.5)	38 (30.2)	82 (65.1)
P-value			<0.05	<0.05	>0.05	>0.05
Community						
Akinlalu	223	92 (41.3)	39 (17.5)	28 (12.6)	64 (28.7)	131 (58.7)
Ogbagba	51	3 (5.8)	5 (9.8)	20 (39.2)	22 (43.1)	47 (92.2)
Total	274	95 (34.7)	44 (16.4)	48 (17.5)	86 (31.4)	178 (65.0)

prevalence infection among children from three years to sixteen years. This is in conformity with the general pattern of *S. haematobium* infection in endemic areas. Similar findings were reported by previous workers who studied *S. haematobium* infection among school children and stated that individuals aged 5 to 15 years were more likely to be infected with *S. haematobium*. (Okoli and Odaibo, 1999; Ejima and Odaibo, 2010). However, some studies reported a decline in infection from 14 years and attributed the declining trend in prevalence of infection among children aged 15 years and above to probable age acquired immunity (Satayathum et al., 2006).

In this study, the prevalence of infection was higher among the male pupils (48.6%) than in female pupils (47.6%), however, no statistically significant difference in the prevalence of infection between both sexes. This may be an indication that both male and female pupils are equally exposed to infection through water contacts. Previous studies have also reported higher prevalence among male pupils than among their female counterparts (Akinwale et al., 2010; Ejima and Odaibo, 2010; Ekpo et al., 2010). The intensity of infection was also higher among male pupils; this is suggestive that male pupils carry a greater worm burden than the females. Similar findings were reported in previous studies (Ariyo et al., 2004; Ejima and Odaibo, 2010).

The prevalence of microhaematuria in this study (65.3%) shows no statistically significant difference between the sexes. This is due to the fact that microhaematuria is a characteristic symptom of urinary schistosomiasis in endemic communities where its prevalence correlated positively with urinary schistosomiasis infection (Anosike et al., 2001). Visible haematuria was recorded in this study and was an on the spot indication of the

Table 3. Prevalence (%) of visible haematuria in relation to age and sex of the children in the two communities.

Age (years)	No. examined	No. and (%) with visible haematuria
3 – 4	30	2 (6.7)
5 – 6	42	6 (14.3)
7 – 8	57	7 (12.3)
9 – 10	55	7 (12.7)
11 – 12	40	9 (22.5)
13 – 14	35	6 (17.1)
15 – 16	15	7 (46.7)
P-value		< 0.05
Sex		
Male	148	32 (21.6)
Female	126	12 (9.5)
P-value		<0.05
Community		
Akinlalu	223	35 (15.7)
Ogbagba	51	9 (17.6)
Total	274	44 (16.1)

presence of urinary schistosomiasis in the study communities. The prevalence of visible haematuria recorded in this study (16.1%) compares favourably with 17.9% from Ishielu Amagunze, Enugu state Nigeria (Nwaorgu et al., 1998). It was noted that even though the disease is associated with river water, a high level of illiteracy, ignorance and traditional beliefs as well as unavailability of pipe borne water and insufficient boreholes have contributed to the endemicity of *S. haematobium* in the two communities. In both communities it is still believed by some that rivers "Amula" and "Ooye" situated in Akinlalu and Ogbagba, respectively have healing and medicinal powers. Provision of basic health education and portable water in the study communities will reduce contact of humans with infested water and thereby reduce transmission. Periodic mass treatment will reduce morbidity and gradually eradicate the disease in the two communities.

ACKNOWLEDGEMENTS

The authors wish to thank the community heads, parents/ guardians and the school authorities of Akinlalu and Ogbagba communities for their willingness to participate in the study and cooperation during the collections of the urine samples.

REFERENCES

Akinwale OP, Ajayi MB, Akande DO, Gyang PV, Adeleke MA, Adeneye AK, Adebayo MO, Dike AA (2010). Urinary Schistosomiasis around Oyan Reservoir, Nigeria: Twenty years after the first Outbreak. Iranian J. Pub. Health 39(1):92-95.

Anosike JC, Nwoke BEB, Njoku AJ (2001). The validity of haematuria in the community diagnosis of urinary schistosomiasis infections. J. Helminth. 75:223-225.

Ariyo O, Olofintoye LK, Adeleke RA, Famurewa O (2004). Epidemiological study of urinary Scistosomiasis among primary school pupils in Ekiti State, Nigeria. Afr. J. Clin. Exp. Micr. 5(1):20–29.

Asaolu SO, Holland CV, Crompton DWT (1991). Community control of *Ascaris lumbricoides* in rural Oyo State, Nigeria: Mass targeted and selective treatment with levamisole. Parasit. 103(2): 291–298.

Ejima IAA, Odaibo AB (2010). Urinary Schistosomiasis in the Niger – Benue Basin of Kogi State, Nigeria. Int. J. Trop. Med. 5(3):73–80.

Ekpo UF, Laja-Deile A, Oluwole AS, Sam-Wobo SO, Mafiana CF (2010). Urinary Schistosomiasis among preschool children in rural community near Abeokuta, Nigeria. Parasite, Vec. 3:58 doi: 10.1186/1756-3305-3-58.

Gryseels B, Polman K, Clerinix J, Kestens L (2006). Human Schistosomiasis. Lancet 368(9541):1106 – 1118.

Hotez PJ (2009). "Mass drug administration and integrated control for the world's high- prevalence neglected tropical diseases. Clin. Pharm. Therap. 85: 659-664.

Hotez PJ, Kamath A (2009). Neglected tropical diseases in Sub – Saharan Africa: review of their prevalence, distribution, and disease burden. PLoS Neglect. Trop. Dise. 3,e412.

Mafiana CF, Ekpo UF, Ojo DA (2003). Urinary Schistosomiasis in preschool children in settlements around Oyan reservoir in Ogun State, Nigeria : implication for control. Trop. Med. Inte. hlth. 8, 78 – 82.

Montresor A, Crompton DWT, Hall A, Bundy DAP, Savioli L (1998). Guidelines for the evaluation of soil-transmitted helminthiasis and schistosomiasis at community level. World Health Organisation, Geneva.

Nwaorgu OC, Okeibunor J, Madu E, Amazingo U, Onyegegbu N, Evans D (1998). A school – based Schistosomiasis and intestinal helminthiasis control program in Nigeria: acceptability to community members. Trop. Med.Int. Health 3(10)842-849.

Ofoezie IE (2002). Human health and sustainable water resources development in Nigeria: Schistosomiasis in artificial lakes. Natur. Resou. For. 26: 150-160.

Okoli EI, Odaibo AB (1999). Urinary Schistosomiasis among school children in Ibadan, an urban community in Southwestern Nigeria. Trop. Med. Int. Health 4: 308-315.

Oladejo SO, Ofoezie IE (2006). Unabated schistosomiasis transmission in Erinle River Dam, Osun State, Nigeria: evidence of neglect of environmental effects of development projects. Trop. Med. Int. Health 11: 843-850.

Opera KN, Udoidung NI, Ukpong IG (2007). Genitourinary schistosomiasis among pre – primary school children in rural community within the Cross River Basin, Nig. J. Helmin. 81:393 – 394.

Satayathum SA, Muchiri EM, Ouma JH, Whalea CC, King CH (2006). Factors affecting infection or reinfection with *Schistosoma haematobium* in coastal Kenya: survival analysis during a nine year, school – based treatment program. Am. J. Trop. Med. Hyg. 75(1): 83-92.

Ugbomoiko US (2000). The prevalence, incidence and distribution of human urinary Schistosomiasis in Edo State, Nigeria. Nig. J. Parasit. 21: 3-14.

Ugbomoiko US, Ofoezie IE, Okoye IC, Heukelbach J (2010). Factors associated with urinary schistosomiasis in two peri – urban communities in Southwestern Nigeria. Ann. Trop. Med. Parasit. 104(5): 409-419.

World Health Organization (1983). Diagnostic Techniques in Schistosomiasis Control. Document WHO/SCHISTO/83.69. Geneva: WHO.

World Health Organisation (2002). Prevention and Control of Schistosomiasis and Soil-transmitted Helminthiasis, Report of a WHO Expert Committee. Technical Report Series No. 912. Geneva: WHO.

Co-infection of HIV and malaria parasites in pregnant women attending major ante-natal health facilities in Akure, Ondo State, Nigeria

Olusi, T. A. and Abe, A. F.

Department of Biology, Federal University of Technology, P M B 704, Akure Ondo State, Nigeria.

Malaria and human immunodeficiency virus (HIV) in pregnancy are the major factors contributing to adverse maternal and perinatal outcome. HIV increases pregnant women's chances of contracting malaria, increases the risk of developing anaemia, delivering a low birth weight infant and premature delivery. This study was designed to investigate the level of co-infection of malaria parasite and HIV in pregnant women in State Specialist Hospital, Akure. 616 pregnant women aged 15 to 46 years who attended major referral health facilities for ante-natal purposes between February and April, 2012 were included for the study. 'Determine' and Uni-Gold rapid diagnostic tests kits were used to determine HIV status, giemsa stained thick blood smear of patients were examined for presence of the asexual stages of *Plasmodium* parasite. Out of 616 pregnant women, 28 (4.55%) were HIV positive and 597 (96.92%) had malaria parasite. Among HIV negative women, 569 (96.8%) had malaria parasite infection while 28 (100%) of the HIV positive women had the infection. This study revealed that 92.37% had malaria alone, none of the women had HIV only and 28 (4.55%) were co-infected with both pathogens, indicating that all HIV positive women also harbor malaria parasite. There was moderate correlation between HIV and malaria parasite $p < 0.05$ which is suggestive that women infected with HIV are most likely to be infected with malaria due to their compromised immune system which makes them more susceptible to malaria infection. The rate of malaria infection was generally high in the sampled population though majority was infected at low parasiteamia. This is therefore attributed to the endemic nature of the disease in Ondo state of Nigeria

Key words: Human immunodeficiency virus (HIV), malaria, co-infection HIV-MALARIA, antenatal care, pregnant women.

INTRODUCTION

Human immunodeficiency virus (HIV) and malaria are among the leading causes of morbidity and mortality in sub-Saharan Africa, where they represent common infections in women of childbearing age. Together, they cause more than four million deaths per year (World Health Organization (WHO), 2011). About 90% of the 300 to 500 million annual acute episodes of malaria are reported in this region, and there is also an estimated 30 million HIV-infected cases (Rowland-Jones and Lohman, 2002). The burden of malaria is particularly severe in children under 5 years (who constitute more than 70% of the 1 million deaths due to malaria) and pregnant women. About 55% of HIV infected adults in sub-Saharan African are women of reproductive age and they account for over 80% of the world's HIV infected women. In this region, the prevalence of maternal malaria is 65% and HIV affects 40% of the pregnant population (Ned et al., 2005). Schantz-Dunn and Nawal (2009) recorded that malaria is one of the most devastating infectious diseases, killing more than one million people annually. Forty percent of the world's population lives in endemic areas. Malaria epidemics have devastated large populations. Consequently the disease poses a serious barrier to economic progress in many developing countries. There are an estimated 300 to 500 million cases of clinical disease per year, with 1.5 to 2.7 million deaths.

Abu-Raddad et al (2006) estimated that, the interaction of malaria and HIV in one Kenyan district alone had caused 980,000 excess malaria episodes and 8,500 excess HIV infections since the emergence of HIV in the 1980s. In areas endemic for malaria, it is estimated that at least 25% of pregnant women are infected with the disease. The highest risk of infection and morbidity is found among the primigravidas, adolescents and those co-infected with HIV (Desai et al., 2007). Malarial infection in HIV-positive women is associated with higher levels of parasitemia, leading to a greater risk of severe anemia. Likewise, HIV viral load is increased, creating opportunity for infection and more severe disease (Brentlinger et al., 2006). As a result of the impaired immune state in pregnant women, HIV infection increases susceptibility to malaria and the morbidity associated with it in co-infected women, resulting in higher incidences of severe anemia and low-birth-weight neonates in co-infected. WHO (2004) explained co-infection of malaria and HIV having disproportionate effects on pregnant women and pose serious risks. Briand et al. (2009) reported that HIV-infected women were approximately 22 times more likely to die than HIV-uninfected women. There was a similar risk for those with malaria. Women with dual infections thus had the greatest risk of death compared with those with only HIV infection or malaria infection. Malaria infection in HIV-positive pregnant women can also increase the risks of mother to child transmission (MTCT) during pregnancy, labor as well as during breastfeeding period due to increase in the level of HIV in the blood, that is increase in viral load (UNICEF, 2003). Thus any interaction between malaria and HIV can be an additional health burden, especially in pregnant women, with resulting poor birth outcomes and the imminent transmission of these intracellular pathogens. Adeoti et al. (2012) studied prevalence of HIV and malaria parasites co-infection in pregnant mothers and their babies post-delivery, clearly showing that the recruitment profile of the samples examined exhibited prevalence of malaria infections in the HIV infected mothers and their infants. The aim of this research work is to determine the prevalence and the level of co-infection of HIV and malaria infection in pregnant women at Akure, Ondo State, Nigeria using State Specialist Hospital Akure.

MATERIALS AND METHODS

An ethical clearance was sought and collected from the Ondo State Ministry of Health for permission to carry out the present research that involved human subject. The study subjects consisted of 616 pregnant women aged 15 to 46 years, who came from different local government areas, for antenatal purposes at Ondo State Specialist Hospital located in Akure, between February and April, 2012. They were selected randomly without the prior knowledge of their clinical and family history.

Sample collection

Venous blood was collected from the upper arm of the pregnant women with the assistance of trained Laboratory Scientists, following the method described by cheesebrough (2005). 5 ml of blood needed for the tests was drawn gently at site selected, using sterile needle and syringe into the anticoagulant (EDTA) bottles. Each sample bottle was labeled to correspond to the age and number of subject at the point of collection.

Determination of human immunodeficiency virus (HIV) status of pregnant women

Standard commercially sourced 'Determine' and "Uni-Gold rapid diagnostic tests kits were used to determine the HIV status of the subjects. 'Determine' HIV Rapid Test kit (For use with whole blood, serum, or plasma) (www.who.int/diagnostics laboratory). Uni-Gold test kits are kits that pick or react to only HIV in the blood sample (The Trinity Biotech *Uni-Gold*™ HIV test).

"Determine diagnostic test kit" procedure

The expiration date of kit was checked to ensure the kit had not expired, after which the protective wrapper was removed from the test device and the kit labelled with the patient identification number. Disposable pipette was filled with blood sample and held over the sample pad marked by an arrow symbol into which two drops of sample was carefully added. Blood was allowed to absorb into the sample pad and result was read and recorded after waiting for 60 min. Any positive result was subjected to Uni-Gold test to confirm the positivity of the test result.

Table 1. Age distribution of HIV antibodies among pregnant women.

Age group (years)	No. examined	No. positive	Percentage positive
15-19	13	1	7.69
20-24	49	2	4.08
25-29	211	9	4.27
30-34	222	11	4.96
35-39	104	5	4.81
40-44	14	0	0
≥45	3	0	0
Total	616	28	4.55

Uni-Gold diagnostic test kit procedure

The expiration date of kits was checked to ensure they had not expired, after which the protective wrapper was removed from the test device and the kit was labelled with the patient identification number. Disposable pipette was filled with sample and held over the sample pot, into which two drops of sample was carefully added. Two drops of appropriate wash agent was also added to the sample pot. Result was read and recorded after waiting for 20 min for reaction to occur. The appearance of a line in the control region indicates a negative test result while a line in both the test and control regions is indicative of a positive result. When no line appears in the control region, the test was regarded as inconclusive and consequently repeated with a fresh kit.

Malaria diagnosis in pregnant women

All thick films were prepared and stained with Giemsa stain using a modification of method described by Cheesebrough (2005). The approximate numbers of parasites were reported using plus sign: 1 to 10 parasites per 100 high power fields as +, 11 to 100 parasites per 100 high power fields as ++, 1 to 10 parasites in every high power field as +++, and if no parasites were found after examined as NPF (Cheesbrough, 2005). This was done by a trained and experienced laboratory scientist.

Analysis of data

Analysis of data was carried out using MS-Excel and SPSS version 16.0 (SPSS Chicago Inc., IL, and U.S.A). Pearson correlation analysis was used to investigate possible relationship or correlation between HIV and malaria positivity. Statistical significance was set at p value < 0.05.

RESULTS

Prevalence was calculated by comparing the number of positive individuals with the total number examined. Also, for the analysis, the population was grouped into seven age categories and for each age group of pregnant mothers, the number positive and infection rate per category was estimated.

Results of HIV screening in different age groups of pregnant women

Result in Table 1 showed that out of the six hundred and sixteen sampled populations, 28 women representing 4.55% were positive to HIV infection. The prevalence of HIV antibodies occurred most among women aged 15 to 19 (7.69%) followed by those aged 30 to 34 years (4.96%) while none of the women aged 40 years and above was found to be infected by the virus. The distribution of infection among other age groups of women did not follow any particular pattern. For instance, women aged 20 to 24 had (4.08%), 25 to 29 had (4.27%), 30 to 34 had (4.96%), and 35 to 39 years had (4.81%). Result also shows that HIV infection rate decrease as age of women increases, with the exception of mothers aged 25 to 29 and 30 to 34 years, where infection rate increased slightly from 4.27 to 4.96%.

Prevalence of malaria parasitaemia in the different age groups of pregnant women

In Table 2, a total of 597 pregnant women representing 96.92% of those tested had malaria infection. The infection was generally high in the sampled population. Women aged 15 to 19 and 40 years and above were all infected with malaria parasite. The rate of infections increased steadily through the age group, starting from age 20 to 24, with 95.92%, up to age 35 to 39 years, with 99.04%.

Characterization of malaria parasite density in relation to age of positive pregnant women

The density of malaria parasite in the different age groups of the 597 infected pregnant women was categorized as follows: (+) low, (++) moderate and (+++) high parasiteamia. Result indicated that malaria with very high

Table 2. The prevalence of malaria parasite among the different age groups of pregnant women.

Age group (years)	No. examined	No. positive	Percentage positive
15-19	13	13	100
20-24	49	47	95.92
25-29	211	203	96.21
30-34	222	214	96.39
35-39	104	103	99.04
40-44	14	14	100
≥45	3	3	100
Total	616	597	96.92

Table 3. Characterization of malaria parasite density in relation to age of positive pregnant women.

Age group (years)	Malaria parasite density		
	Low (+)	Moderate (++)	High (+++)
15-19	12	1	0
20-24	42	5	0
25-29	187	16	0
30-34	201	12	1
35-39	97	6	0
40-44	14	0	0
≥45	3	0	0
Total	556 (93.17%)	40 (6.7%)	1 (0.17%)

Age long medically accepted mode of classification of malaria parasite density in patients.

high parasiteamia was found only in pregnant women aged 30 to 34 (0.17%), moderate parasiteamia was found in mothers aged 15 to 39 (6.7%) while the older ages were not infected, and low parasiteamia was found in all the age groups (93.17%) (Table 3).

Result of co-infection of HIV and malaria parasites among pregnant women

Result obtained on the co-infections of HIV and malaria parasites in this present study ascertain the proportion of pregnant women carrying concurrent infections of both HIV and malaria. This study revealed 28 (4.55%) positivity to both pathogens, 92.37% had malaria alone and none of the women had HIV only (Table 4). Pearson's correlation coefficient (r) of 0.5 shows moderate correlation. There was a significant correlation ($P < 0.05$) between HIV and malaria parasite infection in the pregnant women (Table 5).

DISCUSSION

The present 4.55% prevalence of HIV antibodies in Akure, Ondo State was not surprising based on the number of AIDS cases reported from Nigeria (WHO, 1995). In addition, personal observation showed that practices identified as high risk factors in the contraction and transmissions of HIV infections (Olusi et al., 1996; Olusi, 1998) are common in the state. Such practices include prostitution, keeping of multiple sex partners, use of unsterilized or partially sterilized needle and syringes by qualified and unqualified health workers, especially in mission houses. As a result of these, the majority of persons in the study group can be assumed to have a history of risky practices.

The differences in the pattern of HIV infection in the age group of pregnant mothers in this study has to do with the known epidemiological features of HIV infection (Olusi et al., 1996). The absence of HIV antibodies among women aged 40 years and above suggested that

Table 4. Prevalence of HIV and Malaria parasite Co-infections among pregnant women in Akure.

Infection	No. examined	No. positive	Percentage positive
HIV and Malaria	616	28	4.55
Malaria alone	616	569	92.37
HIV alone	616	0	0

Table 5. Correlation of malaria parasite and HIV.

Parameter		Value	Asymptotic standard error (a)	Approximation T (b)	Approxinmate significance
Interval by Interval	Pearson's R	0.500	0.074	14.078	0.000 (c)
No. of valid cases		597	-	-	-

[a]Not assuming the null hypothesis. [b]Using the asymptotic standard error assuming the null hypothesis. [c]Based on normal approximation.

most of the women in this age group had formal education about the infection and took extra care by keeping to one sex partner and more so that they are in their late reproductive ages, while other age groups could probably have contracted the HIV antibodies through heterosexual intercourse and carefree attitude, most especially among the young pregnant women aged 15 to 19 years that are still sexually active; these therefore probably indicates the reason for the prevalence and distribution of HIV infection among pregnant women in Akure, Ondo State.

The present work is in contrast to the research work done at Cameroon by Nkuo-Akenji et al. (2011) who observed the prevalence rate of 21.1% HIV infection and a higher number (60.7%) of HIVSP patient in ages 26 to 35 and 14.1, 20.0 and 0% in ages 15 to 25, 36 to 45 and > 45, respectively. Perrault (2009) reported mean HIV prevalence in pregnant women attending antenatal clinics in South Africa, where he noted prevalence rate of 15 to 40% in South Africa but a lesser prevalence of 5 to 10% in East Africa. Adeoti et al. (2012) also exhibited 19 (12.8%) prevalence of HIV infection among 149 pregnant mothers examined.

The result obtained in the present study indicates that 96.92% of the population of pregnant women examined had malaria parasite, though majority (93.17%) of the mothers were infected at low parasitaemia, which is suggestive of the reason why they did not come down with malaria despite being infected. This indicated that pregnant women could be carriers of malaria parasites without showing any symptoms to the infection. The high infection rate in this study therefore is attributed to the endemic nature of the disease in Ondo state of Nigeria.

The results obtained in this work and those of previous studies support the endemic nature as well as in other part of the country. For instance, Martra et al. (1993) found *Plasmodium falciparum* affecting 97.27% cases of pregnant women examined. Falade et al. (2008) found 89% asymptomatic malaria parasite among pregnant women attending antenatal in a Secondary health care facility in Ibadan, Nigeria. Adefioye et al. (2007) also recorded 72% prevalence rate and Chukwurah et al. (2003) recoded 63.5% of prevalence of malaria parasite in pregnant women in Akwa Ibom, Nigeria. The high infection rate observed among women aged 15 to 19 and 40 years and above is probably an indication of low immunity to malaria infection due to primigravidae status of pregnant women (first pregnancy), and so they are more susceptible to malaria infection than others (Martra et al., 1993). Immunity to malaria infection diminishes significantly in pregnancy, particularly in primigravidae (Oforie et al., 2009). In first and second pregnancies, women are especially vulnerable to malaria infection. Mcgregor (1984) identified the factors responsible for susceptibility of primigravidae to malaria as inhibition of type 1 cytokine responses (interferon, interleukin 2 and 12 and TNF). Brabin (1991) also confirmed that the primigravidae were more susceptible to malaria infection than the multigravidae. Opare Addo et al. (2002) reported that primigravidae, alongside those women in their second pregnancy, were more vulnerable to malaria parasitemia. This work is in contrast to the work of McGregor (1983) who reported a decline in malaria prevalence as age increase, as well as with improved host immunity, thus reducing susceptibility in later years.

The percentage of women carrying both infections in the study indicated that all the pregnant women that were positive to HIV antibodies also habour malaria parasites. This research work compared with other recent studies, Adeoti et al. (2012), observed that the prevalence of malaria infections in HIV infected mothers was due to the impaired immune status of HIV infected pregnant women to control *P. falciparum*. It was also observed that the pregnant women that are infected with the HIV severely suffered from malaria. Previous studies indicate that people with haemoglobin genotype AS are not prone to malaria susceptibility when they have high level of immunity, but from this study high parasitaemia (+++) was observed from a pregnant woman with genotype AS. The findings in this study further support previous findings that dual infection with HIV and malaria may have serious consequences on the health of pregnant women. The dual infected pregnant women could be susceptible to many other infections. This is the evidence that where malaria and HIV occur together, infections interact. HIV worsens malaria infection by lowering the immunity of the infected pregnant women while malaria increases the risk of transmitting HIV from mother to child and also HIV viral load in the infected pregnant women is increased.

Desia et al. (2007) reported that in the areas endemic for malaria, at least 25% of pregnant women are infected with malaria, with the highest risk for infection and morbidity in primigravidas, adolescents, and those co-infected with HIV. Ned et al. (2005) also observed that the HIV co-infection has its impact on disease presentation, with an increased risk of complicated and severe malaria and death. Malarial infection in HIV-positive women is associated with higher levels of parasitemia, leading to a greater risk of severe anemia. Likewise, HIV viral load is increased, creating opportunity for infection and more severe disease (Briand et al., 2009). The moderate correlation that occurs between malaria and HIV implies that mothers infected with HIV are most likely to be infected with malaria due to the compromised immune system which makes them more susceptible to malaria infection.

RECOMMENDATION

It is therefore recommended that all pregnant women attending antenatal clinic in State Specialist Hospital Akure be tested for malaria parasite every month, in order to detect early, in case of any symptomatic or asymptomatic malaria infection, so as to be treated appropriately and prevent further morbidity and mortality associated with malaria. Intermittent preventive treatment should be provided for all pregnant women with preventive doses of an effective anti-malarial drug during routine antenatal clinic visits. Mosquito treated net should be made available free for the pregnant women. Training on how to use and maintain the net must be given as this will decrease the exposure to infective mosquito bites. More impressive public announcement and awareness should be made to the pregnant women in the public to attend good antenatal clinic for proper handling of their health and pregnancy from malaria and HIV.

ACKNOWLEDGEMENT

The authors are grateful to the management of Ondo State Specialist Hospital for granting us audience to collect the sample for this research work. We also appreciate Mr. Lawrence Ogundoju of Federal University of Technology, Akure Health Centre where the laboratory analyses of the samples gotten were carried out.

Conflict Interests

The authors declare that there is no conflict of interests

REFERENCES

Abu-Raddad LJ, Patnaik P, Kublin JG (2006). Dual infection with HIV and malaria fuels the spread of both diseases in sub-Saharan Africa. Science 314(5805):1603-06.

Adefioye OA, Adeyeba OA, Hassan WO, Oyeniran OA (2007). Prevalence of Malaria Parasite Infection among Pregnant Women in Osogbo, Southwest, Nigerian. American-Eurasian J. Sci. Res. 2(1):43-45.

Adeoti OM, Anumudu CI, Nwuba RI, Awobode HI, Olaniyan MF, Olayiwola O, Fagbade O (2012). Prevalence of HIV and Malaria parasites co-infection in pregnant mothers and their babies post-delivery. J. Biol. Agric. Healthcare 2(6):59.

Brabin BJ (1991). The risks and severity of malaria in pregnant women. In: Applied field research in malaria. Report no. 1. Geneva WHO. pp. 1-34.

Brentlinger PE, Behrens CB, Micek MA (2006). Challenges in the concurrent management of malaria and HIV in pregnancy in Sub-Saharan Africa. Lancet Infect. Dis. 6:100-111.

Briand V, Badaut C, Cot M (2009). Placental malaria, maternal HIV infection and infant morbidity. Ann. Trop. Paediatr. 29:71-83.

Cheesbrough M (2005). District Laboratory Practice in Tropical Countries, 2nd Edition. Cambridge University Press, United Kingdom. pp. 244-251.

Chukwura EI, Okpala EE, Ani I (2003). The prevalence of malaria parasites in pregnant women and other patients in Awka Urban, Anambra State. J. Biomed. Investig. 1:48-52.

Desai M, Kuile FO, Nosten F (2007). Epidemiology and burden of malaria in pregnancy. Lancet Infect. Dis. 7:93-104.

Falade CO, Olayemi O, Dada-Adegbola HO, Aimaku CO, Ademowo OG, Salako LA (2008). Prevalence of malaria at booking among antenatal client in a Secondary health care facility in Ibadan, Nigeria. Afr. J. Reprod. Health 12:141-52.

Maitra N, Joshi M, Hazra M (1993). Maternal manifestation of malaria in pregnancy- A review. Indian J. Matern. Child Health 4(4):98-102.

McGregor IA, Wilson ME, Billewiez WZ (1983). Malaria infection of the placental in the Gambia West Africa. Its incidence and relationship to still birth weight and placental weight. Trans. R. Soc. Trop. Med. Hyg. 77(2):232-44.

McGregor LA (1984). Epidemiology, malaria and pregnancy. Am. J. Trop. Med. Hyg. 33:517-525.

Ned RM, Moore JM, Chaisavaneeyakorn S, Udhayakumar V (2005). Modulation of immune responses during HIV-malaria coinfection in pregnancy. Trends Parasitol. (21):284-91.

Nkuo-Akenji T, Tevoufouet EE, Nzang F, Fon E, Ebong IN (2011). HIV/AIDS and malaria in pregnant women from Cameroon Short running title: HIV, malaria in pregnancy. Afr. J. Health Sci. 18:105-109.

Oforie MF, Ansah E, Agyepong I (2009). Pregnancy-associated malaria in a rural community of Ghana. Ghana Med. J. 43:13-18.

Olusi T, Gross U, Ajayi J, (1996). High incidence of Toxoplasmosis during pregnancy in Nigeria. Scand. J. Infect. Dis. 28(6):645-6.

Olusi TA (1998). Studies on the prevalence and seroepidemiology of *Toxoplasma gondii* and Human Immunodeficiency Virus (HIV) on parts of Benue State, Nigeria. Ph.D Theisis University of Jos, Jos, Nigeria.

Opare AHS, Odoi AT (2002). Malaria in pregnancy. In: Kwawukueme EY, Emuveyan EE (eds), Comprehensive Obstetrics in the tropics, 1st ed. Accra, Ghana: Asante and Hittscher Printing Press Publishers. pp. 250-60.

Perrault SD (2009). Human immunodeficiency virus Co-infection increases placental parasite density and transplacental malaria transmission in Western Kenya. Am. J. Trop. Med. Hyg. 80(1):119-125.

Rowland-Jones S, Lohman B (2002). Interaction between malaria and HIV infections-An emerging public health problem. Microbes Infect. 4(12):1265-70.

Schantz-Dunn J, Nour NM (2009). Malaria and pregnany: A Global Health Perspective. Rev. Obstet. Grnecol. 2(3):186-192.

United Nations International Children's Emergency Fund (UNICEF) (2003). Malaria and HIV. Technical Note No 6.

World Health Organization (WHO) (2004). Malaria and HIV Interactions and their Implications for Public Health Policy: Conclusions of a Technical consultation. pp. 23-25. WHO, Geneva.

World Health Organization (WHO) (2011). Malaria and HIV interactions and their implications for public health policy. WHO, Geneva.

Prevalence of *Schistosoma mansoni* and geo-helminthic infections among patients examined at Workemeda Health Center, Northwest Ethiopia

Tadesse Hailu and Mulat Yimer

Department of Microbiology, Immunology and Parasitology [DMIP], College of Medicine and Health Science, Bahir Dar University, P. O. Box 79, Bahir Dar, Ethiopia.

Schistosoma mansoni **and geo-helminths are considerable medical and public health problems in Ethiopia. However, information is limited on the epidemiology of these parasitic infections in different areas, it is very important to plan effective prevention and control measures. Therefore, the aim of this study was to determine the prevalence of *S. mansoni* and geo-helmenthic infections among patients examined at Workmeda Health Center. Institution based retrospective cross sectional study was conducted to determine the prevalence of *S. mansoni* and geo-helmenthic infections among patients from September 2012 to August 2013. A total of 604 participants (56.6% males and 43.4% females) within the age range of six to eighteen were included. A single stool sample was collected and processed using direct microscope. The analysis of the generated data revealed that the overall prevalence of intestinal parasites infection was 32%. The prevalence of *S. mansoni and* hookworm species were 7.3 and 20.7%, respectively. Patients in the age range of 10 to 14 years had higher (14.7%) rate of infection. Prevalence of *S. mansoni* (12%) infection was higher in the age range 10 to 14 years. High prevalence of *S. mansoni* and hookworm species need periodic deworming programme and health education urgent to reduce morbidity and mortality.**

Key words: Prevalence, *Schistosoma mansoni*, geo-helminths, Workmeda.

INTRODUCTION

Intestinal parasites affect human's health in many parts of the world. Their problem is magnificent, particularly in developing nations (WHO, 1981). Several reports showed that the overall high prevalence of intestinal helminths in the world mainly attributed to infections emanating from environmental contamination by human excreta. Thus, helminthic infections as a whole can be viewed as the best indicators of the sanitation level of a community (Mahfouz et al., 1997).

The most prevalent and important helminths in developing countries are Schistosomes (Rozendaal, 1998) and geo-helminths also called soil-transmitting helminthes (STHs) including *Ascaris lumbricoides*, *Trichuris trichiura* and Hookworms (WHO, 1962). Recent global prevalence estimate shows that *A. lumbricoides* infect 1.221 billion, *T. trichiura* infect 795 million and hookworm infect 740 million people worldwide (Silva et al., 2003).

Schistosomiasis also remains one of the most prevalent parasitic diseases in the world and is endemic in 74 tropical countries worldwide. Approximately, 500 to 600 million people are at risk of schistosomal infection, and 80% of the 200 million people are infected worldwide live in sub-Saharan Africa where *Schistosoma mansoni* and *Schistosoma haematobium* are widely distributed (WHO, 2002; Davis et al., 2003). Schistosomiasis is also endemic in many parts of Ethiopia. The intestinal form of schisosomiasis caused by *S. mansoni* is widely distributed in the country, however, the urinary form caused by *S. hamatobium* is limited to some lowland areas of Ethiopia (Lo et al., 1988; Tedla et al., 1998).

Intestinal schistosomiasis and geo-helminths infections are the major causes of morbidity and mortality in different parts of Ethiopia (Aklilu et al., 1986; Leykun, 2001) causing a series of public health problems such as malnutrition, anaemia, and growth retardation as well as higher susceptibility to other infections (Silva et al., 1997). Although many studies previously conducted in Ethiopia indicated that the distribution of different intestinal parasites on different altitudes in different community groups such as preschool children, school children and other groups confined to camps and refugees, the prevalence of *S. mansoni* and STHs infections was not well addressed in different parts of Ethiopia including our study area. In addition, the scarcity of reports on the distribution and prevalence of *S. mansoni* and STHs infections in Ethiopia promoted investigation of the conditions in Northwest Ethiopia.

Therefore, the aim of this study was to determine the prevalence of *S. mansoni* and STHs infections among patients examined at Workmeda Health Center, Northwest Ethiopia. The research finding could provide baseline data on the distribution and prevalence of intestinal parasites and assisting in proposing strategies to protect parasitic infections.

MATERIALS AND METHODS

Study population, area and period

Institution based retrospective cross sectional study was conducted to determine the prevalence of *S. mansoni* and geo-helminths among patients examined at Workmeda Health Center, Jawe Woreda, Northwestern Ethiopia from September 2012 to August 2013. The area has an elevation of about 1000 to 1050 m above sea level. The study subjects engaged in this study were 604 clinically suspected cases of intestinal parasite infection within age range 6 to 18 who visited Outpatient Department (OPD) of Workmeda Health Center. Eligible study subjects for the study were those patients clinically suspected of intestinal parasitosis and had stool examination from 6 to 18 years. Children less than 5 years and adults greater than 18 years with diarrhea and suspected of intestinal parasitosis were excluded from this study.

Clinical and laboratory diagnosis

One year retrospective data of intestinal parasitosis were collected from Workmeda Heath Center. In this Health Center, direct stool examination of well-prepared smears is used in confirming the presence of Schistosomiasis and other intestinal parasites infection. In Ethiopia, detection of intestinal parasites in stool is conducted according to a standard operating procedure (SOP) in each health center throughout the country. The direct smear method is done by mixing 2 mg of stool with either one drop of saline (diarrhea) or iodine (formed) on a slide, covered with cover slide and sealed first with low power, then high power (WHO, 2004). Therefore, for this study objective, one year (September 2012 to August 2013) of *S. mansoni* and Geo-helminths data, Socio-demographic information and environmental related factors at Workmeda Health Center in October 2013 had been collected by reviewing the patient's card and laboratory log book.

Statistical methods

Data were entered in to excel and transported to SPSS. Analysis was performed by SPSS version 16 statistical software package. Frequency and percentage were calculated for the study variable. Chi-square, p-value and two tail Fisher's exact test were used to calculate and determine significance. In all statistical tests, the differences were considered to be statistically significant if p-value less than 0.05.

Ethical consideration

The department ethical review committee of Microbiology, Immunology and Parasitology, College of Medicine and Health Science, Bahir Dar University approved the project. The researchers obtained informed consent from the Workmeda Health Center.

RESULTS

Socio-demographic characteristics of study subjects

A total of 604 clinically suspected intestinal parasitosis cases who attended at Workmeda Health Centre were enrolled in this study. The mean age of the attendants was 12.24 years with a standard deviation (SD) of 4.05 ranging from 6 to 18. There were more males (56.6%) than females (43.4%). Nearly 9% of the cases were under 7 years, 56% of the cases were between 7 and 14 years, the rest 35% were great than 14 years old. And majority of the study group were greater than 10 years old (Table 1).

The overall prevalence of parasitic infection was 32%. Of which, hookworm species accounted for (20.7%) parasites infection followed by *S. mansoni* (7.3%) (n=604) (Table 2). About 3.6% of the cases were co-infected with hookworm species and *S. mansoni* (n = 193). However, no triple, quadruple and multiple infections were found (Table 2).

Prevalence of *S. mansoni* and geo-helminths

The overall prevalence of *S. mansoni* and geo-helminths

Table 1. Prevalence of intestinal parasitosis infections based on their socio-demographic characters in Northwest Ethiopia [n, %].

Result	Age			Sex		Address	
	6-9	10-14	15-18	M	F	Urban	Town
Positive	45 (25.4)	92 (42.6)	56 (26.5)	114 (33.3)	79 (30.2)	183 (31.9)	10 (33.3)
Negative	132 (74.6)	124 (57.4)	155 (73.5)	228 (66.7)	183 (69.8)	391 (68.1)	20 (66.7)
Total	177 (29.3)	216 (35.8)	211 (34.9)	342 (56.6)	262 (43.4)	574 (95)	30 (5)
χ^2, P		29.09, 0.04		8.53, 0.481		7.839, 0.550	

Table 2. Distribution of intestinal parasites in relation with their address in Northwest Ethiopia [n, %].

Result	Address						
	Asech	Cimida	Dire	Kava	Workmeda	Wobo	Total
Hookworm	22 (17.6)	24 (19.2)	13 (10.4)	45 (36)	8 (6.4)	13 (10.4)	125 (4.8)
A. lumbricoides	0 (0)	1 (33.3)	1(33.3)	1 (33.4)	0 (0)	0 (0)	3 (.11)
T. trichiura	0 (0)	0 (0)	0 (0)	0 (0)	0 (0)	0 (0)	0 (0)
S. mansoni	3 (8.1)	9 (24.3)	8 (21.6)	14 (37.9)	1 (2.7)	2 (5.4)	37 (19.2)
Strongyloides stercoralis	1 (16.7)	2 (33.3)	1 (16.7)	2 (33.3)	0 (0)	0 (0)	6 (31)
H. nana	0 (0)	0 (0)	0 (0)	2 (66.7)	1 (33.3)	0 (0)	3 (.11)
Hookworm + *S. mansoni*	1 (14.2)	0 (0)	3 (42.9)	3 (42.9)	0 (0)	0 (0)	7 (36.3)
Hookworm + *A. lumbricoides*	0 (0)	1 (100)	0 (0)	0 (0)	0 (0)	0 (0)	1 (0.5)
Entamoeba histolytica	1 (12.5)	0 (0)	0 (0)	5 (62.5)	0 (0)	2 (25)	8 (41.5)
Giardia lamblia	0 (0)	0 (0)	2 (66.7)	1 (33.3)	0 (0)	0 (0)	3 (1.55)
Total positive	28 (5.4)	37 (35.9)	28 (35.9)	73 (29.3)	10 (33.3)	17 (26.2)	185 (30.6)
Total	79 (3.1)	103 (17)	78 (12.9)	249 (41.2)	30 (5)	65 (10.8)	604
χ^2, P				47.622, 0.366			

was 30%. The prevalence of *S. mansoni*, hookworm species, *A. lumbicolds* and *T. trichuria* in this study was found to be 7.3, 22, 0.7, and 0%, respectively (n = 604). The prevalence of *S. mansoni* which was 2.8% found in Kava was high followed by 1.8% in Dire Kebele. Similarly, the prevalence of hookworm species (7.9%) was high in Kava Kebele, followed by 4.1% in Cimida (Table 3).

Prevalence of *S. mansoni*

The prevalence of *S. mansoni* obtained from rural and urban was 34 (5.6%) and 10 (1.7%) (n=604), respectively but not statistically associated (Table 4). About 8.2% of male (n= 342) and 6.1% of female (n= 262) cases were found to be infected with *S. mansoni* (Table 4). The prevalence of *S. mansoni* infection was high (12%) in age group 10 to 14 (n = 216), followed by 5.1% in age group 6 to 9 (n = 177). *S. mansoni* infection was statistically associated with high prevalence in males and age group 10 to 14 (Table 4).

DISCUSSION

The prevalence of *S. mansoni* and geo-helminths among school age segments of the population of Jawe Woreda was determined. The overall prevalence of parasitic infection in this study was 32% and the infection rates were very high in children age range of 10 to 14. Prevalence of parasitic infections obtained in the present study was comparable with those previously reported (34.4%) in South Western Ethiopia (Awole et al., 2003) and 36.6% in North Western Tanzania (Mazigo et al., 2010). In contrast to the present study, high prevalence of parasitic infection (50.6%) was reported in Gorgora Town, Northwest Ethiopia (Assefa et al., 2013).

In Ethiopia, most *S. mansoni* infections and transmission sites are in agricultural communities along streams between 1300 and 2000 m altitude (Kloos et al., 1988). The prevalence of *S. mansoni* (7.3%) in the present study was comparable with 10.1% in North West Ethiopia (Zinaye et al., 2013), and 5.6% in North Western Tanzania (Mazigo et al., 2010), but higher than 3% in South Western Ethiopia (Awole et al., 2003). In contrast to the present study, high prevalence of *S. mansoni* infections, 73.7% in Southern Ethiopia (Terefe et al., 2011), 27.1% in Northern Ethiopia (Dejenie et al., 2009), and 23.1% in South Cote De voire (Coulibaly et al., 2013) was reported.

The low prevalence of *S. mansoni* in this study may be due to the direct diagnosis method applied in the Ethiopian health system. The best diagnosis method for

Table 3. Prevalence of *S. mansoni* and geo-helminths in Northwest Ethiopia [n, %].

Result	Address						Total
	Asech	Cimida	Dire	Kava	Workmeda	Wobo	
Hookworm	23 (3.8)	25 (4.1)	16 (2.7)	48 (7.9)	8 (1.3)	13 (2.2)	133 (22)
A. lumbricoides	0 (0)	2 (0..3)	1 (0.2)	1 (0.2)	0 (0)	0 (0)	4 (0.7)
T. trichuria	0 (0)	0 (0)	0 (0)	0 (0)	0 (0)	0 (0)	0 (0)
S. mansoni	4 (0.7)	9 (1.5)	11 (1.8)	17 (2.8)	1 (0.2)	2 (0.3	44 (7.3)
Total positive	27 (4.50)	36 (6)	28 (4.6)	66 (10.9)	9 (1.5)	15 (2.5)	181(30)
Total cases	80 (13.3)	102 (16.9)	78 (12.9)	246 (40.7)	31 (5.1)	67 (11.1)	604

Table 4. Distribution of *S. mansoni* infection based on patients age, sex and address in Northwest Ethiopian [n, %].

Result	Total	Age			Sex		Address	
		6-9	10-14	15-18	M	F	Rural	Urban
Positive	44 (7.3)	9 (5.1)	26 (12)	9 (4.3)	28 (8.2)	16 (6.1)	34 (5.6)	10 (33.3)
Negative	560 (92.7)	168 (94.9)	190 (88)	202 (95,7)	314 (91.8)	246 (93.9)	540 (94.1)	20 (66.7)
Total	604	177 (29.3)	216 (35.8)	211 (34.9)	342 (56.6]	262 (43.4)	574 (95)	30 (5)
χ^2, P		10.527, 0.005			4.156, 0.028		0.708, 0.400	

S. mansoni is the Kato Katz test which is not used as a routine diagnosis method in Ethiopia (Endris et al., 2012). On the other hand, the prevalence of intestinal parasites in the present study may be attributable to walking on bare foot, unhygienic conditions, insufficient provision of safe water, in appropriate utilization of latrine, crossing the river when going to their field work and use river water for washing, swimming and playing. The magnitude of the problem emphasizes the need to take immediate intervention measures. Combined mass chemotherapy and focal snail control using primary health care systems may have an effect on the prevalence and intensity of parasitic infections in the study area.

The distribution of *S. mansoni* infection among age groups showed that there was high prevalence of *S. mansoni* (12%) in age groups 10 to 14. This study was in agreement with previous study reports (28.5%) in South West Ethiopia (Mengistu et al., 2007), 28.53% in North West Ethiopia (Zinaye et al., 2013), and 30.71% in Northern Ethiopia (Assefa et al., 2013) which showed high prevalence of *S. mansoni* in age group 10 to 14.

The prevalence of *S. mansoni* infection was high among males than females. A similar high prevalence of *S. mansonai* in males than females was reported previously in South West Ethiopia (Dejenie et al., 2008). This might be due to males are mostly engaged to manipulate the farming activity so that they might be exposed to river water in washing, crossing and swimming more frequently than females. In addition, there are two main rivers (Burabur and Asewe) which are used for a source of water in the study area. Males do not have any problem to swim and wash in these rivers, but females are culturally influenced by the community not to swim and to swim and wash in the rivers.

Among the geo-helminths, the prevalence of hookworm species (22%) was high in this study. Similar previous high hookworm species prevalence (25.2%) in Tanzania (mazingo et al., 2010) and 28.4% in Southern Ethiopia (Kloos et al., 1988) was reported. In contrast to this study, lower hookworm prevalence reports 12% in South Western Ethiopia (Awole et al., 2003) and 8.2% in North West Ethiopia (Endris et al., 2012).

Though, multiple infections were common in Ethiopia (Mengistu et al., 2007), only 1 to 2 species of parasites per individual were observed in this study and majority of the co-infections were between *S. mansoni* and hookworm species. Prevalence of co-infection of *S. mansoni* and hookworm (3.6%) in this study was in agreement with previous 7.4% in North Ethiopia (Kloos et al., 1988), and 74.6% in Eastern Uganda (Moira et al., 2012). This may be because of the higher prevalence of each parasites and/or their similar mode of transmission which favors dual infections and could also be the different stool examination method.

Conclusion

This study showed that high prevalence of hookworm species and *S. mansoni* were an important health problem in Jawe Woreda North West Ethiopia. Factors such as low awareness of schistosomiasis, swimming, washing, bathing and crossing the river water, and walking on bare foot might be associated risk factors for hookworm species and *S. mansoni* infection in the study area. This calls for periodic deworming programme to reduce

reduce transmission, worm burden and morbidity. Deworming for both *S. mansoni* and STHs should be supplemented with improved sanitation and access to clean water, appropriate health education and environmental measures to have a lasting impact on transmission. The impact of each measure would be maximized through a health education program directed to school age children in particular, and to communities in general.

ACKNOWLEDGEMENT

The authors would like to acknowledge Workmeda Health Center for permission to conduct the research. They would also like to extend their thanks to the laboratory staffs of Workmeda Health Center for their help in the research process.

Conflict Interests

The authors declare that there is no conflict of interests regarding the publication of this article.

REFERENCES

Aklilu L, Demisse M, Bahta M (1986). Parasitiological survey of Addis Ababa and Debre Zeit school children with special emphasis on Bilharziasis. Ethiop. Med. J. 6:1-7.

Assefa A, Dejenie T, Tomass Z (2013). Infection prevalence of Schistosoma mansoni and associated risk factors among schoolchildren in suburbs of Mekelle city, Tigray, Northern Ethiopia. Mek. Univ. 5(1):174-188.

Awole M, Gebre-Selassie S, Kassa T, Kibru G (2003). Prevalence of Intestinal Parasites in HIV-Infected adult Patients in Southwestern Ethiopia. Ethiop. J. Health Dev. 17:71-78.

Coulibaly JT, N'Gbesso YK, Knopp S, N'Guessan NA, Silue KD, Van Dam GJ, NGoran EK, Utzinger J (2013). Accuracy of Urine Circulating Cathodic Antigen Test for the Diagnosis of Schistosoma mansoni in Preschool-Aged Children before and after Treatment. PLOS Neg. Trop. Dis. 7(3):e2109.

Davis A, Cook C, Zumla A (2003). Schistosomiasis: Manson's Tropical Diseases, London. pp. 1431-1469.

Dejenie T, Asmelash T (2008). Impact of irrigation on the prevalence of intestinal parasite infections with emphasis on schistosomiasis in Hintallo-Wejerat, North Ethiopia. Ethiop. J. Health Sci. 18(2):33-38.

Dejenie T, Asmelash T, Teferi M (2009). Intestinal Helminthes Infections and Re-Infections with Special Emphasis on Schistosomiasis Mansoni in Waja, North Ethiopia. Meke. Univ. 1(2):4-16.

Endris M, Tekeste Z, Lemma W, Kassu A (2012). Comparison of the Kato-Katz, Wet Mount, and Formol-Ether Concentration Diagnostic Techniques for Intestinal Helminth Infections in Ethiopia. ISRN Parasitol. 2012:5.

Kloos H, Lo CT, Birrie H, Ayele T, Tedla S, Tsegay F (1988). Schistosomiasis in Ethiopia. Soc. Sci. Med. 26(8):803-827.

Leykun J (2001). Soil Transmitted Helminthic Infection and S. mansoni in School Children from Chilga District, North West Ethiopia. Ethiop. J. Health Sci. 1:79-87.

Lo CT, Kloos H, Hailu B (1988). Schistosomiasis: The Ecology of Health and Disease in Ethiopia. Addis Ababa., EMPDA Press. pp. 196-207.

Mahfouz AAR, El-Morshedy H, Farghaly A, Khalil A (1997). Ecological determinants of intestinal parasitic infections among pre-school children in an Urban Squatter Settlement of Egypt. J. Trop. Ped. 43:341-334.

Mazigo HD, Ambrose EE, Zinga M, Bahemana E, Mnyone LL, Kweka EJ, Heukelbach (2010). Prevalence of intestinal parasitic infections among patients attending Bugando Medical Centre in Mwanza, north-western Tanzania: a retrospective study. Tanzan. J. Health Res. 12:2.

Mengistu A, Gebre-Selassie S, Kassa T (2007). Prevalence of intestinal parasitic infections among urban dwellers in southwest Ethiopia. Ethiop. J. Health Dev. 21(1):12-17.

Moira APD, Jones FM, Wilson S, Tukahebwa E , Fitzsimmons CM, Mwatha JK, Bethony JM, Kabatereine NB, Dunne DW (2012). Childhood schistosome and hookworm co-infections: effects of treatment on IgE responses against parasite allergen-like proteins and immunity to re infection. Infect. Immun. doi:10.1128/IAI.00748-12

Rozendaal JR (1997). Vector Control Methods for use by Individuals and Communities. WHO, Geneva, No. 337-356.

Silva NRDe, Brooker S, Hotez PJ, Montresor A, Engles D, and Savioli L (2003). Soil-transmitted helminth infections: updating the global picture. Trends Parasitol. 19(12):547-551.

Silva NRDe, Guyatt HL, Bundy DA (1997). Morbidity and mortality due to ascaris-induced intestinal obstruction. Trans. R. Soc. Trop. Med. Hyg. 91(1):31-36.

Tedla S, Tilahun G, Burie H (1998). Parasitology. Schistosomiasis in Ethiopia and Eritrea, 2nd ed. Institute of Pathobiology., AAU. pp. 9-28.

Terefe A, Shimelis T, Mengistu M, Hailu A, Erko B (2011). *Schistosomiasis mansoni* and soil-transmitted helminthiasis in Bushulo village, southern Ethiopia. Ethiop. J. Health Dev. 25(1):46-50.

World Heallth Organization (WHO) (1962). Soil transmitted helminthes, WHO Tec Rep ser, Geneva, No. 277.

World Heallth Organization (WHO) (1981). World Health Organization Scientific Group Intestinal protozoa and helminths infection, WHO Tech Rep Ser. 666, Geneva No 7-43.

World Heallth Organization WHO (2002). Prevention and Control of Schistosomiasis and Soil-Transmitted Helminthiasis. WHO, Technical Report Series Geneva.

World health Organization (WHO) (2004). Schistosomiasis and intestinal parasites unit division of control of tropical diseases. WHO, Training manual on diagnosis of intestinal parasites, Geneva.

Zinaye T, Yeshambel B, Amare GH, Beyene M, Meseret W, Getnet A, Misganaw M, Afework K (2013). Epidemiology of intestinal schistosomiasis and soil transmitted helminthiasis among primary school children in Gorgora, Northwest Ethiopia. Asian Pac. J. Trop. Dis. 3(1):61-64.

Epidemiological significance of major hemoparasites of ruminants in and around Debre-Zeit, Central Ethiopia

Tirualem Sitotaw[1], Fikru Regassa[1], Fikre Zeru[2] and Abraha Gebregziabher Kahsay[3,4]

[1]College of Veterinary Medicine and Agriculture, Addis Ababa University, Debre Zeit, Ethiopia.
[2]School of Veterinary Medicine, Semera University, Semera, Ethiopia.
[3]School of Veterinary Medicine, Wollo University, P.O. Box 1145, Dessie, Ethiopia.
[4]Institute of Tropical Medicine, Sint Rochusstraat 21, Antwerp, Belgium.

Prevalence of hemoparasites of ruminants (cattle, sheep and goat) in and around Debre-Zeit was determined using diagnostic records from November, 2010 to April, 2011. A total of 384 blood samples collected from ruminants present in Debre-Zeit town and surrounding peasant associated areas were examined by Giemsa stained, thin blood smears and Diff-quick method for hemoparasite. Packed cell volume (PCV) for each sample was estimated and hematocrit centrifugation (HCT) technique was used to determine the presence of hemoflagellates. An overall prevalence of hemoparasites was 6.3% which consisted of *Anaplasma* species (*Anaplasma ovis* 2.1%, *Anaplasma marginale* 1.6%, *Anaplasma centerale* 0.3%), *Babesia* species (*Babesia bigeemina*, *Babesia bovis* and *Babesia ovis* total of 0.9%) and *Trypanosome* species (only *Trypanosome vivax* 1.6%). The hemoparasite detected alone or in concurrent infection had a significant (P<0.05) effect on PCV of infected animals. The result of this study shows these hemoparasites are not endemic in ruminants in the study area, which may results in serious disease conditions when such animals are subjected to stress full condition.

Key words: Debre-Zeit, hemoparasites, prevalence, ruminants.

INTRODUCTION

Ethiopia is one of the countries with the largest number of livestock in Africa and livestock production plays a major role in the development of Ethiopia's agriculture. Nevertheless, cattle productivity is low (Central Statistics Authority (CSA), 2009). This may be due to improper management, disease, nutritional deficiencies, harsh environment and genetic factors. The presence of diseases caused by hemoparasites is broadly related to the presence and distribution of their vectors. These diseases cause negative effects on the health of the livestock including production and productivity. Arthropod transmitted hemoparasitic disease of cattle is caused by the *Trypanosomae, abesia*, *Theleria* and *Anaplasma* species. The extracellular species belonging to the genus *Trypanosomae* and intracellular species are *Anaplasma*, *Theleria* and *Babesia* in animals (Alekaw, 2000).

Arthropod transmitted hemoparasitic diseases are economically important vector-borne diseases of tropical and subtropical parts of the world including Ethiopia. *Bovine babesiosis* caused by an apicomplexan haemoprotozoan parasite, *Babesia bigemina* (family Babesiidae, order Piroplasmida), is transmitted by brevirostrate tick, *Rhipicephalus microplus*, causing significant morbidity and mortality in cattle and buffaloes. *Trypanosoma evansi*, the causative agent of the "Surra," is mechanically transmitted by tabanid flies (Sumba et al., 1998). Dairy animals, especially bovines, which are bearing production stress along with other diseases, are potential viable host to these infections. Bovines act as reservoir hosts of Surra as the cause of disease remains subclinical (McLeod and Kristjanson, 1999).

Animal trypanosomosis and its vectors occur in vast areas of the Sub-Saharan Africa with devastating impact on livestock productivity posing a serious threat to the livestock and communities. It constitutes the greatest single constraint to livestock and crop production by directly contributing to hunger, poverty, protein malnutrition and suffering of entire communities in Africa (WHO, 2002). Of the 165 million cattle found in Africa, only 10 million are found within the tsetse fly free belt, and these are mostly low producing breeds which are maintained on high drug management regimes to keep trypanosomosis (Jones and Dávila, 2001). The disease has economic importance due to loss of condition, reduction in milk yield, decrease capacity of work (Reghu et al., 2008).

Prevalence of the disease depends on the rate of exposure, availability of infected animals, the insect reservoir and seasons (Mottelib et al., 2005). In Ethiopia, quite a number of epidemiological studies have been conducted on cattle trypanosomosis and these epidemiological studies were carried out using conventional parasitological techniques such as dark phase buffy coat, thin and thick smear (Dagnachew et al., 2005; Sinshaw et al., 2006). It is the major constraints on livestock production and general agricultural development over an estimated area of 98,000 km^2 in the country (Solomon, 1980).

Babesia species are intra erythrocytic protozoal organisms spread by arthropods like ticks and biting flies transplacentally and by blood transfusion. Babesiosis is a tick born disease of domestic, wild and laboratory animals as well as humans caused by the genus *Babesia*. More than 100 known *Babesia* spp. have been identified which infect many types of mammalian host, out of these, 18 cause disease in domestic animals notably in cattle, sheep, goats, horses, pigs, dogs and cats. Almost any mammals, that serves as a host for a *Babesia* infected tick is a potential reservoir (Homer et al., 2000).

Anaplasmosis is a vector born infectious blood disease in cattle caused by the rickettsial parasites, *Anaplasma marginale* and *A. centrael*. It occurs primarily in worm tropical and subtropical areas. The disease is not contagious but transmitted most commonly by ticks. It can also be transmitted via contaminated surgical instruments, biting flies and mosquitoes. The intracellular parasites destroy the red blood cells. It causes anemia, fever, weight loss, breathlessness, uncoordinated movements, abortion and death (Centers for Disease Control and Prevention (CDC), 2009).

Development and transmission of hemoparasites by tick vectors are phenomena closely synchronized with the tick feeding cycle. In all known life cycles, initial infection of tick tissues occurs in midgut epithelial cells and transmission is effected as ticks feed after parasites have developed and multiplied in salivary glands. Many factors reviewed affect development and transmission of hemoparasites by ticks including age of ticks, artificial temperature, climate and/or season, tick stage or sex, hemoparasite variation, concurrent infection of ticks with other pathogens, host cell susceptibility, transovarial transmission, effect of hemoparasites on tick biology, and the effect of infecting parasitemia level in cattle on infection rates in ticks. Four hemoparasites of cattle, *A. marginale*, *Cowdria ruminantium*, *Theileria parva*, and *Babesia* spp., are all dependent on ticks for biological transmission. *Babesia* is transmitted trans-ovarially, whereas the other three are transmitted transracially. Mechanical transfer of infective blood via formites and mouthparts of biting arthropods is also a major means of transmission for *A. marginale*, but not of the others. Potential control methods for hemoparasites include tick control, vaccines (against ticks and parasites), and drugs (against ticks and parasites). Successful application of control strategies will be dependent upon thorough understanding of parasite developmental cycles, biology of the tick vectors and the immune response of cattle to ticks and to hemoparasites (Kocan, 1995). The most effective control measures are targeted against both ticks and the hemoparasites.

In Ethiopia, various surveys have been carried out on distribution, abundance and prevalence of hemoparasite species on livestock in different regions of the country by various investigations (Adam et al., 1971) including trypanosomosis of ruminants, babesiosis in donkeys, and canine bebasiosis (Tolosa, 2010). Such investi-gations have been conducted in the different regions of the country like Ghibe, Asela, Wolayta, Bahirdar, Gondar, Debre Markos, Arbaminch, Tigray (Mekele), Kefa, West Gojjam, Upper Didesa valley, North Omo, Arjo, Wonago, Gore, Mizan Tefere, Debre-Zeit. The prevalence of trypanosomes in some studies, ranged from 20 to 40% and animal mortality was 16% (Tolosa, 2010). However, the detailed status of the hemoparasitic diseases in ruminants is not thoroughly studied in and around Debre-Zeit and information is so far scanty.

Therefore, the objectives of this study were to study the epidemiological significance of the main hemo-parasitic diseases in and around Debre-Zeit and to indicate the risk factors associated with distribution of hemoparasites in the study area.

MATERIALS AND METHODS

Study area

Debre-Zeit is located in Central Oromia regional state, at a distance of 47 km of the South Eastern part of Addis Ababa. It is the main city of Ada'a-Liban district, situated on two international trade routes, which are connected by Franco-Ethiopia Djibouti railway and Addis-Moyale Nairobi, intercontinental asphalted road transport route. This is the only gate for import/export of livestock and agricultural commodities besides other commercial goods, from or to international market (Central Agricultural Census Commission (CACC), 2003).

Debre-Zeit lies between 9° N latitude and 39° E longitude and an

Table 1. Prevalence of hemoparasitosis in ruminants on the basis of species of animals in and around Debre-Zeit, Ethiopia.

Animal	Laboratory test results	
	No. of animals examined (%)	No. of animals positive and prevalence (%)
Bovine	177	15 (3.9)
Ovine	180	2 (0.5)
Caprine	27	7 (1.8)
Total	384	24 (6.3)

altitude of 1860 meter above sea level (m.a.s.l). It has three agro-ecological locations, 6% Dega greater than 2000 m.a.s.l, 89% Woinadega greater than 1600 m.a.s.l, and 5% Kola greater than 1000 m.a.s.l. It gets an annual rainfall of 871 mm of which 80% is received during long rainy season starting from June to September and the remaining in short rainy season extending from March to May, and the dry season from October to February. The mean annual maximum and minimum temperature are 26 and 14°, respectively with a minimum relative humidity of 63.8% (NMSA, 2005).

Study animals

A total of 384 animals (local and cross breed) were randomly selected for infected and suspected cases, irrespective of their age, breed, sex and agro-ecological set. Mainly concerned animals came to Addis Ababa University, School of Veterinary Medicine open air clinic from and around Debre-Zeit town.

Study design

A cross-sectional study was conducted on prevalence rates of hemoparasite infection of ruminants from November, 2010 to April, 2011 in and around Debre-Zeit.

Sample collection

Blood sample was collected randomly from 177 bovines (local and cross breed), 27 caprine and 180 ovine, from ear-vein, at Addis Ababa University, School of Veterinary Medicine open air clinic.

Parasitological examination

Blood samples collected directly from an ear-vein in a heparinized micro hematocrit capillary tube, thin blood smears were drawn, air dried, fixed (methanol 3 to 5 min) and stained by immersing in jar containing diluted 10% Giemsa solution for 30 min (Kaufman, 1996), and by Diff-quick staining for only 15 s, 5 times dip for each of the 3 types of staining containing jar and then examined under oil immersion ×100 magnification to identify the morphology and type of hemoparasites (Coles, 1986).

Hematological examination

Both parastaemic and aparastaemic blood samples were taken from the ear-vein in the heparinized capillary tubes (75 × 1.2 mm). Each of the two tubes was filled to a height of 66 mm and one end sealed with crystaseal (Hauskyley Ltd. England). After centrifugation, 13000 rpm for 5 min packed cell volume was determined by using micro hematocrit reader, to estimate PCV for checking anemia (Woo, 1971). A separated Buffy coat smear was examined. The capillary tube was put at buffy coat junction with plasma and RBC and buffy coat was taken out on the slide, covered by cover slip and examined under low magnification or by WOO technique (Woo, 1971).

Body condition evaluation and estimation of age

On subjective basis, body condition score of animals were evaluated during sample collection. They were classified as emaciated, moderate and good based on anatomical parts and the flesh and fat cover at different body parts (Nicholson and Butterworth, 1986). The ages of the animals were estimated by history and as described by De-lahunta and Habel (1986).

Data analysis

Data entry was made of Microsoft excel spread and the analysis was performed using chi-square of independence which was employed in the prevalence of hemoparasitic disease in ruminants with respect to sex, age as well as breed to determine the associated risk factors. Analysis of variance (ANOVA) (Thrusfield, 2005) was used to assess the difference in mean PCV% of the animals in hemoparasites infected and non-infected.

RESULTS

Out of 384 blood samples examined (177 bovine, 27 caprine and 180 ovine species) which included 359 samples from the local breeds, 24 samples were found infected on different species of hemoparasites. Among the 25 samples from exotic cross breeds, none was found infected, this may be due to less number of animals included in the present study. The overall prevalence rate of hemoparasites in and around Debre-Zeit is 6.3%. This included 7 different types of hemoparasites. Out of 177 bovine samples, 15 were positive (3.9%), *A. marginale* 6 (1.6%), *A. centrale* 1 (0.3%), *Babesia bigemina* 1 (0.3%), *B. bovis* 1 (0.3%) and *Theleria vivax* 6 (1.6%); out of 27 caprine samples, 2 were positive *Anaplasma ovis* (0.3%) *Babesia ovis* 1 (0.3%) and 180 ovine samples *A. ovis* 7 (1.8%). The mean PCV of parasitized animals (21.4 ± 7.1) was significantly lower ($P<0.05$) than those of non-parasitized animals (27.7 ± 6.9) during dry season (Table 8).

The overall prevalence of hemoparasitosis on the basis of species was 3.9, 0.5 and 1.8% in bovine, caprine and ovine species, respectively (Table 1). There was no statistically significant difference ($P>0.05$) in the prevalence of hemoparasites among the different species of animals examined. Table 2 indicates the species of hemoparasitosis detected and their relative frequencies. There was no statistically significant difference ($P>0.05$) between the prevalence of different species of hemoparasites encountered and species of animals examined.

Prevalence of hemoparasites in ruminants on the basis of age, sex, breeds and body conditions in the study area is indicated in Table 3. The species of hemoparasite and their relative frequencies in ruminants on sex basis in

Table 2. Species of parasites and their relative frequencies of hemoparasitosis in ruminants in and around Debre-Zeit, Ethiopia.

Animal species (No. examined)	Species of parasites						
	A. marginale (%)	*A. central* (%)	*A. ovis* (%)	*B. bigemina* (%)	*B. bovis* (%)	*B. ovis* (%)	*T. vivax* (%)
Bovine (177)	6 (1.6)	1 (0.3)	0 (0)	1 (0.3)	1 (0.3)	0 (0)	6 (1.6)
Ovine (180)	0 (0)	0 (0)	7 (1.8)	0 (0)	0 (0)	0 (0)	0 (0)
Caprine (27)	0 (0)	0 (0)	1 (0.3)	0 (0)	0 (0)	1 (0.3)	0 (0)
Total	6 (1.6)	1 (0.3)	8 (2.1)	1 (0.3)	1 (0.3)	1 (0.3)	6 (1.6)

Table 3. Prevalence of hemoparasitosis in ruminants on basis of age, sex, breed and body conditions in and around Debre-Zeit, Ethiopia.

Parameter		No. of animals examined	No. of animals positive and prevalence (%)
Age category	Adult	291	15 (3.9)
	Young	93	9 (2.3)
Sex	Female	241	11 (2.9)
	Male	170	13 (3.4)
Breed	Cross	25	0 (0)
	Local	359	24 (6.25)
Body condition	Emaciated	83	14 (3.6)
	Moderate	167	5 (1.3)
	Good	134	5 (1.3)
Total		384	24 (6.25)

Table 4. Species of hemoparasites and their relative frequencies in ruminants on sex basis in and around Debre-Zeit, Ethiopia.

Sex and No. examined	Species of parasites						
	A. marginale (%)	*A. central* (%)	*A. ovis* (%)	*B. bigemina* (%)	*B. bovis* (%)	*B. ovis* (%)	*T. vivax* (%)
Female (214)	1 (0.3)	0 (0)	5 (1.3)	0 (0)	1 (0.3)	0 (0)	4 (1)
Male (170)	5 (1.3)	1 (0.3)	3 (0.8)	1 (0.3)	0 (0)	0 (0)	2 (0.5)
Total (384)	6 (1.6)	1 (0.3)	8 (2.1)	1 (0.3)	1 (0.3)	1 (0.3)	6 (1.6)

and around Debre-Zeit is shown in Table 4. There was no statistically significant difference (P>0.05) between the species of hemoparasites and their relative frequencies in ruminants on the basis of sex of examined animals. The species of hemoparasite and their relative frequencies in ruminants on breed basis in and around Debre-Zeit is shown in Table 5. There was statistically significant difference (P<0.05) between the species of hemoparasites and their relative frequencies in ruminants on the basis of breed of examined animals.

The species of hemoparasite and their relative frequencies in ruminants on body condition basis in and around Debre-Zeit is shown in Table 6. There was a statistically significant difference (P<0.05) between the species of hemoparasites and their relative frequencies in ruminants on the basis of body condition of examined animals. The mean PCV of animals infected with different species of hemoparasites is shown in Table 7. There was a statistically significant difference (P<0.05) in the mean PCV of animals infected with different species of hemoparasites. A comparative value of mean PCV of parasiteamic and aparasiteamic ruminants in and around around Debre-Zeit is depicted in Table 8. There was a statistically significant difference (P<0.05) in the mean

Table 5. Species of hemoparasites and their relative frequencies in ruminants on the basis of breeds in and around Debre-Zeit, Ethiopia.

Breed and No. examined	Species of parasites						
	A. marginale (%)	*A. central* (%)	*A. ovis* (%)	*B. bigemina* (%)	*B. bovis* (%)	*B. ovis* (%)	*T. vivax* (%)
Cross (25)	0 (0)	0 (0)	0 (0)	0 (0)	0 (0)	0 (0)	4 (1)
Local (359)	6 (1.6)	1 (1.3)	8 (2.1)	1 (1.3)	1 (0.3)	6 (0)	2 (0.5)
Total (384)	6 (1.6)	1 (0.3)	8 (2.1)	1 (0.3)	1 (0.3)	1 (1.6)	6 (1.6)

Table 6. Species of hemoparasites and their relative frequencies in ruminants based on body condition of the examined animals in and around Debre-Zeit, Ethiopia.

Body condition score and No. examined	Species of parasites						
	A. marginale (%)	*A. central* (%)	*A. ovis* (%)	*B. bigemina* (%)	*B. bovis* (%)	*B. ovis* (%)	*T. vivax* (%)
Emaciated (83)	6 (1.6)	1 (0.3)	0 (0)	1 (0.3)	0 (0)	1 (0.3)	5 (1.3)
Good (167)	0 (0)	0 (0)	5 (1.3)	0 (0)	0 (0)	0 (0)	0 (0)
Moderate (134)	0 (0)	0 (0)	3 (0.8)	0 (0)	1 (0.3)	0 (0)	1 (0.3)
Total (384)	6 (1.6)	1 (0.3)	8 (2.1)	1 (0.3)	1 (0.3)	1 (1.6)	6 (1.6)

Table 7. Mean PCV of animals affected with different species of hemoparasites in and around Debre-Zeit, Ethiopia.

Hemoparasite species	Mean (%)	No. of animals examined	Standard deviation
A. marginale	18.1	6	5.2
A. central	15	1	-
A. ovis	23.8	8	10.9
B. bigemina	21	1	-
B. bovis	22.5	1	-
B. ovis	18	1	-
T. vivax	23.1	6	2.7
Total	27.4	24	7.1

Table 8. Mean PCV of parasiteamic and aparasiteamic ruminants in and around Debre-Zeit, Ethiopia.

Laboratory test result	Mean PCV (%)	No. of animals examined	Standard deviation
Aparasitemic	27.7	360	6.9
Parasitemic	21.4	24	7.1
Total	27.4	384	7.1

PCV of parasiteamic and aparasiteamic ruminants.

DISSCUSION

In this study, the overall prevalence rate of hemoparasites in ruminants was found to be 6.3%. These results disagree with 25.7% in North-Central Nigeria by Kamani et al. (2010). Trypanosomosis 1.6% consist of only *T. vivax* and tick born hemoparasitosis 4.9% including *A. marginale* (1.6%), *A. centerale* (0.3%), *A. ovis* (2.1%), *B. bigemina* (0.3%), *B. bovis* (0.3%), *B. ovis* (`0.3%). The present results are comparable with different figures reported from different tsetse infected regions of Ethiopia. Trypanosomosis, 51.1% in West Ethiopia (Terefa, 2008),

21% in Southern rift valley areas (Bekele et al., 2010), 17.2% in Metekele (Afework et al., 2000), 17.5% in the Upper Dedessa valley (Afework et al., 2000) and tick born hemoparasitosis 2.1% (Terefa, 2008) and 17% in other African country, Nigeria (Kamani et al., 2010). Furthermore, the study indicates that *A. ovis* 2.1% was the predominant hemoparasite species detected by Giemsa staining and Diff-quick methods as compared to other hemoparasite in the study animals. *A. ovis* infection rate in small ruminants more in sheep (1.8%) as compared to goats (0.3%) in this report.

In the present study, the prevalence rate of babesiosis was lower in the study area (only 0.6%). So results of the present study are not in agreement with previous study by Yismashewa (2005) in Decha Woreda Southern Ethiopia and Seyoum (2007) in Kobo and Girana valley in Amhara region. They reported high prevalence in those areas. The earlier studies were conducted during the late dry and early rainy season but in the present study all sample collected at the dry season. This implies that crucial factors influenced the rate and seasonal variation of vectors, high humidity and temperature. In general, prevalence intensity rate of tick born hemoparasitic disease infestation were generally low during dry season and higher in rainy season (Solomon et al., 2003).

The prevalence rate of hemoparasitism was generally slightly higher in male than female animals in the present study and this can be due to the fact that male animals work exposed in different areas which tends to increase chance of contact with the vectors of these disease as compared to female, disagree with Kamani et al. (2010). The lower prevalence in young animals compared to adults can be attributed to restricted grazing of young animals which tends to reduced their chance of contact with the vectors of these diseases. There was significant difference between the prevalence in local and exotic cross breed of cattle in management practices, where the exotic cross breeds receive more attention in terms of ecto parasite control and feed supplementation and it can be less number of exotic cross breed animals included in this study. The mean PCV of non-infected animals was significantly higher (p<0.05) than that of animals with one or more parasites during dry season, which may be related to confounding factors like scarcity of good quality feeds and stress of trekking experienced by most pastoral animals during the dry season of feeds (Obi and Anosa, 1980); this report agree with the previous study (Kamani et al., 2010).

Hemoparasitosis infection in the study animals was statistical significant on basis of body condition of the studied animals and breeds of animals. Also, there was no significant difference between age, sex and species of animals. Moreover, as one of the predominant clinical infestations of hemoparasitosis is anemia and had significant difference lower mean PCV as compared to non-infected animals by hematocrit PCV% determination. This finding indicates that anemia is good indicator for infection with hemoparasites (*Trypanosomae, Babesia, Anaplasma*) than the body condition score of the animals. The anemia could be due to the hemopasites as it has been also indicated by (Laha and Sasmal, 2008) that anemia could be apparent in infected animals as a result of production of haemolysin by trypanosomes resulting into haemolysis of RBCs and extra vascular destruction of RBCs. This destruction may be through the erythro-phagocytosis or may be immune mediated. Depression of erythropoiesis and non-specific factors, which increase red cell fragility, may also be responsible for anemia.

The body condition of infected animals were statically significant (P<0.05) as compared to non-infected animals. Loss of body condition not always associated with this disease, can be other chronic disease of other parasitic, bacterial, viral and nutrition deficiency or poor management systems of the farm animals. This study was partly conducted in and around Debre-Zeit town. Thus, it is an area where no tsetse fly was reported and the trypanosome species detected in this area is mainly *T. vivax*, one of the trypanosome species that can be trans-mitted mechanically by biting flies and cyclically by tsetse flies. In the present study, this *Trypanosoma* species is present solely in bovine, but the small ruminants were suspected by this blood parasite. However, in the study area no tsetse flies were reported.

The present report confirms the presence of carrier populations of hemoparasite infected animals which both serve as reservoir of infection for *Trypanosoma* and tick vector and susceptible livestock, and has the potential for clinical relapse under stressful conditions, this report agree with the previous studies conducted in other countries (Sharma et al., 2013; Kamani et al., 2010; Singla et al., 2007).

ACKNOWLEDGEMENTS

The authors appreciate the support by Addis Ababa University. The willingness and cooperation of flock owners were an indispensable input for the accomplishment of this work. All contributions and supports are gratefully acknowledged.

REFERENCES

Adam KMG, Paul J, Zaman V (1971). Medical and veterinary protozology. Churchill Livingstone. Edinburgh and London.

Afewerk Y, Clausen PH, Abebe G, Tilahun G, Mehlitz D (2000). Multiple-drug resistant *Trypanosoma congolense* populations in village cattle of Metekel District, North-West Ethiopia. Acta Trop. 76(3):231-238.

Alekaw S (2000). Distribution of ticks and tick born diseases at Metekel ranch, Ethiopia. J. Ethiop. Vet. Assoc. 4:40-60.

Bekele J, Asmare K, Abebe G, Ayelet G, Gelaye E (2010). Evaluation of Deltamethrin applications in the control of tsetse and trypanosomosis in the Southern rift valley areas of Ethiopia. Vet. Parasitol. 168(3-4):177-184.

Central Agricultural Census Commission (CACC) (2003). Statistical report on livestock farm implements. Ethiopian agricultural sample enumeration, 2001/2, Addis Ababa.

Centers for Disease Control and Prevention (CDC) (2009). Anaplasmosis, question and answer. Tick born rickettisial disease. Centre for diseases control and prevention. pp. 8-18.

Coles H (1986). Erythrocytes. In: Veterinary clinical pathology. 4th Ed. W.B. Saunders Company. pp. 10-79.

Central Statistics Authority (CSA) (2009). Central Statistics Authority, Agricultural sample survey (2008/09). Statistical bulletin 302, Addis Ababa.

Dagnachew S, Sangwan K, Abebe G (2005). Epidemiology of bovine trypanosomosis in the Abay (Blue Nile) Basin areas of Northwest Ethiopia. Elev. Med. Vet. Pays Trop. 58:151-157.

De-lahunta A, Habel E (1986). Teeth. In: Applied veterinary Anatomy. W.B. Saunders Company, pp. 4-16.

Homer J, Aquiler E, Telford III R, Krause J, Pressing H (2000). Babesiosis. Clin. Microbiol. Rev. 13(3):451–469.

Jones TW, Dávila AM (2001). *Trypanosoma vivax*--out of Africa. Trends Parasitol. 17(2):99-101.

Kamani J, Sannus E, Egulu K, Dogo I, Tanko J, Kenza J, Tafariki E, Ghise S (2010). Prevalence and Significance of Haemoparasitic Infections of Cattle in North-Central, Nigeria. Vet. World 3(10):44-45.

Kocan KM (1995). Targeting tick for control of selected hemoparasite disease of cattle. Vet. Parasitol. 57(1-3):121-151.

Kaufman J (1996). Parastic infections of domestic animals: A diagnostic manual. Boston: Birkhauser. pp. 55-61.

Laha R, Sasmal K (2008). Endemic status of *Trypanosoma evansi* infection in a horse stable of eastern region of India--a field investigation. Trop. Anim. Health Prod. 40:357–361.

McLeod R, Kristjanson P (1999). Final Report of Joint ESYS/International Livestock Research Institute/Australian Centre for International Agricultural Research Tick Cost Project-Economic Impact of Ticks and Tick-Borne Diseases to Livestock in Africa, Asia and Australia. Nairobi, Kenya: International Livestock Research Institute.

Mottelib A, Hosin I, Moul I, Ei-sherif M and Abo Z (2005). Comparative evaluation of various diagnostic techniques for *T. evansi* in naturally infected camel SAH- Warsia Poland.

Nicholson MJ, Butterworth MH (1986). A guide to condition scoring of zebu cattle. ILRI (aka ILCA and ILRAD),Addis Ababa, Ethiopia.

NMSA (2005). National Metrological Service Agency; Rain fall and temperature data. Addis Ababa, Ethiopia.

Obi V, Anosa O (1980). Haematological studies on domestic animals in Nigeria IV. Clinico-haematological features of bovine trypanosomiasis, theileriosis, anaplasmosis, eperythrozoonosis and helminthiasis. Zlblatt Vet. Med. B27: 789-797.

WHO, World Health Organization (2002). Pan African tsetse and trypanosomiasis eradication campaign, Fifty-fifth World Health Assembly. Provisional agenda item. 13:17, 1-4.

Reghu R, Jammi R, Rao K, krishan M, Nagjan B, Cheeyan C, Sachivothman R (2008). *Trypanosoma evansi* in camel, donkey and dog in India, comparison of PCR and light microscopy for detection short communication. Vet. Arch. 78:89-94.

Seyoum Z (2007). Bovine babesiosis and vectors at Kobo and Girana valley in Amhara region. M.Sc. Thesis, Addis Ababa University, FVM, Debre-Zeit, Ethiopia.

Sharma A, Singla LD, Tuli A, Kaur P, Batth BK, Javed M, Juyal PD (2013). Molecular prevalence of *Babesia bigemina* and *Trypanosoma evansi* in dairy animals from Punjab, India by duplex PCR: A step forward to detection and management of concurrent latent infections. Biomed. Res. Int. 2013:893862.

Singla LD, Aulakh GS, Juyal PD (2007). Haemato-biochemical and clinico-pathological observations on haemoprotists in cattle and buffaloes. In: Proceedings of National Seminar on Recent Diagnostic Trends and Control Strategies for Haemoprotozoan Infections in Livestock held at Sardarkrushinagar. pp 107-110.

Sinshaw A, Abebe G, Desquesnes M, Yoni W (2006). Biting flies and mechanically *Trypanasoma vivax* infection in three highland bordering Lake Tana, Ethiopia. Vet. Parasitol. 142:35-46.

Solomon G, Nigist M, Kassa B (2003). Seasonal variation of ticks on calves at Sebeta in Western Shoa zone Ethiopia. Ethiop. Vet. J. 7:17-27.

Solomon M (1980). Animal health review in Ethiopia (1972-1979). Addis Ababa: Department of Veterinary Service Division, Ministry of Agriculture.

Sumba AL, Mihok S, Oyieke FA (1998). Mechanical transmission of *Trypanosoma evansi* and *T. congolense* by *Stomoxys niger* and *S. taeniatus* in a laboratory mouse model. Med. Vet. Entomol. 12(4):417-422.

Terefa W (2008). Studies on bovine trypanosomiasis and therapeutic efficacy of selected trypanocidal drug in Birbir valley of Gawo, Dale district West Oromia. M.Sc. Thesis, Addis Ababa University, FVM, Debre-Zeit, Ethiopia.

Thrusfield M (2005). Veterinary Epidemiology, 2nd edition, Black well scientific, London. Pp. 225-228.

Tolosa S (2010). A study on hemoparasites of cattle and major tick species in and around Assela, Oromia region, Ethiopia.

Woo K (1971). Evaluation of the haematocrit centrifuge technique for the diagnosis of African trypanosomiasis. Acta Trop. 27:384–386.

Yismashewa W (2005). Epidemiology of tick and tick born protozaol disease of cattle in Decha Woreda, Southern Ethiopia. M.sc. Thesis, Addis Ababa University, FVM, Debre-Zeit, Ethiopia.

Evaluation of larvicidal activity of the different extracts against important species of mosquito: *Anopheles stephensi*

Babita R. Malik[1] , Mukesh K. Malik[1], N. Balakrishnan[2] and B.suresh[1]

[1]Department of Pharmacognosy, J.S.S. College of Pharmacy Rockland, Ooty-643 001, T.N., India.
[2]National Institute of Communicable Diseases, Southern India branch Conoor-643101, T.N., India.

The aim of this work was to compare larvicidal activity of insecticidal plants against malaria vector mosquito *Anopheles stephensi.* The larvicidal activity of *Lantana camara* Linn. and *Bauhinia racemosa* Lam., extracted in petroleum ether, chloroform and ethyl acetate were tested against mosquito larvae of *A. stephensi*. Late third or early fourth instar larvae were used for the screening. These extracts were used for determining the larvicidal activity using World Health Organization (WHO) method for evaluation of new larvicidal agents. The petroleum ether extract of *L. camara* showed highest larvicidal activity in comparison to petroleum ether extract of *B. racemosa*, and ethyl acetate extract of *B. racemosa* showed highest larvicidal activity in comparison to chloroform extract of *L. camara* against the mosquito vector *A. stephensi*. No mortality was observed in control. The result suggests the use of the plants in insect control as an alternative method for minimizing the noxious effect of some pesticide compounds on the environment. Thus, the extracts of whole plant of *L. camara* Linn. and leaf extracts of *B. racemosa* Lam. may deliver promising, more selective and biodegradable agents.

Key words: *Lantana camara*, *Bauhinia racemosa, Anopheles stephensi.*

INTRODUCTION

Mosquitoes are one of the most medicinally significant vectors and they transmit parasites and pathogens which continue to have devastating impact on human beings (Maheswaran et al., 2008). Several numbers of species belong to genera *Anopheles, Culex, Aedes* and vectors for the pathogens of various diseases like malaria, filaria, Japanese encephalitis, dengue, and yellow fever. Thus, one of the approaches for control of these vector-borne diseases is the interruption of disease transmission by killing mosquitoes or preventing mosquito bites (Das, 1989). Herbal products which have proven potential as insecticides or replicants can play an important role in the interruption of the transmission of mosquito-borne diseases both at the individual and community level. However the discovery, development and use of synthetic organic insecticidal chemicals with persistent residual action not only over shadowed the use of herbal products as insecticides of choice against mosquitoes but also become the major weapon for mosquito control (Sakthivadivel and Daniel, 2008). The extensive use of synthetic organic insecticides during the last decade has resulted in environmental hazards and also in the development of physiological resistance in most vector species. This has necessitated the need for research and development of environmentally safe, biodegradable, low cost indigenous methods for vector control, which can be used with minimum care by individual and communities in specific situations (Singh et al., 2006). The plant *Lantana camara* Linn. (Verbenaceae) and *Bauhinia racemosa* Lam. (Caesalpiniaceae) are described in Ayurveda and Siddha, as a potent drug against a variety of ailments. These plants are widely distributed and cultivated in various parts of India (Yoganarasimhan, 2000; Varier, 2006; Prajapati et al., 2007; Ivan, 1999; Singh and Himadri, 2005; Raveendra and Martin, 2006).

MATERIALS AND METHODS

Collection of plant material

The plants were collected during flowering stage in the month of July to August from Nilgiris. Then, their identification was established with the aid of an expertise botanist and by making comparison with herbarium sheets of the authentic sample. Many of the defensive components are biodegradable, with non-residual effect on the biological environment hence; an attempt has been made in the present investigation to identify plants with the potential to control vector mosquitoes.

Extraction

The plant *L. camara* Linn. and *B. racemosa* Lam. were powdered and extracted in soxhlet, with petroleum ether, chloroform and ethyl acetate. The extracts were concentrated under reduced pressure to a semisolid mass. These extracts were used for determining the larvicidal activity against mosquito larvae.

Biological assay

Larvicidal activity was evaluated in accordance with World Health Organization (WHO) for the evaluation of new larvicidal agents (WHO, 1985). The larvae of *Anopheles stephensi* was obtained and reared from the neonates in National Institute of Communicable Diseases, Southern India branch field station located at Mettupalayam (District Coimbatore of Tamil Nadu State), at 28 ± 2°C with a photoperiod of 12 h light and dark, and 80 ± 10% relative humidity. A brewer's yeast powder mixed with an equal quantity (w/w) of ground dog biscuit was used in the laboratory as food for the larvae. The late third or early fourth instar larvae were collected according to larval size and the degree of chitinization of respiratory siphon (Cheng et al., 2003). Different concentrations of the extracts were prepared in 1 ml of acetone for each experiment. All experimental exposure was done in 500 ml glass beaker in triplicate. Twenty-five (25) larvae were collected with a pasture pipette, placed on a filter paper for removal of excess of water and placed in 250 ml dechlorinated tap water containing various concentration of crude extracts. Three controls in triplicate were setup, one with acetone (1 ml), the other with distilled water (250 ml).

The beakers were covered with muslin cloth to avoid entry of any foreign material. Sufficient control was also kept for each extracts. The observed mortality (crude mortality) was recorded at 24 h of exposure to test solution. From this crude mortality, percentage crude mortality was obtained. Subsequently, controlled mortality, if any, was recorded and percentage crude mortality was obtained. The percentage crude mortality was corrected by using Abbot's formula. The corrected probit mortality and expected mortality was also obtained but no control mortality was recorded during the experiment, so Abbot's formula was not used.

Statistical analysis

Lethal concentration (LC_{50} and LC_{90}) values and their 95% confidence limits were estimated by fitting a probit regression model to the observed relationship between percentage mortality of larvae and logarithmic concentration of the substance. Separate probit models were fitted for each extract (Finnely, 1971). All analysis was carried out using the statistical package for social sciences (SPSS) software, version 13.0.

RESULTS AND DISCUSSION

Seven different concentrations of test solution ranging from 50 to 350 ppm for petroleum ether extract and six concentrations of test solution ranging from 50 to 300 ppm for chloroform extract for *L. camara* Linn. and six different concentrations of test solution ranging from 50 to 300 ppm for petroleum ether extract and five different concentrations of test solution ranging from 40 to 200 ppm for ethyl acetate extract were subjected to 24 h bioassay, using late 3rd or early 4th instar larvae of *A. stephensi*. Based on observations made in the 24 h bioassay studies among the different plant extracts, the petroleum ether extract of *L. camara* Linn. was more potent than petroleum ether extract of *B. racemosa* Lam. and ethyl acetate extract of *B. racemosa* was more potent than chloroform extract of *L. camara* Linn. for vector mosquito and were identified as efficient against them. The results from the *A. stephensi* larvicidal bioassay using different extracts of two different plants, the most active extract against late third or early fourth instar larvae of *A. stephensi*, were the petroleum ether extract of *L. camara* and Ethyl acetate extract of *B. racemosa*. The results of susceptibility of larvae for the extracts were given in Tables 1 and 2; and Figures 1 and 2.

Conclusion

The use of the plants in insect control offers a safer alternative to synthetic chemicals and can be obtained by individuals and communities easily at a very low cost. Moreover, these results could be useful in the search for newer, more selective and biodegradable larvicidal natural compounds.

ACKNOWLEDGEMENTS

The authors are thankful to the J.S.S. College of Pharmacy, Ooty, Tamil Nadu and National Institute of

Table 1. Observed and expected mortality of *Anopheles stephensi* larvae exposed to *Lantana camara* with petroleum ether and chloroform extracts. Expected mortality is based on probit regression analysis.

Concentration	No. of Larvae		Mortality (%)		Expected mortality			Probit (mortality)	χ^2, DF, P-value	LC_{50}	LC_{90}
(µg/ml)	Exposed	Dead	Crude	Corrected	Probit	Dead	%	= a + b × concentration		(95% CI)	(95% CI)
					Petroleum extract						
50	75	17	22.7	22.7	-0.80	15.8	21.1	-1.3270+0.0105×conc.	χ^2=1.56, DF=5, P=0.9	126.7 (112.0-139.7)	248.9 (231.2-271.8)
100	75	30	40.0	40.0	-0.28	29.3	39.1				
150	75	44	58.7	58.7	0.25	44.8	59.8				
200	75	56	74.7	74.7	0.77	58.5	78.0				
250	75	67	89.3	89.3	1.30	67.7	90.3				
300	75	73	97.3	97.3	1.82	72.4	96.6				
350	75	75	100.0	100.0	2.35	74.3	99.1				
					Chloroform extract						
50	75	8	10.7	10.7	-1.95	1.9	2.6	-2.7408+0.0158×conc.	χ^2=39.9, DF=4, P<0.005	173.8 (120.2-235.3)	255.0 (205.5-435.0)
100	75	9	12.0	12.0	-1.16	9.2	12.3				
150	75	13	17.3	17.3	-0.37	26.7	35.5				
200	75	44	58.7	58.7	0.42	49.7	66.2				
250	75	73	97.3	97.3	1.21	66.5	88.7				
300	75	75	100.0	100.0	2.00	73.3	97.7				

DF = degrees of freedom, conc. = concentration, CI = confidence interval.

Table 2. Observed and expected mortality of *Anopheles stephensi* larvae exposed to *Bauhinia racemosa* with petroleum ether and ethyl acetate extracts. Expected mortality is based on probit regression analysis.

Concentration	No. of Larvae		Mortality (%)		Expected Mortality			Probit (mortality)	χ^2, DF, P-value	LC_{50}	LC_{90}
(µg/ml)	Exposed	Dead	Crude	Corrected	Probit	Dead	%	= a + b x concentration		(95% CI)	(95% CI)
					Petroleum ether extract						
50	75	11	14.7	14.7	-1.21	8.4	11.2	-1.8238+0.0122×conc.	χ^2= 9.7, DF=4, P=0.046	149.3 (121.2-175.3)	254.3 (220.5-316.0)
100	75	24	32.0	32.0	-0.60	20.5	27.3				
150	75	31	41.3	41.3	0.01	37.7	50.2				
200	75	49	65.3	65.3	0.62	54.8	73.1				
250	75	69	92.0	92.0	1.23	66.7	89.0				

Table 2. Contd

300	75	75	100.0	100.0	1.84	72.5	96.7				
						Ethyl acetate extract					
40	75	12	16.0	16.0	-1.14	9.5	12.7				
80	75	25	33.3	33.3	-0.43	25.1	33.5				
120	75	42	56.0	56.0	0.28	45.9	61.2	-1.8513+0.0178×conc.	χ^2=5.9, DF=3, P=0.11	104.0 (95.2-112.5)	176.1 (163.3-193.1)
160	75	60	80.0	80.0	1.00	63.0	84.1				
200	75	75	100.0	100.0	1.71	71.7	95.6				

DF = degrees of freedom, conc. = concentration, CI = confidence interval.

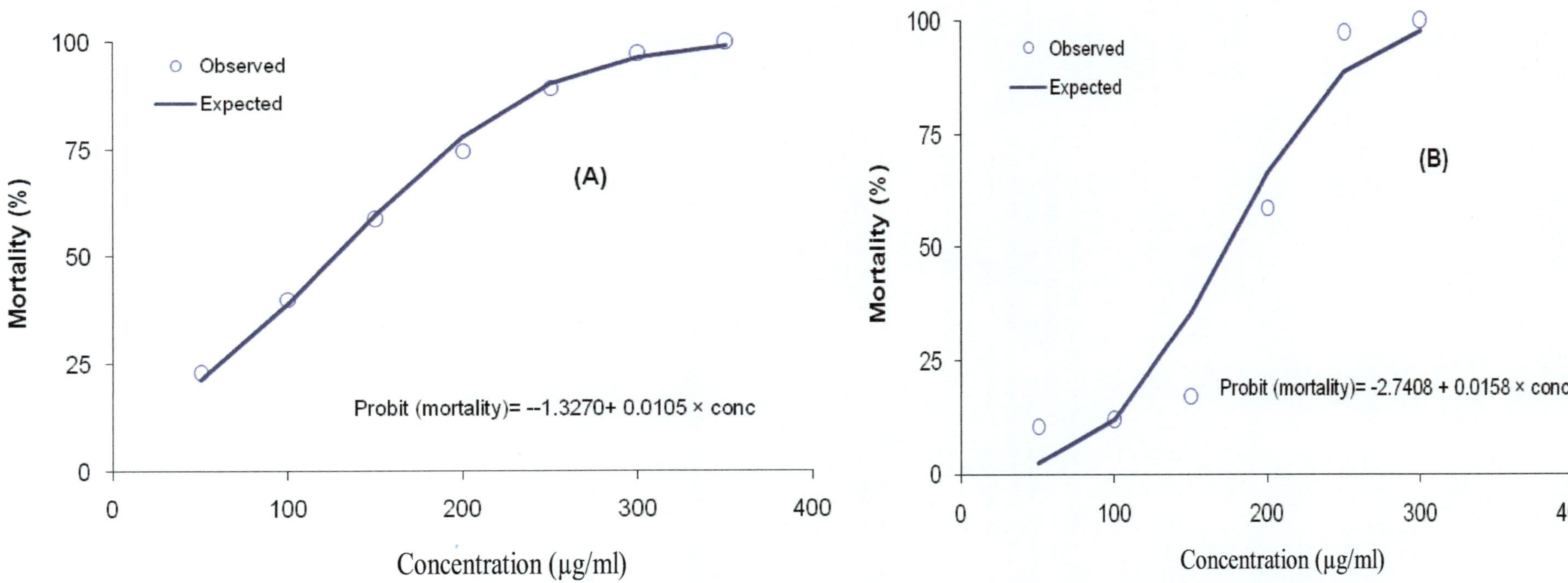

Figure 1. Relationship between *Anopheles stephensi* larval mortality and concentration of *Lantana camara* with (A) petroleum ether and (B) chloroform extracts. Expected values are based on probit regression analysis. Conc = concentration.

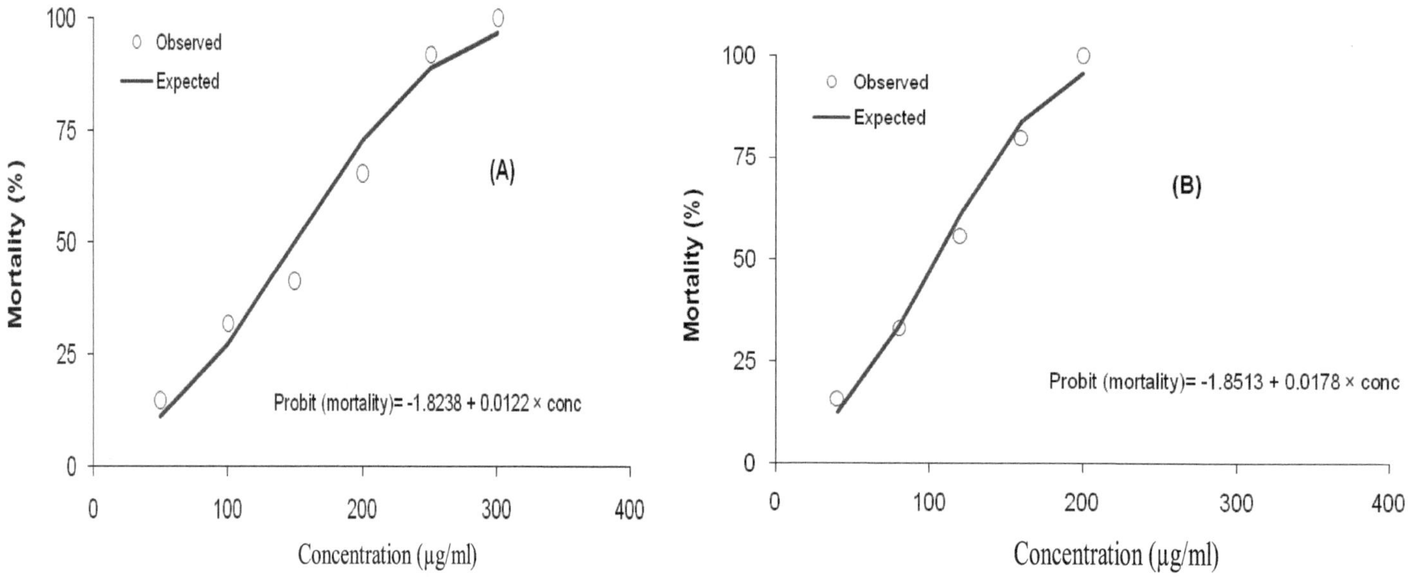

Figure 2. Relationship between *Anopheles stephensi* larval mortality and concentration of *Bauhinia racemosa* with (A) petroleum ether and (B) ethyl acetate extracts. Expected values are based on probit regression analysis. Conc = concentration.

Communicable Diseases, (Coonoor), Southern India branch field station located at Mettupalayam (District Coimbatore of Tamil Nadu State) for providing all necessary facilities required FOR completing this research work.

REFERENCES

Cheng SS, Chang HT, ST Chang, Tsai KH, Wei JC (2003). Bioactivity of selected plant essential oils against the yellow fever mosquito *Aedes aegypti* larvae. Bio. Resour. Tech. 89:99-102.

Das PKKM (1989). A module for chemical control of mosquito vectors, V.C.R.C Pondicherry.

Finnely DJ (1971). Probit Analysis, 3rd ed. Cambridge: Cambridge Univ.press; p 333.

Ivan AR (1999). Medicinal Plants of the World Chemical Constituents Traditional and Modern medicinal Uses. NJ: Published by Humana Press Inc. Totowa. pp. 179-180.

Maheswaran R, Kingsley S, Ignacimuthu S (2008). Larvicidal and Repellent activity of Clerodendron phlomides against Culex quinquefasciatus Say (Diptera: Culicidae). proceed. Recent Trends Insect Pest Manage. pp.240-243.

Prajapati ND, Purohit SS, Sharma AK, Kumar T (2007). Handbook of Medicinal Plants: A complete source book. India: Published by Agrobis. pp. 306-307.

Raveendra KR, Martin P (2006). Ethnomedicinal Plants. Published by Agrobis (India), Jodhpur, p 80.

Sakthivadivel M, Daniel T (2008). Evaluation of certain insecticidal plants for the control of vector mosquitoes viz. *Culex quinquefasciatus*, *Anopheles stephensi* and *Aedes aegypti*. Appl. Entomol. Zool. 43(1):57-63.

Singh MP, Himadri P (2005). Medicinal herbs with their formulations. Published by Daya Publishing House, Delhi, Volume I, p. 158.

Singh RK, Dhiman RC, Mittal P K (2006). Mosquito larvicidal properties of *Momordica charantia* Linn. J. Vect. Borne Dis. 43:88-91.

Varier PS (2006). Indian Medicinal Plant: A Compendium of 500 Species, Orient Longman Arya Vaidya Sala. Chennai: Orient Longman Private Limited; pp. 300-301.

World Health Organisation (WHO) (1985). Bioassay method for the titration of *Bacillus sphaericus* preparations with RB standard. Informal consultations on the development of *Bacillus sphaericus* as a microbial larvicide. World Health Organization, Geneva. (TDR/BVC/SPHAERICUS/ 85.3).

Yoganarasimhan SN (2000). Medicinal Plants of India. Tamil Nadu: 2:315-373.

Using health education intervention to improve knowledge and practice of prevention of avian influenza among bird handlers in Sokoto, Nigeria

M. O. Oche[1], A. U. Junaidu[2], A. S. Mainasara[3] and M. A. Ndakotsu [3]

[1]Department of Community Health, Usman Danfodiyo University, Sokoto, Nigeria.
[2]Department of Veterinary Public Health, Usman Danfodiyo University, Sokoto, Nigeria.
[3]Department of Pathology/Microbiology, Usman Danfodiyo University, Sokoto, Nigeria.

Zoonotic diseases are under-reported in most parts of the world particularly in developing countries. Approximately 75% of emerging diseases are zoonotic which include avian influenza (AI), Lassa fever and toxoplasmosis. The persistence and spread of avian/fatal avian influenza A (A/H5NI) in poultry and people in Asia, Europe and Africa has led the World Health Organization (WHO) to state that the world is now in a pandemic alert phase. For these reasons, an integrated and cross-sectorial plan was needed which ensures that Nigeria was prepared to address both the causes and consequences of avian and human influenza pandemic. This study was conducted to assess the impact of educational intervention on the knowledge and practice of prevention of AI among bird handlers in Sokoto. A total of 100 male bird handlers in Sokoto metropolis were enrolled for the quasi-experimental study with pre- and post-intervention components in the same group. A set of comprehensive and pre-tested questionnaire was administered to the respondents for information on the socio-demographic characteristics, knowledge and practices regarding AI. The pre-intervention phases involved the application of the questionnaires to the respondents. The intervention measures instituted included health education on causes of AI, clinical features, communicability, prevention, morbidity/mortality patterns and a demonstration session. One month after the intervention, the same set of questionnaires were administered to the study subjects. Scores obtained for knowledge were subsequently graded as adequate (≥50%) and inadequate (<50%). At baseline, 51% of the respondents scored ≥50% with a mean knowledge score of 64±7.0, while at post-intervention, 86% of the respondents scored ≥50% with a mean knowledge score of 91±17.64 (p<0.0001). The commonest form of protection adopted by the study subjects was the use of protective clothing, including face masks.

Key words: Avian influenza, bird handlers, health education, preventive measures.

INTRODUCTION

The human population has been on the rise with attendant urbanization resulting in more humans sharing wild life and livestock habitats at the same time developing closer interactions with domestic animals. Increasing globalization, international trade and international movement of humans, goods and animals impact on the global health from one geographical location to the other. Therefore, protecting the public health where one lives means improving the overall global health (Pappaioanou, 2004). Avian influenza (AI) is a highly contagious disease primarily of birds, and caused by influenza A viruses. It is one of the greatest concerns for public health that has

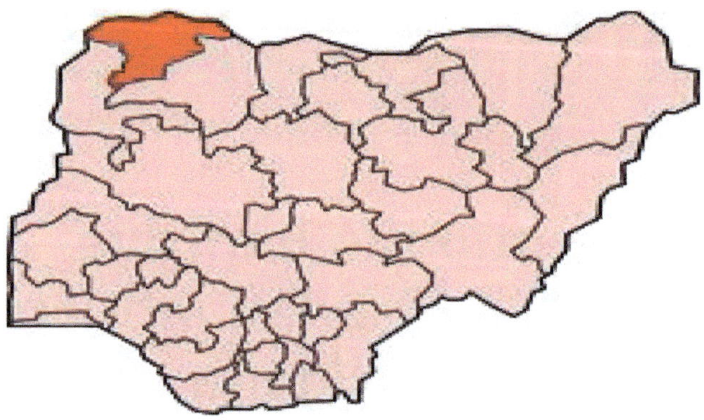

Figure 1. Map of Nigeria showing Sokoto state in extreme North West (Red color).

emerged from the animal reservoir (Brown et al., 2006; Capua and Marangon, 2007). The spread of the highly pathogenic avian influenza (HPAI) to countries in which hygienic standards are deficient increases the virus' pandemic potential and raises concerns about food security particularly in rural villages (Capua and Marangon, 2007). Aquatic birds are the sources of AI viruses (Pantin-Jackwood and Swayne, 2009; Krauss and Webster, 2010).

Infection of poultry with influenza A (subtype H5N1) virus is responsible for outbreaks in birds and a human case-fatality rate of 58% (WHO, 2006). The most likely means of transmission is from infected birds to humans and from the environment to humans, but evidence for human-to-human transmission is limited (WHO, 2005). This virus can be transmitted if a person has direct contact with infected poultry or surfaces and objects contaminated by poultry droppings.

Workers in the poultry industry, who commonly have contact with live, sick, or dying poultry, are at higher risk for AI. These workers are at increased risk, because of food handling and preparation of raw poultry meat and products. Concern exists that AI could be transmitted from uncooked birds or bird products to humans (Bridges et al., 2002; Swayne, 2006).

Nigeria with an estimated poultry population of 140 million birds has about 60% of poultry production taking place in small backyard flocks. Outbreaks of AI have been reported from commercial farms in the states of Kano, Kaduna, Plateau, Katsina, Bauchi and Abuja area and to date four patients have been diagnosed with respiratory symptoms and a history of exposure to diseased poultry have been investigated for possible infection. This number includes a woman who died of an acute respiratory illness traced to infection with AI virus (WHO, 2006). Current surveillance for human cases has resulted in the identification of individuals with influenza-like illness who have had a history of exposure to sick or dying birds. An integrated and cross sectoral plan is needed to ensure that Nigeria is prepared to address both the causes and consequences of avian and human pandemic influenza. Literature search has shown that no interventional studies have been carried out in North Western Nigeria since the advent of avian influenza to improve the knowledge and preventive practices of bird handlers and the public in general. Since the outbreak of the disease, the Sokoto State Ministry of Animal Resources has carried out a series of enlightenment campaigns in the media aimed at creating public awareness on the dangers posed by HPAI virus. However, there is the need to identify bird handlers in strategic market places who are at greater risk of acquiring the deadly virus with the aim of equipping them with preventive measures to stop further spread of the influenza virus since they come in contact with live birds from different geographical localities.

This study is therefore aimed at assessing the effect of health education intervention on the knowledge and practice of preventive measures against AI among bird handlers in Sokoto metropolis.

The research hypothesis was that at the end of the health education intervention, there will be at least 20% improvement in the knowledge and practices related to AI.

MATERIALS AND METHODS

Sokoto state is located in the extreme North West of Nigeria, close to the confluence of Sokoto river and Rima river, sharing borders with Niger Republic to the north, Zamfara state to the east, kebbi state to the south east and Benin republic to the west (Figure 1). As of 2010, it has an estimated population of more than 4.2 million. Sokoto city is the modern day capital of Sokoto state (and its predecessor, the Northwestern State). The name Sokoto (which is the modern/anglicized version of the local name, Sakkwato) is of Arabic origin, representing suk, 'market'. It is also known as Sakkwato, Birnin Shaihu da Bello or "Sokoto, Capital of Shaihu and Bello").

Being the seat of the former Sokoto caliphate, the city is predominantly Muslim and an important seat of Islamic learning in Nigeria. The Sultan who heads the caliphate is effectively the spiritual leader of Nigerian Muslims. The metropolis is made up of four local government areas namely Sokoto North and South, Dange Shuni and Wamakko. Farming is the main stay of the economy of the state. It has an estimated population of 3 million cattle, 3 million sheep, 5 million goats, 4,600 camels and variable species of poultry including chickens, guinea fowls, ducks and turkeys (Sokoto State Profile, 2013).

This study was a quasi-experimental type with pre- and post-intervention components in the same group. The study population comprised all bird handlers in Sokoto metropolis. Using the formula for comparison of proportions in same group (Kirkwood and Sterne, 2003) and a 24% practice of prevention of AI in a previous study (Abbate et al., 2006), a sample size of 100 was obtained. Cluster sampling technique was applied in the selection of the 100 study subjects. The metropolis has 12 clusters where the bird handlers are found, out of which 6 were selected by simple random sampling technique using roll of papers. The selected respondents were given identification numbers and at the same time their telephone numbers were given to the officials of the Market Poultry sellers association for ease of communication when their presence was needed. Advocacy visits were paid to the officials of the Poultry Association of Nigeria, Sokoto state branch during which the objectives of the study were explained to them and also to solicit for

Table 1. Age distribution of respondents (n=100).

Age (years)	No. (%)
18-27	19 (19)
28-37	24 (24)
38-47	40 (40)
48-57	11 (11)
58-67	6 (6)
Total	100 (100)

Table 2. Educational status of respondents.

Educational level	No. (%)
None	18 (18)
Qur'anic only	45 (45)
Primary	28 (28)
Secondary	9 (9)
Tertiary	0 (0)
Total	100 (100)

Table 3. Length of time handling birds.

Length of time (years)	No. (%)
1-5	27 (27)
6-10	34 (34)
11-15	29 (29)
16-20	10 (10)
Total	100 (100)

Mean=9.1±1.14.

Table 4. Sources of information about avian influenza.

Source of information	No. (%)
Radio/Television	60 (60)
Public enlightenment/lectures	23 (23)
Friends	10 (10)
Others	7 (7)
Total	100 (100)

their members' cooperation in ensuring the success of the study. A set of comprehensive, pretested questionnaire was administered by 6 trained research assistants on the respondents. Observational checklists were also used to assess the practices of the bird handlers with respect to preventive measures. The questionnaires sought such information as the socio-demographic characteristics, knowledge and practices regarding AI. The study instrument was pretested in a poultry market of another local government area outside the metropolis.

The members of the association in the 6 identified clusters were informed of the research and a day was set aside for the application of the study instruments. The study subjects were grouped into 2 for the purpose of the intervention. Each group was headed by 2 principal researchers. Informed verbal consent was obtained from individual study subjects and they were also given the option of opting out of the study when they so wished. Institutional ethical clearance was obtained from the ethical committee of the Usman Danfodiyo University, Sokoto. The pre-intervention phase of the study involved the application of the study instruments (questionnaires) to the respondents which was followed shortly by the intervention phase. The intervention which was basically health education had a demonstration segment. There were 2 sessions of the health education, each facilitated by 2 principal researchers. The health education sessions took place in the secretariat of the Market Poultry Association, Sokoto state branch located a stone throw from the market, each session lasting three hours. The contents of the health education included the cause of AI, clinical features, differentials, communicability, treatment and prevention, morbidity and mortality pattern all over the world including Nigeria.

The demonstration session involved the use of personal protective apparels. There was also a question and answer session that lasted for forty minutes to enable respondents get clarification on some grey areas. Information, education and communication materials used for the health education included posters, flip charts, pictures, etc. The health education sessions were repeated a week after to reinforce the knowledge acquired. One month after the intervention, the same set of questionnaires was administered to the study subjects in the same venue to obtain the post-intervention data. Check list was used in the markets and poultry farms to observe preventive measures by the subjects.

Data collected using the questionnaires was entered into and analyzed using EPI INFO version 3.5.1 (2008) computer software. Cross tabulation of variables was done and level of statistical significance was set at $p<0.05$. Each correct answer on the knowledge questions was awarded a mark with no marks awarded for wrong answers. Knowledge was graded with scores <50 and ≥50% adjudged inadequate and adequate knowledge, respectively.

RESULTS

A total of one hundred (n=100) respondents participated in the study. The ages of the respondents ranged from 18 to 63 years, 38 to 47 years being the modal class and those above 58 years constituted 6%. The mean age (±2 standard deviation (SD)) of the respondents was 37±3.34 years (Table 1). All the study subjects were males and Muslims. A total of 45% of them had only Qur'anic education, 18% had no education, while only 9% had secondary education (Table 2). Most of the respondents, 34% had been handling birds for between 6 and 10 years, while only 10% of them handled birds for over 16 years (Table 3). Sixty percent of the respondents have heard of AI, with the media being the commonest source of information followed by public enlightenment campaigns (Table 4). At baseline, 51% of the respondents scored ≥50% on the knowledge and prevention of AI (adequate knowledge), with a mean (±2 SD) knowledge score of 64±7.01. At post-intervention, 86% of the respondents scored ≥50% (adequate knowledge) with a mean knowledge score of 91±17.64 ($p<0.0001$) (Table 5). The commonest form of personal protection adopted by the study subjects was the use of protective clothing including face masks (Table 6). Results from the

Table 5. Knowledge of AI.

Knowledge of AI	Study phase		Test statistics
	Pre-intervention	Post-intervention	
	No. (%)	No. (%)	
Adequate	51 (51)	86 (86)	
Inadequate	49 (49)	14 (14)	χ^2=26.79; df=1; p<0.0001
Total	100 (100)	100 (100)	

Table 6. Respondents' ways of protection against AI.

Protective measures*	No. (%)
Protective clothing	27 (27)
Face masks	53 (53)
Hand gloves	16 (16)
Boots	21 (21)
Goggles	12 (12)
Washing hands after handling birds	25 (25)
Bathing after handling baths	8 (8)
Use of aprons	17 (17)

*Multiple answers allowed.

application of checklists showed that of the 50 bird handlers in 50 farms surveyed, only 10 used some form of protective clothing with 10 wearing boots. Five out of the 50 farms had their workers wearing masks, hand gloves and aprons, while only one farm had a few of their workers wearing goggles. Forty of these farms did not as a matter of routine practice provide any form of protective clothing for their workers. Although 20 farms claimed that they use soap and water to wash their hands after work, there was really no evidence on ground to justify that claim.Thirty farms used ordinary water for washing after attending to birds. The sanitary condition of most of the farms is everything, but good as only five farms had facilities for proper waste management. Only 3 farms had dips for vehicles and 25 had for humans. It was observed that only 4 farms had good methods of egg collection which prevents the birds coming in contact with the eggs after they are laid.

DISCUSSION

The poultry farmer by virtue of his close contact with birds is among individuals most at risk of contracting AI. The mean age of the respondents in this study was 37±3.3 years, which is similar to the mean age obtained in the study from Nepal, India (Neupane et al., 2012). However, the mean age in our study was lower than 43 years observed in Italy (Abbate et al., 2006) but higher than figures observed in other studies from Zaria and Oyo state of Nigeria (Idris et al., 2009; Fatiregun and Saani, 2008). Findings from this study showed that the mean duration of handling of birds was 9 years which is lower than 12 and 16 years observed in studies from Zaria and Hong Kong, respectively (Idris et al., 2009; Kim et al., 2011). The results of the study showed that 60% of the respondents were aware of the existence of AI, which is lower than the 93 and 97% awareness recorded in similar studies from Oyo state, Nigeria and Nepal, India, respectively (Fatiregun and Saani, 2008; Neupane et al., 2012). In contrast to the level of awareness recorded in this study, other centers have observed lower rates (Eastwood et al., 2009; Blendon et al., 2008). The level of awareness recorded in our study may not be unconnected to the low morbidity and mortality associated with the infection since it was first noticed in Nigeria, compared to that seen in other countries. The commonest source of information on AI in this study was the media (60%), particularly radio and television. The radio is a common household appliance, which is found useful in the information, education and entertainment of the populace. Equally, our respondents are avid listeners of international Hausa service radio stations e.g. British Broadcasting Corporation (BBC), Voice of America (VOA), Radio France International (RFI), De-Welle, Russia, China, etc. This finding is in consonance with those observed in other studies (Abbate et al., 2006; Eastwood et al., 2009; Neupane et al., 2012; Idris et al., 2009; Fatiregun and Saani, 2008). However, the study from Hong Kong, China (Kim et al., 2011) recorded the internet and health talks as the commonest sources of information about AI which may not be unrelated to the level of education of their subjects as compared to this study where majority (63%) of the subjects had no formal education.

At baseline, 51% of the study subjects had adequate knowledge of the cause, mode of transmission and prevention of AI, with a mean knowledge score of 64±7.01. The level of knowledge exhibited by the respondents may not be unrelated to their exposure to foreign mass media that usually broadcast in the local language, Hausa in the wake of the outbreak of the infection and this had necessitated the setting up of AI desks in all the Local Government Areas of the state. Adequate knowledge of the mode of transmission of AI will invariably assist individuals in protecting themselves from the menace of the infection.

After the intervention, 86% of the subjects had adequate knowledge of AI with a mean knowledge score of 91±17.64, and this increase in the level of knowledge post intervention was found to be statistically significant ($p<0.0001$). In agreement with our findings, the study conducted in Vietnam showed high knowledge for AI. However, low knowledge scores were observed in the studies from Zaria, Nigeria and Hong Kong (Idris et al., 2009; Kim et al., 2011).

In the assessment of the respondents' practices aimed at protecting themselves against AI, the commonest measure was the use of face masks (53%). In the wake of the outbreak of AI, the state government, Sokoto state branch of the Veterinary Medical Association (NVMA) and other donor agencies distributed face masks free of charge to the people most at risk, which could have accounted for high proportion of our subjects who used it. Cultures and habits are often difficult to change overnight which accounted for near average use of face masks by the respondents as most opined that for long they have been practicing their trade without contracting any infections. Similar to our findings, the use of face masks was the commonest preventive measure adopted in the study from Italy (Abbate et al., 2006).

However, the use of other personal protective equipment (PPE) was not a routine among our respondents and this is in consonance with the findings from other studies (Fatiregun and Saani, 2008; Idris et al., 2009; Kim et al., 2011; Neupane et al., 2012). Sometimes the low profit margin of these bird handlers makes it difficult to acquire these PPE. It has been observed that the distribution of promotional material during outbreaks of epidemics will instill confidence and demonstrate to the public that health authorities are transparent about the risk and have containment plans in place.

This in turn, has the added benefit of allowing people to accept socially unfamiliar control measures such as wearing masks and home quarantine, before a pandemic occurs. The strategy should also include the promotion of personal infection control practices, such as hand washing and sick leave which are all lifestyle activities that would protect individuals and the community against any communicable disease (Eastwood et al., 2009). Despite World Health Organization (WHO) recommendations on ways to prevent infection with AI virus (WHO, 2006), there has always been low adherence or outright disregard to these recommendations. Hand washing with soap and water was practiced in only 20 out of the 50 farms visited. This practice is in consonance with findings from other centers (Neupane et al., 2012; Kim et al., 2011; Abbate et al., 2006; Idris et al., 2009).It was observed that 30 farms used ordinary water for washing after attending to birds. The sanitary condition of most of the farms is everything but good as only five farms had facilities for proper waste management. Only 3 farms had dips for vehicles and 25 had for humans.

It was observed that only 4 farms had good methods of egg collection which protects the birds against the spread of infection as the birds do not come in contact with eggs after they are laid. In the market place, the disposal of waste is also poor and birds of different species are mixed together by majority off the sellers. Only few of the sellers separated sick birds from the healthy ones. This underscores the need for regular inspection of farms and other points of sale of birds to ensure adequate sanitation and adherent to safety guidelines. After intervention, it was observed that there was tremendous improvement both in the farms and the markets. Consequent upon the intervention, 30 farms bought boots, gloves, aprons and masks for their workers. As is common with practices in most work places, most workers chose to ignore safety rules thereby increasing their chances of contracting infections.

After the intervention, the sanitary conditions of most farms had improved considerably with most farms initiating a weekly environmental sanitation exercise in the farms and their environs. The provision of dips for vehicles was not done as all the farms claimed it was capital intensive. The training intervention showed considerable increase in the level of knowledge about AI, its cause, transmission and prevention while the subjects exhibited marked improvement in the practices of prevention of AI. Improvement of knowledge of transmission and the use of preventive measures is an important public health strategy for the reduction of the effects of AI and its complications amongst bird handlers.

The mixing of birds of different species reduced considerably; sick birds were often separated from healthy ones, even though the market environmental waste management was yet to be corrected as the disposal was still a major problem.

Conclusions

The health education intervention study has demonstrated increased awareness of AI among the study subjects as 86% of them had a mean knowledge score of 91±17.64 after the introduction of the intervention measures as compared to 51% with a mean knowledge score of 64±7.01 at baseline. The use of media particularly radio was found to be an effective and veritable tool for raising the awareness of bird handlers on topical issues in this environment such as AI. The findings from this study will help relevant government agencies come up with sustainable policies that will nip in the bud lapses in preventive measures in case of future outbreaks. Given the necessary education, bird handlers are more likely to take more proactive measures against the spread of diseases like AI.

ACKNOWLEDGEMENTS

The authors wish to acknowledge the efforts and cooperation of their respondents who volunteered to participate in the study, the Poultry Association of Nigeria

(PAN) Sokoto state branch, the research assistants, management of Usmanu Danfodiyo University and the NFELTP 7/FMOH/CDC facilitators, for their motivation to undertake this research.

REFERENCES

Abate R, Di Giuseppe G, Marinelli P, Angelillo IF (2006). Knowledge , attitudes and practices of Avian Influenza, Poultry workers, Italy. Emerg. Infect. Dis.12(11):1762-1765.

Blendon RJ, Koonin LM, Benson JM, Centron MS, Pollard WE, Mitchell EW et al (2008). Public response to community mitigation measures for Pandemic Influenza. Emerg Infect Dis. 14:778-86.

Bridges CB, Lim W, Hu-Primmer J, Sims L, Fukuda K, Mak KH, Rowe T, Thompson WW, Conn L, Lu X, Cox NJ, Katz JM (2002). Risk of influenza A (H5N1) infection among poultry workers, Hong Kong, 1997–1998. J. Infect. Dis;185:1005–10.

Brown JD, Stallknecht D, EBeck JR, Suarez DL, Swayne DE (2006). Susceptibility of North American ducks and gulls to H5N1 highly pathogenic avian influenza viruses, Emerg. Infect. Dis.12 (11): 1663–1670.

Capua I, Marangon S (2007). Control and prevention of avian influenza in an evolving scenario, Vaccine 25(30): 5645–5652.

Eastwood K, Durrheim D, Francis JL, d'Espaignet ET, Duncan S, Islam F, Speare R (2009). Knowledge about Pandemic influenza and compliance with containment measures among Australians. Bulletin of the World Health Organization, 87:588-594.doi:10.2471/BLT.08.060772.

Fatiregun AA, Saani MM (2008). Knowledge, Attitudes and compliance of poultry workers with preventive measures for avian Influenza in Lagelu, Oyo state, Nigeria. J. Infect. Dev. Ctries; 2(2):130-4

Idris AM, Olugbenga , Abdulrazaq HG (2009). Knowledge, Attitudes and Practices of Avian Influenza among Poultry traders in Nigeria . Int. J. Infect. Dis. Volume 8 number 2. DOI:10.5580/89e.

Kim JH, Lo FK, Cheuk KK, Kwong MS, Goggins WB, Cai YS, Lee SS, Griffiths S (2011). Knowledge of Avian Influenza (H5N1) among poultry workers, Hong Kong, China. Emerg. Infect. Dis. 17(12):2319-2321.

Kirkwood BR, Sterne JAC (2003). Essentials of medical statistics, 2nd edition, Massachusetts: Blackwell science Ltd. Pp191.

Krauss S, Webster RG (2010). "Avian influenza virus surveillance and wild birds: past and present," Avian Dis. 54(1): 394–398.

Neupane D, Khanal V, Ghimire K, Aro AR, Leppin A (2012). Knowledge, Attitudes and Practices related to avian Influenza among poultry workers in Nepal: a cross sectional study. BMC Infectious Diseases;12:76 http://www.biomedcentral.com/1471-2334/12/76

Pantin-Jackwood MJ, Swayne DE (2009). "Pathogenesis and pathobiology of avian influenza virus infection in birds," Revue Scientifique et Technique, 28(1): 113–136.

Pappaioanou M (2004). Veterinary medicine protecting and Promoting the Public's Health and Well being. Prevent. Vet. Med. 62:153-163

Sokoto state profile,2013. Available from www.sokotostate.gov.ng (accessed on 5th August, 2013).

Swayne DE (2006). Occupational and consumer risks from avian influenza viruses. Dev. Biol. Basel. 124:85–90.

The Writing Committee of the World Health Organization (WHO)Consultation on Human Influenza A/H5. Avian influenza A (H5N1) infection in humans. New Engl. J. Med. 2005;353:1374–85.

World Health Organization. Cumulative number of confirmed human cases of avian influenza A/(H5N1) reported to WHO. 2006 [cited 2006 Aug 2]. Available from http://www.who.int/csr/disease/avian_influenza/country/cases_table_2006_07_26/en/index.html.

WHO Global Alert and Response (GAR): Avian influenza situation in Nigeria –update 2006. World Health organization Geneva 2006).

World Health Organization. Public health interventions for prevention and control of avian influenza. A manual for improving biosecurity in the food supply chain: focusing on live animal markets. New Delhi: Regional Office for South-East Asia. 2006.

Feeding habits of culicine mosquitoes in the Cameroon lowland forests based on stable isotopes and blood meal analyses

Kevin Yana Njabo[1]*, Thomas B Smith[1,2] and Elizabeth Yohannes[3]

[1]Center for Tropical Research UCLA Institute of the Environment and Sustainability, Los Angeles CA USA.
[2]Department of Ecology and Evolutionary Biology, University of California, Los Angeles, Los Angeles, CA 90095, USA.
[3]University of Constance, Limnological Institute, Stable Isotope Laboratory,Mainaustr 252, D-78464, Germany.

Mosquito blood feeding behavior is a very significant component of pathogen transmission and determinant of disease epidemiology. Yet, knowledge of foraging ecology of mosquitoes often depends on the presence of undigested blood in the mosquito mid gut. Approximately 36 h after feeding, the blood meal is sufficiently digested to make identification by molecular techniques difficult, leaving a very narrow window in which these methods can be utilized. Here, we investigated the feeding habits of wild caught culicine mosquitoes from four genera, *Aedes, Anopheles, Coquillettidia* and *Mansonia* of the lowland rainforests of Cameroon based on the isotopic ratios of nitrogen ($\delta^{15}N$), carbon ($\delta^{13}C$) and sulfur ($\delta^{34}S$). Results showed that unfed mosquitoes had a lower $\delta^{13}C$, $\delta^{15}N$ and middle $\delta^{34}S$ values than mosquitoes fed with $\delta^{13}C$ appearing to be the best element to differentiate between mosquito species that fed on different host species. Isotopic analyses show that the different mosquito genera may be separated based on their diets, suggesting that linking stable isotope-based assays and DNA analysis may be a powerful new tool to investigate mosquito feeding ecology and the dynamics of vector-borne pathogens.

Key words: Carbon ($\delta^{13}C$), nitrogen ($\delta^{15}N$), sulfur ($\delta^{34}S$), isotope, feeding habits, mosquitoes, polymerase chain reaction (PCR).

INTRODUCTION

Over the past decades, there has been an increase in the incidence of infectious diseases throughout the world, with serious implications for human and wildlife populations (Kilpatrick et al., 2006; Baker et al., 2012). Due to their diversity and abundance, demonstrated vector competence (Sardelis et al., 2002) and frequent infection in nature (Molaei et al., 2006; Burkett-Cadena et al., 2008a, b), mosquitoes are regarded as one of the most important vectors of disease.

The behavioral characteristics of mosquitoes in disease transmission differ vastly between different regions and species (Fontenille and Lochouarn, 1999; Kent and Norris, 2006), and can only be understood in a local context of vector-host interactions. An important aspect of these interactions is that of host preference, with each vector species feeding on a limited range of host species. Vector species that switch between hosts are particularly considered important in disease transmission because they have the potential to act as bridge vectors, by transferring pathogens from the reservoir to humans and/or domestic animals (Chandler et al., 1975; Rasgon, 2008). For example, West Nile Virus (WNV) is primarily an infection of birds, but can be transferred to humans or horses by mosquitoes that feed on both birds and mammals (Hamer et al., 2008). To accurately describe an

*Corresponding author. E-mail: kynjabo@hotmail.com, kynjabo@ucla.edu.

arthropod-borne pathogen transmission cycle, it is critical to know which available host species are preferred by the corresponding vectors. Most studies evaluate the host-feeding patterns of mosquitoes by identifying the source of the vertebrate blood meals through sequencing portions of the cytochrome *b* gene of mitochondrial DNA. However, there is a very limited time period in which these methods can be utilized because these approaches only provide information on recent feeding events. After approximately 36 hours of feeding, the blood meal is sufficiently digested to make the identification by immunological or polymerase chain reaction (PCR)-based techniques particularly difficult, thus effective methods to assess feeding habits over longer periods of time are necessary.

Stable isotope techniques offer the opportunity to overcome some of these constraints by obtaining information on nutrients assimilated over extended periods of time (Rasgon, 2008). By comparing the isotopic signatures of a given organism or tissue, inferences on its feeding behavior could be done. Stable isotopes of carbon, nitrogen and sulfur have been widely used in ecological research (Wada et al., 1991; Fry, 2007). Stable carbon (δ13C); (δ15N) and nitrogen (δ34S) as indicated in ms are commonly used to extract information about the feeding habitat and carbon dietary sources (McCutchan et al., 2003; Michener and Lajtha, 2007; Yohannes et al., 2008); while nitrogen isotopes are often applied to evaluate the trophic position of a given species (Peterson and Fry, 1987). Based on the natural variation of the composition of stable sulfur isotope, measurement values can provide information on dietary protein sources and geographical origin (Richards et al., 2003). Also, because the rate of metabolism of different tissues determines the turnover of stable isotopes in tissues, it is possible to obtain dietary information on varying time-scales and extended period by sampling tissues with different turnover rates after feeding events (Tieszen et al., 1983; Rasgon, 2008).

The lowland forest areas in Cameroon are known to experience hyper-endemic transmission of *Plasmodium falciparum* mostly spread by the species, *Anopheles gambiae*, but little is known about the ecology of the different mosquito species in this region, and how vector species composition and their relative roles in transmission vary geographically (Akono et al., 2009). Very few studies on wild caught mosquitoes in Cameroon have been published (Rageau and Adam, 1952; Akono et al., 2009; Njabo et al., 2011), and very little information is available on their feeding habits (Snow and Boreham, 1978; Wanji et al., 2003).

In this study, we compared the feeding habits of wild caught blood-engorged mosquitoes and those with no visible blood meals (unfed mosquitoes or completely digested blood meal) that exhibit ornithophilic feeding habits, and characterize their feeding behavior by applying individual $\delta^{13}C$, $\delta^{15}N$ and $\delta^{34}S$. We also coupled our isotopic approach with PCR based molecular techniques and identified vertebrate host blood meals following extraction of DNA from the engorged mosquitoes and PCR amplification of the cytochrome *b* gene.

METHODOLOGY

Sampling sites and habitat characterization

Mosquito samples were collected in August, 2007 and May, 2008 at two lowland forest sites in Cameroon. Locations and dates of fieldwork for these forest sites were: Ndibi (N 2° 43' 50", E 9° 52' 19"), August 21st to 26th, 2007 and May 2nd to 16th, 2008; Nkouak (N 3° 52', E 13° 18'), August 27th to 31st, 2007 and May 17th to 29th, 2008. Ndibi has an elevation of 500 meters above sea level and its habitat is characterized by disturbed lowland forest, and seasonally flooded swamp forest habitat and high human population (exacerbated by its close proximity to the city of Akonolinga), while Nkouak has less human population density and more forest cover and is the source of river Nyong. Rainfall patterns in these areas have been fairly stable for more than 40 years. There is one rainy season each year that lasts from November to April, followed by cool dry (April to August) and hot dry (August to November) seasons.

Mosquito collections

Two main trapping methods were used as described in Njabo et al. (2011). Briefly, six Center for Disease Control (CDC) Miniature Light Traps baited with CO_2 (John W. Hock, Gainesville, FL), and four modified bird-baited Ehrenberg lard cans (Downing and Crans, 1977). Three bird species Feral Pigeon (*Columba livia*), Yellow-whiskered Greenbul (*Andropadus latirostris*) and Greater Swamp Warbler (*Acrocephalus rufescens)* were used as baits in the lard cans.

Traps were set out each day for at least 12 h (06.00 pm to 06.00 am). Following each trapping period, the collection bags were removed from traps and the mosquitoes were transported to the field lab site where they were immobilized with chloroform and/or smoke. On the day of collection and immobilization, mosquitoes were sorted by sex and identified to species, with the aid of a stereomicroscope (× 90) and morphological keys (Hopkins et al., 1941; Service, 1990). The mosquitoes were then separated by gonotrophic condition (unfed, blood-fed, gravid) and enumerated. They were then sorted into four groups: Pigeon fed, Warbler fed, Greenbul fed and unfed (all mosquitoes with no visible blood meals in the abdomen are considered for the purpose of this manuscript as unfed). We used t-test to compare each fed group against the unfed mosquitoes (that is, binary comparisons of Pigeon fed, Warbler fed or Greenbul versus unfed were conducted separately). At both sites, sampling date and species, fed, unfed and gravid mosquitoes were pooled separately into groups of 5 to 20 specimens and placed in 95% alcohol and later stored at -20°C. Blood meals from blood-fed mosquitoes were subsequently tested by PCR, as described below, to determine the birds on which mosquitoes had fed.

Host blood meal identification

Within the pools of blood-fed mosquitoes, five individuals were randomly selected for PCR analyses. Total DNA was extracted from the abdominal contents of the blood-fed mosquitoes individually, using standard proteinase k digestion and phenol chloroform purification. Isolated DNA from the mosquito blood meals served as DNA templates in subsequent PCR reactions. PCR primers (Forward 5'-3'GACTGTGACAAAATCCCNTTCCA, Reverse 5'-3' GGTCTTCATCTYHGGYTTACAAGAC) were based upon previously published multiple alignment of cytochrome *b* sequences of avian species (Ngo and Kramer, 2003). Amplified fragments from the blood-fed mosquitoes were sequenced and compared with sequences in GenBank for host-species identification as described in Molaei and Andreadis (2006) and Hellgren et al. (2008).

Stable isotope analyses

We analyzed stable carbon and nitrogen isotopes from representative whole mosquitoes using pooled samples within each group. We removed lipids from each homogenized sample, using a 2:1chloroform-methanol solution rinse for 24 h. Samples were rinsed using methanol solution and air-dried in a fume hood. Dried and powdered sub-samples, approximately 0.6 mg, were weighed into small tin cups to the nearest 0.001 mg, using a micro-analytical balance. Samples were then combusted in a Eurovector (Milan, Italy) elemental analyser (Limnological Institute, University of Konstanz, Germany). The resulting CO_2 and N_2 were separated by gas chromatography and admitted into the inlet of a Micromass (Manchester, UK) Isoprime isotope ratio mass spectrometer (IRMS) for determination of $^{13}C/^{12}C$, $^{15}N/^{14}N$ ratios. Measurements were reported in δ-notation ($\delta^{13}C$ and $\delta^{15}N$, respectively) where $\delta = 1000 \times (R_{sample}/R_{standard}) - 1$ ‰ relative to the Pee Dee Belemnite (PDB) for carbon and atmospheric N_2 for nitrogen in parts per thousand deviations (‰). Two sulfanilamide (Iso-prime internal standards) and two Casein were used as a laboratory standard for every 10 unknowns in sequence.

We also analyzed sulfur isotopes in mosquitoes. Tin capsules containing reference or sample were loaded into an automatic sampler from where they were dropped in sequence, into a furnace held at 1080°C and combusted in the presence of oxygen, raising the temperature in the region of the sample to ~ 1800°C. The reference material used for analysis was sulfanilamide calibrated and traceable to NBS-127 (barium sulphate, $\delta^{34}S$ = +20.3‰). Hundreds of replicate assays of internal laboratory standards indicate measurement errors (SD) of ± 0.05, 0.15 and 0.05‰ for $\delta^{13}C$, $\delta^{15}N$ and $\delta^{34}S$, respectively.

For each isotopic element separately, unfed versus fed mosquitoes of *Coquillettidia aurites* were compared using t-test. A second t-test was applied to compare elemental isotopic differences between blood-engorged *Coquillettidia pseudoconopas* and those with no visible blood meals. A one-way analysis of variance (ANOVA) followed by a Scheffe´ Post-hoc-test was used to compare differences in $\delta^{13}C$ and $\delta^{15}N$ between Warbler-fed, Greenbul-fed and unfed *Mansonia uniformis*

RESULTS

Five mosquito species belonging to four genera *Aedes* (*Aedes mcintoshi*), *Anopheles* (*Anopheles coustani*), *Coquillettidia* (*C. aurites and C. pseudoconopas*), and *Mansonia* (*M. uniformis*) were collected in the light traps while only three species (*C. aurites, C. pseudoconopas* and *M. uniformis*) were collected in the lard cans. All the species collected from the CDC light traps were unfed (had no visible blood meals) while all species from the bird-baited lard cans were blood-engorged.

Host blood meal identification

To confirm the applicability of the blood meal-specific PCR to Cameroon lowland forest mosquitoes and their avian hosts, the assay was tested for its ability to amplify cytochrome *b* sequences from several native Cameroon lowland forest bird species (results published elsewhere (Chasar et al., 2009; Njabo et al., 2011). The specific amplification conditions were found to support the amplification of detectable PCR products from all the bird species tested. The PCR diagnostic successfully identified all blood meals from engorged mosquitoes collected from the baited lard cans. No other species were identified and none of the mosquitoes fed on multiple hosts, indicating that the bias towards bird feeding was not indicative of a specific host preference for birds but a reflection of relative host availability. All unfed mosquitoes tested were negative.

Mosquito stable isotope analyses

The $\delta^{13}C$, $\delta^{15}N$ and $\delta^{34}S$ values of unfed mosquitoes and fed mosquitoes are shown in Table 1. In general, unfed mosquitoes have relatively lower $\delta^{13}C$ and $\delta^{15}N$ compared to the blood-engorged mosquitoes. Mean $\delta^{13}C$ values from unfed mosquitoes ranged from ca.-20.0 to -34.5‰, while $\delta^{15}N$ ranged from 3.1 to 4.8‰, and $\delta^{34}S$ ranged from 6.4 to 10‰. Mean $\delta^{13}C$ values from fed mosquitoes ranged from -19.8 to -24.5‰, while $\delta^{15}N$ ranged from 4.3 to 7.1‰, and $\delta^{34}S$ ranged from 5.5 to 8‰.

The mean ± SE of $\delta^{13}C$ and $\delta^{34}S$ of each mosquito species are given in Figures 1 and 2, and Table 1 separately. Fed and unfed *C. aurites* showed a significantly different $\delta^{13}C$ (t-test, p = 0.02) signature (Figure 1). However, there was no difference between fed and unfed $\delta^{34}S$ (Figure 2) or $\delta^{15}N$ (Figure 1). Blood-engorged *C. pseudoconopas* and those with no visible blood meals showed significant difference at both $\delta^{13}C$ (t-test, p < 0.0001) and $\delta^{34}S$ (p = 0.01, equal variance not assumed).

ANOVA revealed that there was a significant diet effect on $\delta^{13}C$ and $\delta^{15}N$ values of *M. uniformis* ($\delta^{13}C$: $F_{15,98} = -7.19$, $p < 0.001$; $\delta^{15}N$: $F_{12,97} = -3.65$, $p < 0.001$). Post-hoc analysis of $\delta^{13}C$ for Warbler-fed and Greenbul-fed *M. uniformis* species indicated no significant difference (p > 0.05), and thus we pooled these data in further analyses and compared unfed and fed *M. uniformis* using t-test. On the contrary, *M. uniformis* that fed on Warbler were significantly different in $\delta^{15}N$ (t-test: t =14.44, p = 0.003 equal variance not assumed due to unbalanced sample size) than those that fed on Greenbul (Figure 1 and Table 1).

Bi-directional error bar graph (Figure 1) for $\delta^{13}C$ and $\delta^{15}N$, for fed (grouped together in a rectangle frame) and unfed groups show that unfed *A. mcintoshi* survives on

Table 1. Mean ± SE (n) values of nitrogen, carbon and sulfur Isotopes of unfed and fed mosquitoes.

Species	Diet	$\delta^{15}N$ (‰)	t-test	$\delta^{13}C$ (‰)	t-test	$\delta^{34}S$ (‰)	t-test
C. aurites	Unfed	4.5±0.31(8)	t=0.23	-30.4±3.5 (8)	t=-2.52	6.4±0.46 (3)	t=0.80
*C. aurites**	Pigeon	4.3±0.78 (9)	p=0.82	-21.8±2.47 (9)	p=0.02*	5.9±0.59 (9)	p=0.44
C. pseudoconopas	Unfed	4.8±0.37 (6)	t=-0.90	-34.7±1.07 (6)	t=-0.77	6.8±0.34 (6)	t=0.33
*C. pseudoconopas**	Pigeon	5.2±0.11 (3)	p<0.0001*	-19.8±1.05 (3)	p=0.46	5.4±0.28 (3)	p=0.01*
M. uniformis	Unfed	3.1±0.23 (90)	t=-14.25	-29.9±0.48 (90)	t=-7.52	8.4±0.36 (3)	t=-1.29
M. uniformis[1]	Greenbul	7.1±0.51 (6)	p<0.0001*	-24.5±1.80 (6)	p<0.0001*	8.9±0.36 (3)	p=0.27
M. uniformis[2]	Warbler	5.8±0.23 (3)	t =-11.21	-23.7±1.11 (3)	p<0.0001*	8.9±0.36 (3)	p=0.27
A. Mcintoshi	Unfed	3.9±0.34 (17)	p<0.0001*	-20.0±0.64 (17)	p<0.0001*	8.0±0.47 (3)	p<0.0001*
A. coustani	Unfed	4.0±0.63 (3)	p<0.0001*	-26.8±0.52 (3)	p<0.0001*	10.3±0.09 (3)	p<0.0001*

Diet: *Feral Pigeon, [1]Yellow-whiskered Greenbul and [2]Greater Swamp Warbler fed mosquitoes in the lowland forests of Cameroon. Results of the Student's t-test are given for each comparison. *Statistically significant differences ($p<0.05$). $\delta^{13}C$ (‰) in unfed *M. uniformis* was compared using pooled Greenbul and Warbler fed *M. uniformis*. Sample sizes for $\delta^{34}S$ are lower than for $\delta^{15}N$ or $\delta^{13}C$ due to missing samples.

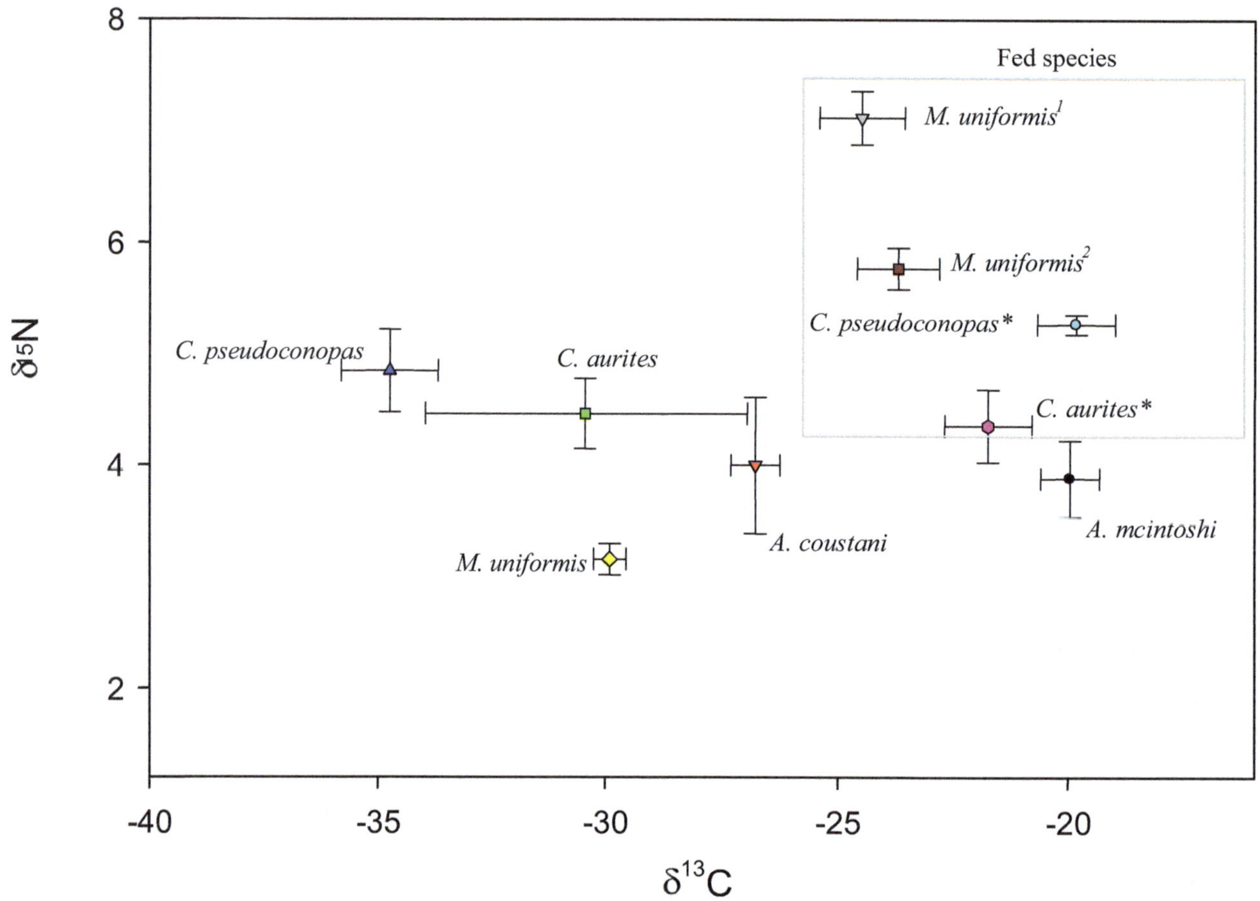

Figure 1. Bi-directional mean (± SE) values of carbon ($\delta^{13}C$) and nitrogen ($\delta^{15}N$) of unfed and blood-engorged mosquitoes in the lowland forests of Cameroon. Mosquitoes were fed with *Feral Pigeon, [1]Yellow-whiskered Greenbul and [2]Greater Swamp Warbler.

coustani, C. aurites and *C. pseudoconopas*, respectively. These values for *A. mcintoshi* are very similar to those of *Coquillettidia* species that fed on pigeons. As shown in Figure 1, the highest $\delta^{34}S$ in these study species was recorded for *A. coustani*. This species also showed an intermediate $\delta^{13}C$ values (ca. -27‰) and $\delta^{15}N$ (ca. 4‰), indicating a unique sulfur dietary sources in this species Interestingly, *M. uniformis* which fed on warblers has a comparable $\delta^{34}S$ (ca. 9‰) to that of *A. coustani* (ca. 9‰) signature. This implies the two mosquito species rely on

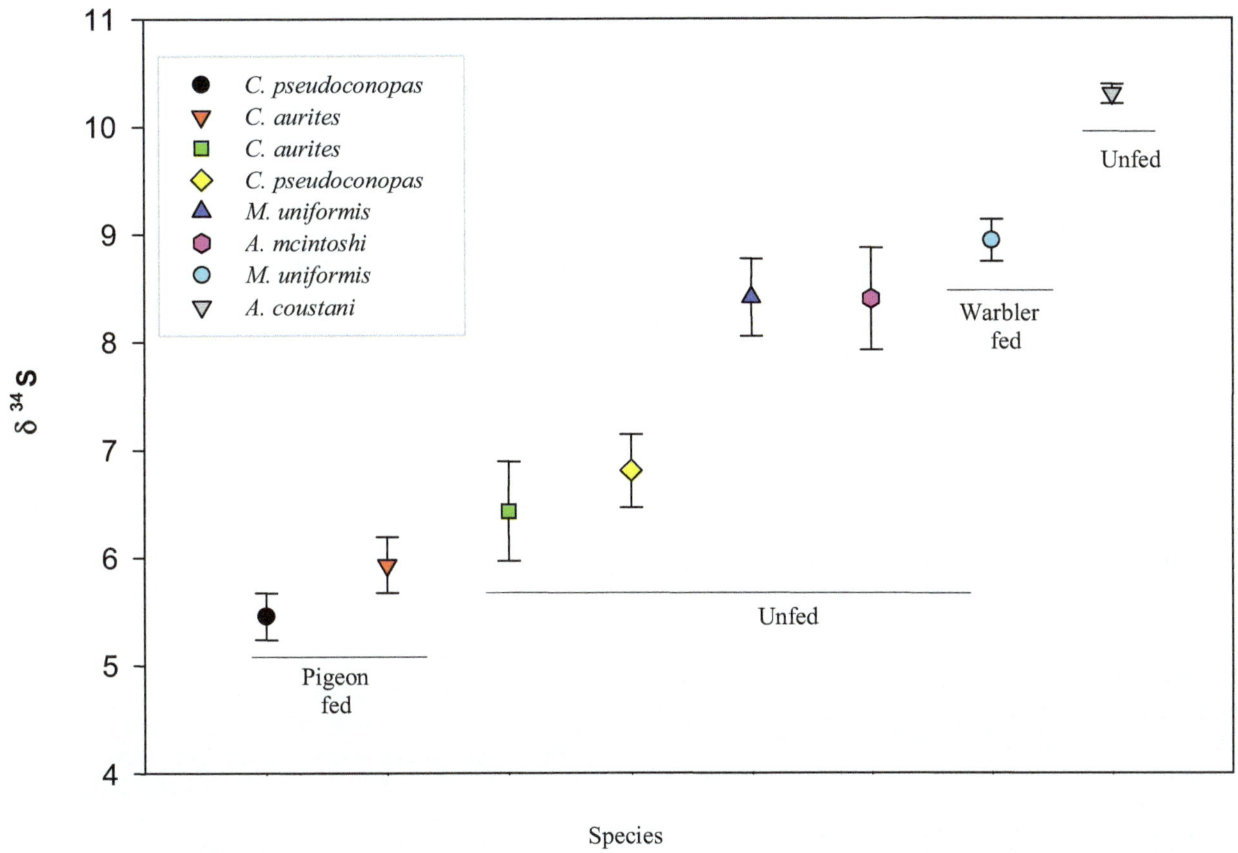

Figure 2. Mean (± SE) values of carbon ($\delta^{34}S$) for unfed and fed (Pigeon and Warbler fed) mosquitoes collected from the lowland forests of Cameroon.

similar sulfur dietary source.

DISCUSSION

In this study, four mosquito genera, *Aedes, Anopheles, Coquillettidia* and *Mansonia* and three different host types, were examined for triple isotopic elements: $\delta^{13}C$, $\delta^{15}N$, and $\delta^{34}S$.

Stable carbon isotopes

Our results suggest that the five mosquito species studied exhibit a wide-range of $\delta^{13}C$ signature spanning from ca -49 to -15‰, indicating a variety of host-carbon source in the region. These results support the findings of Hassan et al. (2003) who showed that ornithophilic mosquitoes typically feed significantly more or less on available bird species than predicted based on biomass, surface area, or relative abundance.

Although these mosquitoes seem to potentially cover a rather wider carbon source, each species shows a more specific and distinctive isotopic niche-width. For instance, *A. mcintoshi* apparently feeds on a host with a relatively higher stable carbon isotope value. Using mean $\delta^{13}C$ values reported for C_4 (-13‰) and C_3 (-27‰) plants (Michener and Lajtha, 2007) as endpoint tissue, $\delta^{13}C$ of *A. mcintoshi* departs from a dietary source based on C_3 plant and supports a C_4 dominated carbon source incorporated in its food chain. Relatively higher $\delta^{13}C$ values, as those reported for this species, can also be linked to C_3 plants of drier areas and xeric condition in response to water-use efficiency (Ehleringer and Cooper, 1988).

The $\delta^{13}C$ values of *A. mcintoshi* also complement those of *Coquillettidia* spp. that fed on pigeons, indicating that *A. mcintoshi* may also feed on pigeons or on other avian/vertebrate hosts with similar food habits. Pigeons are known to forage mainly in agricultural landscapes, as these areas represent an important and well exploited source of food (Hetmanski et al., 2005). There were no significant differences in isotopic values in $\delta^{13}C$ ($p > 0.05$) for Warbler-fed and Greenbul-fed *M. uniformis*. Although we observed slight differences in *M. uniformis* that fed on Warblers to those that fed on Greenbul (of $\delta^{13}C$ and also $\delta^{15}N$), the wide-range of isotopic signatures from both fed and unfed individuals (for example, mean $\delta^{13}C$ range: ca.

-24 to -30‰ of this species) suggest that *M. uniformis* seems to be a generalist that feeds on available host species of different isotopic values.

Indeed, *Mansonia* mosquitoes are major vectors of filariasis caused by *Brugia malayi* filarial nematodes and feed on a wide range of vertebrate hosts in nature (Laurence, 1960; Phumee et al., 2011). A larger range of $\delta^{13}C$ was also observed for unfed *C. aurites*. -21 to -48‰, implying the availability of multiple host species or hosts that rely on wider isotope biomes. Analysis of additional isotopes elements such as deuterium or oxygen may improve the resolution of nutrient sources in blood meal identification, especially in situations of many potential host species.

Stable nitrogen isotopes

The $\delta^{15}N$ signatures in consumer tissues are primarily used to assess trophic position in food webs. It has been shown that $\delta^{15}N$ values can be associated with plant and animal tissues grown in relatively low rainfall locations and arid environmental conditions (Ambrose, 1991; van der Merwe et al., 1993; Sealy et al., 1995). Our study illustrates that almost all unfed mosquito species exhibited similar nitrogen isotope signature. However, Greenbul-fed *M. uniformis* showed a much more enriched (up to Δ + 4‰) $\delta^{15}N$ than unfed conspecific species indicating a higher trophic position than the other mosquito species. We also measured a slight difference in $\delta^{15}N$ of *M. uniformis* that fed on Warblers and those fed on Greenbul.

Stable sulfur isotopes

A. coustani seems to depend on a higher sulfur isotopic niche. *M. uniformis* which fed on warblers has a clearly separated nitrogen (Figure 1) and sulfur isotope (Figure 2) compared to Pigeon-fed species. The similar $\delta^{34}S$ in Warbler-fed and unfed *M. uniformis,* as well as to unfed *A. mcintoshi* imply that these two mosquito species might feed on similar hosts. By implication, while others occupy a unique specific niche, others do share host sources and rely on similar nutrient sources. In summary, our results illustrate the advantages of stable isotope analyses for the study of mosquito host feeding and niche segregation analyses. Using both fed and unfed mosquitoes collected in the field, we showed that the stable isotopes-based assays could play an essential role as tracers when applied on specimens unsuitable for PCR, such as gravid individuals or mosquitoes with digested or no observable blood meal to reconstruct the history of previous feeding events and dietary sources.

While stable isotope profiling is more sensitive in tracing element and nutrient sources, it is less specific in identifying digested blood meals to specific species as PCR techniques are. Thus the best results are obtained when both methods, stable isotope profiling and PCR, are used in concert. PCR may be used on fresh blood meals to gain a recent snapshot of current feeding habits, while stable isotope analysis may be used to reconstruct dietary history, trophic status and niche width of vectors over the longer time frames.

In general, field situations are likely to be more complicated, with multiple potential hosts present in the environment. Additional analyses of stable isotope ratios through more extensive sampling of potential hosts may provide further insights into seasonal changes in host preference and individual species' dietary strategies. To fully get the complete picture of field situations therefore, initial stable isotope profiling will have to be performed against all potential hosts in the study area, and temporal-spatial variation in stable isotope profiles within populations and within individual mosquitoes taken into account.

ACKNOWLEDGEMENT

We are grateful to Tanga Mbi and Eric Djomo Nana for assistance in the field, Birgit Beck and Katharina Greiner-Perth for assistance in stable isotope analyses of mosquitoes, and Tyffany Chen for assistance with PCR techniques. Dr Anton Cornel helped with the identification of all the mosquito species. We thank the Government of Cameroon for providing permits for field research. The present study was supported by the joint NSF-NIH Ecology of Infectious Diseases Program award EF-0430146, the Rufford Small Grants for Nature Conservation, and by the University of Konstanz (start-up-grant, Zukunftskolleg). We sincerely acknowledge the unreserved support by Karl-Otto Rothhaupt.

REFERENCES

Akono NP, Bakwo EM, Belong P, Kekeunou S, Foko G, Messi J (2009). Abondance et diversite de la faune culicidienne a Yaounde (Cameroun). Entomol. faunistique 62:(3)115-124.

Ambrose SH (1991). Effects of diet, climate and physiology on nitrogen isotope abundances in terrestrial food webs. J. Archaeol Sci. 18:293-317.

Baker MG, Barnard LT, Kvalsvig A, Verrall A, Zhang J, Keall M, Wilson N, Wall T, Howden-Chapman P (2012). Increasing incidence of serious infectious diseases and inequalities in New Zealand: A national epidemiological study. Lancet 379 (9821):1112-1119.

Burkett-Cadena ND, Eubanks MD, Unnasch TR (2008a). Preference of female mosquitoes for natural and artificial resting sites. J. Am. Mosq. Control. Assoc. 24(2):228-235.

Burkett-Cadena ND, Graham SP, Hassan HK, Guyer C, Eubanks MD, Katholi CR, Unnasch TR (2008b). Blood feeding patterns of potential arbovirus vectors of the genus *Culex* targeting ectothermic hosts. Am. J. Trop. Med. Hyg. 79(5):809-815.

Chandler JA, Boreham PFL, Highton RB, Hill MN (1975). A study of the host selection patterns of the mosquitoes of the Kisumu area of Kenya. Trans. R. Soc. Trop. Med. Hyg. 69:415-425.

Chasar A, Loiseau C, Valkiūnas G, Iezhova T, Smith TB, Sehgal RN (2009). Prevalence and diversity patterns of avian blood parasites in degraded African rainforest habitats. Mol. Ecol. 18:4121-4133.

Downing JD, Crans WJ (1977). The Ehrenburg pigeon trap as a sampler of *Culex* mosquitoes for St. Louis encephalitis surveillance. Mosq. News 37:48-53.

Hopkins GHE, Evans AM, Edwards FW (1941). Mosquitoes of the Ethiopian Region: Culicine Adults and Pupae. In: Edwards FW (Ed.), Mosquitoes of the Ethiopian Region, order of the Trustees, sold at the British Museum (Nat. Hist.) B. Quaritch, Limited, London.

Ehleringer, JR, Cooper TA (1988). Correlations between carbon isotope ratio and microhabitat in desert plants. Oecologia 76:562-566.

Fontenille D, Lochouarn L (1999). The complexity of the malaria vectorial system in Africa. Parassitologia (Rome) 41:267-271.

Fry B (2007). Coupled N, C, and S isotope measurements using a dual column GC system. Rapid Commun. Mass Spectrom. 21:750-756.

Hamer GL, Uriel DK, Jeffrey DB, Scott RL, Marilyn OR, Tony LG, Edward DW (2008). *Culex pipiens* (Diptera: Culicidae): A bridge vector of West Nile virus to humans. J. Med. Entomol. 45(1):125-128.

Hassan HK, Cupp EW, Hill GE, Katholi CR, Klingler K, Unnasch TR (2003). Avian host preference by vectors of eastern equine encephalomyelitis virus. Am. J. Trop. Med. Hyg. 69:641-647.

Hellgren O, Bensch S, Malmqvist B (2008). Bird hosts, parasites and their vectors-associations uncovered by molecular analyses of blackfly blood meals. Mol. Ecol. 17:1605-1613.

Hetmanski T, Wolk E (2005). The effect of environmental factors and nesting conditionson clutch overlap in the feral pigeon *Columba livia* f. *urbana* (Gm.). Pol. J. Ecol. 53(4):523-534.

Kent RJ, Norris DE (2006). Identification of mammalian blood meals in mosquitoes by a multiplexed polymerase chain reaction targeting cytochrome *b*. Am. J. Trop. Med. Hyg. 73:336-342.

Kilpatrick AM, Kramer LD, Jones MJ, Marra PP, Daszak P (2006). West Nile virus epidemics in North America are driven by shifts in mosquito feeding behavior. PLoS Biol. 4:e82.

Laurence BR (1960). The biology of two species of mosquito, *Mansonia africana* (Theobald) and *Mansonia uniformis* (Theobald), belonging to the subgenus *Mansonioides* (Diptera, Culicidae). Bull Entomol. Res. 51:491-517.

McCutchan JH Jr, Lewis WM Jr, Kendall C, McGrath CC (2003). Variation in trophic shift for stable isotope ratios of carbon, nitrogen, and sulfur. Oikos 102:378-390.

Michener R, Lajtha K (2007). Stable Isotopes in ecology and environmental science (2 Ed.), Blackwell Publishing Ltd. 594p.

Molaei G, Andreadis TA, Armstrong PM, Anderson JF, Vossbrinck CR (2006). Host feeding patterns of *Culex* mosquitoes and West Nile virus transmission, northeastern United States. Emerg. Infect. Dis. 12:468-474.

Ngo KA, Kramer LD (2003). Identification of mosquito bloodmeals using polymerase chain reaction (PCR) with order-specific primers. J. Med. Entomol. 40:215-222.

Njabo KY, Anthony JC, Camille B, Erin T, Sehgal RNM, Valkiunas G, Andrew FR, Smith TB (2011). Non-specific patterns of vector, host and avian malaria parasite associations in a Central African Rainforest. Mol. Ecol. 20:1049-1061.

Peterson BJ, Fry B (1987). Stable isotopes in ecosystem studies. Ann. Rev. Ecol. Syst. 18:293-320.

Phumee A, Kanok P, Kanyarat K, Usavadee T, Apiwat T, Yutthana P, Padet S (2011). Morphology and protein profiles of salivary glands of filarial vector mosquito *Mansonia uniformis*; Possible relation to blood feeding process. Asian Biomed. 5(3):353-360.

Rageau J, Adam JP (1952). Culicinae du Cameroun. Ann. Parasitol. Hum. Comp. 32:610-635.

Rasgon JL (2008). Stable isotope analysis can potentially identify completely-digested Blood meals in mosquitoes. PLoS ONE 3(5):e2198.

Richards MP, Fuller BT, Sponheimer M, Robinson T, Aliffe L (2003). Sulphur isotopes in palaeodietary studies: A review and results from a controlled feeding experiment. Int. J. Osteoarchaeol. 13:37-45.

Sardelis MR, Turell MJ, O'Guinn ML, Andre RG, Roberts DR (2002). Vector competence of three North American strains of *Aedes albopictus* for West Nile virus. J. Am. Mosq. Control Assoc. 18:284-289.

Sealy J, Armstrong R, Schrire C (1995). Beyond lifetime averages: Tracing life histories through isotopic analysis of different calcified tissues from archaeological human skeletons. Antiquity 69:290-300.

Service MW (1990). Handbook to the Afrotropical toxorhynchitine and culicine mosquitoes, excepting *Aedes* and *Culex*. British Museum (Natural History), London.

Snow WF, Boreham PFL (1978). The host-feeding patterns of some culicine mosquitoes (Diptera: Culicidae) in Gambia. Bull. Entomol. Res. 68:695-706.

Tieszen LL, Boutton TW, Tesdahl KG, Slade NA (1983). Fractionation and turnover of stable carbon isotopes in animal tissues: implications for δ13C analysis of diet. Oecologia 57:32-37.

van der Merwe NJ, Lee-Thorp JA, Raymond JS (1993). Light, stable isotopes and the subsistence base of formative cultures at Valdivia, Ecuador," in Prehistoric Human Bone: Archaeology at the molecular level. In: Lambert JB, Grupe G, Berlin: Springer-Verlag pp. 63-97.

Wada E, Mizutani H, Minagawa M (1991). The use of stable isotopes for food web analysis. Crit Rev. Food Sci. Nutr. 30:361-371.

Wanji S, Tanke T, Atanga SN, Ajonina C, Nicholas T, Fontenille D (2003). Anopheles species of the mount Cameroon region: Biting habits, feeding behaviour and entomological inoculation rates. Trop. Med. Int. Health 8(7):643-9.

Yohannes E, Hansson B, Lee WR, Waldenström J, Westerdahl H, Åkesson S, Hasselquist D, Bensch S (2008). Isotope signatures in winter moulted feathers predict malaria prevalence in a breeding avian host. Oecologia 158:299-306.

Two years impact of single praziquantel treatment on urinary schistosomiasis in the Barombi Kotto focus, South West Cameroon

L. Nkengazong[1] , F. Njiokou[1] and T. Asonganyi[2]

[1]Laboratory of General Biology and Physiology, Faculty of Science, P. O. Box 812, Yaounde, University of Yaounde I, Cameroon.
[2]Faculty of Medicine and Biomedical Science, University of Yaounde I, P. O. Box 1364, Cameroon.

To evaluate the impact of a single dose praziquantel on urinary schistosomiasis in the Barombi Kotto focus, urine samples were collected from 306 participants (279 school children and 27 volunteer parents) of the Barombi Kotto health area from May, 2007 to May, 2009 and were examined using the filtration technique. A malacological survey was conducted to identify and follow up the infection rates of the intermediate snail host. The overall prevalence (69.17%) of *Schistosoma haematobium* was significantly different between boys (73.22%) and girls (66.66%; P = 0.03), same as between the island (84.3%) and mainland (60.45%; P = 0.0001) quarters. The prevalence reduced significantly by 55.03% (from 69.17 to 31.10%; P = 0.0001) and intensity of infection by 76.16% (from 212.1 to 50.56 eggs/10 ml urine; P = 0.01) at 2 years post treatment. Heavy *S. haematobium* infections in school children decreased from 23.31 to 2.12% at 2 years post treatment. The infection rates of the intermediate snail hosts (*Bulinus truncatus* and *Bulinus camerunensis)* identified reduced from 3.7 to 0.9%. These results show a significant impact of a single dose praziquantel in reducing *S. haematobium* infections after two years. The general uptrend of the prevalence and intensity of infection observed requires a continued monitoring of the disease transmission, repeated treatment and availability of adequate sanitation facilities in the Barombi Kotto focus.

Key words: *Schistosoma haematobium*, prevalence, intensity, praziquantel, Barombi Kotto focus, Cameroon.

INTRODUCTION

Three main forms of human schistosomiasis exist in Africa (intestinal, rectal and urinary), caused by the species *Schistosoma mansoni*, *Schistosoma intercalatum*, and *Schistosoma haematobium,* respectively. Schistosomiasis remains a significant health burden for many parts of the world, especially where health resources are limited (King, 2010).

Out of the 239 million people with active *Schistosoma* infection in 2009 (King et al., 2011), 85% lived in sub-Saharan Africa, where about 112 and 54 million are infected with urinary and intestinal schistosomiasis, respectively and the number of subjects at risk of infection stands superior to 600 million (Martyne et al., 2007). Estimated data show that 200.000 deaths/year where caused by schistosomiasis (Zhang et al., 2007).World Health Organization (WHO) recommendation on schistosomiasis control is based on morbidity control through chemotherapy with praziquantel (WHO, 2002, 2006). Schistosome morbidity is mainly caused by eggs trapped in various parts of the human body; therefore, the fundamental aim of morbidity control is to reduce intensity o f infection by drug treatment

(Touré et al., 2008). Although praziquantel has been available as an effective treatment for *Schistosoma* infection for nearly 30 years; treatment may not be fully curative and questions remain about the best possible timing and frequency of praziquantel dosing for optimal control of infection and morbidity (King et al., 2011). It has been observed in some studies that preventive chemotherapy using praziquantel has resulted in significant reduction of infection and morbidity in one year post treatment (Koukounari et al., 2006; Zhang et al., 2007; Nkengazong et al., 2009) and two years post treatment (Saathoff et al., 2004; Nsowah-Nuamah et al., 2004; Touré et al., 2008). Previous findings suggest that the WHO recommended treatment strategies of schistosomiasis should be adopted with some degree of flexibility according to the different epidemiological and geographical settings (Nsowah-Nuamah et al., 2004; Touré et al., 2008).

In Cameroon, the three forms of human schistosomiasis exist with unequal distribution foci. Intestinal and urinary schistosomiasis is more prevalent in the northern part of the country, while only a few foci are found in the southern part (Ratard et al., 1990). Schistosomiasis due to *S. haematobium* is sparsely distributed in the South West Cameroon. Lake Barombi Kotto in Barombi Kotto village and Barombi Mbo in Kumba serve as the main transmission foci, where transmission is assured by *Bulinus truncatus* and *Bulinus camerunensis* (lake Barombi Kotto) and *B. truncatus* (lake Barombi Mbo). Multiple outbreaks of the disease have been observed in the Barombi Kotto focus despite the control done in the mid-1970s (Duke and Moore, 1976) and the treatment given after each survey, with the highest prevalence and infection intensity observed in children living around the transmission site (Moyou-Somo et al., 1987; Ratard et al., 1990; Ndamukong et al., 2001; Nkengazong et al., 2009). Past control done in this focus has mainly focused on the treatment of school-age children (Moyou-Somo et al., 1987; Ndamukong et al., 2001), but it has not been well defined for adults resident in the same vulnerable area.

The study reported here aimed to evaluate the parasitological impact of a single treatment on urinary schistosomiasis in the Barombi Kotto health area after two years by: (1) assessing the prevalence and parasite intensity at base line, prevalence and intensity reduction at one and two years post treatment among school children of 3 to 22 years; (2) identifying and following-up the infection rates of the intermediate snail hosts following the treatment of the human population. The results obtain will be vital in determining the timing of treatment in the area and other endemic areas.

MATERIALS AND METHODS

Study area

The study was conducted in the Barombi Kotto health area, where the principal transmission site (Lake Barombi Kotto) was found. The Barombi Kotto village (4° 28' 4"N; 9° 15', 2"E) is located in Mbonge subdivision (South West Cameroon). The Crater Lake Barombi Kotto (altitude of 400 m) is situated at about 41 km from Kumba, the head quarter of the division (Moyou-Somo et al., 1987). This village belongs to the equatorial forest zone, Cameroon-type climate having one long rainy season (9 months) and one short dry season (3 months). The total annual rain fall varies between 2000 and 10000 mm. Most water bodies are perennial, while few others dry up for only brief periods (Greer et al., 1990; Ratard et al., 1990). The Crater Lake Barombi Kotto has a width of about 1 km on its largest diameter. Two small streams meet in the lake, one empting into it (the inlet) and the other flowing out of it (outlet). Some of the inhabitants in Barombi Kotto village live in a small island situated in the middle of the lake, while the majority lives around the lake and along the road leading to Kumba through other villages in the middle of the equatorial forest. This village is characterised by the absence of good water sources (wells, taps, forages), poorly disposed garbage and inadequate sanitation facilities; this means that most human activities necessitating water (laundry, bathing, fishing) are done only in the lake.

The residents of the island cross the lake every day for their activities (Farming, marketing, schooling, etc) in the other villages. This village was qualified as a highly endemic area, because of the great number of patients' diagnosed (Moyou-Somo et al., 1987; Ndamukong et al., 2001; Nkengazong et al., 2009).

Malacological study

Three water points were chosen for the collection of snails: Point I, situated at the mainland shoreline of the lake; Point II, with multiple sub-points situated at the island shoreline of the lake; and Point III, situated at about 1.5 km away from the lake in a cocoa plantation (Figure 1). These water points were the ones used by the population for multiples activities (laundry, bathing, fishing, and farming).

Snails were collected at each point following the method described by Njiokou et al. (2004) during 7 surveys (first, second, third, forth, fifth, sixth and seventh), respectively in the months of May and October, 2007, January, February, March, September and November, 2008. Snails collected were brought to the laboratory for test of natural infestation and for species identification following the morphology of their shells (Mimpfoundi and Ndassa, 2005). The test of cercarial shedding and observation was done during one month according to Njiokou et al. (2004).

Study subjects

The study was conducted from May, 2007 to May, 2009. Out of the 380 school children contacted, 279 (153 girls and 126 boys), 197 (109 girls and 88 boys) and 151 (81 girls and 70 boys) of ages between 3 and 22 years, participated, respectively for the initial, one and two year post treatment surveys. An additional 21 volunteer parents participated in the longitudinal survey. The infection status in these parents should give information on the quality of community-based treatment in the focus.

Subject consent

Administrative authorities (Chief of health District, school directors and traditional leaders) were informed about the project and they gave their verbal consent for the study to be undertaken. A written informed consent that met the standards of the National Ethical Commission was obtained from the pupils or the guardians of the young children that accepted to participate in the study.

Figure 1. Map showing the different collection points of snails.

Parasitological study

One urine sample was collected from each participant in a 10 ml plastic screw-cap vials between 10 am and 2 pm. The samples were transported to the General Biology Laboratory (Faculty of Science) at the University of Yaounde I and were examined using the filtration technique (Plouvier et al., 1975). The urine was filtered, stained with 1% Lugol solution, eggs counted under a light microscope at 10× magnification and their number in 10 ml urine recorded (e/10 ml). The entire population (children and parents: ≥ 3 years) living in the Barombi Kotto health area were treated two weeks after the initial survey with a single dose of praziquantel (40 mg/kg of body weight). The drug impact was evaluated one and two years later by collecting urine samples from all those who participated during the initial survey. The samples were examined using the same method described earlier

Data analysis

The parameters assessed during the study were the prevalence (Margolis et al., 1982), egg load, egg reduction rate (Touré et al., 2008; Nkengazong et al., 2009), the level of infection intensity (Kihara et al., 2007) and the natural infection of intermediate snail hosts.

The Chi-square test was used to compare the prevalence in relation to sex, quarters and between the different survey periods, while one way analysis of variance (ANOVA) or Kruskal-Wallis (Sokal and Rohlf, 1981) tests were used to compare the egg load in relation to sex, quarters, egg load and the level of infection intensity during the different survey periods. The level of statistical significance was at 95% ($P < 0.05$). Participants who dropped out or missed either of the two follow-up surveys were not included in the longitudinal analysis.

RESULTS

Malacology

During the malacological survey of a total of 2847 snails, all intermediate snail hosts of *S. haematobium* were collected. This number included 1309 (46.0%) *B. truncatus* and 1538 (54.02%) *B. camerunensis*. Snails collected during the first (May 2007) and the fifth (March 2008) surveys from the Barombi Kotto lake emitted *S. haematobium* forked-tailed cercariae with their respective infection rates being 3.7 and 0.9%. Those collected during the other surveys did not emit forked-tailed cercariae during one month period of exposure in the laboratory (Table 1).

Parasitology

Baseline results

The global prevalence of *S. haematobium* among 279 children examined at baseline was 69.17%. Children

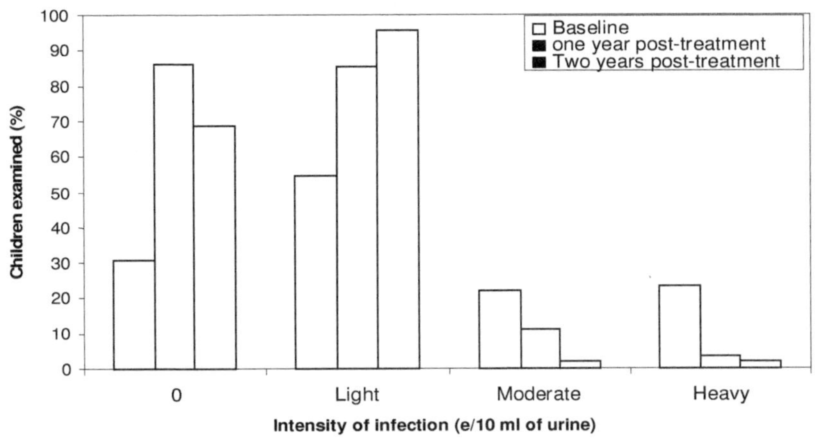

Figure 2. Categories of *S. haematobium* intensity before (n = 279), one year post treatment (n = 197) and two years post treatment (n = 151): 0 = number of persons uninfected.

from the island (84.3%) were significantly more infected than those from the mainland (60.45%; P = 0.0001). A significant difference of prevalence was observed between boys (73.22%) and girls (66.66%) (P = 0.03). The global intensity of infection was 212.1 e/10 ml urine. This value was significantly higher in children from the island (280.1 e/10 ml urine) than the mainland (130.2 e/10 ml urine) (P= 0.01). The intensity of infection did not vary significantly between boys and girls (Table 2).

The proportion of light (≤ 99 e/10 ml urine), moderate (100 to 399 e/10 ml urine) and heavy (≥ 400e/10ml urine) infections of the school children examined were 54.4 %, 22.27 % and 23.31 % respectively (Figure 1). The level of heavy infection did not vary significantly by sex and quarter of residence. Of the 27 volunteer parents examined, six (22.22%) were positive for *S. haematobium* eggs among whom four (14.81%) were heavily infected.

Post treatment results

One round of mass Praziquantel treatment significantly reduced the prevalence in the cohort children from 69.17% at baseline to 13.7 and 31.10% at one year and two years post treatment, respectively (P = 0.0001) with the corresponding overall reduction rates of 80.19 and 55.03%.

A significant decreased of prevalence was observed at baseline as compared to one and two years post treatment, with a significant increase from one to two years post treatment (P = 0.01). The overall intensity of infection significantly reduced from 212.1 at baseline to 32.65 and 50.56 e/10 ml urine at one and two years post treatment, respectively (P = 0.01), with the corresponding overall reduction rates of 84.60 and 76.16% (Table 2). No significant difference, though a high prevalence was observed in females, and two years post treatment with a high intensity observed in males. At one and two yearspost-treatment,nosignificantdifferencewasobserved.

Importantly, heavy infections decreased to 3.7% at 1 year post treatment and remained at 2.12% at two years post treatment. Moderate infection was maintained at 11.10 and 2.12% and light infections at 85.20 and 95.74%, respectively at one and two years post treatment (Figure 2). Four parents were found infected at one and two years post treatment among which one who was negative at base line was heavily infected during the two follow-up periods.

DISCUSSION

The malacological survey conducted enabled us to identify two water points in the Barombi Kotto village harbouring the intermediate snail hosts of *S. haematobium* (*B. truncatus* and *B. camerunensis*) known since the discovery of the focus. *B. camerunensis* was more abundant than *B. truncatus* during our survey confirming the results observed by Duke and Moore (1976), Moyou-Somo et al. (1987) and Greer et al. (1990). This shows stability in the Gastropod population in the Barombi Kotto focus. The cercariae observed from snails collected during the first and fifth survey from the lake confirm that the lake remains the main transmission site of urinary schistosomiasis in the Barombi Kotto health area. The absence of cercariae during the second, third and forth surveys could likely be due to the fact that both children and parents were treated, leading to a reduction in the level of snail infection. The sixth and seventh surveys were conducted during the rainy season, when the human population had another water source (rain water), that could limit the frequency of contact with the transmission site. This is in conformity with the result of Sturrock et al. (1979) who showed that the rate of natural infestation of snails in a transmission focus of schistosomiasis could be linked to the pressure of snail infestation in a transmission site or to the frequency of water contact by infected people. However, most water

Table 1. Distribution and infection rates of *Bulinus* species from lake Barombi Kotto, South West Cameroon.

Collection period	Species and sites*						Total*			Overall total	Infection rate (%)
	B. truncatus			*B. camerunensis*							
	I	II	III	I	II	III	I	II	III		
May 2007	-	-	-	52 (2)	112 (4)	-	52 (2)	112 (4)	-	164 (6)	3.7
October 2007	5	117	2	25	187	6	30	304	8	342	0.0
January 2008	2	298	-	51	48	10	53	346	10	409	0.0
February 2008	-	6	-	81	256	21	81	262	21	364	0.0
March 2008	18 (1)	43 (2)	-	49	312 (2)	32	67 (1)	355 (3)	32	454 (4)	0.9
September 2008	11	489	3	9	171	15	20	660	18	698	0.0
November 2008	-	315	-	26	68	7	26	383	7	416	0.0
Total	36	1268	5	293	1154	91	329	2422	96	2847	0.4

I, II and III: Collection sites for mainland, island and in cocoa plantation, respectively. *Values in parentheses indicate the number of snails infected.

Table 2. Prevalence and intensity of *S. haematobium* infection before and after treatment in Barombi Kotto focus, South West Cameroon.

Variable	Baseline	1 year post-treatment	2 years post-treatment	Overall reduction (%)	
				1 year post-treatment	2 years post-treatment
Prevalence (%)					
Overall prevalence	69.17 (193: n = 279)	13.7 (27: n = 197)	31.10 (47: n = 151)	80.19	55.03
By quarter					
Mainland	60.45 (107: n =177)	8.8 (11: n = 125)	27.05 (n = 85)	85.44	55.33
Island	84.3 (86: n = 102)	22.2 (16: n = 72)	33.36 (n = 66)	73.66	60.42
By sex					
Boys	73.22 (91: n = 126)	14.80 (13: n = 88)	28.60 (20: n = 70)	79.80	60.93
Girls	66.66 (102: n = 153)	12.84 (14: n = 109)	33.33 (27: n = 81)	80.73	50.0
Egg load (e/10 ml urine)					
Overall mean	212.1 (n = 193)	32.65 (n = 27)	50.56 (n = 47)	84.60	76.16
By quarter					
Mainland	130.2 (n = 107)	41.8 (n = 11)	20.96 (n = 23)	67.90	83.90
Island	280.1 (n = 86)	24.31 (n = 16)	61.41 (n = 24)	91.32	78.08
By sex					
Boys	213.7 (n = 91)	36.07 (n = 13)	62.61 (n = 18)	83.12	70.70
Girls	210.53 (n = 102)	29.23 (n = 14)	29.13 (n = 29)	86.11	86.16

n: Total.

periods, as contamination is determined by human behaviour for specific purposes. This makes snails infestation to be a discontinuous process, taking place only at particular moments in such a way that a given snail population can be uninfected completely. This may explain the case of snails collected in the third water point located in a cocoa plantation where people come into contact with water point only during cocoa harvesting and clearing of the farm. In line with previous studies, high prevalence and intensity of *S. haematobium* was found at baseline in the Barombi Kotto health area (Moyou-Somo et al., 1987; Ndamukong et al., 2001) with the prevalence rate of 76.0 and 75.9%, respectively. Even though our prevalence obtained was slightly lower as compared to that of previous authors, it confirms the high endemicity of schistosomiasis in this area. These differences may reflect the difference in the number of years spent since the last mass treatment before each of the three surveys.

Our results are similar to those of King (2006) who showed that villages without pipe-borne water access maintained a high level of infection. Our results confirm the well documented observation that children living near the transmission site of schistosomiasis are more infected (Moyou-Somo et al., 1987; Ndamukong et al., 2001; Saathoff et al., 2004; Kabatereine et al., 2011), and that boys are generally more active in water collection than girls (Tchuem Tchuenté et al., 2003; Njiokou et al., 2004).

Following treatment, the proportion of heavy infections was greatly reduced, which is particularly important as high intensity of *S. haematobium* infection has been shown to contribute to morbidity, including anaemia in children (Koukounari et al., 2007; Green et al., 2011; Sousa-Figueiredo et al., 2012). In previous studies conducted on *S. haematobium* control in Eastern Africa, an annual treatment strategy was predominantly used, with varying results (Magnussen, 2003; Satayathum et al., 2006). However, Nsowah-Nuamah et al. (2004) in Southern Ghana and Touré et al. (2008) in Burkina Faso observed an important reduction of prevalence and parasite intensities at two years post treatment, while Garba et al. (2004) in Niger observed a significant reduction of prevalence and parasite intensities after three years post treatment. Our results are in line with the results of these studies, suggesting that a more spaced treatment strategy is highly effective on *S. haematobium* infections, even in highly endemic areas. However, the effectiveness of such strategy may depend on the local epidemiological and geographical settings of each area.

Though one dose of praziquantel reduced the level of infection significantly at one and two years post treatment; it has been shown that between infection intensity and morbidity risk, even light infection may impose risk of serious disease (Wamachi et al., 2004). Given the association between schistosome infection and anaemia and other disability-related outcomes, infection at any level may impose a significant burden on local health (King et al., 2005). Our results showed that one parent was found heavily infected at one and two years post treatment.

This may be a reflection of the observation that targeted school-age treatment can suppress transmission in some, but not all *S. haematobium* infected communities, because one individual usually uses multiple water sites, and a single infected person can maintain local transmission for an indefinite period (Woolhouse et al., 1997, 1998).

The significant increase of prevalence observed from one to two years post treatment as compared to baseline and one year post treatment may reflect the period of treatment which was in the rainy season (June); the availability of rain water probably reduces human contact with the transmission site during the first year post treatment as compared to the second year. Nevertheless, the general uptrend of the prevalence and intensity of infection showed that two years post treatment could be a sign of a potential rebound of *S. haematobium* should drug distribution be interrupted, as reported by Touré et al. (2008); it could also be a reflection of the fact that multiple villages interact with multiple water contact sites, thus favouring transmission of infection or contamination of water sites (King, 2006). In general, our results highlight the importance of continued effort in monitoring schistosomiasis transmission and repeated treatment in the Barombi Kotto focus, as already shown in some studies (Mduluza et al., 2001; King et al., 2011).

Conclusions

This study has shown: (1) a high prevalence and intensity of *S. haematobium* infection in the Barombi Kotto health area, (2) a significant impact of a single dose of praziquantel in reducing prevalence and intensity of *S. haematobium* after two years, and (3) a high pace of re-infection and the presence of other water sites harbouring the intermediate snail hosts of *S. haematobium* beside the main transmission site (lake Barombi Kotto). It is therefore necessary to: (1) carry out a continuous monitoring of the disease transmission to maintain infection prevalence and intensity at a low level, (2) repeat treatment once every year at the start of the dry season, involving both children and parents with high coverage, (3) organise a survey regarding knowledge, attitude, and practices (KAP) of the people living in the area, and (4) improve the socio-economic conditions of the area (availability of good water sources in all villages).

ACKNOWLEDGEMENTS

The authors thank the headmasters and teachers of Presbyterian and Government primary schools of Barombi Kotto, the Barombi Kotto Chief of District Health Centre for the facilities given during this work, Mr Nchonganyi Ignatius and Mrs Forcha Florence for their financial support.

REFERENCES

Duke BO, Moore PJ (1976). The use of a molluscicide in conjunction with chemotherapy to control *Schistosoma haematobium* at the Barombi Lake foci in Cameroon. II. Urinary examination methods, the use of niridazole to attack the parasite in man, and the effect on transmission from man to snail. Tropenmed. Parasitol. 27(4):489-504.

Garba A, Campagne G, Tassie JM, Barkire A, Vera C, Sellin B, Chippaux JP (2004). Evaluation à long terme d'un traitement de masse par le praziquantel sur la morbidité due à *Schistosoma haematobium* dans deux villages hyperendémiques du Niger. Bull. a Soc. Pathol. Exot. 97:7-11.

Green HK, Sousa-Figueiredo JC, Basanez MG, Betson M, Kabatereine NB, Fenwick A, Stothard JR (2011). Anaemia in Ugandan preschool-aged children: The relative contribution of intestinal parasites and malaria. Parasitology, 12:1534-1545.

Greer GJ, Mimpfoundi R, Malek EA, Joky A, Ngonseu E, Ratard RC (1990). Human schistosomiasis in Cameroon. II. Distribution of the snail hosts. Am. J. Trop. Med. Hyg. 42(6):573-580.

Kabatereine NB, Standley CJ, Sousa-Figueiredo JC, Fleming FM, Stothard JR, Talisuna A, Fenwick A (2011). Integrated prevalence mapping of schistosomiasis, soil-transmitted helminthiasis and malaria in lakeside and island communities in Lake Victoria, Uganda. Parasit. Vectors 4:232.

Kihara JH, Muhoho N, Njomo D, Mwobobia IK, Josyline K, Awazawa T, Amano T, Mwandawiro C (2007). Drug efficacy of praziquantel and albendazole in school children in Mwea Division, Central Province, Kenya. Acta Trop. 102(3):165-171.

King CH (2006). Long-term outcomes of school-based treatment for control of urinary schistosomiasis: A review of experience in Coast Province, Kenya. Mem. Inst. Oswaldo Cruz 101(1):299-306.

King CH (2010). Parasite and poverty: The case of schistosomiasis. Acta Trop. 113(2):95-104.

King CH, Dickman K, Tisch DJ (2005). Reassessment of the cost of chronic helmintic infection: A meta-analysis of disability-related outcomes in endemic schistosomiasis. Lancet 365(9470):1561-1569.

King CH, Olbrych SK, Soon M, Singer ME, Carter J, Colley DG (2011). Utility of repeated praziquantel dosing in the treatment of schistosomiasis in high risk communities in Africa: A systematic review. PLoS Negl. Trop. Dis. 5(9):e1329.

Koukounari A, Fenwick A, Whawell S, Kabatereine NB, Kazibwe F, Tukahebwa EM, Stothard JR, Donnelly CA, Webster JP (2006). Morbidity indicators of *Schistosoma mansoni*: Relationship between infection and anemia in Uganda school children before and after praziquantel and albendazol chemotherapy. Am. J. Trop. Med. Hyg. 75(2):278-286.

Magnussen P (2003). Treatment and re-treatment strategy for schistosomiasis control in different epidemiological settings: a review of 10 years experience. Acta Trop. 86:243-254.

Margolis L, Esch W, Holmes JC, Kuris AM, Schad GA (1982). The use of ecological terms in parasitology (Report of an adhoc committee of American Society of parasitologists). J. Parasitol. 68(1):131-133.

Martyne TS, Essame O, Ratard RC (2007). High risk behaviours and schistosomiasis infection in Kumba, South-West Province, Cameroon. Int. J. Environ. Res. Pub. Health 4(2):101-105.

Mduluza T, Ndhlovu PD, Madziwa TM, Midzi N, Zinyama R, Turner CMR, Chandiwana SK, Nyazema N, Hagan P (2001). The impact of repeated treatment with praziquantel of schistosomiasis in children under six years of age living in an endemic area for *Schistosoma haematobium* infection. Mem. Inst. Oswaldo Cruz 96:157-164.

Mimpfoundi R, Ndassa A (2005). Studies on the morphology and compatibility between *Schistosoma haematobium* and the *Bulinus* spp. Complex (Gastropoda: Planorbidae) in Cameroon. Afr. J. Biotechnol. 4(9):1010-1016.

Moyou-Somo R, Tagni Zukam D, Kouamouo J, Enyong P, Ripert C (1987). Epidemiologic and radiologic study of urinary bilharziasis in the focus of Barombi Lake (Meme Department), Cameroon. Bull. Soc. Pathol. Exot. 80(5):813-825.

Ndamukong KJ, Ayuk MA, Dinga JS, Akenji TN, Ndiforchu VA, Titanji VP (2001). Prevalence and intensity of urinary schistosomiasis in primary school children of the Kotto Barombi Health Area, Cameroon. East Afr. Med. J. 78(6):287-289.

Njiokou F, Onguene AR, Tchuem Tchuente LA, Kenmogne A (2004). Bilharziose urbaine au Cameroun : étude longitudinale de la transmission dans un nouveau site d'extension du foyer de bilharziose de Melen, Yaoundé. Bull. Soc. Pathol. Exot. 97:37-40.

Nkengazong L, Njiokou F, Teukeng F, Enyong P, Wanji S (2009). Reassessment of endemicity level of urinary schistosomiasis in the Kotto Barombi focus (South West Cameroon) and impact of mass drug administration (MDA) on the parasitic indices. J. Cell Anim. Biol. 3(9):159-164.

Nsowah-Nuamah NN, Aryeetey ME, Jolayemi ET, Wagatsuma Y, Mensah G, Dontwi IK (2004). Predicting the timing of second praziquantel treatment and its effect on reduction of egg counts in Southern Ghana. Acta Trop. 90:263-270.

Plouvier S, Leroy JC, Collette J (1975). A propos d'une technique simple de filtration des urines dans la diagnostic de la bilharziose urinaire en enquête de masse. Méd. Trop. 35(3):229-230.

Ratard CR, Kouemeni LE, Bessala MMES, Ndamkou CN, Greer GJ, Spilsbur J, Cline BC (1990). Human schistosomiasis in Cameroon. I. Distribution of schistosomiasis. Am. J. Trop. Med. Hyg. 42(6):561-572.

Saathoff E, Olsen A, Kvalsvig JD, Appleton CC (2004). Patterns of *Schistosoma haematobium* infection, impact of praziquantel treatment and re-infection after treatment in school children from rural Kwazulu-Natal, South-Africa. BMC Infect. Dis. 4(40):1-14.

Satayathum SA, Muchiri EM, Ouma JH, Whalen CC, King CH (2006). Factors affecting infection or reinfection with *Schistosoma haematobium* in coastal Kenya: survival analysis during a nine year school based treatment program. Am. J. Trop. Med. Hyg. 75: 83-92.

Sokal RR, Rohlf FJ (1981). Biometry, 4th ed. WH Freeman and Co., New York. 937p.

Sokal RR, Rohlf FJ (1981). Biometry: The principles and practice of statistics in biological research (2nd ed.). Freeman & Co., New York. P 659.

Sousa-Figueiredo JC, Gamboa D, Pedro JM, Fançony C, Langa AJ, Magalhaes JS, Stothard JR, Nery SV (2012). Epidemiology of malaria, geohelminths, anaemia and malnutrition in the context of a demographic surveillance system in Northern Angola. PLoS One 7(4):1-9.

Sturrock RF, Karam sadkar SJ, Ouma JH (1979). Schistosome infection rate in field snails. *Schistosoma mansoni* in *Biomphalaria pfeifferi* from Kenya. Ann. Trop. Med. Parasitol. 73:365-371.

Tchuem Tchuenté LA, Behnke JM, Gilbert FS, Southgate VR, Vercruysse J (2003). Polyparasitism with *Schistosoma haematobium* and soil-transmitted helminth infections among school children in Loum, Cameroon. Trop. Med. Int. Health 8(11):975-986.

Touré S, Zhang Y, Bosqué-Oliva E, Césaire KY, Amado O, Koukounari A, Gabrielli AF, Sellin B, Joanne PW, Alan F (2008). Two year impact of single praziquantel treatment on infection in the national control programme on schistosomiasis in Burkina Faso. Bull. World Health Organ. 86(10):780-787.

Wamachi AN, Mayadev JS, Mungai PL, Magak PL, Ouma JH, Magambo JK, Muchiri EM, Koech DK, King CH, King CL (2004). Increased ratio of tumor necrosis factor-alpha to interleukin-10 production is associated with *Schistosoma haematobium*-induced urinary-tract morbidity. J. infect. Dis. 190:2020-2030.

WHO (2002). Prevention and control of schistosomiasis and soil-transmitted helminthiasis. WHO Tech. Rep. Ser. 912:1-57.

WHO (2006). Preventive chemotherapy in human helminthiasis. Coordinated use of antihelminthic drugs I control interventions: A manual for health professionals and programme managers. World Health Organisation, Geneva.

Woolhouse ME, Dye C, Etard JF, Smith T, Charlwood JD, Garnett GP, Hagan P, Hii JL, Ndhlovu PD, Quinnell RJ, Watts CH, Chandiwana SK, Anderson RM (1997). Hetergeneities in the transmission of infectious agents: implications for the design of control programs. Proc. Natl. Acad. Sci. USA, 94:338-342.

Woolhouse ME, Etard JF, Dietz K, Ndhlovu PD, Chandiwana SK (1998). Hetergeneities in schistosome transmission dynamics and control. Parasitoloy 117:475-482.

Zhang Y, Koukounari A, Kabaterine N, Fleming F, Kazibwe F, Tukahebwa E, Stohard R, Webster JP, Fenwick A (2007). Parasitological impact of 2-year preventive chemotherapy on schistosomiasis and soil-transmitted helminths in Uganda. BMC Med. 5:27-43.

Cockroach and rodents infestation in Benue State University students' hostel in Makurdi and their epidemiological implications

Edward Agbo Omudu and Gabriel Akosu

Department of Biological Sciences, Benue State University, Makurdi, Nigeria.

Adverse housing conditions in homes and hostels are strongly associated with increased odds of both rodent and cockroach infestation. This study was designed to determine the species of rodents and cockroaches infesting university hostels, the extent of infestation and the methods of control practiced by students. Cockroaches and rodents were trapped using the Hercules mouse glue board. An average of 1.72 rats per room was collected. The number of rats caught from boys' hostel (Unity campus) and girls' hostel (Technical block) were significantly higher than those caught from the other hostel blocks (χ^2 = 39.3, df = 4, P < 0.05). Three species of rodents were identified, *Mus musculus* (24.1%), *Rattus norvegicus* (34.0%) and *Rattus rattus* (41.9%). 502 cockroaches were caught from 142 (54.6%) of the hostel rooms inspected, and an average 3.53 cockroaches were caught from a room. Girls' hostel technical block accounted for the highest cockroach infestation rate (32.3%) and this differs significantly from the other hostel blocks (χ^2 = 11.4, df = 4, P < 0.05). This study shows that overcrowding, cooking in the hostel and refuse accumulation provide conducive breeding ground for cockroaches and rodents. The severe health implications arising from cockroach and rodent infestation justify the need to provide hygiene education to both students and hostel management staff to be able to link cockroach and rodent infestation to health and disease.

Key words: Cockroach, rodents, infestation, university hostels, Makurdi, Nigeria.

INTRODUCTION

Pest infestation is a recognized residential hazard that has been associated with higher risk of asthma symptoms in the case of cockroach (Wang et al., 2008; Sarinho et al., 2004) and Lassa fever, in the case of rodents (Bonner et al., 2007; Bradman et al., 2005). Due to inadequate hostel accommodation for students and the lack of adequate maintenance of existing facilities, a disproportionately high incidence of pest infestation occurs, arising from poor hygiene and improper storage and disposal of waste (Bradman et al., 2005; Wang et al., 2008; Bamigboye, 2006). Poor housing conditions resulting to overcrowding of limited hostel facilities in many Nigerian Universities, provide ample habourage for pest infestation due to unsanitary stacking of food items, as well as congestion of rooms with varieties of articles (Majekodunmi et al., 2002; Onyido et al., 2009). The high incidence of cockroaches in public housing apartments poses numerous public health risks, including exposure to allergens through the air and via food contamination, resulting in an increased incidence of health problems, mainly asthma (Wang et al., 2008; Lwebuga-Mukasa et al., 2002). Cockroaches are also capable of transmitting many pathogens including bacteria, viruses, fungi, protozoa and pathogenic helminthes that threaten human health (Tawatsin et al., 2001; Mlso et al., 2005; Pai et al., 2004; Ghosh and Gayen, 2006). Rodents infestation on the other hand has long been associated with risk of Lassa fever epidemics and the transmission of many serious diseases including plague, hantavirus pulmonary syndrome, murine typhus, rat bite fever, salmonella and leptospirosis (Fisher and Miller, 2003; Bonner et al., 2007; Bradman et al., 2005). Lassa fever is endemic in West Africa and has been reported from Sierra Leone, Guinea, Liberia, and Nigeria. Some studies indicate that 300,000 to 500,000 cases of Lassa fever and 5000 deaths, occur yearly across West Africa (Ogbua et al., 2007). The synergistic health and economic consequences of rodents and cockroach infestation is resulting to the use of home pesticide for their control, a situation that increases indoor pesticide residue (Arlian, 2002).

That adverse housing condition in homes and hostels is strongly associated with increased odds of both rodent and cockroach infestation is not in doubt.

The extent of infestation, destruction caused to property, the species involved and their perceived importance, is yet to attract the required research interest in Nigeria. The objectives of this study therefore, were to determine the extent of rodents and cockroach infestation in a University hostel, species of these pests involved and methods of control practiced by the students. The findings of this study will be used to design and develop pest control strategies that will involve education and students' participation.

MATERIALS AND METHODS

Description of study area

The study was conducted at the Benue State University (BSU), Makurdi, between October, 2009 and March, 2010. Makurdi is the capital of Benue State in Nigeria; the State covers an area of about 34,059 square kilometers with a population of over 4.2 million people (National Population Commission, 2006). Benue State has two seasons called the rainy and dry season, as rainfall is the real climatic variable.

The rainy season is from April to October (7 months, with 800 to 1,100 mm of rain) while the dry season runs from November to March (5 months, with 200 to 400 mm of rain). Temperature ranges from 26 to 29.5°C in the dry season and 19.5 to 24°C in the rainy season, with mean relative humidity of 78%.

The town lies between latitude 7° 30' to 8° 00' N and longitude 8° 30' to 9° 00' E and situated in the guinea savanna vegetative region in Nigeria. The University was established 17 years ago and has 13 halls of residence, each averaging about 48 rooms to serve a student population of about 18,000. Each room was originally designed for 2 students but now officially has between 3 to 4 students per room due to large increase in student admission without corresponding increase in hostel facilities. A common feature of most Nigerian university hostels is the stocking of foodstuffs by individual students inside their rooms, especially under their beds and inside the wardrobe (Bamigboye, 2006). Significant among the problems created by cooking in the hostel are the increased volume of garbage generated and poor strategies for disposal and hygiene.

Selection of rooms and participants

Twenty (20) rooms were randomly selected in each hall of residence, and in each of the selected rooms the first occupant was identified to provide information on students' attitude and practices towards cockroach and rodent infestation. Advocacy visits were made to each of the selected rooms to solicit occupants' cooperation and participation throughout the period of the study. A semi-structured questionnaire was administered to the room occupants. The questions investigated activities of cockroaches and rodents in the rooms and methods of control employed by students.

Methods of trapping cockroaches and rodents

Cockroaches and rodents were trapped using the Hercules mouse glue board, measuring 30 cm in length and 20 cm in width. After thorough visual inspection of the room to identify cockroach and rodents' tracks, droppings and eggs, the glue board was then placed on the path. Pieces of crayfish or smoked fish were placed in the middle of the board as bait. The glue board was left overnight and trapped cockroaches and rodents were collected in the morning and taken to the Zoology laboratory for preservation and identification. While rodents were identified using keys provided in Fisher and Miller (2003), the cockroaches were identified using identification keys in Runstrum and Bennet (1990).

Data analysis

Data was analyzed using percentages and frequencies while chi-square was used to test for significance of infestation rates in the different halls of residence.

RESULTS

The study inspected 260 students' hostel rooms for rodents and cockroach infestation, and a total of 203 rats were caught from 118 (45.5%) rooms (an average of 1.72 rats per room). The number of rats caught from Boys' hostel on the unity campus and girls' hostel Technical block were significantly higher than those caught from the other hostel blocks (χ^2 = 39.3, df = 4, $P < 0.05$). Three species of rodents were identified, *Mus musculus* (24.1%), *Rattus norvegicus* (34.0%) and *Rattus rattus* (41.9%) (Table 1). On the other hand, 502 cockroaches were caught from 142 (54.6%) of the hostel rooms inspected, and an average 3.53 cockroaches were caught from a room. Girls' hostel Technical block accounted for the highest cockroach infestation rate (32.3%) and this differs significantly from the other hostel blocks (χ^2 = 11.4, df = 4, $P < 0.05$) (Table 2). The American cockroach, *Periplaneta americana* was the most predominant cockroach species encountered, accounting for 50.6% of total cockroaches caught. Other species were the German cockroach, *Blattala germanica* (34.7%) and *Supella longipalpa* (14.7%).Students reported several evidences of rat infestation in their rooms, the three most reported evidences were rat runs noticed in room corners (24.6%), rat droppings in wardrobes (18.5%) and rat burrows in boxes and cupboards (15.8%)

Table 1. Common species of rodents trapped from BSU students' hostels in Makurdi, Nigeria.

Hostel	Number of room inspected	Number of rooms with rodents (%)	Number of rodents caught	*Mus musculus*	*Rattus norvegicus*	*Rathus rattus*
Boys' hostel main campus	20	7 (35.0)	16 (7.9)	2 (12.5)	6 (37.5)	8 (50.0)
Girls' hostel main campus	40	17 (42.5)	26 (12.8)	6 (23.0)	8 (30.8)	12 (46.2)
Boys' hostel unity campus	60	38 (63.3)	62 (30.5)	14 (22.5)	28 (45.2)	20 (32.3)
Girls' hostel unity campus	80	21 (26.3)	28 (13.8)	16 (57.1)	11(39.3)	1 (3.6)
Girls' hostel technical block	60	35 (58.3)	71 (35.0)	11 (15.5)	16 (22.5)	44 (62.0)
Total	260	118 (45.4)	203	49 (24.1)	69 (34.0)	85 (41.9)

(X^2 = 39.3, df = 4, P < 0.05).

Table 2. Common species of cockroach trapped from BSU students' hostels in Makurdi, Nigeria.

Hostel	Number of room inspected	Number of rooms with cockroach (%)	Number of cockroaches caught	*Periplaneta americana*	*Blattela germanica*	*Supella longipalpa*
Boys' hostel main campus	20	18 (90.0)	125 (25.0)	79 (63.2)	41 (92.8)	5 (4.0)
Girls' hostel main campus	40	19 (47.5)	58 (11.6)	23 (39.6)	29 (50.0)	6 (10.4)
Boys' hostel unity campus	60	28 (46.7)	84 (16.7)	46 (54.8)	21 (25.0)	17 (20.2)
Girls' hostel unity campus	80	35 (43.8)	73 (14.5)	31 (42.5)	22 (30.1)	20 (27.4)
Girls' hostel technical block	60	42 (70.0)	162 (32.3)	75 (46.3)	61 (37.7)	26 (16.0)
Total	260	142 (54.6)	502	257 (50.6)	174 (34.7)	74 (14.7)

(X^2 = 11.4, df = 4, P< 0.05).

Table 3. Respondents' noticeable evidence of rodents' infestation in BSU hostels in Makurdi, Nigeria.

Indicators of rodents infestation	Frequency	Percentage
Rat runs noticed in corners of rooms	64	24.6
Rate droppings found on floor and in wardrobes	48	18.5
Hole/burrow in wardrobes/cupboards	41	15.8
Live rats often sighted inside cupboards/wardrobes	26	10.0
Reduction in amount of food stuff	24	9.2
Rat noise heard	21	8.1
No evidence of rat infestation	36	13.8
Total	260	100

(Table 3). Some of these noticeable evidences of rat infestation were reported even in rooms where rats were not caught. The three most reported evidences of cockroach infestation were sighting live cockroaches (38.8%), cockroach body parts seen in wardrobes/ cupboard (27.3%) and cockroach odour perceived in cupboards (13.5%) (Table 4). Respondents were more concerned with rat infestation than cockroach infestation; 64.7% of respondents' preferred cockroach infestation to rats.

The most commonly reported damages to properties resulting from cockroach and rat infestation were contamination and consumption of food stuff (29.6%), books (21.1%) and bags (9.6%) (Table 5).

Students in the hostels have resorted to a wide range of activities targeted at controlling cockroaches and rats. The most commonly used method is the use of insecticides for cockroach (44.6%) and a variety of rodenticides popularly known as rat killers (26.1%). This usually combined with physical killing for both rats and cockroaches (Table 6).

The findings from this study demonstrate that students in the hostel have to cope with both physical health risks and the nuisance of cockroach and rat infestation. Both rodents and cockroaches were caught from 20.7% of the rooms inspected; this combined infestation could have additional consequences.

Table 4. Respondents' noticeable evidence of cockroach infestation in BSU hostels in Makurdi, Nigeria.

Indicator of cockroach infestation	Frequency	Percentage (%)
Live cockroach often sighted in room/cupboard	104	38.8
Cockroach body parts seen in wardrobes/cupboard	71	27.3
Cockroach odour perceived in wardrobes/cupboards	35	13.5
Cockroach eggs seen in cupboards/wardrobes	28	10.8
cockroach droppings seen in cupboards	9	3.5
Cockroach noise heard in room	3	1.1
No evidence of cockroach infestation	13	5.0
Total	260	100

Table 5. Damage to items caused by rodents and cockroaches in BSU students hostels in Makurdi, Nigeria.

Item	Frequency	Percentage
Food stuff	77	29.6
Books	55	21.1
Bags/ boxes	25	9.6
Textiles	27	8.5
Shoes	21	8.1
Carpets rugs	16	6.2
Mattresses/pillows	11	4.2
No item damaged	33	12.7
Total	260	100

DISCUSSION

The incidence of cockroaches and rodents in students' hostels poses numerous public health risks. The outcome of this study indicates that overcrowding, cooking in the hostel and low level sanitation and hygiene encourage the breeding of rodents and cockroaches. Species of cockroaches and rodents encountered in university hostels in this study seem to be widespread. *R. rattus* and *R. norvegicus* have been frequently reported in residential buildings in both urban and rural areas in many parts of the world (Stojcevic et al., 2004; Bradman et al., 2005; Omudu and Ati, 2009). The American cockroach, *P. americana* and the German cockroach, *B. germanica,* on the other hand are the commonest roaches infesting buildings worldwide (Tawatsin et al., 2001; Sarinho et al., 2004; Mlso et al., 2005; Bradman et al., 2005; Omudu and Eyumah, 2008). These species of rodent and cockroaches are known to reproduce faster and thrive in habitats with availability and abundance of diverse food materials, suitable refuge, and lack of comprehensive control efforts.

Several studies have linked adverse housing conditions in residential apartments to an increased odds of both rodent and cockroach infestations (Bradman et al., 2005; Bonner et al., 2007; Wang et al., 2008). Our finding corroborates earlier studies in university hostels and residential apartments in Nigeria (Mbanong et al., 2002; Bamigboye, 2006; Omudu and Eyumah, 2008; Omudu and Ati, 2009). In addition to poor sanitation and lack of proper maintenance, Nigerian universities are known to be largely overcrowded, with more than twice the number of students expected to occupy the rooms. Also contributing is the fact that many students cook in the hostel, these factors substantially provide a wide range of food items, increase the amount of waste generated and escalate the breeding of rodents and cockroaches (Bamigboye 2006; Wang et al., 2008; Onyido et al., 2009).

Poor sanitary conditions in the hostel and overcrowding provide enabling environment for cockroach and rodent infestations, both of which are allergenic and can carry infectious diseases. A large number of microorganisms have been isolated from cockroaches captured, either from residential apartments, hospitals or other buildings. Recent studies in Iran reported a high percentage of test cockroaches (98%) carrying various microorganisms (bacteria, fungi and parasltes) in and on their bodies, some of them of medical importance (Salehzadeha et al., 2007).Exposure to cockroach allergens through air and via food contamination results to an increased incidence of health problem, mainly asthma and cockroaches are potential to vector dysentery and gastroenteritis (Bradman et al., 2005; Wang et al., 2008). Studies have reported robust association between cockroach infestation and development of asthma as relevant allergens have been identified in the body and secretions from cockroach's body (Sarinho et al., 2004). On the other hand, rodent infestation is associated with the occurrence of Lassa fever. Evidences of rodent infestation like rodent burrows have been strongly associated with Lassa fever incidence in homes (Davis et al., 2005; Bonner et al., 2007). Other health implications of rodent infestation are that they are reservoirs for a variety of zoonotic diseases like *Toxoplasma goondi* (Stojcevic et al., 2004; Murphy et al., 2007). The involvement of rodent and cockroaches in transmission of life-threatening infections is epidemiologically significant and could be responsible for increasing incidence of emerging infectious diseases. Refuse disposal and food handling practices in many Nigerian

Table 6. Respondents' methods of rodent/cockroach control in BSU hostels in Makurdi, Nigeria.

Adopted control methods	Frequency	Percentage
Rodents		
Physical killing	96	37.0
Rat poison	68	26.1
Use of traps (glue board, snap trap)	44	17.0
blockage of burrows and tracks	17	6.5
Proper storage of food stuff	10	3.8
Never did anything	25	9.6
Total	260	100
Cockroaches		
Use of insecticides (Chlorpyrifos DDVP, Permethrin, Dichlorvos)	116	44.6
Physical killing	93	35.8
Use of adhesive gum	18	70
Proper storage of food stuff	9	3.5
Never did anything	24	9.2
Total	260	100

universities hostels provide ideal habitats for cockroaches and rodents to breed. The Girls' hostel technical block and the boys' hostel Unity campus are surrounded by open fields overgrown with weeds providing a mosaic of agricultural and village areas. These may be the reason for the significantly high infestation rates recorded in these hostel blocks. It has been reported that human environment, such as shrubs and refuse dumps, attract rodents for shelter and refuge (Onyido et al., 2009). Poor external hygiene may act as a risk factor for rodent and cockroach infestation of residential homes and transmission of lassa virus and precipitation of asthmatic attacks (Sarinho et al., 2004; Bonner et al., 2007).

Additional health consequences of rodent and cockroach infestation are the resultant increase in use of pesticides. Frequent pesticide use results in environmental pollution and poor air quality within rooms. The high percentage of student resorting to rodenticides and insecticide to control infestation could cause accidental contamination of foods stuff inside students' rooms and increase indoor pesticide residue (Majekodumi et al., 2002; Bamigboye, 2006; Wang et al., 2008). Control strategies should therefore be directed to emphasis of knowledge of the biology and ecology of target rodent and cockroaches, improvement in sanitation and hygiene and limited pesticide use.

Conclusion

This study shows that overcrowding, cooking in the hostel and accumulating refuse provide conducive breeding ground for cockroaches and rodents. The severe health implications arising from cockroach and rodent infestation justify the need to provide hygiene education to link cockroach and rodent infestation to health and disease.

Controlling pest infestation in hostel settings will require a comprehensive approach targeted at reducing overcrowding of rooms, improvement of hygiene and sanitation, and general maintenance of hostel facilities on campus.

REFERENCES

Arlian J (2002). The risk of human health from cockroach infestations. J. Environ. Health Res. 5 (1):23-29.

Bamigboye AP (2006). Rat infestation in a typical Nigerian University Halls of residence: Implications for hygiene and sanitation. Res. J. Agric. Biol. Sci. 2(6):328-331.

Bonner PC, Schmidt WP, Belmain SR, Oshin B, Baglode D, Borchet M (2007). Poor housing quality increase risk of rodent infestation and Lassa fever in refugee camps of Sierra Leone. Am. J. Trop. Med. Hyg. 77:169-175.

Bradman A, Chevrier J, Tager I, Lipsett M, Weldon K (2005). Association of housing disrepair indicators with cockroach and rodent infestation in a cohort of pregnant Latino women and their children. Environ. Health Perspect.11:21-29.

Davis S, Calvet E, Leirs H (2005). Fluctuating rodent population and the risk to human from rodent borne zoonoses. Vector Borne Zoonotic Dis. 5(4):305-314.

Fisher ML, (2003). School integrated pest management for rats and mice. Department of Entomology Publication, No. 444. Virginia Tech, USA. pp. 21-29.

Ghosh J, Gayen A (2006). The protozoan fauna living in the digestive system of *Periplaneta americana* in kolkata, West Bengal, India. J. Parasit. Dis. 30(1):76-80.

Lwebuga-Mukasa JS, Wojcik R, Dunn-Georgiou E, Johnson C (2002). Home environmental factors associated with asthma prevalence in

two Buffalo inner-city neighbourhoods. J. Health Care Poor Underserved 13(2):214-228.
Majekodunmi A, Howard MT, Shah V (2002). The perceived importance of cockroach (*Blatta orientalis* and *Blattela germanica*) infestation to social housing residents. J. Environ. Health Res. 1(2):27-34.
Mbanong WF, Nock IH, Ezealor AU (2002). A survey of ecto-and endoparasites of murid rats in Samaru, Zaria, Nigeria. J. Trop. Biosci. 2(1):106-112.
Mlso WR, Qureshi AH, Khan IA, Hussain S (2005). Frequency of different species of cockroaches in tertiary care hospital and their role in transmission of bacterial pathogens. Pak. J. Med. Res. 44(4):25-32.
Murphy RG, William RH, Hughes JM, Ford NJ, Hide G, Oldbury DJ (2007). The urban house mouse (*Mus domesticus*) as reservoir of infection for the human parasite *Toxoplasma gondi*: An unrecognized public health issue? Int. J. Environ. Health Res.17:45-51.
Ogbua O, Ajuluchukwu E, Uneke CJ (2007). Lassa fever in West African sub-region: An overview. J. Vector Borne Dis. 44:1-11.
Omudu EA, Eyumah AO (2008). A study of cockroach (Dictyoptera: Blattidae) infestation in different types of residential buildings in Makurdi, Nigeria. Niger. J. Entomol. 25:11-17.
Omudu EA, Ati TT (2009). A survey of rats trapped in residential apartments and their ectoparasites in Makurdi, Nigeria. Res. J. Agric. Biol. Sci. 8(1):7-13.
Onyido AE, Okolo PO, Obiukwu MO, Amadi ES (2009). Survey of vectors of public health diseases in un-disposed refuse dumps in Awka town, Anambra State, Nigeria. Res. J. Parasitol. 4(1):22-27.
Pai HH, Chen WC, Peng CF (2004). Cockroaches as potential vectors of nosocomial infestations. Infect. Control Hosp. Epidemiol. 11:979-984.
Sarinho E, Schor D, Veloso MA, Rizzo JA (2004). There are more asthmatics in homes with high cockroach infestation. Braz. J. Med. Biol. Res. 37(4):503-510.
Salehzadeha A, Tavacolb P, Mahjubc H (2007). Bacterial, fungal and parasitic contamination of cockroaches in public hospitals of Hamadan, Iran. J. Vector Borne Dis. 44:105-110.
Stojcevic D, Mihaijevic Z, Marinculic A (2004). Parasitological survey of rats in rural regions of Croatia. Vet. Med. 49(3):70-74.
Tawatsin A, Thavara U, Chompoosri J, Chansang C, Poosriwong S (2001). Cockroach surveys in 14 provinces of Thailand. J. Vector Ecol. 26(2):1-7.
Wang C, Abou El-Nour MM, Bennett GW (2008). Survey of pest infestation, asthma and allergy in low-income housing. J. Community Health 33(1):31-39.

Permissions

All chapters in this book were first published in JPVB, by Academic Journals; hereby published with permission under the Creative Commons Attribution License or equivalent. Every chapter published in this book has been scrutinized by our experts. Their significance has been extensively debated. The topics covered herein carry significant findings which will fuel the growth of the discipline. They may even be implemented as practical applications or may be referred to as a beginning point for another development.

The contributors of this book come from diverse backgrounds, making this book a truly international effort. This book will bring forth new frontiers with its revolutionizing research information and detailed analysis of the nascent developments around the world.

We would like to thank all the contributing authors for lending their expertise to make the book truly unique. They have played a crucial role in the development of this book. Without their invaluable contributions this book wouldn't have been possible. They have made vital efforts to compile up to date information on the varied aspects of this subject to make this book a valuable addition to the collection of many professionals and students.

This book was conceptualized with the vision of imparting up-to-date information and advanced data in this field. To ensure the same, a matchless editorial board was set up. Every individual on the board went through rigorous rounds of assessment to prove their worth. After which they invested a large part of their time researching and compiling the most relevant data for our readers.

The editorial board has been involved in producing this book since its inception. They have spent rigorous hours researching and exploring the diverse topics which have resulted in the successful publishing of this book. They have passed on their knowledge of decades through this book. To expedite this challenging task, the publisher supported the team at every step. A small team of assistant editors was also appointed to further simplify the editing procedure and attain best results for the readers.

Apart from the editorial board, the designing team has also invested a significant amount of their time in understanding the subject and creating the most relevant covers. They scrutinized every image to scout for the most suitable representation of the subject and create an appropriate cover for the book.

The publishing team has been an ardent support to the editorial, designing and production team. Their endless efforts to recruit the best for this project, has resulted in the accomplishment of this book. They are a veteran in the field of academics and their pool of knowledge is as vast as their experience in printing. Their expertise and guidance has proved useful at every step. Their uncompromising quality standards have made this book an exceptional effort. Their encouragement from time to time has been an inspiration for everyone.

The publisher and the editorial board hope that this book will prove to be a valuable piece of knowledge for researchers, students, practitioners and scholars across the globe.

List of Contributors

Sahidou SALIFOU
Laboratoire de Recherche en Biologie Appliquée, Université d'Abomey-Calavi, 01 BP 2009 Cotonou, Bénin

Dadjo Florian DAGA
Département de Production Animale, Faculté des Sciences Agronomiques, Université d'Abomey-Calavi, 01 BP 526 Cotonou, Bénin

Sabbas ATTINDEHOU
Laboratoire de Recherche en Biologie Appliquée, Université d'Abomey-Calavi, 01 BP 2009 Cotonou, Bénin

Rodolph DEGUENON
Laboratoire de Recherche en Biologie Appliquée, Université d'Abomey-Calavi, 01 BP 2009 Cotonou, Bénin

Cyprien Félix BIAOU
Laboratoire de Recherche en Biologie Appliquée, Université d'Abomey-Calavi, 01 BP 2009 Cotonou, Bénin

T. Auta
Department of Biological Sciences, Federal University, Dutsin-Ma, Katsina State, Nigeria

J. R. Wartu
Department of Microbiology, Kaduna State University, Kaduna, Nigeria

J. A. Bawa
Department of Biological Sciences, Federal University, Dutsin-Ma, Katsina State, Nigeria

A. M. Jabbi
Department of Biological Sciences, Police Academy, Wudil, Kano State, Nigeria

D. E. Nwele
Department of Applied Biology, Faculty of Biological Sciences, Ebonyi State University, Abakaliki, Nigeria

A. C. Uhuo
Department of Applied Biology, Faculty of Biological Sciences, Ebonyi State University, Abakaliki, Nigeria

E. C. Okonkwo
Department of Applied Microbiology, Faculty of Biological Sciences, Ebonyi State University Abakaliki, Nigeria

G. A. Ibiam
Department of Medical Laboratory Science, Faculty of Health Sciences, Ebonyi State University Abakaliki, Nigeria

C. S. Onwe
Department of Applied Biology, Faculty of Biological Sciences, Ebonyi State University, Abakaliki, Nigeria

J. I Ugwu
Department of Medical Laboratory Science, Faculty of Health Sciences, Ebonyi State University Abakaliki, Nigeria

J. C. Onwe
Department of Medical Laboratory Science, Faculty of Health Sciences, Ebonyi State University Abakaliki, Nigeria

A. C. Nwuzo
Department of Applied Microbiology, Faculty of Biological Sciences, Ebonyi State University Abakaliki, Nigeria

Tadesse Hailu Jember
Department of Microbiology, Immunology and Parasitology [DMIP], College of Medicine and Health Science, Bahir Dar University, P. O. Box 79, Bahir Dar, Ethiopia

A. B. Akeredolu
Department of Zoology, Obafemi Awolowo University, Ile-Ife, Osun State, Nigeria

O. A. Sowemimo
Department of Zoology, Obafemi Awolowo University, Ile-Ife, Osun State, Nigeria

Yitagele Terefe
Faculty of Veterinary Medicine, Haramaya University, Ethiopia

Ketema Tafess
Faculty of Veterinary Medicine, Gondar University, Gondar, Ethiopia

Getasew Fekadie
Faculty of Veterinary Medicine, Haramaya University, Ethiopia

Nigatu Kebede
Aklilu Lemma Institute of Pathobiology, Addis Ababa University, Addis Ababa, Ethiopia

Enyaru C. K. John
Department of Biochemistry and Sports Sciences, Makerere University, P. O. Box 7062, Kampala, Uganda

Njuguna James
Astel Diagnostics, Uganda

Alibu Vincent Pius
Department of Biochemistry and Sports Sciences, Makerere University, P. O. Box 7062, Kampala, Uganda

Matovu Enock
College of Veterinary Medicine and Biosafety, Makerere University, P. O. Box 7062, Kampala, Uganda

Malele I. Imna
Vector and Vector Borne Diseases Institute, Tanga, Tanzania

Chisi E. John
College of Medicine, University of Malawi, Private bag 360, Blantyre, Malawi

Mbongo Nicolas
Laboratoire national de Sante Publique, Brazzaville, Congo

Mansinsa Philemon
National Program of Human African Trypanosomosis, Kinshasa, Congo

Intisar E. l. Rayah
Tropical Medicine Research Institute, Khartoum, Sudan

Mohammed Yassir
Tropical Medicine Research Institute, Khartoum, Sudan

Mubarak Mustafa
Tropical Medicine Research Institute, Khartoum, Sudan

Ochi Erneo
Ministry of Animal Resources and Fisheries, Juba, South Sudan

Nantulya Vinand
Astel Diagnostics, Uganda

Girmaye Benti
Department of Biology, Dilla University College of Natural and Computational Science, P. O. Box-419 Dilla, Ethiopia

Fekadu Gemechu
Department of Biology, Dilla University College of Natural and Computational Science, P. O. Box-419 Dilla, Ethiopia

Doaa Abdelhafeez Yones
Department of Medical Parasitology, Faculty of Medicine, Assiut University, Assiut, Egypt

Enas Abdelhameed Mahmoud
Department of Medical Parasitology, Faculty of Medicine, Assiut University, Assiut, Egypt

Ragaa Ali Othman
Department of Medical Parasitology, Faculty of Medicine, Assiut University, Assiut, Egypt

Mohamed Korany
Department of General Surgery, Faculty of Medicine, Assiut University, Assiut, Egypt

Tarek Hassan El-Metwally
Department of Medical Biochemistry, Faculty of Medicine, Assiut University, Assiut, Egypt

E. Eyob
Department of Parasitology, Yabello Regional Veterinary Laboratory, Ethiopia

L. Matios
Department of Parasitology, Yabello Regional Veterinary Laboratory, Ethiopia

Sahidou Salifou
Laboratory of Applied Biology Research, University of Abomey-Calavi, 01 BP 2009 Cotonou, Benin

Dadjo Florian Daga
Department of Animal Production, Faculty of Agricultural Sciences, University of Abomey, 01 BP 526 Cotonou, Benin

Sabbas Attindehou
Laboratory of Applied Biology Research, University of Abomey-Calavi, 01 BP 2009 Cotonou, Benin

Boon Allwin
Department of Wildlife Science, Madras Veterinary College, Chennai, India

M. G. Jayathangaraj
Department of Wildlife Science, Madras Veterinary College, Chennai, India

M. Palanivelrajan
Department of Wildlife Science, Madras Veterinary College, Chennai, India

S. Gomathinayagam
Department of Veterinary Parasitology, Madras Veterinary College, Chennai, India

M. Raman
Programme director, TRVPB, TANUVAS, Chennai, India

S. T. Bino sundar
Department of Veterinary Parasitology, Madras Veterinary College, Chennai, India

Thomas Hile
Laboratory of Parasitology, Department of Biomedical Science, Cosendai Adventist University, Eboko, Cameroon

J. P. Assam Assam
Laboratory of Parasitology, Department of Biomedical Science, Cosendai Adventist University, Eboko, Cameroon
Biotechnology Center, University of Yaoundé I, Cameroon
Department of Biochemistry, Faculty of Science, University of Douala

E. Amali
Department of Biological Science, University of Makurdi, Benue State, Nigeria

E. Amuta
Department of Biological Science, University of Makurdi, Benue State, Nigeria

Javid Ahmad Dar
Department of Zoology, University of Kashmir, Srinagar-190 006, India

Syed Tanveer
Department of Zoology, University of Kashmir, Srinagar-190 006, India

A. E. Ogunkanbi
Department of Zoology, Obafemi Awolowo University, Ile - Ife, Osun State, Nigeria

O. A. Sowemimo
Department of Zoology, Obafemi Awolowo University, Ile - Ife, Osun State, Nigeria

Rock Aïkpon
Centre de Recherche Entomologique de Cotonou (CREC), Cotonou, Benin
Faculte des Sciences et Techniques de l'Université d'Abomey Calavi, Benin

Razaki Ossè
Centre de Recherche Entomologique de Cotonou (CREC), Cotonou, Benin
Faculte des Sciences et Techniques de l'Université d'Abomey Calavi, Benin

Renaud Govoetchan
Centre de Recherche Entomologique de Cotonou (CREC), Cotonou, Benin
Faculte des Sciences et Techniques de l'Université d'Abomey Calavi, Benin

Arthur Sovi
Centre de Recherche Entomologique de Cotonou (CREC), Cotonou, Benin
Faculte des Sciences et Techniques de l'Université d'Abomey Calavi, Benin

Frédéric Oké-Agbo
Centre de Recherche Entomologique de Cotonou (CREC), Cotonou, Benin

Martin C. Akogbéto
Centre de Recherche Entomologique de Cotonou (CREC), Cotonou, Benin
Faculte des Sciences et Techniques de l'Université d'Abomey Calavi, Benin

Bosly A. Hanan
Biology Department, Faculty of Science, Jazan University, Jazan, Saudia Arabia

Achenef Melaku
Department of Veterinary Pharmacy and Biomedical Sciences, Faculty of Veterinary Medicine, University of Gondar, P. O. Box: 196, Gondar, Ethiopia

N. Ouchene
Département des Sciences Vétérinaires, Université d'El Tarf, BP73, 36000, El Tarf, Algérie

N. A. Ouchene-Khelifi
Département des Sciences Vétérinaires, Université d'El Tarf, BP73, 36000, El Tarf, Algérie

F. Zeroual
Département des Sciences Vétérinaires, Université d'El Tarf, BP73, 36000, El Tarf, Algérie

A. Benakhla
Département des Sciences Vétérinaires, Université d'El Tarf, BP73, 36000, El Tarf, Algérie

K. Adjou
UMR Bipar, Unité de pathologie du bétail. École Nationale Vétérinaire d'Alfort7, Avenue du Général-de-Gaulle 94704 Maisons-Alfort, France

Mengistu Wale
Ethiopian Biodiversity Institute, P. O. Box 30726 Addis Ababa, Ethiopia

Melaku Wale
College of Science, Bahar Dar University P. O. Box 26 Bahir Dar, Ethiopia

Tesfu Fekensa
Ethiopian Biodiversity Institute, P. O. Box 30726 Addis Ababa, Ethiopia

Nazaire Aïzoun
Centre de Recherche Entomologique de Cotonou (CREC), 06 BP 2604, Cotonou, Bénin
Faculté des Sciences et Techniques, Université d'Abomey Calavi, Calavi, Bénin

Rock Aïkpon
Centre de Recherche Entomologique de Cotonou (CREC), 06 BP 2604, Cotonou, Bénin
Faculté des Sciences et Techniques, Université d'Abomey Calavi, Calavi, Bénin

Roseric Azondekon
Centre de Recherche Entomologique de Cotonou (CREC), 06 BP 2604, Cotonou, Bénin
University of Massachusetts Amherst, Amherst, Massachusetts, USA

Virgile Gnanguenon
Centre de Recherche Entomologique de Cotonou (CREC), 06 BP 2604, Cotonou, Bénin
Faculté des Sciences et Techniques, Université d'Abomey Calavi, Calavi, Bénin

Razaki Osse
Centre de Recherche Entomologique de Cotonou (CREC), 06 BP 2604, Cotonou, Bénin
Faculté des Sciences et Techniques, Université d'Abomey Calavi, Calavi, Bénin

Gil Germain Padonou
Centre de Recherche Entomologique de Cotonou (CREC), 06 BP 2604, Cotonou, Bénin
Faculté des Sciences et Techniques, Université d'Abomey Calavi, Calavi, Bénin

Martin Akogbéto
Centre de Recherche Entomologique de Cotonou (CREC), 06 BP 2604, Cotonou, Bénin
Faculté des Sciences et Techniques, Université d'Abomey Calavi, Calavi, Bénin

Ebere V. Ugwoke
Department of Veterinary Public Health and Preventive Medicine, Ahmadu Bello University, Zaria, Kaduna State, Nigeria

Jarleth U. Umoh
Department of Veterinary Public Health and Preventive Medicine, Ahmadu Bello University, Zaria, Kaduna State, Nigeria

Emmanuel C. Okolocha
Department of Veterinary Public Health and Preventive Medicine, Ahmadu Bello University, Zaria, Kaduna State, Nigeria

Idris A. Lawal
Department of Veterinary Parasitology and Entomology, Faculty of Veterinary Medicine, Ahmadu Bello University, Zaria, Kaduna State, Nigeria

Samia E. Etewa
Department of Medical Parasitology, Faculty of Medicine, Zagazig University, Zagazig, Egypt

Naglaa F. Abd El-Aal
Department of Medical Parasitology, Faculty of Medicine, Zagazig University, Zagazig, Egypt

Sara A. Abdel-Rahman
Department of Medical Parasitology, Faculty of Medicine, Zagazig University, Zagazig, Egypt

Mahmoud A. El-Shafey
Department of Clinical Pathology, Faculty of Medicine, Zagazig University, Zagazig, Egypt

Ashenafi Teklemariam
Department of Biology, College of Natural and Computational Science, Welkitie University, Ethiopia

Tadesse Dejenie
Department of Biology, College of Natural and Computational Science, Mekelle University, Ethiopia

Zewdneh Tomass
Department of Biology, College of Natural and Computational Science, Wolaita Sodo University, Ethiopia

E. M. El Hassan
Department of Microbiology and Parasitology, King Faisal University, P.O. Box 400, Al-Ahsa 31982, Kingdom of Saudi Arabia

O. M. Agbolade
Department of Plant Science and Applied Zoology, Parasitology and Medical Entomology Laboratory, Olabisi Onabanjo University, P.M.B. 2002, Ago-Iwoye, Ogun State, Nigeria

O. W. Lawal
Department of Plant Science and Applied Zoology, Parasitology and Medical Entomology Laboratory, Olabisi Onabanjo University, P.M.B. 2002, Ago-Iwoye, Ogun State, Nigeria

K. A. Jonathan
Department of Plant Science and Applied Zoology, Parasitology and Medical Entomology Laboratory, Olabisi Onabanjo University, P.M.B. 2002, Ago-Iwoye, Ogun State, Nigeria

Eyob Eshetu
Department of Parasitology, Yabello Regional Veterinary Laboratory, Oromia Regional State, Yabello, Ethiopia

Getachew Dinede
Oromia Pastoral Areas Technical and Vocational Education and Training College, Oromia Regional State, Yabello, Ethiopia

Matios Lakew
Oromia Pastoral Areas Technical and Vocational Education and Training College, Oromia Regional State, Yabello, Ethiopia

Tadele Tolosa
Department of Veterinary Public Health and Veterinary Microbiology, Jimma University College of Agriculture and Veterinary Medicine, Jimma, Ethiopia

Bayleyegn Gizachew
College of Veterinary Medicine, Haramaya University, P.O Box 138, Dawa, Ethiopia

Fikadu Kibru
College of Veterinary Medicine, Haramaya University, P.O Box 138, Dawa, Ethiopia

Biruhtesfa Asrade
College of Veterinary Medicine, Haramaya University, P.O Box 138, Dawa, Ethiopia

Khadija Lahouiti
Laboratory of Microbial Biotechnology, Department of Biology, Faculty of Sciences and Technology, University Sidi Mohamed Ben Abdellah, Fez, Morocco

Abdelhakim El Ouali Alami
Regional Laboratory of Epidemiological Diagnosis and Environmental Health, Entomology Unit, Hospital Al Ghassani, Fez, Morocco

Asmae Hmamouch
Laboratory of Microbial Biotechnology, Department of Biology, Faculty of Sciences and Technology, University Sidi Mohamed Ben Abdellah, Fez, Morocco

Khadija Bekhti
Laboratory of Microbial Biotechnology, Department of Biology, Faculty of Sciences and Technology, University Sidi Mohamed Ben Abdellah, Fez, Morocco

M. K. Bouyou-Akotet
Department of Parasitology-Mycology, Faculty of Medicine, Université des Sciences de la Santé, Libreville, Gabon

D. P. Mawili Mboumba
Department of Parasitology-Mycology, Faculty of Medicine, Université des Sciences de la Santé, Libreville, Gabon

V. Guiyedi
Department of Parasitology-Mycology, Faculty of Medicine, Université des Sciences de la Santé, Libreville, Gabon

M. Pemba Mihindou
Department of Parasitology-Mycology, Faculty of Medicine, Université des Sciences de la Santé, Libreville, Gabon

M. Maryvonne Kombila
Department of Parasitology-Mycology, Faculty of Medicine, Université des Sciences de la Santé, Libreville, Gabon

James C. Dunford
Navy and Marine Corps Public Health Center Detachment, Centers for Disease Control and Prevention, Atlanta, GA, 30341, USA

Robert A. Wirtz
Centers for Disease Control and Prevention, Atlanta, GA, 30341, USA

William G. Reifenrath
Stratacor, Inc., Richmond, CA, 94804, USA

Aneika Falconer
Centers for Disease Control and Prevention, Atlanta, GA, 30341, USA

Laura N. Leite
Centers for Disease Control and Prevention, Atlanta, GA, 30341, USA

William G. Brogdon
Centers for Disease Control and Prevention, Atlanta, GA, 30341, USA

T. A. Babatunde
Department of Zoology, Obafemi Awolowo University, Ile-Ife, Nigeria

S. O. Asaolu
Department of Zoology, Obafemi Awolowo University, Ile-Ife, Nigeria

O. A . Sowemimo
Department of Zoology, Obafemi Awolowo University, Ile-Ife, Nigeria

T. A. Olusi
Department of Biology, Federal University of Technology, P M B 704, Akure Ondo State, Nigeria

A. F. Abe
Department of Biology, Federal University of Technology, P M B 704, Akure Ondo State, Nigeria

Tadesse Hailu
Department of Microbiology, Immunology and Parasitology [DMIP], College of Medicine and Health Science, Bahir Dar University, P. O. Box 79, Bahir Dar, Ethiopia

Mulat Yimer
Department of Microbiology, Immunology and Parasitology [DMIP], College of Medicine and Health Science, Bahir Dar University, P. O. Box 79, Bahir Dar, Ethiopia

Tirualem Sitotaw
College of Veterinary Medicine and Agriculture, Addis Ababa University, Debre Zeit, Ethiopia

Fikru Regassa
College of Veterinary Medicine and Agriculture, Addis Ababa University, Debre Zeit, Ethiopia

Fikre Zeru
School of Veterinary Medicine, Semera University, Semera, Ethiopia

Abraha Gebregziabher Kahsay
School of Veterinary Medicine, Wollo University, P.O. Box 1145, Dessie, Ethiopia
Institute of Tropical Medicine, Sint Rochusstraat 21, Antwerp, Belgium

Babita R. Malik
Department of Pharmacognosy, J.S.S. College of Pharmacy Rockland, Ooty-643 001, T.N., India

Mukesh K. Malik
Department of Pharmacognosy, J.S.S. College of Pharmacy Rockland, Ooty-643 001, T.N., India

N. Balakrishnan
National Institute of Communicable Diseases, Southern India branch Conoor-643101, T.N., India

B.suresh
Department of Pharmacognosy, J.S.S. College of Pharmacy Rockland, Ooty-643 001, T.N., India

M. O. Oche
Department of Community Health, Usman Danfodiyo University, Sokoto, Nigeria

A. U. Junaidu
Department of Veterinary Public Health, Usman Danfodiyo University, Sokoto, Nigeria

A. S. Mainasara
Department of Pathology/Microbiology, Usman Danfodiyo University, Sokoto, Nigeria

M. A. Ndakotsu
Department of Pathology/Microbiology, Usman Danfodiyo University, Sokoto, Nigeria

Kevin Yana Njabo
Center for Tropical Research UCLA Institute of the Environment and Sustainability, Los Angeles CA USA

Thomas B Smith
Center for Tropical Research UCLA Institute of the Environment and Sustainability, Los Angeles CA USA
Department of Ecology and Evolutionary Biology, University of California, Los Angeles, Los Angeles, CA 90095, USA

Elizabeth Yohannes
University of Constance, Limnological Institute, Stable Isotope Laboratory,Mainaustr 252, D-78464, Germany

L. Nkengazong
Laboratory of General Biology and Physiology, Faculty of Science, P. O. Box 812, Yaounde, University of Yaounde I, Cameroon

F. Njiokou
Laboratory of General Biology and Physiology, Faculty of Science, P. O. Box 812, Yaounde, University of Yaounde I, Cameroon

T. Asonganyi
Faculty of Medicine and Biomedical Science, University of Yaounde I, P. O. Box 1364, Cameroon

Edward Agbo Omudu
Department of Biological Sciences, Benue State University, Makurdi, Nigeria

Gabriel Akosu
Department of Biological Sciences, Benue State University, Makurdi, Nigeria